Lecture Notes in Computer Science 12141

More information about this series at http://www.springer.com/series/7407

Valeria V. Krzhizhanovskaya ·
Gábor Závodszky · Michael H. Lees ·
Jack J. Dongarra · Peter M. A. Sloot ·
Sérgio Brissos · João Teixeira (Eds.)

Computational Science – ICCS 2020

20th International Conference
Amsterdam, The Netherlands, June 3–5, 2020
Proceedings, Part V

 Springer

Editors
Valeria V. Krzhizhanovskaya ⓘ
University of Amsterdam
Amsterdam, The Netherlands

Michael H. Lees
University of Amsterdam
Amsterdam, The Netherlands

Peter M. A. Sloot ⓘ
University of Amsterdam
Amsterdam, The Netherlands

ITMO University
Saint Petersburg, Russia

Nanyang Technological University
Singapore, Singapore

João Teixeira
Intellegibilis
Setúbal, Portugal

Gábor Závodszky ⓘ
University of Amsterdam
Amsterdam, The Netherlands

Jack J. Dongarra ⓘ
University of Tennessee
Knoxville, TN, USA

Sérgio Brissos
Intellegibilis
Setúbal, Portugal

ISSN 0302-9743   ISSN 1611-3349   (electronic)
Lecture Notes in Computer Science
ISBN 978-3-030-50425-0   ISBN 978-3-030-50426-7   (eBook)
https://doi.org/10.1007/978-3-030-50426-7

LNCS Sublibrary: SL1 – Theoretical Computer Science and General Issues

This Springer imprint is published by the registered company Springer Nature Switzerland AG
The registered company address is: Gewerbestrasse 11, 6330 Cham, Switzerland

Preface

Twenty Years of Computational Science

Welcome to the 20th Annual International Conference on Computational Science (ICCS – https://www.iccs-meeting.org/iccs2020/).

During the preparation for this 20th edition of ICCS we were considering all kinds of nice ways to celebrate two decennia of computational science. Afterall when we started this international conference series, we never expected it to be so successful and running for so long at so many different locations across the globe! So we worked on a mind-blowing line up of renowned keynotes, music by scientists, awards, a play written by and performed by computational scientists, press attendance, a lovely venue... you name it, we had it all in place. Then corona hit us.

After many long debates and considerations, we decided to cancel the physical event but still support our scientists and allow for publication of their accepted peer-reviewed work. We are proud to present the proceedings you are reading as a result of that.

ICCS 2020 is jointly organized by the University of Amsterdam, NTU Singapore, and the University of Tennessee.

The International Conference on Computational Science is an annual conference that brings together researchers and scientists from mathematics and computer science as basic computing disciplines, as well as researchers from various application areas who are pioneering computational methods in sciences such as physics, chemistry, life sciences, engineering, arts and humanitarian fields, to discuss problems and solutions in the area, to identify new issues, and to shape future directions for research.

Since its inception in 2001, ICCS has attracted increasingly higher quality and numbers of attendees and papers, and 2020 was no exception, with over 350 papers accepted for publication. The proceedings series have become a major intellectual resource for computational science researchers, defining and advancing the state of the art in this field.

The theme for ICCS 2020, "Twenty Years of Computational Science", highlights the role of Computational Science over the last 20 years, its numerous achievements, and its future challenges. This conference was a unique event focusing on recent developments in: scalable scientific algorithms, advanced software tools, computational grids, advanced numerical methods, and novel application areas. These innovative novel models, algorithms, and tools drive new science through efficient application in areas such as physical systems, computational and systems biology, environmental systems, finance, and others.

This year we had 719 submissions (230 submissions to the main track and 489 to the thematic tracks). In the main track, 101 full papers were accepted (44%). In the thematic tracks, 249 full papers were accepted (51%). A high acceptance rate in the thematic tracks is explained by the nature of these, where many experts in a particular field are personally invited by track organizers to participate in their sessions.

ICCS relies strongly on the vital contributions of our thematic track organizers to attract high-quality papers in many subject areas. We would like to thank all committee members from the main and thematic tracks for their contribution to ensure a high standard for the accepted papers. We would also like to thank Springer, Elsevier, the Informatics Institute of the University of Amsterdam, the Institute for Advanced Study of the University of Amsterdam, the SURFsara Supercomputing Centre, the Netherlands eScience Center, the VECMA Project, and Intellegibilis for their support. Finally, we very much appreciate all the Local Organizing Committee members for their hard work to prepare this conference.

We are proud to note that ICCS is an A-rank conference in the CORE classification.

We wish you good health in these troubled times and hope to see you next year for ICCS 2021.

June 2020

Valeria V. Krzhizhanovskaya
Gábor Závodszky
Michael Lees
Jack Dongarra
Peter M. A. Sloot
Sérgio Brissos
João Teixeira

Organization

Thematic Tracks and Organizers

Advances in High-Performance Computational Earth Sciences: Applications and Frameworks – IHPCES

Takashi Shimokawabe
Kohei Fujita
Dominik Bartuschat

Agent-Based Simulations, Adaptive Algorithms and Solvers – ABS-AAS

Maciej Paszynski
David Pardo
Victor Calo
Robert Schaefer
Quanling Deng

Applications of Computational Methods in Artificial Intelligence and Machine Learning – ACMAIML

Kourosh Modarresi
Raja Velu
Paul Hofmann

Biomedical and Bioinformatics Challenges for Computer Science – BBC

Mario Cannataro
Giuseppe Agapito
Mauro Castelli
Riccardo Dondi
Rodrigo Weber dos Santos
Italo Zoppis

Classifier Learning from Difficult Data – CLD^2

Michał Woźniak
Bartosz Krawczyk
Paweł Ksieniewicz

Complex Social Systems through the Lens of Computational Science – CSOC

Debraj Roy
Michael Lees
Tatiana Filatova

Computational Health – CompHealth

Sergey Kovalchuk
Stefan Thurner
Georgiy Bobashev

Computational Methods for Emerging Problems in (dis-)Information Analysis – DisA

Michal Choras
Konstantinos Demestichas

Computational Optimization, Modelling and Simulation – COMS

Xin-She Yang
Slawomir Koziel
Leifur Leifsson

Computational Science in IoT and Smart Systems – IoTSS

Vaidy Sunderam
Dariusz Mrozek

Computer Graphics, Image Processing and Artificial Intelligence – CGIPAI

Andres Iglesias
Lihua You
Alexander Malyshev
Hassan Ugail

Data-Driven Computational Sciences – DDCS

Craig C. Douglas
Ana Cortes
Hiroshi Fujiwara
Robert Lodder
Abani Patra
Han Yu

Machine Learning and Data Assimilation for Dynamical Systems – MLDADS

Rossella Arcucci
Yi-Ke Guo

Meshfree Methods in Computational Sciences – MESHFREE

Vaclav Skala
Samsul Ariffin Abdul Karim
Marco Evangelos Biancolini
Robert Schaback

Rongjiang Pan
Edward J. Kansa

Multiscale Modelling and Simulation – MMS

Derek Groen
Stefano Casarin
Alfons Hoekstra
Bartosz Bosak
Diana Suleimenova

Quantum Computing Workshop – QCW

Katarzyna Rycerz
Marian Bubak

Simulations of Flow and Transport: Modeling, Algorithms and Computation – SOFTMAC

Shuyu Sun
Jingfa Li
James Liu

Smart Systems: Bringing Together Computer Vision, Sensor Networks and Machine Learning – SmartSys

Pedro J. S. Cardoso
João M. F. Rodrigues
Roberto Lam
Janio Monteiro

Software Engineering for Computational Science – SE4Science

Jeffrey Carver
Neil Chue Hong
Carlos Martinez-Ortiz

Solving Problems with Uncertainties – SPU

Vassil Alexandrov
Aneta Karaivanova

Teaching Computational Science – WTCS

Angela Shiflet
Alfredo Tirado-Ramos
Evguenia Alexandrova

Uncertainty Quantification for Computational Models – UNEQUIvOCAL

Wouter Edeling
Anna Nikishova
Peter Coveney

Program Committee and Reviewers

Ahmad Abdelfattah
Samsul Ariffin
  Abdul Karim
Evgenia Adamopoulou
Jaime Afonso Martins
Giuseppe Agapito
Ram Akella
Elisabete Alberdi Celaya
Luis Alexandre
Vassil Alexandrov
Evguenia Alexandrova
Hesham H. Ali
Julen Alvarez-Aramberri
Domingos Alves
Julio Amador Diaz Lopez
Stanislaw
  Ambroszkiewicz
Tomasz Andrysiak
Michael Antolovich
Hartwig Anzt
Hideo Aochi
Hamid Arabnejad
Rossella Arcucci
Khurshid Asghar
Marina Balakhontceva
Bartosz Balis
Krzysztof Banas
João Barroso
Dominik Bartuschat
Nuno Basurto
Pouria Behnoudfar
Joern Behrens
Adrian Bekasiewicz
Gebrai Bekdas
Stefano Beretta
Benjamin Berkels
Martino Bernard

Daniel Berrar
Sanjukta Bhowmick
Marco Evangelos
  Biancolini
Georgiy Bobashev
Bartosz Bosak
Marian Bubak
Jérémy Buisson
Robert Burduk
Michael Burkhart
Allah Bux
Aleksander Byrski
Cristiano Cabrita
Xing Cai
Barbara Calabrese
Jose Camata
Mario Cannataro
Alberto Cano
Pedro Jorge Sequeira
  Cardoso
Jeffrey Carver
Stefano Casarin
Manuel Castañón-Puga
Mauro Castelli
Eduardo Cesar
Nicholas Chancellor
Patrikakis Charalampos
Ehtzaz Chaudhry
Chuanfa Chen
Siew Ann Cheong
Andrey Chernykh
Lock-Yue Chew
Su Fong Chien
Marta Chinnici
Sung-Bae Cho
Michal Choras
Loo Chu Kiong

Neil Chue Hong
Svetlana Chuprina
Paola Cinnella
Noélia Correia
Adriano Cortes
Ana Cortes
Enrique
  Costa-Montenegro
David Coster
Helene Coullon
Peter Coveney
Attila Csikasz-Nagy
Loïc Cudennec
Javier Cuenca
Yifeng Cui
António Cunha
Ben Czaja
Pawel Czarnul
Flávio Martins
Bhaskar Dasgupta
Konstantinos Demestichas
Quanling Deng
Nilanjan Dey
Khaldoon Dhou
Jamie Diner
Jacek Dlugopolski
Simona Domesová
Riccardo Dondi
Craig C. Douglas
Linda Douw
Rafal Drezewski
Hans du Buf
Vitor Duarte
Richard Dwight
Wouter Edeling
Waleed Ejaz
Dina El-Reedy

Amgad Elsayed
Nahid Emad
Chriatian Engelmann
Gökhan Ertaylan
Alex Fedoseyev
Luis Manuel Fernández
Antonino Fiannaca
Christos
    Filelis-Papadopoulos
Rupert Ford
Piotr Frackiewicz
Martin Frank
Ruy Freitas Reis
Karl Frinkle
Haibin Fu
Kohei Fujita
Hiroshi Fujiwara
Takeshi Fukaya
Wlodzimierz Funika
Takashi Furumura
Ernst Fusch
Mohamed Gaber
David Gal
Marco Gallieri
Teresa Galvao
Akemi Galvez
Salvador García
Bartlomiej Gardas
Delia Garijo
Frédéric Gava
Piotr Gawron
Bernhard Geiger
Alex Gerbessiotis
Ivo Goncalves
Antonio Gonzalez Pardo
Jorge
    González-Domínguez
Yuriy Gorbachev
Pawel Gorecki
Michael Gowanlock
Manuel Grana
George Gravvanis
Derek Groen
Lutz Gross
Sophia
    Grundner-Culemann

Pedro Guerreiro
Tobias Guggemos
Xiaohu Guo
Piotr Gurgul
Filip Guzy
Pietro Hiram Guzzi
Zulfiqar Habib
Panagiotis Hadjidoukas
Masatoshi Hanai
John Hanley
Erik Hanson
Habibollah Haron
Carina Haupt
Claire Heaney
Alexander Heinecke
Jurjen Rienk Helmus
Álvaro Herrero
Bogumila Hnatkowska
Maximilian Höb
Erlend Hodneland
Olivier Hoenen
Paul Hofmann
Che-Lun Hung
Andres Iglesias
Takeshi Iwashita
Alireza Jahani
Momin Jamil
Vytautas Jancauskas
João Janeiro
Peter Janku
Fredrik Jansson
Jirí Jaroš
Caroline Jay
Shalu Jhanwar
Zhigang Jia
Chao Jin
Zhong Jin
David Johnson
Guido Juckeland
Maria Juliano
Edward J. Kansa
Aneta Karaivanova
Takahiro Katagiri
Timo Kehrer
Wayne Kelly
Christoph Kessler

Jakub Klikowski
Harald Koestler
Ivana Kolingerova
Georgy Kopanitsa
Gregor Kosec
Sotiris Kotsiantis
Ilias Kotsireas
Sergey Kovalchuk
Michal Koziarski
Slawomir Koziel
Rafal Kozik
Bartosz Krawczyk
Elisabeth Krueger
Valeria Krzhizhanovskaya
Pawel Ksieniewicz
Marek Kubalcík
Sebastian Kuckuk
Eileen Kuehn
Michael Kuhn
Michal Kulczewski
Krzysztof Kurowski
Massimo La Rosa
Yu-Kun Lai
Jalal Lakhlili
Roberto Lam
Anna-Lena Lamprecht
Rubin Landau
Johannes Langguth
Elisabeth Larsson
Michael Lees
Leifur Leifsson
Kenneth Leiter
Roy Lettieri
Andrew Lewis
Jingfa Li
Khang-Jie Liew
Hong Liu
Hui Liu
Yen-Chen Liu
Zhao Liu
Pengcheng Liu
James Liu
Marcelo Lobosco
Robert Lodder
Marcin Los
Stephane Louise

Frederic Loulergue
Paul Lu
Stefan Luding
Onnie Luk
Scott MacLachlan
Luca Magri
Imran Mahmood
Zuzana Majdisova
Alexander Malyshev
Muazzam Maqsood
Livia Marcellino
Tomas Margalef
Tiziana Margaria
Svetozar Margenov
Urszula
    Markowska-Kaczmar
Osni Marques
Carmen Marquez
Carlos Martinez-Ortiz
Paula Martins
Flávio Martins
Luke Mason
Pawel Matuszyk
Valerie Maxville
Wagner Meira Jr.
Roderick Melnik
Valentin Melnikov
Ivan Merelli
Choras Michal
Leandro Minku
Jaroslaw Miszczak
Janio Monteiro
Kourosh Modarresi
Fernando Monteiro
James Montgomery
Andrew Moore
Dariusz Mrozek
Peter Mueller
Khan Muhammad
Judit Muñoz
Philip Nadler
Hiromichi Nagao
Jethro Nagawkar
Kengo Nakajima
Ionel Michael Navon
Philipp Neumann

Mai Nguyen
Hoang Nguyen
Nancy Nichols
Anna Nikishova
Hitoshi Nishizawa
Brayton Noll
Algirdas Noreika
Enrique Onieva
Kenji Ono
Eneko Osaba
Aziz Ouaarab
Serban Ovidiu
Raymond Padmos
Wojciech Palacz
Ivan Palomares
Rongjiang Pan
Joao Papa
Nikela Papadopoulou
Marcin Paprzycki
David Pardo
Anna Paszynska
Maciej Paszynski
Abani Patra
Dana Petcu
Serge Petiton
Bernhard Pfahringer
Frank Phillipson
Juan C. Pichel
Anna
    Pietrenko-Dabrowska
Laércio L. Pilla
Armando Pinho
Tomasz Piontek
Yuri Pirola
Igor Podolak
Cristina Portales
Simon Portegies Zwart
Roland Potthast
Ela Pustulka-Hunt
Vladimir Puzyrev
Alexander Pyayt
Rick Quax
Cesar Quilodran Casas
Barbara Quintela
Ajaykumar Rajasekharan
Celia Ramos

Lukasz Rauch
Vishal Raul
Robin Richardson
Heike Riel
Sophie Robert
Luis M. Rocha
Joao Rodrigues
Daniel Rodriguez
Albert Romkes
Debraj Roy
Katarzyna Rycerz
Alberto Sanchez
Gabriele Santin
Alex Savio
Robert Schaback
Robert Schaefer
Rafal Scherer
Ulf D. Schiller
Bertil Schmidt
Martin Schreiber
Alexander Schug
Gabriela Schütz
Marinella Sciortino
Diego Sevilla
Angela Shiflet
Takashi Shimokawabe
Marcin Sieniek
Nazareen Sikkandar
    Basha
Anna Sikora
Janaína De Andrade Silva
Diana Sima
Robert Sinkovits
Haozhen Situ
Leszek Siwik
Vaclav Skala
Peter Sloot
Renata Slota
Grazyna Slusarczyk
Sucha Smanchat
Marek Smieja
Maciej Smolka
Bartlomiej Sniezynski
Isabel Sofia Brito
Katarzyna Stapor
Bogdan Staszewski

Jerzy Stefanowski
Dennis Stevenson
Tomasz Stopa
Achim Streit
Barbara Strug
Pawel Strumillo
Dante Suarez
Vishwas H. V. Subba Rao
Bongwon Suh
Diana Suleimenova
Ray Sun
Shuyu Sun
Vaidy Sunderam
Martin Swain
Alessandro Taberna
Ryszard Tadeusiewicz
Daisuke Takahashi
Zaid Tashman
Osamu Tatebe
Carlos Tavares Calafate
Kasim Tersic
Yonatan Afework
Tesfahunegn
Jannis Teunissen
Stefan Thurner

Nestor Tiglao
Alfredo Tirado-Ramos
Arkadiusz Tomczyk
Mariusz Topolski
Paolo Trunfio
Ka-Wai Tsang
Hassan Ugail
Eirik Valseth
Pavel Varacha
Pierangelo Veltri
Raja Velu
Colin Venters
Gytis Vilutis
Peng Wang
Jianwu Wang
Shuangbu Wang
Rodrigo Weber
dos Santos
Katarzyna
Wegrzyn-Wolska
Mei Wen
Lars Wienbrandt
Mark Wijzenbroek
Peter Woehrmann
Szymon Wojciechowski

Maciej Woloszyn
Michal Wozniak
Maciej Wozniak
Yu Xia
Dunhui Xiao
Huilin Xing
Miguel Xochicale
Feng Xu
Wei Xue
Yoshifumi Yamamoto
Dongjia Yan
Xin-She Yang
Dongwei Ye
Wee Ping Yeo
Lihua You
Han Yu
Gábor Závodszky
Yao Zhang
H. Zhang
Jinghui Zhong
Sotirios Ziavras
Italo Zoppis
Chiara Zucco
Pawel Zyblewski
Karol Zyczkowski

Contents – Part V

Computational Optimization, Modelling and Simulation

Computational Science in IoT and Smart Systems

Computer Graphics, Image Processing and Artificial Intelligence

Computational Optimization, Modelling and Simulation

Information Theory-Based Feature Selection: Minimum Distribution Similarity with Removed Redundancy

Yu Zhang[ID], Zhuoyi Lin, and Chee Keong Kwoh[✉]

School of Computer Science and Engineering, Nanyang Technological University, 50 Nanyang Avenue, Singapore 639798, Singapore
{YUOO7, ZHUOYIOO1, ASCKKWOH}@ntu.edu.sg

Abstract. Feature selection is an important preprocessing step in pattern recognition. In this paper, we presented a new feature selection approach in two-class classification problems based on information theory, named minimum Distribution Similarity with Removed Redundancy (mDSRR). Different from the previous methods which use mutual information and greedy iteration with a loss function to rank the features, we rank features according to their distribution similarities in two classes measured by relative entropy, and then remove the high redundant features from the sorted feature subsets. Experimental results on datasets in varieties of fields with different classifiers highlight the value of mDSRR on selecting feature subsets, especially so for choosing small size feature subset. mDSRR is also proved to outperform other state-of-the-art methods in most cases. Besides, we observed that the mutual information may not be a good practice to select the initial feature in the methods with subsequent iterations.

Keywords: Feature selection · Feature ranking · Information theory · Redundancy

1 Introduction

In many pattern recognition applications, the original dataset can be in a large feature size and may contain irrelevant and redundant features, which would be detrimental to the training efficiency and model performance [1, 2]. In order to reduce the undesirable effect of the curse of dimensionality and to simply the model for parsimony [3], an intuitive way is to determine a feature subset, and this process is known as feature selection, or variable selection.

The ideal situation for feature selection is to select the optimal feature subset that maximize the prediction accuracy, however, this is impractical due to the intractable computation caused by the exhausted searching over the whole feature space, especially when the prior knowledge is limited and the dependency among features remains unknown. Therefore, varieties of suboptimal feature selection algorithms have been proposed, and they are mainly divided into three categories according to the evaluation metric: Wrapper, Embedded, and Filter methods [3].

V. V. Krzhizhanovskaya et al. (Eds.): ICCS 2020, LNCS 12141, pp. 3–17, 2020.
https://doi.org/10.1007/978-3-030-50426-7_1

Both Wrapper and Embedded methods are dependent on the classifiers. Wrapper methods score the feature subsets via the error rate on a given classifier, and certain searching strategies are employed to generate the next feature subset to avoid NP-hard problem [4, 5]. Such kinds of methods take the interactions among features into consideration and always can find the best subset for a particular learning algorithm, however, they are computation complex and prone to over-fitting, additionally, the subset needs to be reselected when changing the classifiers [6, 7]. In the meantime, embedded methods select the features during the model construction processes and thus they are more efficient than Wrappers [3], but the main limitation is that they rely heavily on the hypotheses the classifier makes [6, 8].

Unlike Wrapper and Embedded methods, Filter methods are independent of the classifiers, therefore they can better expose the relationships among features. Besides, Filter methods are simpler, faster and more scalable than Wrapper and Embedded methods, as they rank the features according to the proximity measures, such as the mutual information (MI) [3], correlation [10], chi-square [11] and relief-based algorithms [12]. The determination of the best feature subset of Filter methods is to select a cut-off point on their ranked features via the cross validation. But the drawback of Filter methods is that they cannot investigate the interaction between the features and classifiers [3, 5].

In Filter methods, information theory-based measure which exploiting not only the relationships between features and labels, but also the dependencies among features, plays a dominant role [9, 13]. Battiti [9] proposed Mutual Information Feature Selection (MIFS) method, which finds feature subset via greedy selection according to the MI between feature subsets and labels. After that, varieties of methods are presented to improve MIFS. Kwak and Choi developed MIFD-U method by considering more about the MI between features and labels [14], Peng *et al.* proposed minimal-redundancy-maximal-relevance (mRMR) framework by providing the theoretical analysis and combining with wrappers [15], Estévez *et al.* proposed Normalised MIFS (NMIFS) which replaces the MI with normalized MI [16], and Hoque *et al.* developed MIFS-ND by considering both MI of feature-feature and feature-label [17].

Consequently, a lot of information theory-based measures have been designed and adopted in Filter methods. Joint MI (JMI) [18], Interaction Capping (ICAP) [19], Interaction Gain Feature Selection (IGFS) [20] and Joint MI Maximisation (JMIM) [7] were raised by taking the joint MI into consideration, from which ICAP and IGFS depend on the feature interaction; Conditional MI Maximization (CMIM) criterion [21], Conditional Infomax Feature Extraction (CIFE) [22] and Conditional MIFS (CMIFS) [23] were proposed by adopting conditional MI; Double Input Symmetrical Relevance (DISR) method was developed by using symmetrical relevance as the objective function [24].

Intuitively, the common procedure for the above information theory related Filter methods is, selecting the initial feature with maximum MI, then increasing the feature subset size on previously defined features according to an objective function. Therefore, the selection of the initial feature will influence the determination of the final feature subset. As a result, a good initial feature may lead to a smaller feature subset size as well as a good classification performance. However, the maximum MI between the features and labels may not be a good criteria to determine the initial feature. We argue that the initial feature found in this way may have a lower classification ability than many other features, which will further lead to relatively large feature subset size or a poor performance that actually can be avoided.

In this paper, we proposed a new approach to select feature subset based on information theory in two-class classification problems, named minimum Distribution Similarity with Removed Redundancy (mDSRR). Different from previous methods which use greedy search approach in adding the features into the feature subset, we rank the features according to their distribution similarity and then remove the redundancy from the ranked feature subset. Furthermore, we compared mDSRR with other state-of-the-art methods on 11 public datasets with different classifiers, results show that mDSRR is superior to other methods, which can achieve high performance with only a few features. Additionally, by comparing the classification performance and the initial feature defined by mDSRR and other methods, we demonstrated that using MI to determine the initial feature may not be a good practice.

This paper is organized as follows: Sect. 2 provides the background of information theory, theoretical analysis and the implementation of the proposed method, Sect. 3 demonstrates the results of the experiments and discuss the results, and Sect. 4 concludes this work.

2 Methods

2.1 Information Theory

This section briefly introduces the related concepts of information theory that will be used in this work. Suppose random variables X and Y represent feature vectors, and random variable C denotes the class label.

The entropy is a measure of the amount of uncertainty before a value of random variable is known, for a discrete random variable which takes value x from the alphabet χ, i.e. $x \in \chi$, with probability $p_X(x)$, the entropy is defined as

$$H(X) = -\sum_{x \in \chi} p_X(x) \log(p_X(x)) \tag{1}$$

The entropy is positive and bounded, i.e. $0 \le H(X) \le \log(|X|)$. Similarly, when taking two discrete random variables X and Y and their joint probability $p(x, y)$ into consideration, the joint entropy can be represented as

$$H(X, Y) = -\sum_{x \in \chi} \sum_{y \in y} p(x, y) \log(p(x, y)) \tag{2}$$

The conditional entropy of X given that C is a measure of the average additional information in X when C is known, which is defined as

$$H(X|C) = -\sum_{c \in C} \sum_{x \in \chi} p(x, c) \log(p(x|c)) \tag{3}$$

where $p(x|c)$ is the conditional probability for x given that c, and according to the chain rule,

$$H(X,C) = H(X|C) + H(C) \tag{4}$$

Mutual information of X and Y is the average amount of information that we get about X from observing Y, or in other word, the reduction in the uncertainty of X due to the knowledge of Y, it is represented as

$$I(X;Y) = H(X) - H(X|Y) = H(X) + H(Y) - H(X,Y) \tag{5}$$

The mutual information is symmetrical and non-negative, it equals to 0 only when the variables are statistically independent. Similarly, the conditional mutual information of X and Y given that C is represented as

$$I(X;Y|C) = H(X|C) - H(X|Y,C) \tag{6}$$

Kullback-Leibler divergence, also known as relative entropy, is another important concept that will be used in this work. It measures the difference between two probability mass vectors, if we denote the two vectors as p and q, then the relative entropy between p and q is defined as

$$D(p||q) = \sum_{x \in \mathcal{X}} p(x) \log\left(\frac{p(x)}{q(x)}\right) \tag{7}$$

Obviously, it is asymmetric between **p** and **q**.

2.2   Minimum Distribution Similarity Feature Ranking

We try to rank the features according to their distribution similarities between classes to separate the objects, where the distribution similarity can be measured by the relative entropy.

Recalling the mathematic representation of relative entropy, Eq. (7) in Sect. **2.1**, D $(p||q)$ represents the "distance" between the probability mass vectors p and q, and it also can be regarded as the measurement of information loss of q from p. Here, we regard x as the event representing the instances in one type of feature whose values located in a certain range, or a small bin, X as the event set, and p and q as the distributions of x for two classes separately. In this work, we use small bins rather than using the exact value of feature, because when $q(x) = 0$, $p(x) \neq 0$, $\log(\frac{p(x)}{q(x)}) \to \infty$; when $p(x) = 0$, $q(x) \neq 0$, $p(x) \log\left(\frac{p(x)}{q(x)}\right) = 0$. Obviously, if there is no overlapped feature values in two classes, $p(x) \log\left(\frac{p(x)}{q(x)}\right)$ would either equal to ∞ or 0, under this circumstance, $D(p||q)$ would equal to ∞ or 0. It is worth noting that if a large number of features do not have overlapped values in different classes, all these features would be assigned value of infinity or zero and we cannot know which one is more important.

Although introducing small bins can solve the above problem, if the bin number is selected too small, the difference for two classes cannot be captured accurately; if it is too large, the relative entropy value would tend to be the same as when taking exact value stated above. Therefore, several choices of bin number determined according to the corresponding total instance amount are evaluated to select the proper solution here. Furthermore, since $D(p\|q) \neq D(q\|p)$, we use D = $D(p\|q) + D(q\|p)$ as the measure to sort the features.

Another problem is how to deal with the circumstance when D $\rightarrow \infty$ in practical. In this work, we use $q(x) = \frac{1}{instance\ number\ of\ class\ II}$ in $D(p\|q)$ to replace $q(x) = 0$ to avoid D becomes infinity. This practice, although may not so accurate, can reflect the trend of D. On one side of the spectrum, when the number of total instances in a certain bin is large, the measure of distribution similarity when there is only one sample of class II in this bin is almost the same as that when there is no sample of class II, both of the two circumstance means an extremely low similarity between the distribution for two classes. At another extreme, when the instance number in a certain bin is small and there is only samples of class I in this bin, due to its small sample number, $p(x)$ tends to be 0, $p(x) \log\left(\frac{p(x)}{q(x)}\right)$ would be an extremely small value and would not contain too much information. Therefore, replacing $q(x) = 0$ with $q(x) = \frac{1}{instance\ number\ of\ class\ II}$ in the practical calculation is rational. This practice can reflect the real distribution similarity trend, especially under the truth that this circumstance would not happen too much with the application of bin. The above algorithm has been briefly described in a previous work [25].

2.3   Minimum Distribution Similarity with Removed Redundancy (mDSRR)

In Sect. 2.2, we already got a feature list sorted according to their potential importance to classification, where the irrelevant features will be ranked in the back. However, the redundancy may still exist. The combination of the features redundant to each other may not contribute to higher performance but lead to overfitting. Therefore, removing redundant features is necessary to improve model's performance and efficiency Different from previous works which use MI or conditional MI between features and label, in this work, we consider the conditional MI between features under the condition of label as the criteria to remove redundancy. In particular, we adopt conditional MI rather than the MI because according to, features have high mutual information may have different information within the class, hence taking labels into consideration is essential.

The complexity to calculate the conditional MI between two features within a feature set with feature size n is proportional to $(n - 1) + (n - 2) + \cdots + 1 = \frac{n^2-n}{2}$, as the increase of n, the calculation complexity will grow as n^2. In case of the feature number is large, it would be impractical to find all conditional MI as the time consumption would be large. Hence, we only calculate the conditional MI for the first m ranked features.

From Eq. (4) and (6), we can obtain

$$I(X;Y|C) = H(X|C) - H(X|Y,C) = H(X|C) - H(X,Y|C) + H(Y|C)$$
$$= H(X|C) + H(Y|C) + H(C) - H(X,Y,C) \tag{8}$$

to calculate the conditional MI between two features under the condition of labels. The relationship among the items in Eq. (8) is illustrated in Fig. 1. From the Venn diagram, the redundancy between feature X and Y under the condition of class C can be measured as the percentage of $I(X;Y|C)$ taken in $H(X,Y|C)$, which can be represented as the ratio r between $I(X;Y|C)$ and $H(X|C) + H(Y|C) - I(X;Y|C)$, that is

$$r = \frac{I(X;Y|C)}{H(X|C) + H(Y|C) - I(X;Y|C)} \tag{9}$$

where $0 \leq r \leq 1$. In Eq. (9), $r = 0$ means X and Y are independent under the condition of C; the larger the r is, the larger the redundancy between X and Y; when $r = 1$, X and Y are totally dependent, or totally redundant under the condition of the label. And once r exceeds the predefined threshold, the feature which is ranked in later position between two redundant features will be removed from the feature subset.

Algorithm 1. mDSRR method

---

1 **Input:** class I data feature set P, class II data feature set Q, bin number n.
2 **Initialize:** initial feature set F.
3 **Begin**
4   for $i = 1, \dots, |P|$*:
5      $D_i = D(p_i||q_i, n) + D(q_i||p_i, n)$
6   end
7   sort D_i from largest value to smallest value, get a new feature set F_{new}.
8   take the first m items in F_{new}, $\rightarrow F_m$.
9   for $j = 1, \dots, m$:
10   calculate $H(F_m(j)|C)$, $H(F_m(j+1)|C)$, and $I(F_m(j); F_m(j+1)|C)$
11     $r_{j,j+1} = \frac{I(F_m(j);F_m(j+1)|C)}{H(F_m(j)|C) + H(F_m(j+1)|C) - I(F_m(j);F_m(j+1)|C)}$
12   if $r_{j,j+1} > r_{th}$:
13   set $F_m \leftarrow F_m\{f_m(j+1)\}$
14   end
15   end
16 **End**
17 **Output:** F_m.

---

* $|P| = |Q|$ is the number of features.

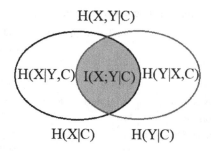

Fig. 1. Venn diagram of the relationship among items in Eq. (8).

Combining the feature ranking algorithm in Sect. 2.2 and the remove redundancy step described above, our feature selection method is finally developed and named as mDSRR (minimum Distribution Similarity with Removed Redundancy). The algorithm is summarized in Algorithm 1.

3   Results

To highlight the effectiveness of mDSRR in feature selection, five state-of-the-art methods, including mRMR [15], DISR [24], ICAP [19], CIFE [22], and CMIM [21] are used for comparison. These five approaches are chosen because that, (i) they are all Filter methods based on information theory; (ii) they cover information theory measures like MI, joint MI and conditional MI; (iii) they are classic and popular feature selection methods which have been applied widely in diversity of areas. All methods are evaluated on 11 public datasets and 4 kinds of classifiers, including Supporting Vector Machine (SVM), Decision Tree (DT), Random Forest (RF) and Naïve Bayes (NB). The average classification accuracy (Acc) for the 10-fold cross validation, which can reflect the general performance while avoiding the bias, is recorded and works as the criteria to evaluate different methods.

3.1   Datasets

Eleven public datasets from UCI Repository [26] are used for comparison, they were all in two classes with multivariate and integer or real attributes, and covering a diversity of areas, such as life, economic, chemistry and biology, medical, computer, artificial and physical. These datasets vary in instance numbers and feature numbers, hence can be used for fair comparisons, the related information for these datasets are briefly listed in Table 1.

Table 1. Datasets used in this work.

Dataset name	Number of instances	Number of features
Arcene	200	10000
Audit	776	17
Biodegradation	1055	41
Breast Cancer Wisconsin	683 (origin 699)	10
Breast Cancer Coimbra	116	9
Diabetic Retinopathy Debrecen	1151	19
Madelon	2600	500
Musk	7074	166
Parkinson	756	751
Sonar	208	60
Spambase	4601	57

3.2 Parameter Determination in mDSRR

In mDSRR, the parameter bin number needs to be assured, to achieve it, we did experiments on 11 datasets to see the impacts of the selection of bin number. Following the rules what we described in Sect. 2.2, the bin number cannot be selected too large and too small, we consider the circumstance when bin number equals to $\frac{1}{10}, \frac{1}{15}, \frac{1}{20}, \frac{1}{30}$ and $\frac{1}{50}$ of the total number of instance for dataset with a relatively large instance amount, i.e. > 600, and $\frac{1}{2}, \frac{1}{5}, \frac{1}{10}, \frac{1}{15}$ and $\frac{1}{20}$ of the total number of instance for dataset with a relatively small instance amount, i.e. < 300. The comparison of the Acc achieved by feature subsets with different choices of bin number for 11 datasets are plotted in Supplementary Figure S1 and S2. The best choice of bin number is selected as the one that achieves the best Acc most times in four classifiers, if two or more

Table 2. The best choice of bin number for 11 datasets.

Datasets	Number of instances	Best bin number in portion*
Breast Cancer Coimbra	116	1/5
Arcene	200	1/15
Sonar	208	–**
Breast Cancer Wisonsin	683	–
Parkinson	756	1/20
Audit	776	1/20 or 1/30
Biodegradation	1055	1/50
Diabetic Retinopathy Debrecen	1151	1/50
Madelon	2600	1/50
Spambase	4601	1/50
Musk	7074	1/50

*Best bin number in portion is the portion of bin number taken up in total instance number.
**Means the choice of best bin number percentage is hard to define according to the existing results.

choices of bin number realize the best Acc equal times, then we consider their performance under the same feature subset size. The times for different choices of bin number which achieving the best Acc are counted in Supplementary Table S1, and the summarize of the best choices of bin number for these datasets are listed in Table 2.

From the experiment results, the portion that the bin number taking up in the total instances decreased as the total number of instances increases. For datasets with instance number at around 100, we use bin number roughly as 1/5 of the total instance number; for total instance number around 200–500, we use its 1/15 as bin number; for total instance number around 500–1000, we use 1/20; for total instance number larger than 1000, we use 1/50.

Actually, with the histogram idea, there is no best choice of bin number without a strong assumption about the shape of the distribution, and the parameter we chose is a suboptimal one, but in later sections, we will show the good performance of mDSRR with such choices of bin number.

The other parameter in mDSRR is the redundancy remove threshold r_{th}. Although sometimes the redundancy between two features may be large, they may still contain useful information and should not be deleted, hence we consider the circumstances when r_{th} = 0.9, 0.99 and 0.9999. The feature subset sizes after removing redundancy with different selection of r_{th} are recorded in Supplementary Table S2 and their comparisons are plotted in Supplementary Figure S3. 0.9999 is chosen as the final value of r_{th} which achieves the best performance most times, although in some cases no feature is removed with such a high threshold.

3.3   Performance Comparison on Datasets with Large Feature Size

To compare the performance of different feature selection methods on datasets with a relatively large feature set size, we plotted the average Acc of 10-fold cross validation for feature subsets with different sizes in Fig. 2. For datasets whose feature size exceeding 100, e.g. Arcene, Madelon, Musk and Parkinson, the results of feature subsets with size from 1 to 50 are shown, and for the others (Biodegradation, Spambase and Sonar), the results of feature subsets with all possible sizes are shown. The bin numbers are chosen following the rules we concluded in Sect. 3.2.

The application of mDSRR method on dataset Arcene, Parkinson and Sonar has obvious advantages over other methods, the feature subset determined by mDSRR can achieve the highest value with only a small feature subset size. For example, in Arcene dataset, the feature subset with size 10 selected by mDSRR realize 85% Acc with SVM classifier, while other methods never reach this value no matter how many features are added; although mDSRR does not realize the best Acc in this dataset with NB classifier, the feature subset it determined with only 2 features achieve Acc as high as 71%, which is almost the same as the best Acc achieved by ICAP (71.5%), with feature subset size 32. Furthermore, with the same feature subset size which are less than 25, 40 and 10 in dataset Arcene, Parkinson and Sonar, respectively, the performance of mDSRR are much better than other methods in most cases.

Although the general performances of mDSRR on the remaining 4 datasets is not as outstanding as the above datasets, the advantages of mDSRR still can be found. For Biodegradation dataset, the best Acc are achieved by mRMR with SVM and DT

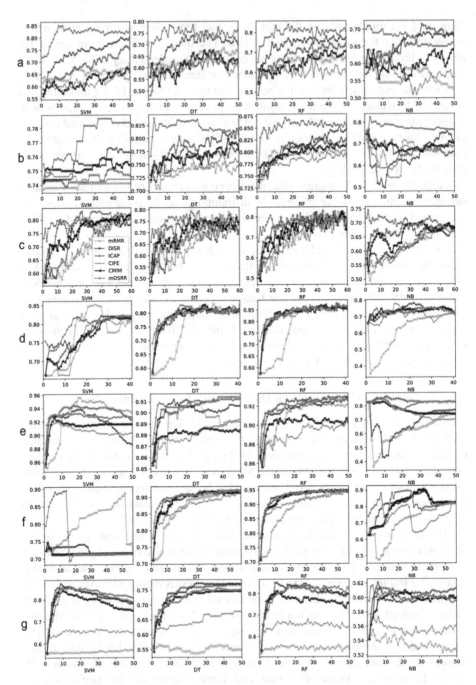

Fig. 2. The average Acc of different feature subset sizes for different feature selection methods with classifier SVM, DT, RF and NB on dataset (a) Arcene, (b) Parkinson, (c) Sonar, (d) Biodegradation, (e) Musk, (f) Spambase, and (g) Madelon.

classifier, but when the feature number is smaller than 16 and 18 in two classifiers separately, the Acc value of mDSRR are higher than that of mRMR with up to 9.7% and 18.7% within the same feature subset sizes. Additionally, mDSRR reaches 75.1% Acc with only 5 features in NB classifier, while DISR, the one realizes the best Acc, reaches the same value with 17 features. For Musk dataset, mDSRR achieves Acc of 93% and 90.9% in SVM and DT separately with only 5 features, while mRMR and ICAP, the methods that achieve the best Acc in two classifiers, reach the same value with 10 and 17 features separately. For Spambase dataset, mDSRR achieves the best Acc of 89.5% with only 14 features in SVM classifier, and when the feature subset sizes are smaller than 18 and 29 in DT and NB classifiers, mDSRR leads to much higher performances than other methods. And for Madelon dataset, the feature subset determined by mDSRR realizes the second-highest Acc value with only 3 features in NB classifier, which far exceeding the performance with the same feature subset size defined by other methods.

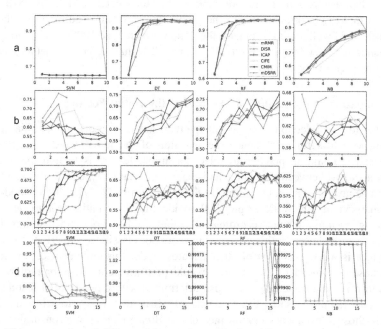

Fig. 3. The average Acc of different feature subset sizes for different feature selection methods with classifier SVM, DT, RF and NB on dataset (a) Breast Cancer Wisconsin, (b) Breast Cancer Coimbra, (c) Diabetic Retinopathy Debrecen, and (d) Audit.

3.4 Performance Comparison on Datasets with Small Feature Size

For datasets with a relatively small feature number, the impact of a single feature can be significant. We plotted the average Acc achieved by 10-fold cross validation obtained by the feature subsets with different sizes selected by different methods for four datasets whose feature set sizes are relatively small, as shown in Fig. 3.

The overall performance of mDSRR is much better than the other methods on all datasets except Audit, where mDSRR achieves the highest Acc with only several features in most cases. For instances, in Breast Cancer Wisconsin dataset, the Acc for feature subsets with size from 1 to 9 selected by mDSRR remain at a high Acc value, i.e. no less than 92%, while the feature subsets determined by other methods only reach this value when the subset size is larger than 4 for DT and RF classifiers, but never exceed 90% for SVM and NB classifiers. In Breast Cancer Coimbra dataset, the remove redundancy step in mDSRR removes 2 to 5 features in 10-fold split datasets, hence we only kept the first 4 features for plotting. The feature subset selected by mDSRR with size 3 in SVM and size 1 in NB classifier achieve the best values, while the feature subsets determined by other methods never reach the same values no matter how many features are used. Similarly, in Diabetic Retinopathy Debrecen dataset, mDSRR removes one feature and it achieves the best Acc with only 8 features in SVM classifier, and with 6 features in DT and RF classifiers, which is superior to other methods with better performance but smaller feature subset size. In addition, for the above three datasets, when feature subset sizes are same and less than a certain value, mDSRR keep leading to higher performance than other methods. As to the Audit dataset, the first feature selected by all methods remains the same which realize 100% Acc, although the performance for later feature subsets may vary, all feature selection methods can be regarded as performing equally.

3.5   Evaluation of the Ability of MI to Find the Initial Feature

Most Filter feature selection methods enlarge their feature subsets on the initial feature according to a loss function, hence the selection of the initial feature is extremely important. The common practice to select the initial feature is to find the one that maximizes the MI between the feature and classes, i.e. $\max\{I(X_i; C)\}$, as this solution is proved to be near the optimal to minimize the Bayes error [22, 27]. All 5 methods used for comparison in this work employ MI to identify their initial feature. However, determining the initial feature according to MI may not be a good practice in practical, the initial feature chosen in this way may have a poor classification performance or its performance is worse than that of other features, under this circumstance, the feature subset chosen followed by this feature would perform poor either or require more features to reach the same result as those determined with a good initial feature.

A good example to show the weakness of MI as the criteria to select the initial feature in those subsequent iteration methods is Breast Cancer Wisconsin dataset. In this dataset, mDSRR determines the 4th feature and the 1st feature in the original feature set as the first and the last feature to be added to the feature subset separately, while the other 5 methods select the 1st feature in original feature set as the initial feature. However, combining the results in Sect. 3.4, the 1st feature in original feature set leads to poor performances with SVM and NB classifiers, whose Acc are around 65% and 53% separately, and when this feature is added to the feature subset found by mDSRR, the high performances of other features, i.e. over 90% Acc, decrease significantly. Particularly, when the 1st feature exists, the accuracies of SVM and NB classifiers can never exceed 65% and 90% separately. Hence, when using the initial feature identified by MI in Breast Cancer Wisconsin dataset, the performance with SVM and NB classifier can hardly be improved. The finding here proved that it is important to properly select the initial feature for methods which based on the subsequent iteration, and the criteria of maximum MI may not be a good choice to determine the initial feature.

4   Conclusion

In this paper, we proposed a new feature selection method based on information theory: minimum Distribution Similarity with Removed Redundancy (mDSRR). mDSRR employs the concept of relative entropy, combined with the histogram idea to rank the features, and the redundancy is removed according to the conditional MI between two features under the condition of classes. The idea of mDSRR is different from previous information theory related Filter feature selection method, which usually follow the practice of determining an initial feature and then enlarging the feature subset on the initial feature according to a loss function.

The comparison results between mDSRR and five state-of-the-art methods on 11 public datasets with four kinds of classifiers in this work show that mDSRR is a valuable method to select effective feature subset. mDSRR leads to better performance than other methods under the same sizes of the feature subsets especially when the sizes are small. Besides, by taking Breast Cancer Wisconsin dataset as an example, we also demonstrate that MI may not be a good practice to determine the initial feature in those subsequent iteration methods.

However, one limitation of mDSRR is that it can only be utilized to two-class classification problems, while are not suitable for multi-class classification problems. Because the relative entropy only can calculate the "distance" between two distributions. For multi-class classification problems, if we measure the distribution similarity of any two classes and then integrate the results, the workload would be heavy.

In a nutshell, mDSRR is a good method to select feature subset, especially to select small size feature subset due to its high efficiency in ranking the features according to their potential contribution to distinct the classes. The successful applications of mDSRR on a range of datasets in different fields with different classifiers highlight its value in feature selection.

Supplementary Data

Supplementary data are available at https://github.com/yuuuuzhang/feature-selection/blob/master/fs_supplementary.docx.

References

1. Yan, H., Hu, T.: Unsupervised dimensionality reduction for high-dimensional data classification. Mach. Learn. Res. **2**, 125–132 (2017)

2. Gu, S., Cheng, R., Jin, Y.: Feature selection for high-dimensional classification using a competitive swarm optimizer. Soft. Comput. **22**(3), 811–822 (2018)
3. Guyon, I., Elisseeff, A.: An introduction to variable and feature selection. J. Mach. Learn. Res. **3**, 1157–1182 (2003)
4. Chandrashekar, G., Sahin, F.: A survey on feature selection methods. Comput. Electr. Eng. **40**(1), 16–28 (2014)
5. Wah, Y.B., Ibrahim, N., Hamid, H.A., Abdul-Rahman, S., Fong, S.: Feature selection methods: case of filter and wrapper approaches for maximising classification accuracy. Pertanika J. Sci. Technol. **26**(1), 329–340 (2018)
6. Jain, D., Singh, V.: Feature selection and classification systems for chronic disease prediction: a review. Egypt. Inform. J. **19**(3), 179–189 (2018)
7. Bennasar, M., Hicks, Y., Setchi, R.: Feature selection using joint mutual information maximisation. Expert Syst. Appl. **42**(22), 8520–8532 (2015)
8. Hira, Z.M., Gillies, D.F.: A review of feature selection and feature extraction methods applied on microarray data. Adv. Bioinform. **2015**, 1–13 (2015)
9. Battiti, R.: Using mutual information for selecting features in supervised neural net learning. IEEE Trans. Neural Netw. **5**(4), 537 (1994)
10. Liu, H., Li, J., Wong, L.: A comparative study on feature selection and classification methods using gene expression profiles and proteomic patterns. Genome Inform. **13**, 51–60 (2002)
11. Jin, X., Xu, A., Bie, R., Guo, P.: Machine learning techniques and chi-square feature selection for cancer classification using SAGE gene expression profiles. In: Li, J., Yang, Q., Tan, A.-H. (eds.) BioDM 2006. LNCS, vol. 3916, pp. 106–115. Springer, Heidelberg (2006). https://doi.org/10.1007/11691730_11
12. Urbanowicz, R.J., Meeker, M., La Cava, W., Olson, R.S., Moore, J.H.: Relief-based feature selection: introduction and review. J. Biomed. Inform. **85**, 189–203 (2018)
13. Torkkola, K.: Feature extraction by non-parametric mutual information maximization. J. Mach. Learn. Res. **3**, 1415–1438 (2003)
14. Kwak, N., Choi, C.-H.: Input feature selection for classification problems. IEEE Trans. Neural Netw. **13**(1), 143–159 (2002)
15. Peng, H., Long, F., Ding, C.D.: Feature selection based on mutual information: criteria of max-dependency, max-relevance, and min-redundancy. IEEE Trans. Patter Anal. Mach. Intell. **1**(8), 1226–1238 (2005)
16. Estévez, P.A., Tesmer, P.A., Perez, C.A., Zurada, J.M.: Normalized mutual information feature selection. IEEE Trans. Neural Netw. **20**(2), 189–201 (2009)
17. Hoque, N., Bhattacharyya, D.K., Kalita, J.K.: MIFS-ND: a mutual information-based feature selection method. Expert Syst. Appl. **41**(14), 6371–6385 (2014)
18. Yang, H., Moody, J.: Feature selection based on joint mutual information. In: Proceedings of International ICSC Symposium on Advances in Intelligent Data Analysis (1999)
19. Jakulin, A.: Machine learning based on attribute interactions. Univerza v Ljubljani (2006)
20. Akadi, A.E., Ouardighi, A.E., Aboutajdine, D.: A powerful feature selection approach based on mutual information. Int. J. Comput. Sci. Netw. Secur. **8**(4), 116 (2008)
21. Fleuret, F.: Fast binary feature selection with conditional mutual information. J. Mach. Learn. Res. **5**, 1531–1555 (2004)
22. Lin, D., Tang, X.: Conditional infomax learning: an integrated framework for feature extraction and fusion. In: Leonardis, A., Bischof, H., Pinz, A. (eds.) ECCV 2006. LNCS, vol. 3951, pp. 68–82. Springer, Heidelberg (2006). https://doi.org/10.1007/11744023_6
23. Cheng, G., Qin, Z., Feng, C., Wang, Y., Li, F.: Conditional mutual information-based feature selection analyzing for synergy and redundancy. ETRI J. **33**(2), 210–218 (2011)

24. Meyer, P.E., Bontempi, G.: On the use of variable complementarity for feature selection in cancer classification. In: Rothlauf, F., et al. (eds.) EvoWorkshops 2006. LNCS, vol. 3907, pp. 91–102. Springer, Heidelberg (2006). https://doi.org/10.1007/11732242_9

25. Zhang, Y., Jia, C., Fullwood, M.J., Kwoh, C.K.: DeepCPP: a deep neural network based on nucleotide bias and minimum distribution similarity feature selection for RNA coding potential prediction. Brief. Bioinform. (2020). https://doi.org/10.1093/bib/bbaa039

26. Dua, D., Graff, C.: UCI Machine Learning Repository (2019). http://archive.ics.uci.edu/ml

27. Vasconcelos, N.: Feature selection by maximum marginal diversity. In: Advances in Neural Information Processing Systems, pp. 1375–1382 (2003)

On the Potential of the Nature-Inspired Algorithms for Pure Binary Classification

Iztok Fister Jr.[1(✉)], Iztok Fister[1], Dušan Fister[2], Grega Vrbančič[1], and Vili Podgorelec[1]

[1] Faculty of Electrical Engineering and Computer Science, University of Maribor, Koroška cesta 46, Maribor, Slovenia
iztok.fister1@um.si
[2] Faculty of Economics and Business, University of Maribor, Razlagova 14, Maribor, Slovenia

Abstract. With the advent of big data, interest for new data mining methods has increased dramatically. The main drawback of traditional data mining methods is the lack of comprehensibility. In this paper, the firefly algorithm was employed for standalone binary classification, where each solution is represented by two classification rules that are easy understandable by users. Implicitly, the feature selection is also performed by the algorithm. The results of experiments, conducted on three well-known datasets publicly available on web, were comparable with the results of the traditional methods in terms of accuracy and, therefore, the huge potential was exhibited by the proposed method.

Keywords: Firefly algorithm · Data mining · Binary classification

1 Introduction

Data Mining is the most complex part of the Knowledge Discovery from Data (KDD) process that is comprised of: Data selection and creation, preprocessing (i.e., data cleaning and transformation), data mining, and evaluation [9]. Typically, the data preprocessing captures the feature extraction and feature selection. The aim of the former is to derive a new, less complex dataset, while the latter to find the best subset of features from a set of features. Classification and clustering are two of the most widely studied tasks of data mining, where the classification is referred to a prediction of the class labels on the basis of test observations during the process of learning [16].

Mainly, the traditional classification methods are Decision Trees [14], Bayesian networks [6], Neural Networks [7], and Support Vector Machines [8]. Although these methods are able to find the local optimal classification models in some situations, the majority of them are not very comprehensible, and thus hard to handle by casual users. Usually, they are time consuming too. Fortunately, searching for the best classification model of all the possible candidates

V. V. Krzhizhanovskaya et al. (Eds.): ICCS 2020, LNCS 12141, pp. 18–28, 2020.
https://doi.org/10.1007/978-3-030-50426-7_2

can be defined as an optimization problem appropriate for solving with stochastic nature-inspired population-based algorithms. Here, the quality of solutions can be evaluated according to classification accuracy and comprehensibility. The majority of these algorithms represents the classification model in terms of "If-then" rules, and are, therefore, close to human comprehension.

Stochastic nature-inspired population-based algorithms have frequently been applied to data mining in the last three decades. For instance, Srikanth et al. in [18] proposed the Genetic Algorithm (GA) for clustering and classification. In the Swarm Intelligence (SI) domain, Particle Swarm Optimization (PSO) and Ant Colony Optimization (ACO) attracted many scientists to use them for solving the problems in data mining. For instance, Sousa et al. [17] compared the implemented PSO algorithms for data mining with the GA, while Ant-Miner, developed by Parpinelly et al. [11] using ACO, was proposed to discover classification rules. The more complete surveys of using EAs in data mining can be found in [4,5,15], while the review of papers describing SI-based algorithms in data mining was presented in [10]. Recently, a Gravitational Search (GS) has achieved excellent results in discovering classification models, as reported by Peng et al. [13].

This paper tries to answer the question whether the stochastic nature-inspired population-based algorithms can be competitive tool for pure binary classification compared with the classical data mining methods. Here, the "pure" means that algorithms perform classification task standalone, i.e., without any interaction with traditional methods. The Firefly Algorithm (FA) [20] for binary classification was proposed, capable of discovering the classification models and evaluating their quality according to a classification accuracy. In our opinion, the main advantage of the FA against the PSO algorithm lays in the principle of FA working, because particles in this algorithm are not dependent on the global best solution as in PSO only, but also on the more attractive particles in the neighborhood. On the other hand, the model in the proposed FA consists of two classification rules, i.e., one for True Negative (TN) and the other for True Positive (TP) classification results. Prior to classification, feature selection is executed implicitly by FA.

The proposed FA was applied to three well-known datasets for binary classification that are publicly available on the web. The obtained results showed big potential in binary classification field that could also be applied for general classification.

The main goals of this paper are as follows:

- to develop the new classification method based on real-coded FA,
- to encode two classification rules simultaneously, and decode by the new genotype-phenotype mapping,
- to perform the feature selection implicitly by the classification,
- to evaluate the proposed method on some binary classification datasets.

In the remainder of the paper, the structure is as follows: Sect. 2 introduces fundamentals of the FA. In Sect. 3, the proposed classification method is described in detail. The experiments and results are presented in Sect. 4, while the paper is concluded with Sect. 5, in which directions for the future work are also outlined.

2 Fundamentals of the Firefly Algorithm

The inspiration for the Firefly Algorithm (FA) was fireflies with flashing lights that can be admired on clear summer nights. The light is a result of complex chemical reactions proceeding in a firefly's body and has two main purposes for the survival of these small lightning bugs: (1) To attract mating partners, and (2) To protect against predators. As follows from a theory of physics, the intensity of the firefly's light decreases with increasing the distance r from the light source, on the one hand, and the light is absorbed by the air as the distance from the source increases, on the other.

Both physical laws of nature are modeled in the FA developed by Yang at 2010 [20], as follows: The FA belongs to a class of SI-based algorithms, and therefore operates with a population of particles representing solutions of the problem in question. Thus, each solution is represented as a real-valued vector, in other words:

$$\mathbf{x}_i^{(t)} = \{x_{i,1}^{(t)}, \dots, x_{i,D}^{(t)}\}, \quad \text{for } i = 1, \dots, N, \tag{1}$$

where N denotes the population size, D a dimension of the problem to be solved, and t is a generation number. Here, the elements are initialized according to the following equation:

$$x_{i,j}^{(0)} = U(0,1) \cdot (Ub_j - Lb_j) + Lb_j, \quad \text{for } i = 1, \dots, N \wedge j = 1, \dots, D, \tag{2}$$

where $U(0,1)$ denotes the random number drawn from uniform distribution in interval $[0,1]$, and Ub_j and Lb_j are the upper and lower bounds of the j-th element of the vector.

The physical laws of a firefly flashing are considered in the FA by introducing the light intensity relation, as follows:

$$I(r) = I_0 \cdot \exp^{-\gamma r^2}, \tag{3}$$

where I_0 denotes the light intensity at the source, and γ is a light absorption coefficient. Similar to the light intensity, the attraction between two fireflies, where the brighter is capable of attracting a potential mating partner more, is calculated according to the following equation:

$$\beta(r) = \beta_0 \cdot \exp^{-\gamma r^2}, \tag{4}$$

where β_0 is the attraction at $r = 0$.

The distance $r_{i,j}^{(t)}$ between two fireflies $\mathbf{x}_i^{(t)}$ and $\mathbf{x}_j^{(t)}$ is expressed as an Euclidian distance, as follows:

$$r_{i,j}^{(t)} = \|\mathbf{x}_i^{(t)} - \mathbf{x}_j^{(t)}\| = \sqrt{\sum_{k=1}^{D} x_{i,k}^{(t)} - x_{j,k}^{(t)}}. \tag{5}$$

The variation operators are implemented as a move of a definite virtual firefly i towards the more attractive firefly j according to the following equation:

$$\mathbf{x}_i^{(t+1)} = \mathbf{x}_i^{(t)} + \beta_0 \cdot \exp^{-\gamma r_{i,j}^{(t)^2}} \left(\mathbf{x}_j^{(t)} - \mathbf{x}_i^{(t)} \right) + \alpha \cdot \epsilon_i, \qquad (6)$$

which consists of three terms: The current position $\mathbf{x}_i^{(t)}$ of the i-th firefly, the social component determining the move of the i-th firefly towards the more attractive j-th firefly, and a randomization component determining the random move of the same firefly in the search space.

Typically, the step size scaling factor α is proportional to the characteristics of the problem, while the randomization factor ϵ_i is a random number drawn from Gaussian distribution with mean zero and standard deviation one, denoted as $N(0,1)$. Contrarily, the uniform distribution $U(0,1)$ from interval $[0,1]$ was used in our study instead of the normal distribution.

The quality of the solution, expressed by the fitness function, is, in the FA, proportional to the light intensity as $I(\mathbf{x}_i) \propto f(\mathbf{x}_i)$. The pseudo-code of the FA is illustrated in the Algorithm 1, from which it can be seen that this consists of the following components: (1) Representation of a solution, (2) Initialization (line 1), (3) Termination condition (line 4), (4) Move operator (line 8), (5) Evaluation function (lines 2 and 11), and (6) Ranking and finding the best solution (line 13).

Algorithm 1. Pseudo code of the basic Firefly algorithm

Input: Population of fireflies $\mathbf{x} = (\mathbf{x}_1, \ldots, \mathbf{x}_N)$, objective function $f(\mathbf{x}_i)$.
Output: The best solution $\mathbf{x}_{best}$ and its value $f_{min} = \min(f(\mathbf{x}_{best}))$.

 1: generate_initial_population $\mathbf{x}^{(0)} = (\mathbf{x}_1^{(0)}, \ldots, \mathbf{x}_N^{(0)})$;
 2: $f(\mathbf{x}_i^{(0)}) =$ evaluate_new_solution_and_update_light_intensity;
 3: $t = 0$;
 4: **while** $t < MAX\_GEN$ **do**
 5:     **for** $i = 1$ to N **do**
 6:         **for** $j = 1$ to N **do**
 7:             **if** $I_j > I_j$ **then**
 8:                 move_firefly_i_towards_j_using_Gaussian_distribution;
 9:             **end if**
10:         **end for**
11:         $f(\mathbf{x}_i^{(t)}) =$ evaluate_new_solution_and_update_light_intensity;
12:     **end for**
13:     rank_fireflies_and_find_the_best;
14:     $t = t + 1$
15: **end while**

---

3   Proposed Method

The task of the proposed stochastic nature-inspired population-based algorithm is to search for the model appropriate for binary classification of an arbitrary dataset. The model consists of two rules containing features of the dataset for predicting the True Negative and True Positive values. Thus, the learning is divided into a training phase, in which 80% of dataset instances are included, and a test phase, where we operate with the remaining 20% of the instances in the same dataset. The search for a model is defined as an optimization, where the fitness function is defined as a classification accuracy metric that is expressed mathematically as:

$$Acc = \frac{TN + TP}{TN + FN + TP + FP},\tag{7}$$

where TN= True Negative, TP= True Positive, FN= False Negative, and FP= False Positive.

Solutions $\mathbf{x}_i$ for $i = 1, \ldots, N$ in the proposed algorithm are represented as real-valued vectors with elements $x_{i,j} \in [0,1]$ for $j = 1, \ldots, L$, representing features, where L is the length of a solution, to which the binary vector $\mathbf{b}_i = \{b_{i,j}\}$ is attached with elements $b_{i,k} \in \{0,1\}$ for $k = 1, \ldots, M$, and M is the number of features. These are obtained in the preprocessing phase, where the dataset in question is analyzed in detail. The features can be either categorical (i.e., $b_{i,k} = 0$) or numerical (i.e., $b_{i,k} = 1$). The former consists of attributes drawn from a discrete set of feasible values, while the latter of continuous intervals limited by their lower and upper bounds. Each feature has its own predecessor *control*, determining its presence or absence in the specific rule.

In summary, the length of the solution L is calculated as:

$$L = 2 \cdot num\_of\_category\_attr + 3 \cdot num\_of\_numeric\_attr + 1,\tag{8}$$

where $num\_of\_category\_attr$ denotes the number of categorical features, $num\_of\_numeric\_attr$ is the number of numerical features, and one is reserved for *threshold* that determines if the definite feature belongs to the rule or not. Obviously, the feature belongs to the rule when the relation *control* $\geq$ *threshold* is satisfied.

In order to transform the representation of solutions into their problem context, the genotype-phenotype mapping is needed. The genotype-phenotype mapping determines how the genotype $\mathbf{x}_i$ of length L, calculated according to Eq. (8), is mapped into the corresponding phenotype $\mathbf{y}_i$ for $k = 1, \ldots, M$, where the variable M denotes the number of features in a dataset.

There are two ways in which to perform the genotype-phenotype mapping, depending on the type of feature: Actually, the categorical variables demand two, and the numerical even three elements for this mapping. In general, the mapping is expressed mathematically as (Fig. 1):

$$y_{i,k} = \begin{cases} -1, & \text{if } x_{i,0}^{(k)} < x_{i,L}, \\ \left\lfloor |Attr_k| \cdot x_{i,1}^{(k)} \right\rfloor, & \text{if } b_{i,k} = 0, \\ \left[\left\lfloor |D_k| \cdot x_{i,1}^{(k)} \right\rfloor, \left\lfloor |D_k| \cdot x_{i,2}^{(k)} \right\rfloor \right], & \text{if } b_{i,k} = 1, \end{cases}\tag{9}$$

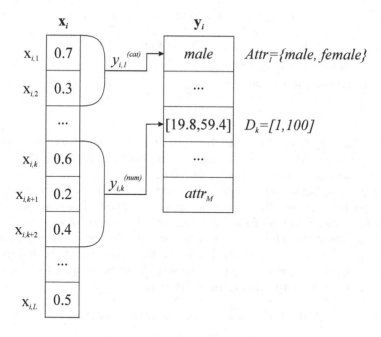

Fig. 1. Genotype-phenotype mapping.

for $k = 1, \ldots, M$, where $|Attr_k|$ denotes the size of the k-th attribute set $Attr_k = \{a_1, \ldots, a_{n_k}\}$, and D_k is a domain of feasible values of attributes, expressed as $(Max_k - Min_k)$, where Max_k and Min_k represent the maximum and minimum values of the numerical feature found in the dataset.

In summary, the phenotype value $y_{i,k}$ can obtain three values after the genotype-phenotype mapping: (1) -1, If the feature is not present in the rule, (2) The attribute of the feature set, if the feature is categorical, and (3) The interval of the feasible values, if the feature is numerical.

4   Experiments and Results

The aim of conducting experiments was twofold: (1) To evaluate the performance of the proposed method on some well-known binary datasets, and (2) To compare the obtained results with the results of some classical classification methods. In line with this, the results of the FA for binary classification were compared with the results obtained by: (1) Random Forest (RF) [3], (2) Multi-Layer Perceptron (MLP) [21], and (3) Bagging [2]. All algorithms in the experiments were applied to three well-known datasets for binary classification taken from the UCI Machine Learning Repository [1], whose characteristics are depicted in Table 1.

Table 1. Datasets used in our experiments.

Dataset name	No. of features	No. of instances
Pima Indians Diabetes dataset	8	768
Haberman survival dataset	3	306
Breast cancer	9	683

As can be seen from the table, binary datasets in question are relatively elementary, because the number of features are less than 10, while the number of instances does not exceed the value of 1,000.

The parameter setting of the FA during the tests is presented in Table 2, from which it can be seen that the maximum number of fitness function evaluations amounts to 5,000. Settings of the other algorithm parameters were taken from the existing literature. This means that no specific optimization of parameter settings was performed in the study. Moreover, the proposed FA did not include any domain-specific knowledge about imposed classification problems incorporated in the sense of adaptation or hybridization.

Table 2. Parameter settings of firefly algorithm.

Parameter	Abbreviation	Value
Maximum number of generations	$nFES$	5,000
Population size	N	90
Step size scaling factor	α	0.5
Attractiveness at $r = 0$	β_0	0.2
Light absorption factor	γ	1.0

Indeed, the proposed FA for binary classification was implemented in the Python programming language using the external NiaPy library [19]. The implementations of the remaining three methods were taken from the scikit-learn Python package [12], where default parameter settings were adopted. Let us emphasize that 25 independent runs were conducted for each method in question, where the achieved classification accuracy was collected after each run. As a result, the quality of the methods was evaluated according to the five aforementioned standard statistical measures: minimum, maximum, average, median, and standard deviation values.

The detailed results of the comparative analysis according to classification accuracy are illustrated in Table 3, where five statistical measures are analyzed according to the used algorithms and the observed datasets. Here, the best results are presented in bold case. As can be seen in the Table 3, the best results were achieved by the RF and MLP classification methods, where the RF gained the better accuracy by classifying the Pima dataset, the MLP was better at the Haberman's, while, at the Breast's, both mentioned methods obtained the same classification accuracy.

Table 3. Detailed results of the binary classification according to accuracy.

Dataset	Algorithm	Min	Max	Mean	Median	Stdev
Pima	FA	0.5844	0.7662	0.6849	0.6818	0.0519
	RF	0.6688	**0.7987**	0.7387	0.7402	0.0342
	MLP	0.5779	0.7467	0.6800	0.6818	0.0437
	Bagging	0.6948	0.7922	**0.7407**	0.7402	0.0273
Haberman	FA	0.6451	0.7903	0.7264	0.7419	0.0368
	RF	0.5645	0.8064	0.7070	0.6935	0.0565
	MLP	0.6612	**0.8709**	**0.7658**	0.7741	0.0587
	Bagging	0.5645	0.7903	0.6741	0.6774	0.0558
Breast	FA	0.8029	0.8978	0.8616	0.8686	0.0293
	RF	0.9270	**0.9854**	**0.9623**	0.9562	0.0162
	MLP	0.9343	**0.9854**	0.9620	0.9635	0.0133
	Bagging	0.9270	0.9781	0.9570	0.9562	0.0146

According to Table 4, where the percent in deviation of the results of the definite method from the best results designated by '‡' in the Table, are calculated, the RF and MLP classification methods exhibit the best percent in general, because they outperformed all the others even three times. The bagging achieved the best mean results by classification of the Pima dataset. Although the FA for binary classification did not achieve the best accuracy in any instance of dataset, its best, as well as mean results, were no worse than 10 percentage points of the best results, except at Breast's, where the accuracy was close to this border value (precisely 89.49 %).

Table 4. Summary results of the binary classification according to accuracy.

Dataset	Measure	FA	RF	MLP	Bagging
Pima	Max [%]	95.93	‡	93.49	99.19
	Mean [%]	92.47	99.73	91.81	‡
Haberman	Max [%]	90.75	92.59	‡	90.75
	Mean [%]	94.86	92.32	‡	88.03
Breast	Max [%]	91.11	‡	‡	99.26
	Mean [%]	89.49	‡	99.92	99.40

Finally, an example of classification rules generated by the proposed FA in classifying the Pima Diabetes dataset is illustrated in Table 5 that is divided into two parts: (1) Feature, and (2) Classification rules. The former consists of three fields: Sequence number, feature name and type. The latter is divided into two rules, i.e., for True Negative TN, and for True Positive TP classifications.

Table 5. An example of classification rule in classifying Pima Diabetes dataset generated by the proposed FA for binary classification.

Feature			Classification rules	
Num.	Name	Class	TN	TP
1	Number of times pregnant	Numeric	[0.79,16.04]	[13.69,16.28]
2	Plasma glucose concentration	Numeric	[25.92,148.08]	n/a
3	Diastolic blood pressure	Numeric	[6.18,84.45]	[53.71,81.74]
4	Triceps skin fold thickness	Numeric	[8.33,52.15]	[15.39,27.88]
5	2-h serum insulin	Numeric	[435.02,730.53]	[759.30,840.51]
6	Body mass index	Numeric	[36.43,37.96]	[31.75,58.41]
7	Diabetes pedigree function	Numeric	n/a	n/a
8	Age	Numeric	[68.45,75.98]	34.29,41.01]

As can be seen from the Table, there are eight features in the dataset. Interestingly, the dataset supports only numerical attributes. These attributes are, therefore, represented as continuous domains of values. Thus, the first rule determines the combination attributes that are classified as True Negative predictions, while the second rule as True Positive predictions.

From this table, it can be concluded that this representation of rules is undoubtedly comprehensive, and, in that way, is easily understandable by the user.

5 Discussion

The huge progress in big data has caused rapid development of new data mining methods that need to satisfy two requests: (1) To process enormous volumes of data, and (2) To ensure enough processing time for their analysis. The classical data mining methods suffer from a lack of comprehensibility that disallows users to use them as effectively as possible. Mainly, the stochastic nature-inspired population-based algorithms are well-known general tools suitable for solving the hardest optimization problems. Recently, this family of algorithms has been applied to data mining field, in order to search for the best model in the model search space.

In this preliminary study, the FA was proposed for the binary classification task, with the following advantages: The FA search process searches for new solutions, not only on basis of the best global solution, but moves each particle in the search space with regard to its neighborhood consisting of the more attractive particles. Furthermore, the original FA operates with real-valued vectors, which represent the solutions of the problem in question. The genotype-phenotype mapping must be performed in order to transform the representation in the genotype space into the solution in the problem context. In our case, the mapping decodes two classification rules from each solution, where the first is dedicated for classification of TN predictions, while the second for classification of TP. Moreover, the algorithm is capable of performing the feature selection implicitly, because only the more important features must be presented in the solution. Finally, the features can be either categorical or numerical. Both types are represented as real values and, therefore, no discretization is necessary.

The proposed FA for binary classification was applied to three well-known datasets publicly available on the web. The obtained results were compared with three classical classification methods: RF, MLP, and boosting. Although the FA did not improve the results achieved by the classical methods, they showed that this has a big potential for improving its results in the future, especially due to the fact that the algorithm was used as is, i.e., no features were implemented to improve it.

In line with this, the improvement of the FA in the sense of adaptation and hybridization should be a reasonable direction for the future. However, testing the behavior of the algorithm on general classification problems could also be challenging.

Acknowledgment. Iztok Fister Jr., Grega Vrbančič and Vili Podgorelec acknowledge the financial support from the Slovenian Research Agency (Research Core Funding No. P2-0057). Dušan Fister acknowledges the financial support from the Slovenian Research Agency (Research Core Funding No. P5-0027). Iztok Fister acknowledges the financial support from the Slovenian Research Agency (Research Core Funding No. P2-0041).

References

1. Asuncion, A., Newman, D.: UCI machine learning repository (2007)
2. Breiman, L.: Bagging predictors. Mach. Learn. **24**(2), 123–140 (1996)
3. Breiman, L.: Random forests. Mach. Learn. **45**(1), 5–32 (2001)
4. Cantú-Paz, E.: On the use of evolutionary algorithms in data mining. In: Data Mining: A Heuristic Approach, pp. 22–46. IGI Global (2002)
5. Freitas, A.A.: A survey of evolutionary algorithms for data mining and knowledge discovery. In: Ghosh, A., Tsutsui, S. (eds.) Advances in Evolutionary Computing. NCS, pp. 819–845. Springer, Heidelberg (2003). https://doi.org/10.1007/978-3-642-18965-4_33
6. Friedman, N., Geiger, D., Goldszmidt, M.: Bayesian network classifiers. Mach. Learn. **29**(2–3), 131–163 (1997)
7. Haykin, S.: Neural Networks: A Comprehensive Foundation, 2nd edn. Prentice Hall PTR, Upper Saddle River (1998)
8. Hearst, M.A.: Support vector machines. IEEE Intell. Syst. **13**(4), 18–28 (1998)
9. Maimon, O., Rokach, L.: Introduction to soft computing for knowledge discovery and data mining. In: Maimon, O., Rokach, L. (eds.) Soft Computing for Knowledge Discovery and Data Mining, pp. 1–13. Springer, Boston (2008). https://doi.org/10.1007/978-0-387-69935-6_1
10. Martens, D., Baesens, B., Fawcett, T.: Editorial survey: swarm intelligence for data mining. Mach. Learn. **82**(1), 1–42 (2011)
11. Parpinelli, R.S., Lopes, H.S., Freitas, A.A.: Data mining with an ant colony optimization algorithm. Trans. Evol. Comput. **6**(4), 321–332 (2002)
12. Pedregosa, F., et al.: Scikit-learn: machine learning in python. J. Mach. Learn. Res. **12**(Oct), 2825–2830 (2011)

13. Peng, L., Yang, B., Chen, Y., Abraham, A.: Data gravitation based classification. Inf. Sci. **179**(6), 809–819 (2009)
14. Ross Quinlan, J.: Induction of decision trees. Mach. Learn. **1**(1), 81–106 (1986)
15. Rekha, S.: A survey of evolutionary algorithms and its use in data mining application. J. Pure Appl. Math. **119**(12), 13593–13600 (2018)
16. Ratnoo, S., Vashishtha, J., Goyal, P., Ahuja, J.: A novel fitness computation framework for nature inspired classification algorithms. Procedia Comput. Sci. **132**, 208–217 (2018)
17. Sousa, T., Silva, A., Neves, A.: Particle swarm based data mining algorithms for classification tasks. Parallel Comput. **30**(5–6), 767–783 (2004)
18. Srikanth, R., George, R., Warsi, N., Prabhu, D., Petry, F.E., Buckles, B.P.: A variable-length genetic algorithm for clustering and classification. Pattern Recogn. Lett. **16**(8), 789–800 (1995)
19. Vrbančič, G., Brezočnik, L., Mlakar, U., Fister, D., Fister Jr., I.: NiaPy: python microframework for building nature-inspired algorithms. J. Open Source Softw. **3**, 613 (2018)
20. Yang, X.-S.: Firefly algorithm, stochastic test functions and design optimisation. Int. J. Bio-Inspired Comput. **2**(2), 78–84 (2010)
21. Zurada, J.M.: Introduction to Artificial Neural Systems, vol. 8. West Publishing Company, St. Paul (1992)

Analytical Techniques for the Identification of a Musical Score: The Musical DNA

Michele Della Ventura$^{(\boxtimes)}$

Department of Information Technology, Music Academy "Studio Musica",
Via Andrea Gritti, 25, 31100 Treviso, Italy
michele.dellaventura@tin.it

Abstract. In the information age, one of the main research field that is being developed is the one related to how to improve the quality of the search engine as regards knowing how to manage the information contained in a document in order to extract its content and interpret it. On the one hand due to the heterogeneity of the information contained on the web (text, image, video, musical scores), and on the other hand to satisfy the user who generally searches for information of a very different type. This paper describes the development and evaluation of an analytical method for the analysis of musical score considered in its symbolic level. The developed method is based on the analysis of the fundamental elements of the musical grammar and takes into account the distance between the sounds (which characterize a melody) and their duration (which makes the melody active and alive). The method has been tested on a set of different musical scores, realizing an algorithm in order to identity a musical score in a database.

Keywords: Musical DNA · Information retrieval · Musical score search engine

1 Introduction

Internet and the Web are an immense information resource. As such, it represents the first global communication network that allows users to transmit, receive, communicate, and make available information contents. Given the complexity and heterogeneity of the information contained on the web [1], search engines are becoming crucial to allow easy navigation through the data.

Search engines are based on specific algorithms for Information Retrieval (IR). The main objective is to make the right information available to the user based on his requests and expectations [2]. In order to be identified by Information Retrieval (IR) Systems, documents are generally transformed into an adequate representation [3]. Each method of information recovery has a different model based on the type of document [4, 5]: text, image, video.

From these considerations one can immediately infer that while in the case of a linguistic text it is easy to create indexes [6, 7], in the case of a musical language significant difficulties emerge. As far as a musical piece is concerned, the indexes are created exclusively in reference to the title, the name of the author, the tonality and other information of a purely textual and informative kind [8]. Recent research

© Springer Nature Switzerland AG 2020
V. V. Krzhizhanovskaya et al. (Eds.): ICCS 2020, LNCS 12141, pp. 29–39, 2020.
https://doi.org/10.1007/978-3-030-50426-7_3

analyzed the audio files, creating Audio Search Engines, through the audio fingerprint technique that may be used not only to identify an audio file [9] but also to synchronize multiple audio files [10, 11].

However, in the case of a musical text examined at a symbolic level (i.e. the musical score), it becomes difficult to create indexes in the absence of specific indications that often lead to approximate results. There are many scientific researches in the ambit of the musical text that have the objective of searching for more accurate systems in order to determine the identifying elements of a composition, as for instance a melody, a motif, a rhythmic structure [12–14] and so forth, that might be used in order to create an indexation of the same composition.

This document presents a method for the analysis of a musical score considered in its symbolic level to analytically represent its distinctive characters. These refer to the score and are unique: each music is different from another music and each person can interpret these differences by listening and segmenting the continuous sonorous. The representation of these characters through a vector permits to define an objective comparison criterion. The method is based on the mathematical formalization of the distinctive elements of the sound: pitch and duration.

This paper is organized as follows.

Section 2 analysis the concept of "sound" and its characteristics. Section 3 explains the method used to obtain the Fingerprint of a musical score. available experimental results are shown that illustrate the effectiveness of the proposed method. Finally, Sect. 5 concludes this paper with a brief discussion.

2 The Concepts of Sound

Melody and rhythm are two fundamental components as far as musical structuring is concerned, two nearly inseparable components: a melody evolves along the rhythm in the absence of which it does not exist [15]: *"melody in itself is weak and quiescent, but when it is joined together with rhythm it becomes alive and active"* [16] (Fig. 1).

Fig. 1. Excerpt from the score of Ravel's "Bolero". The initial notes of the theme are represented on the first staff without any indication with respect to rhythm; the same notes are represented of the second staff together with the rhythm assigned by the composer.

2.1 The Melody

The melody of a musical piece is represented by a number of sounds, each one separated from the next by a number of semitones: the melodic interval.

The various melodic intervals were classified as symbols of the alphabet [17]. The classification of an interval consists in the *denomination* (generic indication) and in the *qualification* (specific indication) [18, 19]. The *denomination* corresponds to the number of degrees that the interval includes, calculated from the lowest one to the highest one; it may be of a 2nd, a 3rd, 4th, 5th, and so on.; the *qualification* is deduced from the number of tones and semi-tones that the interval contains; it may be: perfect (G), major (M), *minor* (m), *augmented* (A), *diminished* (d), *more than augmented* (A+), *more than diminished* (d-), *exceeding (E), deficient* (def).

A melody is usually represented as a sequence S_i of N intervals n_x indexed on the basis of their order of occurrence x [17]:

$$S_i = (n_x)_{x \in [0, N-1]}$$

The musical segment may, therefore, be seen as a vector the elements of which are, respectively, the intervals that separate the various sounds from one another. The corresponding value of every interval equals the number of semi-tones between the i-th note and the preceding one: this value will be respectively positive or negative depending on whether the note is higher or lower than the preceding note (Fig. 2) [17].

$$S_i = <2, 2, 1, -3, 2, -4>$$

Fig. 2. Melodic segment and its related vector.

2.2 The Rhythm

The rhythm is associated with the duration of the sounds: duration intended as the time interval in which sound becomes perceptible, regardless of whether it is due to a single sign or to several signs joined together by a value connection [17, 20, 21].

If we were to analyze a score, the sound duration will not be expressed in seconds but calculated on the basis of the musical sign (be it sound or rest) with the smallest duration existing in the musical piece [17]. The duration of every single sign will therefore be a (integer) number directly proportional to the smallest duration. In the example shown in Fig. 3, the smallest duration sign is represented by the thirty-second note to which the value 1 is associated (automatically): it follows that the sixteenth note shall have the value 2, the eighth note the value 4, …

$$R_i = <2, 4, 4, 4, 4, 4, 4, 4, 4, 1, 1>$$

Fig. 3. Rhythmic segment and its related vector.

On the base of all the above considerations related to the concept of melody, it is possible to deduce that each melody has a succession of sounds that differentiates it from other melodies (i.e. there are different intervals that separate sounds) and each sound has a specific duration that confers meaning to the whole melody. Figure 4 shows two incipits derived from two different songs that have the same sounds but different rhythm.

Fig. 4. A) "Frere Jacque" (French popular song), B) "DO-RE-MI" (from the Musical "The sound of Music").

3 The DNA of a Musical Score

The aim of this study is to mathematically formalize the features of the sound that allow a listener to recognize a particular composition: the pitch and the duration. These elements are unique for each composition and therefore allow to delineate the "DNA" of the composition.

The difference in pitch between two sounds allows you to define the musical interval; the sequence of intervals within the composition allows the definition of the melody. In order to analyze and represent the succession of the interval within the musical piece it has been used the Markov Process (or Markov Stochastic Process – MSP): the choice was made to describe the passage from one sound to the next sound considering the number of semitones between the two sounds (melodic interval), the trend of the interval (a = ascending or d = descending) and the duration of the two sounds.

Table 1 shows an excerpt of a transition matrix: the first column (and first row) indicates the denomination of the interval (classification), the second column (and the second row) presents the number of semi-tones that make up the interval (qualification), the third column (and the third row) displays the ascending (a) or descending (d) movement between two consecutive sounds, and the fourth column (and forth row) presents the time-space between two consecutive sounds.

Table 1. Example of a transitions matrix.

Interval				2° m	2° m	2° m	2° m	...
	Semitones			1	1	1	1	...
		Trend		a	d	a	d	...
			Duration	♪	♪	♪	♪	...
2° m	1	a	♪					...
2° m	1	d	♪					...
2° m	1	a	♪					...
2° m	1	d	♪					...
...	...	...	...	...	...	...	...	...

To determine the DNA of the score (or of the musical research segment) it is necessary to fill in the matrix of the transitions and then for each row it is necessary to add the values present in the matrix: if the result is zero in the DNA column (and in the cell corresponding to the same row) the number zero is written, while if the value is different from zero in the DNA column (and in the cell corresponding to the same row) the number 1 is written.

Table 2. DNA of the musical object in Fig. 1.

1				2°m	2°m	2°m	2°m	2°m	2M	2°M	2°M	3°m	3°m	3°m	3°m	3°M	3M	3°M	3°M	...	
	S			1	1	1	1	2	2	2	2	3	3	3	3	4	4	4	4	...	
		T		a	d	a	d	a	d	a	d	a	d	a	d	a	d	a	d	...	
			D	♪	♪	♪	♪	♪	♪	♪	♪	♪	♪	♪	♪	♪	♪	♪	♪	...	DNA
2°m	1	a	♪									①								...	1
2°m	1	d	♪																	...	0
2°m	1	a	♪																	...	0
2°m	1	d	♪																	...	0
2°M	2	a	♪	①			①									①				...	1
2°M	2	d	♪																	...	0
...	...	...	...	...	...	...	...	...	...	...	...	...	...	...	...	...	...	...	...	...	...

Given a musical segment S_2 of length $N_1 < N_0$, where N_0 is the length of the music score, it is necessary to define its representative vector following the procedure described above, and compare it with the representative vector of the music score. To identify the DNA of the segment S_2 it is necessary to use a transition matrix with the same columns and rows of the music score: only in this way is it possible to obtain a vector with the same length as the vector of the score.

The comparison takes place by making a bit-to-bit difference, in correspondence with the bits with value 1 of the segment S_2: if the resulting difference is zero for each bit, the score was identified (Fig. 5) otherwise the choice is made considering the musical score with the greatest number of zeros (which means a greater number of common elements between the musical segment S_2 and the musical sores) (see Fig. 6). In this case the concept of similarity is taken into consideration, as a discriminating factor [22]. The distinction between similarity and identity of two musical segments is very rigorous [23]: similarity is defined as partial identity, that is, two entities (A and B) are similar if they share some properties, but not necessarily all. Therefore, the similarity between A and B depend on their common features: the more common features they share, the more similar they are (Fig. 6). At the same time, the similarity between A and B depend on the differences between them: the more differences they have, the less similar they are.

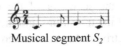

Musical segment S_2

Frére Jacque

DNA	0	1	1	0	0	0	0	0	0	0	1	1	1	0	0	0	0	0	0	0	...
Vector S_2											1			1							...
Check											0			-1							...

DO-RE-MI

DNA	1	1	0	1	1	0	0	0	0	1	1	1	1	1	1	0	0	1	0	1	...
Vector S_2											1				1						...
Check											0				0						...

Fig. 5. Musical score identification.

Musical Score 1

DNA	1	1	0	1	1	0	0	0	0	0	1	1	1	1	1	0	0	1	0	1	...
Vector S_2						1			1	1				1							...
Check						-1			-1	0				0							...

Musical Score 2

DNA	1	1	0	1	1	0	1	0	0	0	1	1	1	1	1	0	0	1	0	1	...
Vector S_2						1			1	1				1							...
Check						0			-1	0				0							...

Fig. 6. Musical score identification.

4 Application and Analysis: Obtained Results

The aim of this research was to develop an analytical method in order to realize a Search Engine able to identify a musical score in a database.

To evaluate the proposed method, it has been realized an algorithm the structure of which takes in consideration each and every single aspect previously described. The algorithm does not provide any limitations with regard the number of notes that can be wrote for the research and therefore the musical score representative matrix that is automatically dimensioned on the base of the features of the musical score. The algorithm presents some limitations related to the duration of the notes used for the research. This is due to the fact that the solidity of the algorithm had to be checked against complex rhythmic structures (typical of the contemporary music).

4.1 Database Preparation

The collection of documents was prepared using musical scores of 18th, 19th and 20th Century of different authors, for a total of 220 musical scores.

All of the musical scores were stored in MIDI format (a symbolic music interchange format), or in the Lilypond notation format (a textual notation for music based on the set of ASCII characters) [24], or in the more recent MusicXML file (a text-based language that permits to represent common Western musical notation) [24].

All the musical scores were for piano, in order to take polyphony into consideration; grace notes (or musical ornaments) were removed from each score (Fig. 7), because musically they are secondary notes and do not characterize the musical structure.

Fig. 7. Database preparation: removing grace motes.

4.2 Musical Score Decoding

In the decoding stage, it is important to underline the fact that in the case of two or more sounds connected together, these are considered as a single sound whose duration is equal to the sum of the durations of the individual sounds [17] (see example in Fig. 8).

Fig. 8. Sound duration.

Each musical score (read from the file) is transformed into a list of numbers (Musical DNA) [17]: each sound is associated with two values, the first one related to the distance with the next sound and the second one related to its duration [25, 26] (Fig. 9).

Fig. 9. Musical DNA.

In case the musical score is composed of several voices placed on different staves (such as the piano score) the list of sounds is composed first of all by the relative numbers of the first staff, then those of the second staff and so on (Fig. 10) [12].

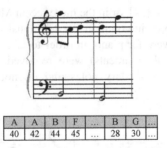

A	A	B	F	...	B	G	...
40	42	44	45	...	28	30	...

Fig. 10. Sequential representation of the score.

4.3   Input Sample Music

After filling in the representative matrix of the musical score, its binary representative vector (Musical DNA) is defined: see example in Table 2.

The search for a score in a database is performed by writing the sounds of a sample musical segment (with their respective durations) within a dialog similar to the dialog of a search engine, with the difference that this is represented by a musical pentagram (Fig. 11).

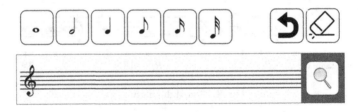

Fig. 11. Music search engine interface.

4.4   Obtained Results

The initial tests were carried out on a set of sample musical segments of two different lengths, 5 and 10 musical notes. For the tests the processing time was not taken into consideration, because it was strictly connected to the type of computer used for the analysis.

It was not important that the durations of the sounds indicated in the sample were the same as the durations present in the scores: the durations could be different under the condition that the mathematical proportion was always respected (Fig. 12).

The results of information retrieval are present in Table 3. The percentage improvement obtained considering 10 notes instead of 5 notes is approximately 5% in the case of musical scores of the 18th Century, 7% in the case of musical scores of the 19th Century and 4% in the case of the musical scores of the 20th Century. This is explained by the fact that: in the 18th Century music, the type of intervals present in a score is not very varied; in 19th Century music the type of intervals increases and therefore a greater number of notes allows to diversify the research results; finally, in the music of the 20th Century there is a particular element, namely the presence of complex rhythms which reduces the similarity between musical segments.

$R_i = <1, 1, 4, 2>$

Fig. 12. Proportional duration.

Table 3. Comparative performance under different length of the sample musical segment.

Period of the musical scores	Averaged recognition rate (%)	
	5 musical notes	10 musical notes
18th century	87	92
19th century	84	91
20th century	63	67

5 Discussion and Conclusions

The Internet and the Web represent an immense and heterogeneous information resource for a vast and heterogeneous public like current Internet users. However, the nature and characteristics of the Internet highlight problems relating to how to search and find information online, in order to satisfy the user's requests.

This study has been able to successfully develop an algorithm (a sort of IR system) that allows the identification of a musical score within a database, through the indication of musical notes. The results show that to obtain a satisfactory result it is necessary to insert a suitable number of notes, capable of analytically describing the identifying characteristics of the musical score.

In addition, the following study areas can be improved in future studies to create a more robust IR:

- Increase in the size of the database size (number of musical scores);
- Representation of irregular musical rhythms (typical of contemporary music).

References

1. Dong, Y., Chawla, N.V., Swami, A.: metapath2vec: scalable representation learning for heterogeneous networks. In: Proceedings of the KDD, pp. 135–144 (2017)
2. Cao, X., Zheng, Y., Shi, C., Li, J., Wu, B.: Meta-path-based link prediction in schema-rich heterogeneous information network. Int. J. Data Sci. Anal. 3(4), 285–296 (2017). https://doi.org/10.1007/s41060-017-0046-1
3. Blummer, B., Kenton, J.M.: Information research and the search process. In: Improving Student Information Search, pp. 11–21. Chandos Publishing (2014)
4. Lewandowski, D.: A framework for evaluating the retrieval effectiveness of search engines. In: Jouis, C., Biskri, I., Ganascia, G., Roux, M. (eds.) Next Generation Search Engines: Advances Models for Information Retrieval, pp. S456–S479. IGI Global, Hershey (2012)
5. Lazonder, A.W., Biemans, H.J.A., Wopereis, I.G.J.H.: Differences between novice and experienced users in searching information on the world wide web. J. Am. Soc. Inf. Sci. 51(6), 576–581 (2000)
6. Chang, G., Healey, M.J., McHugh, J.A.M., Wang, J.T.L.: Multimedia search engines. In: Chang, G., Healey, M.J., McHugh, J.A.M., Wang, J.T.L. (eds.) Mining the World Wide Web. INRE, vol. 10, pp. 51–63. Springer, Boston (2001). https://doi.org/10.1007/978-1-4615-1639-2_4
7. Yuan, X.J., Zhang, X.J., Avery, J.: Seeking information with an information visualization system: a study of cognitive styles. Inf. Res. 16(4), 499 (2011)
8. Zmudzinski, S., Steinebach, M., Butt, M.: Watermark embedding using audio fingerprinting. In: Shi, Y.Q., Katzenbeisser, S. (eds.) Transactions on Data Hiding and Multimedia Security VIII. LNCS, vol. 7228, pp. 63–79. Springer, Heidelberg (2012). https://doi.org/10.1007/978-3-642-31971-6_4
9. Miller, M.L., Rodriguez, M.A., Cox, I.J.: Audio fingerprinting: nearest neighbor search in high dimensional binary spaces. J. VLSI Signal Process. Syst. Signal Image Video Technol. 41(3), 285–291 (2005). https://doi.org/10.1007/s11265-005-4152-2
10. Oostveen, J., Haitsma, J., Kalker, T.: Algorithms for audio and video fingerprinting. In: Verhaegh, W.F.J., Aarts, E., Korst, J. (eds.) Algorithms in Ambient Intelligence. PRBS, vol. 2, pp. 201–219. Springer, Dordrecht (2004). https://doi.org/10.1007/978-94-017-0703-9_11
11. Mau, T.N., Inoguchi, Y.: Audio fingerprint hierarchy searching on massively parallel with multi-GPGPUS using K-modes and LSH. In: Eighth International Conference on Knowledge and Systems Engineering (KSE), pp. 49–54. IEEE (2016)
12. Della Ventura, M.: Musical DNA. ABEditore, Milano (2018). ISBN 978-88-6551-281-4
13. Neve, G., Orio, N.: A comparison of melodic segmentation techniques for music information retrieval. In: Rauber, A., Christodoulakis, S., Tjoa, A.M. (eds.) ECDL 2005. LNCS, vol. 3652, pp. 49–56. Springer, Heidelberg (2005). https://doi.org/10.1007/11551362_5
14. Rodrıguez Lopez, M.E.: Automatic melody segmentation. Ph.D. thesis, Utrecht University (2016)
15. Fraisse, P.: Psychologie du rythme. Puf, Paris (1974)
16. Sioros, G., Davies, M.E., Guedes, C.: A generativemodel for the characterization of musical rhythms. J. New Music Res. 47(2), 114–128 (2018)
17. Della Ventura, M.: The influence of the rhythm with the pitch on melodic segmentation. In: Abraham, A., Jiang, X.H., Snášel, V., Pan, J.-S. (eds.) Intelligent Data Analysis and Applications. AISC, vol. 370, pp. 191–201. Springer, Cham (2015). https://doi.org/10.1007/978-3-319-21206-7_17
18. de la Motte, D.: Manuale di armonia, Bärenreiter (1976)
19. Schoenberg, A.: Theory and Harmony. Univ of California Pr; Reprint edition (1992)

20. Moles, A.: Teorie de l'information et Perception esthetique, Paris, Flammarion Editeur (1958)
21. Louboutin, C.: Multi-scale and multi-dimensional modelling of music structure using polytopicgraphs. Sound [cs.SD]. Université Rennes 1 (2019)
22. Della Ventura, M.: Similarity measures for music information retrieval. In: Le, N.-T., Van Do, T., Nguyen, N.T., Thi, H.A.L. (eds.) ICCSAMA 2017. AISC, vol. 629, pp. 165–174. Springer, Cham (2018). https://doi.org/10.1007/978-3-319-61911-8_15
23. Della Ventura, M.: Speech assessment based on entropy and similarity measures. In: Le Thi, H.A., Le, H.M., Pham Dinh, T., Nguyen, N.T. (eds.) ICCSAMA 2019. AISC, vol. 1121, pp. 218–227. Springer, Cham (2020). https://doi.org/10.1007/978-3-030-38364-0_20
24. http://lilypond.org/
25. Madsen, S., Widmer, G.: Separating voices in MIDI. In: ISMIR, Canada (2006)
26. Della Ventura, M.: Using mathematical tools to reduce the combinatorial explosion during the automatic segmentation of the symbolic musical text. In: Nguyen, T.B., Do, T.V., Le Thi, H.A., Nguyen, N.T. (eds.) Advanced Computational Methods for Knowledge Engineering. AISC, vol. 453, pp. 281–293. Springer, Cham (2016). https://doi.org/10.1007/978-3-319-38884-7_20

Reduced-Cost Constrained Modeling of Microwave and Antenna Components: Recent Advances

Anna Pietrenko-Dabrowska[1] , Slawomir Koziel[1,2(✉)] ,
and Leifur Leifsson[3]

[1] Faculty of Electronics Telecommunications and Informatics,
Gdansk University of Technology, Narutowicza 11/12, 80-233 Gdansk, Poland
anna.dabrowska@pg.edu.pl
[2] Engineering Optimization and Modeling Center,
School of Science and Engineering, Reykjavík University,
Menntavegur 1, 101 Reykjavík, Iceland
koziel@ru.is
[3] Department of Aerospace Engineering, Iowa State University,
Ames, IA 50011, USA
leifur@iastate.edu

Abstract. Electromagnetic (EM) simulation models are ubiquitous in the design of microwave and antenna components. EM analysis is reliable but CPU intensive. In particular, multiple simulations entailed by parametric optimization or uncertainty quantification may considerably slow down the design processes. In order to address this problem, it is possible to employ fast metamodels. Here, the popular solution approaches are approximation surrogates, which are versatile and easily accessible. Notwithstanding, the major issue for conventional modeling methods is the curse of dimensionality. In the case of high-frequency components, an added difficulty are highly nonlinear outputs that need to be handled. A recently reported constrained modeling attempts to broaden the applicability of approximation surrogates by confining the surrogate model setup to a small subset of the parameter space. The said region contains the parameter vectors corresponding to high-quality designs w.r.t. the considered figures of interest, which allows for a dramatic reduction of the number of training samples needed to render reliable surrogates without formally restricting the parameter ranges. This paper reviews the recent techniques employing these concepts and provides real-world illustration examples of antenna and microwave structures.

Keywords: Microwave engineering · Antenna engineering · Electromagnetic simulation · Surrogate modeling · Performance-driven modeling · Kriging interpolation

1 Introduction

Design of contemporary microwave and antenna components has been increasingly dependent on full-wave electromagnetic (EM) simulation tools. EM analysis permits reliable evaluation of arbitrary geometries and taking into account cross-couplings

V. V. Krzhizhanovskaya et al. (Eds.): ICCS 2020, LNCS 12141, pp. 40–56, 2020.
https://doi.org/10.1007/978-3-030-50426-7_4

between system components, dielectric anisotropy, or the effects of installation fixtures and radomes. The growing involvement of EM simulation packages is especially pertinent to parameter tuning also referred to as design closure [1]. This is where the performance parameters of the structure at hand are enhanced subject to the assumed design constraints. It is most often achieved through numerical optimization, which entails significant computational expenses. Handling massive EM analyses is one of the major challenges pertaining to EM-based design processes even if local optimization is of concern. Solving tasks such as global optimization [2], uncertainty quantification or tolerance-aware design [3], may become computationally unmanageable when attempted directly at full-wave EM simulation level.

Reducing the CPU cost of EM-driven design has been the subject of extensive research. One possible option is the development of more efficient numerical algorithms, e.g., incorporation of adjoint sensitivities to speed up gradient-based procedures [4], or the employment of (local) surrogate models [5]. A notable example of the latter is space mapping [6]. Other approaches include response correction methods [7], feature-based optimization [8], or machine learning frameworks, often surrogate-assisted [9]. Another option is an overall replacement of the EM model by its faster surrogate, which permits a rapid execution of all types of simulation-based design procedures. Data-driven surrogates belong to the most popular ones due to their versatility [10]. They are constructed by approximating the data sampled from the original (here, EM) model with no physical insight required. Commonly used modeling techniques include polynomial regression [11], radial basis functions [12], kriging [13], support vector regression (SVR) [14], and polynomial chaos expansion (PCE) [15]. Available alternatives include, among others, hybridization of one of the aforementioned methods. One of these is PC kriging [16], in which polynomial chaos expansion surrogate becomes a trend function, whereas kriging interpolation is employed to account for the residuals.

Despite their merits and popularity, applicability of approximation models is limited by the curse of dimensionality. In the case of high-frequency structures, which often feature nonlinear responses, reliable data-driven surrogates can be constructed for systems described by up to four or five variables. Design utility of such models is of course questionable given the complexity of contemporary devices (both antennas and microwave). A range of methods have been developed to mitigate these problems, including high-dimensional model representation (HDMR) [17], feature-based modeling [18], orthogonal matching pursuit (OMP) [19], as well as variable-fidelity methods (Bayesian model fusion [20], co-kriging [21], or two-stage GPR [22]).

In [23], an alternative approach has been suggested, where the issue of high cost of training data acquisition is addressed by confining the surrogate model domain to a region that contains the designs that are of high quality with respect to the figures of interest relevant for the system at hand. The volume of such a region is dramatically smaller than the conventional box-constrained domain so that restricting the model validity (and, consequently, training data allocation) leads to significant computational savings. Determination of the constrained domain requires additional knowledge, normally in the form of the reference designs pre-optimized with respect to the chosen figures of interest of choice. Several variations of the constrained modeling have been reported, including rudimentary frameworks rendering surrogates that can handle a single operating condition [23], versions that allow accounting for supplementary figures of interest (e.g., substrate permittivity [24]), to techniques that permit arbitrary allocation of the reference designs [25]. The nested kriging of [26] also allows for straightforward uniform domain sampling and surrogate model optimization, which was not possible in the earlier versions.

This paper reviews the recent developments of the constrained modeling, focusing on three approaches: (i) modeling with structured reference design set [23, 24], (ii) triangulation-based modeling [25], as well as (iii) the nested kriging framework [26]. The presented methods are illustrated using real-world microwave and antenna structures, and benchmarked against conventional surrogate modeling methods. A generic formulation of the constrained modeling concept is also provided.

2   Surrogate Modeling with Domain Confinement

We start by formulating the constrained modeling concept. One of the most important components is the confined surrogate model domain. The fundamental criterion deciding upon the domain geometry is design optimality for the assumed design objectives.

2.1   Fundamental Concepts. Design Variables and Figures of Interest

Let us denote by X the parameter space for the design problem at hand. It is delimited by the lower and upper bounds on design variables $l \leq x \leq u$, where $x = [x_1 \ldots x_n]^T$, $l = [l_1 \ldots l_n]^T$, $u = [u_1 \ldots u_n]^T$, or $X = [l_1 \ u_1] \times \ldots \times [l_n \ u_n]$. The relevant figures of interest are f_k, $k = 1, \ldots, N$. Perhaps the most representative example of a performance figure is the operating frequency (or frequencies in the case of multi-band structures) [23]. Another example is the substrate permittivity [24]. The objective space F is defined by the ranges $f_{k.\min} \leq f_k^{(j)} \leq f_{k.\max}, k = 1, \ldots, N$, i.e., $F = [f_{1.\min} \ f_{1.\max}] \times \ldots \times [f_{N.\min} \ f_{N.\max}]$. The design goals for a given target vector $f = [f_1 \ldots f_N]^T$ are encoded in an objective function $U(x, f)$. The optimum design $U_f(f)$ w.r.t. f, is

$$U_f(f) = \arg \min_x U(x, f) \tag{1}$$

An illustration example follows: if f_k are operating frequencies of a multi-band antenna, $U(\cdot)$ may be defined as $-\min\{B_1, \ldots, B_N\}$, where B_j is the fractional bandwidth corresponding to f_j. Thus, minimization of $U_f(f)$ leads to achieving the largest possible bandwidths.

$U_f(F) \subset X$ is an N-dimensional manifold (cf. Fig. 1), which determines the region of interest from the point of view of the figures f_k. It contains the designs that are of high quality w.r.t. f_k as specified by U. Hence, the surrogate constructed in the vicinity of $U_f(F)$ is all one needs to carry out the design tasks where f_k are of concern. Focusing on $U_f(F)$ yields considerable savings in terms of training data acquisition as compared to building the model in X. Yet, some problems arise: (i) how to identify $U_f(F)$, (ii) how to carry out design of experiments, and, (iii) how to employ the surrogate (e.g., for parametric optimization) given geometrical complexity of the domain. Sections 3 through 5 address these issues when discussing particular realizations of the constrained modeling concept.

2.2 Modeling Flow

The modeling flow is shown in Fig. 2. The fundamental step of the process is a definition of the model domain. The latter is based on information acquired from a certain number of reference designs optimized for the selected values of performance figures. A particular way of utilizing this data is what distinguishes different versions of the constrained modeling frameworks.

3 Constrained Modeling with Predefined Reference Point Allocation

The initial versions of constrained modeling framework utilized (structurally) fixed set of reference points. The advantage of this approach was simpler implementation. The downside was limited flexibility. This section describes a specific technique designed to model narrowband antennas with two figures of interest: operating frequency, and relative permittivity of the dielectric substrate [24].

3.1 Constructing the Surrogate

Here, we assume two figures of interest, specifically, the antenna operating frequency f and the relative substrate permittivity ε_r. The goal is to construct the surrogate for $f_{min} \leq f \leq f_{max}$, and $\varepsilon_{min} \leq \varepsilon_r \leq \varepsilon_{max}$. The design that is optimum for particular values of f and ε_r will be denoted as $U_f(f, \varepsilon_r)$.

The domain of the model is a neighborhood of the surface defined using 9 reference points allocated within the discussed ranges of the operating frequency and substrate permittivity. We have $U_f(f^*, \varepsilon_r^*)$, where $f^* \in \{f_{min}, f_0, f_{max}\}$ and $\varepsilon_r^* \in \{\varepsilon_{min}, \varepsilon_{r0}, \varepsilon_{max}\}$, cf. Fig. 3.

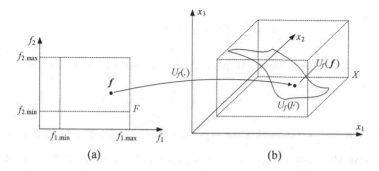

(a)   (b)

Fig. 1. Basic concepts of constrained modeling: (a) the space F of figures of interest, and (b) the parameter space X. Note that the set $U_f(F)$ constitutes an N-dimensional object (surface) in the parameter space. This set contains optimum designs for all $f \in F$. In principle, restricting the modeling process only to $U_f(F)$ is sufficient to maintain the design utility of the surrogate [26].

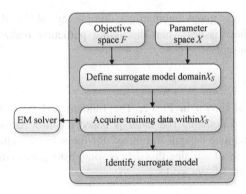

Fig. 2. Performance-driven modeling flow.

We define vectors $v_1 = U_f(f_{min}, \varepsilon_{min}) - U_f(f_0, \varepsilon_{r0})$, $v_2 = U_f(f_{min}, \varepsilon_{r0}) - U_f(f_0, \varepsilon_{r0})$, $v_3 = U_f(f_{min}, \varepsilon_{max}) - U_f(f_0, \varepsilon_{r0})$, $v_4 = U_f(f_0, \varepsilon_{max}) - U_f(f_0, \varepsilon_{r0})$, $v_5 = U_f(f_{max}, \varepsilon_{max}) - U_f(f_0, \varepsilon_{r0})$, $v_6 = U_f(f_{max}, \varepsilon_{r0}) - U_f(f_0, \varepsilon_{r0})$, $v_7 = U_f(f_{max}, \varepsilon_{min}) - U_f(f_0, \varepsilon_{r0})$, and $v_8 = U_f(f_0, \varepsilon_{min}) - U_f(f_0, \varepsilon_{r0})$, see Fig. 3(a). Let M be a manifold spanned by vectors $[v_1, v_2]$, $[v_2, v_3]$, …, $[v_8, v_1]$.

$$M = \bigcup_{k=1}^{8} M_k = \bigcup_{k=1}^{8} \{y = U_f(f_0, \varepsilon_{r0}) + \alpha v_k + \beta v_{k+1} : \alpha, \beta \geq 0, \ \alpha + \beta \leq 1\} \quad (2)$$

For consistency, $v_9 = v_1$. Let z be a point in the parameter space, and $P_k(z)$ be its projection onto M_k (cf. Fig 4(b)). We have

$$\arg\min_{\bar{\alpha}, \bar{\beta}} \left\| z - \left[U_f(f_0, \varepsilon_{r0}) + \bar{\alpha} v_k + \bar{\beta} v_{k+1}^{\#} \right] \right\|^2 \quad (3)$$

where $v_{k+1}^{\#} = v_{k+1} - p_k v_k$ with $p_k = v_k^T v_{k+1}(v_k^T v_k)$. Thus, $v_{k+1}^{\#}$ is a component of v_{k+1} that is orthogonal to v_k. Consider

$$\left[v_k \ v_{k+1}^{\#} \right] \left[\bar{\alpha} \ \bar{\beta} \right]^T = z - U_f(f_0, \varepsilon_{r0}) \quad (4)$$

The least-square solution to (4) (equivalent to the solution of (3)) is given as

$$\left[\bar{\alpha} \ \bar{\beta} \right]^T = \left(V_k^T V_k \right)^{-1} V_k^T \left(z - U_f(f_0, \varepsilon_{r0}) \right) \quad (5)$$

where $V_k = \left[v_k v_{k+1}^{\#} \right]$. In practice, the expansion coefficients with respect to v_k and v_{k+1} are of interest. These are given as $\alpha = \bar{\alpha} - p_k \bar{\beta}, \beta = \bar{\beta}$. Note that $P_k(z) \in M_k$ if and only if $\alpha \geq 0$, $\beta \geq 0$, and $\alpha + \beta \leq 1$.

Let $x_{max} = \max\{U_f(f_0, \varepsilon_{r0}) + v_1, \ldots, U_f(f_0, \varepsilon_{r0}) + v_8\}$ and $x_{min} = \min\{U_f(f_0, \varepsilon_{r0}) + v_1, \ldots, U_f(f_0, \varepsilon_{r0}) + v_8\}$; $dx = x_{max} - x_{min}$ is the range of variation of geometry parameters within M. Using these, we can define the domain X_S by imposing the following conditions: a $y \in X_S$ if

1. y is close to M in the sense that its orthogonal projection belongs to at least one M_k, i.e., we have $K(y) = \{k \in \{1,\ldots,8\} : P_k(y) \in M_k\} \neq \varnothing$;
2. $\min\{\|(y - P_k(y))//dx\| : k \in K(y)\} \leq d_{max}$ (here, //stands for component-wise division); d_{max} is a domain thickness parameter (typically, $0.1 \leq d_{max} \leq 0.2$).

Note that d_{max} in the second condition determines the "perpendicular" size of X_S. The size of X_S is dramatically smaller (volume-wise) than the size of the hypercube containing the reference designs. The surrogate itself is constructed using kriging interpolation of the EM model response R based on the training data allocated in X_S [24]. The design of experiments is based on random sampling within the interval $[x_{min}, x_{max}]$ assuming uniform probability distribution. The samples allocated outside $[x_{min}, x_{max}]$ are rejected.

3.2 Case Study: Ring Slot Antenna

The method is illustrated using a ring slot antenna of Fig. 5(a) [27], implemented on the 0.76-mm-thick substrate. The design parameters are $x = [l_f \, l_d \, w_d \, r \, s \, s_d \, o \, g \, \varepsilon_r]^T$; ε_r represents relative permittivity of the substrate. The feed line width w_f is computed for each ε_r to ensure 50 Ω input impedance. The EM is implemented in CST ($\sim$300,000 cells, simulation time 90 s). The goal is to construct the surrogate for the following ranged of the operating frequency and substrate permittivity: $f_{min} = 2.5$ GHz to $f_{max} = 6.5$ GHz, and $\varepsilon_{min} = 2.0$ to $\varepsilon_{max} = 5.0$. There are nine reference points generated by optimizing the antenna [28] for the pairs $\{f_0, \varepsilon_r\}$ with $f \in \{2.5, 4.5, 6.5\}$ GHz and $\varepsilon_r \in \{2.0, 3.5, 5.0\}$. The optimization objective is matching improvement at f_0.

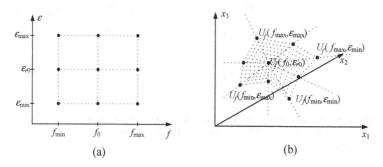

(a)   (b)

Fig. 3. Graphical illustration of reference points: (a) objective space, and (b) parameter space. The domain-defining surface is marked as the dotted area between the reference points [24].

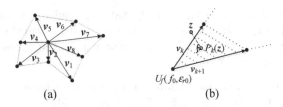

(a)   (b)

Fig. 4. Auxiliary concepts for domain definition: (a) the surface of Fig. 3(b); (b) the kth surface M_k marked along with its corresponding vectors v_k and v_{k+1}. Also shown is an exemplary point z and its projection onto M_k [24].

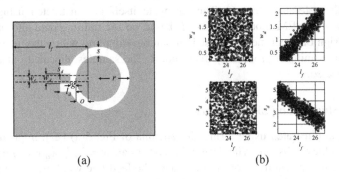

(a)   (b)

Fig. 5. Ring slot antenna: (a) geometry [27], (b) uniform versus constrained sampling for selected two-dimensional projections onto the l_f-w_d plane and the l_f-s_d plane [24].

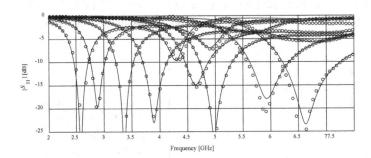

Fig. 6. Reflection characteristics of the ring-slot antenna at the selected test locations. The surrogate constructed with $N = 1000$ training samples: EM simulations (—), constrained surrogate (o) [24].

For the sake of validation, the surrogate model was constructed using the training set of various sizes, from 100 to 1000 samples. In all cases, we used $d_{max} = 0.2$. Model validation was performed using the split-sample method with 100 test designs. The benchmark was kriging interpolation surrogate established in the original domain $X = [x_{min}, x_{max}]$ using 1000 samples. The numerical results are gathered in Table 1, see also Fig. 6. Furthermore, Fig. 5(b) illustrates selected projections of the training data set for conventional (uniform) and proposed design of experiments. It can be observed that the accuracy improvement due to constrained sampling is considerable (by a factor of about 3.5). The constrained surrogate that exhibits the same predictive power as the corresponding conventional model can be obtained using around ten times less data samples.

4   Constrained Modeling Using Domain Triangulation

In [25], a generalization of the technique presented in Sect. 3 has been proposed, which is based on triangulation of the reference designs. This technique does not only allow for arbitrary distribution of the reference points but it also has no limitations in the number of figures of interest that can be handled.

4.1   Constructing the Surrogate

The parameter and objective spaces are defined as in Sect. 2. The reference designs $\boldsymbol{x}^{(j)} = [x_1^{(j)} \ldots x_n^{(j)}]^T, j = 1, \ldots, p$, are optimized with respect to the figure of interest vectors $\boldsymbol{f}^{(j)} = [f_1^{(j)} \ldots f_N^{(j)}]^T$. The reference designs $\boldsymbol{x}^{(j)}$ are subject to Delaunay triangulation [29] to form simplexes $S^{(k)}$, $k = 1, \ldots, N_S$, whose vertices are $S^{(k)} = \{\boldsymbol{x}^{(k.1)}, \ldots, \boldsymbol{x}^{(k.N+1)}\}$, where $\boldsymbol{x}^{(k,j)} \in \{\boldsymbol{x}^{(1)}, \ldots, \boldsymbol{x}^{(N)}\}, j = 1, \ldots, N + 1$ (cf. Fig. 7(a)).

Table 1.   Ring slot antenna: modeling results.

Design space sampling and surrogate modeling technique*	Average relative RMS error
Uniform sampling in the original space, $N = 1000$	7.3%
Constrained sampling, $N = 100$	7.8%
Constrained sampling, $N = 200$	5.5%
Constrained sampling, $N = 500$	3.3%
Constrained sampling, $N = 1000$	2.1%

*In all cases, the surrogate model constructed using kriging interpolation [28].

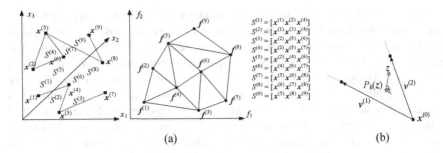

Fig. 7. Surrogate modeling using design reference triangulation: (a) triangulation of the reference designs (left plot) and corresponding objective vectors (right plot); (b) simplex $S^{(k)}$ a, point z and its projection onto the hyper-plane H_k containing $S^{(k)}$.

The model domain X_S is determined as a vicinity of the manifold M defined as

$$M = \bigcup_k \{y = \sum_{j=1}^{N+1} \alpha_j x^{(k,j)} : 0 \le \alpha_j \le 1, \sum_{j=1}^{N+1} \alpha_j = 1\} \tag{6}$$

The region is defined using the distance from the surface M in the orthogonal complements of the subspaces containing $S^{(k)}$. Given a point z, it is necessary to find the distance from it to M, which can be done by considering a projection $P_k(z)$ onto the affine subspace $H_k \supset S^{(k)}$. We also define the simplex anchor $x^{(0)} = x^{(k,1)}$, and its spanning vectors $v^{(j)} = x^{(k,j+1)} - x^{(0)}$, $j = 1, \ldots, N$ (cf. Fig. 7(b)). The projection corresponds to the expansion coefficients w.r.t. $v^{(j)}$ [25]

$$\arg \min_{[\bar{\alpha}^{(1)},\ldots,\bar{\alpha}^{(N)}]} \left\| z - \left[x^{(0)} + \sum_{j=1}^{N} \bar{\alpha}^{(j)} \bar{v}^{(j)} \right] \right\|^2 \tag{7}$$

where the vectors $\bar{v}^{(j)}$ are obtained from $v^{(j)}$ by orthogonalization (i.e., $\bar{v}^{(1)} = v^{(1)}$, $\bar{v}^{(2)} = v^{(2)} - a_{12}v^{(1)}$ where $a_{12} = v^{(1)T}v^{(2)}(v^{(1)T}v^{(1)}$, etc.). In general

$$\bar{V} = \left[\bar{v}^{(1)} \, \bar{v}^{(2)} \, \ldots \, \bar{v}^{(N)} \right] = \left[v^{(1)} \, v^{(2)} \, \ldots \, v^{(N)} \right] A \tag{8}$$

Note that A is a triangular matrix that contains coefficients obtained from the above orthogonalization procedure. The problem (7) is equivalent to

$$\left[\bar{v}^{(1)} \, \bar{v}^{(2)} \, \ldots \, \bar{v}^{(N)} \right] \begin{bmatrix} \bar{\alpha}^{(1)} \\ \vdots \\ \bar{\alpha}^{(N)} \end{bmatrix} = z - x^{(0)} \tag{9}$$

The expansion coefficients can be found analytically as

$$\left[\bar{\alpha}^{(1)} \ldots \bar{\alpha}^{(N)}\right]^T = (\bar{V}^T \bar{V})^{-1} \bar{V}^T (z - x^{(0)}) \tag{10}$$

The critical factor is whether $P_k(z) \in hull(S^{(k)})$ ($hull()$ stands for the convex hull). For that, it is necessary to identify the expansion coefficients $\alpha^{(j)}$ of z w.r.t. $\{v^{(j)}\}$. The latter can be obtained as $[\alpha^{(1)} \ldots \alpha^{(N)}]^T = A[\bar{\alpha}^{(1)} \ldots \bar{\alpha}^{(N)}]^T$. It should be noted that $P_k(z) \in S^{(k)}$ if

1. $\alpha^{(j)} \geq 0$ for $j = 1, \ldots, N$, and
2. $\alpha^{(1)} + \ldots + \alpha^{(N)} \leq 1$.

that is, if $P_k(z)$ is a convex combination of the vectors $v^{(j)}$.

The next step of surrogate modeling domain definition is to define $x_{max} = \max\{x^{(k)}, k = 1, \ldots, p\}$ and $x_{min} = \min\{x^{(k)}, k = 1, \ldots, p\}$. The vector $dx = x_{max} - x_{min}$ is an indication of the geometry parameter variability within the surface M. Using these, the surrogate model domain X_S can be defined as in [24]. More specifically, a vector $y \in X_S$ if

1. $K(y) = \{k \in \{1, \ldots, N_S\} : P_k(y) \in S^{(k)}\} \neq \emptyset$;
2. $\min\{\|(y - P_k(y))//dx\| : k \in K(y)\} \leq d_{max}$ (//stands for component-wise division); d_{max} is a user-defined parameter determining the domain thickness.

The major benefit of this definition is that X_S is considerable smaller than the original domain $X = [x_{min}, x_{max}]$ volume-wise, thus the surrogate can be constructed using a reduced training set. Notwithstanding, the domain still contains the optimum designs (with respect to the selected performance figures) so that the surrogate retains its design utility. This is demonstrated in the next section by modeling a dual-band antenna over broad ranges of both geometry and material parameters.

4.2 Verification Case: Uniplanar Dipole Antenna

For the sake of verification, let us consider a dipole antenna of Fig. 8(a) [30]. The structure is realized on Taconic RF-35 substrate of relative permittivity $\varepsilon_r = 3.5$ and thickness $h = 0.762$ mm. There are six adjustable variables $x = [l_1 \; l_2 \; l_3 \; w_1 \; w_2 \; w_3]^T$. Other parameters are fixed: $l_0 = 30$, $w_0 = 3$, $s_0 = 0.15$ and $o = 5$ are fixed (dimensions in mm). The computational model R is simulated in CST Microwave Studio ($\sim 100,000$ cells; simulation time 1 min).

We aim at constructing the surrogate within the following objective space: 2.0 GHz $\leq f_1 \leq 4.0$ GHz (lower band), and 4.5 GHz $\leq f_2 \leq 6.5$ GHz (upper band). Figure 8(b) shows the allocation of the reference designs. The latter have been generated using variable-fidelity feature-based optimization [28].

Validation has been carried out by rendering the surrogates using various numbers of training samples, from 100 to 1600. In all cases, $d_{max} = 0.05$ was employed. The numerical results are gathered in Table 2. Conventional kriging metamodel is used as a benchmark. The antenna reflection characteristics according to the surrogate and EM simulation have been shown in Fig. 9.

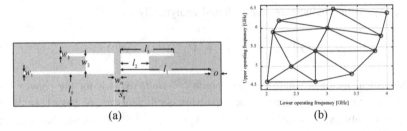

(a)   (b)

Fig. 8. Uniplanar dipole antenna: (a) geometry [30], (b) allocation of the reference designs and their triangulation [25].

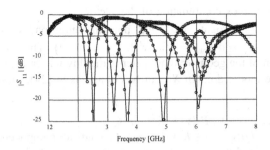

Fig. 9. Reflection characteristics of the antenna of Fig. 8: electromagnetic simulations (—), triangulation-based surrogate rendered with $N = 1600$ training samples (o).

Table 2. Modeling results and benchmarking for antenna of Fig. 8.

Number of training samples[#]	Relative RMS error[*]	
	Conventional surrogate[$]	Triangulation-based surrogate
100	17.2%	4.6%
200	12.7%	3.5%
400	9.3%	2.8%
800	6.9%	2.6%
1600	5.7%	2.3%

[*]In all cases, the surrogate model constructed using kriging interpolation.
[#]The cost of finding the reference designs for constrained modeling is about 400 evaluations of the EM antenna model.
[$]Conventional surrogate established in the parameter space $X = [x_{min}, x_{max}]$.

5 Modeling Using Nested Kriging

The nested kriging framework proposed in [26] employs two kriging metamodels. One of these models is used to establish the surrogate model domain by mapping the figure-of-interest space into the parameter space and to provide the first approximation of the region of interest. The major advantage of [26] is that design of experiments but also model optimization can be implemented in a convenient way.

5.1 Surrogate Model Construction

We use the same definitions as in Sect. 2.1 for the objective and parameter spaces. The reference designs optimized w.r.t. $f^{(j)} = [f_1^{(j)} \dots f_N^{(j)}]$ are denoted as $x^{(j)} = [x_1^{(j)} \dots x_n^{(j)}]^T$, $j = 1, \dots, p$. The technique employs two kriging metamodels. The first-level one $s_I(f)$ transforms F into the parameter space X. The model s_I is set up using the training pairs $\{f^{(j)}, x^{(j)}\}_{j=1,\dots,p}$ (see Fig. 10 for a graphical illustration).

The surrogate model domain X_S is defined to contain the designs that are optimum w.r.t. f_k, $k = 1, \dots, N$. The information obtained from the reference designs only permits for establishing an initial approximation of the optimum design set $U_f(f)$. As the domain should contain the entire $U_f(f)$ (or a vast majority of it), $s_I(F)$ must be enlarged. This is realized by an orthogonal extension of $s_I(F)$ towards its normal vectors. We denote by $\left\{ v_n^{(k)}(f) \right\}$, $k = 1, \dots, n - N$, an orthonormal basis of vectors normal to $s_I(F)$ at f, and define $x_{\max} = [x_{\max.1} \ \cdots \ x_{\max.n}]^T$, $x_{\min} = [x_{\min.1} \ \cdots \ x_{\min.n}]^T$, with $x_{\max.k} = \max\left\{ x_k^{(j)}, j = 1, \dots, p \right\}$, and $x_{\min.k} = \min\left\{ x_k^{(j)}, j = 1, \dots, p \right\}$. We also define $x_d = x_{\max} - x_{\min}$ (parameter variations within $s_I(F)$). Further, extension coefficients are defined as follows:

$$\alpha(f) = [\alpha_1(f) \dots \alpha_{n-N}(f)]^T = 0.5 d_{\max}\left[|x_d v_n^{(1)}(f)| \ \cdots \ |x_d v_n^{(n-N)}(f)| \right]^T \quad (11)$$

Similarly as for the method discussed before, $d_{\max}$ denotes the domain thickness. The coefficients α_k are used to delimit X_S (see also Fig. 10(b)) [26] by defining

$$M_{\pm} = \left\{ x \in X : x = s_I(f) \pm \sum_{k=1}^{n-N} \alpha_k(f) v_n^{(k)}(f) \right\} \quad (12)$$

Using (12), we get

$$X_S = \left\{ \begin{matrix} x = s_I(f) + \sum_{k=1}^{n-N} \lambda_k \alpha_k(f) v_n^{(k)}(f) : f \in F, \\ -1 \leq \lambda_k \leq 1, \ k = 1, \dots, n - N \end{matrix} \right\} \quad (13)$$

The second-level surrogate is a kriging model rendered in X_S based on $\{x_B^{(k)}, R(x_B^{(k)})\}_{k=1,\dots,NB}$, where R is the EM-simulation model of the structure of interest.

Note that the definition of X_S facilitates design of experiments which was a problem for both [23] and [25]. It is implemented using (13) and the mappings from the unit interval $[0, 1]^n$ onto X_S. Let $\{\mathbf{z}^{(k)}\}$, $k = 1, ..., N_B$, where $\mathbf{z}^{(k)} = [z_1^{(k)} \ ... \ z_n^{(k)}]^T$, denote the set of uniformly distributed data points in $[0, 1]^n$ (here, using LHS [31]). The mapping is realized in two stages. First, the function h_1.

$$
\begin{aligned}
\mathbf{y} = h_1(\mathbf{z}) &= h_1([z_1 ... z_n]^T) \\
&= [f_{1.\min} + z_1 (f_{1.\max} - f_{1.\min}) ... f_{N.\min} + z_N (f_{N.\max} - f_{N.\min})] \times [-1 + 2z_{N+1} ... -1 + 2z_n]
\end{aligned}
\tag{14}
$$

transforms the unit hypercube onto a $F \times [-1,1]^{n-N}$ ($\times$ is a Cartesian product). Subsequently, a function h_2 is defined as

$$
\begin{aligned}
\mathbf{x} = h_2(\mathbf{y}) &= h_2([y_1 \ ... \ y_n]^T) \\
&= \mathbf{s}_l([y_1 \ ... \ y_N]^T) + \sum_{k=1}^{n-N} y_{N+k} \alpha_k([y_1 \ ... \ y_N]^T) \mathbf{v}_n^{(k)}([y_1 \ ... \ y_N]^T)
\end{aligned}
\tag{15}
$$

which maps $F \times [-1, 1]^{n-N}$ onto X_S. Hence, uniformly distributed samples $\mathbf{x}_B^{(k)}$ in X_S are obtained as $\mathbf{x}_B^{(k)} = H(\mathbf{z}^{(k)}) = h_2(h_1(\mathbf{z}^{(k)}))$.

The surjective mapping H also allows for implementing surrogate model optimization in its domain X_S. In particular, the optimization process can be formally carried out in the primary domain $F \times [-1, 1]^{n-N}$, whereas the mapping H can be employed to perform evaluation of the structure.

5.2 Verification Case: Miniaturized Impedance Transformer

The modeling technique described in Sect. 5.1 has been validated using a miniaturized impedance matching transformer [32]. The structure is shown in Fig. 11(b). It is realized on Taconic RF-35 substrate of relative permittivity $\varepsilon_r = 3.5$ and thickness $h = 0.762$ mm. The circuit employees compact microstrip resonant cells (CMRCs) shown in Fig. 11(a).

The adjustable variables are $\mathbf{x} = [l_{1.1} \ l_{1.2} \ w_{1.1} \ w_{1.2} \ w_{1.0} \ l_{2.1} \ l_{2.2} \ w_{2.1} \ w_{2.2} \ w_{2.0} \ l_{3.1} \ l_{3.2}$ $w_{3.1} \ w_{3.2} \ w_{3.0}]^T$. The figure-of-interest space contains the operating bands $[f_1 \ f_2]$ with the following ranges: 1.5 GHz $\leq f_1 \leq 3.5$ GHz, and 4.5 GHz $\leq f_2 \leq 6.5$ GHz. Here, the optimum design is understood by minimization of the maximum reflection $|S_{11}|$ within $[f_1 \ f_2]$ (which is a minimax problem). The allocation of the reference designs has been shown in Fig. 11(c).

Surrogate model validation has been carried out for various numbers of training data samples from 50 to 800. All models were generated using $d_{\max} = 0.05$. The modeling error was calculated using the split-sample method with 100 random testing designs. The numerical results are shown in Table 3 (see also Fig. 12). The benchmark includes conventional kriging and RBF metamodels. It can be observed that the nested kriging framework enables two- or even three-fold reduction of the modeling error assuming the same training data set size. It should be emphasized that for this case the predictive power of all conventional surrogates is poor. The primary reason is a large number of geometry parameters and, consequently, a large volume of the conventional domain.

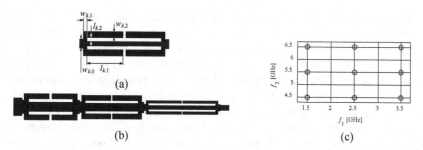

(a)   (b)

Fig. 10. The main components of the nested kriging framework: (a) reference points and the space of figures of interest F; (b) the domain defining surfaces: $s_I(F)$, the exemplary normal vector $v_1^{(k)}$ at $f^{(k)}$; the surfaces M_- and M_+, and the domain X_S. The latter is defined as an extension of $s_I(F)$ according to (13) [26].

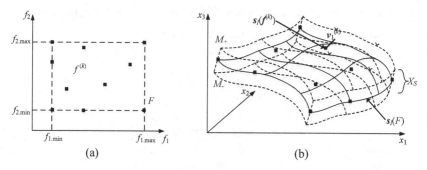

(a)

(b)   (c)

Fig. 11. Compact impedance transformer as verification case study for the nested kriging framework: (a) CMRC cell, (b) impedance transformer structure [32], (c) reference points [26].

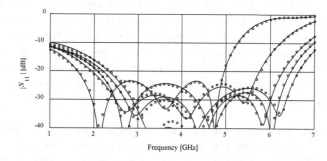

Fig. 12. Reflection characteristics of the circuit of Fig. 11(b): full-wave electromagnetic simulations (—), and the nested kriging surrogate rendered with $N = 800$ training samples (o) [26].

Table 3. Modeling results and benchmarking for impedance matching transformer.

Number of training samples	Relative RMS error		
	Conventional models		Nested Kriging model
	Kriging	RBF	
50	49.1%	56.2%	17.3%
100	31.1%	33.0%	13.9%
200	25.9%	27.5%	10.3%
400	20.4%	23.1%	7.4%
800	15.7%	16.8%	6.1%

$^\$$Conventional surrogate established in the parameter space $X = [x_{min}, x_{max}]$.

6 Conclusions

This paper discussed the recent developments in constrained modeling of high-frequency structures. Domain confinement permits a reduction of the computational overhead related to training data acquisition. At the same time, reliable surrogates can be rendered without formally restricting the ranges of geometry and material parameters as well as the operating conditions. There were three specific realizations of this concept discussed in the work. The main differences between these were in analytical formulation of the surrogate model domain. The simplest technique, discussed in Sect. 3, requires structured allocation of the reference designs. The second method, the triangulation-based modeling of Sect. 4, is more flexible, allows for an arbitrary placement of the reference set, and it can directly handle arbitrary number of operating conditions. The downside is a non-trivial design of experiments. The nested kriging framework is the most comprehensive: the very formulation of the model domain incorporates the means to carry out uniform design of experiments as well as optimization of the surrogate. Notwithstanding, implementation of this approach is more involved. In general, surrogate modeling with domain confinement can be considered a viable workaround dimensionality and parameter range issues, both of which are the fundamental challenges of conventional methods. At the same time, the initial computational overhead required to generate the reference points is well justified when the surrogate is reused, for example, for the purpose of dimension scaling for various operating conditions.

Acknowledgement. The authors would like to thank Dassault Systemes, France, for making CST Microwave Studio available. This work was supported in part by the Icelandic Centre for Research (RANNIS) Grant 174573051 and by National Science Centre of Poland Grant 2017/27/B/ST7/00563.

References

1. Koziel, S., Unnsteinsson, S.D.: Expedited design closure of antennas by means of trust-region-based adaptive response scaling. IEEE Antennas Wirel. Propag. Lett. **17**(6), 1099–1103 (2018)
2. Torun, H.M., Swaminathan, M.: High-dimensional global optimization method for high-frequency electronic design. IEEE Trans. Microw. Theory Tech. **67**(6), 2128–2142 (2019)
3. Hosder, S.: Stochastic response surfaces based on non-intrusive polynomial chaos for uncertainty quantification. Int. J. Num. Model. Num. Optim. **3**(1/2), 117–139 (2012)
4. Ghassemi, M., Bakr, M., Sangary, N.: Antenna design exploiting adjoint sensitivity-based geometry evolution. IET Microw. Antennas Propag. **7**(4), 268–276 (2013)
5. Koziel, S., Bekasiewicz, A.: Reliable multistage optimization of antennas for multiple performance figures in highly dimensional parameter spaces. IEEE Antennas Wirel. Propag. Lett. **18**(7), 1522–1526 (2019)
6. Bandler, J.W., Hailu, D.M., Madsen, K., Pedersen, F.: A space-mapping interpolating surrogate algorithm for highly optimized EM-based design of microwave devices. IEEE Trans. Microw. Theory Tech. **52**(11), 2593–2600 (2004)
7. Koziel, S., Leifsson, L.: Simulation-Driven Design by Knowledge-Based Response Correction Techniques. Springer, Heidelberg (2016). https://doi.org/10.1007/978-3-319-30115-0
8. Koziel, S., Bekasiewicz, A.: Fast simulation-driven feature-based design optimization of compact dual-band microstrip branch-line coupler. Int. J. RF Microw. Comput. Aided Eng. **26**(1), 13–20 (2015)
9. Rasmussen, C.E., Williams, C.K.I.: Gaussian Processes for Machine Learning. MIT Press, Cambridge (2006)
10. Gorissen, D., Dhaene, T., De Turck, F.: Evolutionary model type selection for global surrogate modeling. J. Mach. Learn. Res. **10**, 2039–2078 (2009)
11. Chávez-Hurtado, J.L., Rayas-Sánchez, J.E.: Polynomial-based surrogate modeling of RF and microwave circuits in frequency domain exploiting the multinomial theorem. IEEE Trans. Microw. Theory Tech. **64**(12), 4371–4381 (2016)
12. Kitayama, S., Arakawa, M., Yamazaki, K.: Sequential approximate optimization using radial basis function network for engineering optimization. Optim. Eng. **12**(4), 535–557 (2011)
13. Kleijnen, J.P.C.: Kriging metamodeling in simulation: a review. Eur. J. Oper. Res. **192**(3), 707–716 (2009)
14. Angiulli, G., Cacciola, M., Versaci, M.: Microwave devices and antennas modelling by support vector regression machines. IEEE Trans. Magn. **43**(4), 1589–1592 (2007)
15. Du, J., Roblin, C.: Statistical modeling of disturbed antennas based on the polynomial chaos expansion. IEEE Antennas Wirel. Propag. Lett. **16**, 1843–1846 (2017)
16. Schobi, R., Sudret, B., Wiart, J.: Polynomial-chaos-based kriging. Int. J. Uncertainty Quant. **5**(2), 171–193 (2015)
17. Liu, H., Hervas, J.R., Ong, Y.S., Cai, J., Wang, Y.: An adaptive RBF-HDMR modeling approach under limited computational budget. Struct. Multidisc. Optim. **57**(3), 1–18 (2018)
18. Koziel, S., Bekasiewicz, A.: Computationally feasible narrow-band antenna modeling using response features. Int. J. RF Microw. Comput. Aided Eng. **27**(4), e21077 (2017)
19. Tropp, J.A., Gilbert, A.C.: Signal recovery from random measurements via orthogonal matching pursuit. IEEE Trans. Inf. Theory **53**(12), 4655–4666 (2007)
20. Wang, F., et al.: Bayesian model fusion: large-scale performance modeling of analog and mixed-signal circuits by reusing early-stage data. IEEE Trans. CAD Integr. Circuits Syst. **35**(8), 1255–1268 (2016)

21. Koziel, S., Ogurtsov, S., Couckuyt, I., Dhaene, T.: Variable-fidelity electromagnetic simulations and co-kriging for accurate modeling of antennas. IEEE Trans. Antennas Propag. **61**(3), 1301–1308 (2013)
22. Jacobs, J.P., Koziel, S.: Two-stage framework for efficient Gaussian process modeling of antenna input characteristics. IEEE Trans. Antennas Propag. **62**(2), 706–713 (2014)
23. Koziel, S.: Low-cost data-driven surrogate modeling of antenna structures by constrained sampling. IEEE Antennas Wirel. Propag. Lett. **16**, 461–464 (2017)
24. Koziel, S., Bekasiewicz, A.: On reduced-cost design-oriented constrained surrogate modeling of antenna structures. IEEE Antennas Wirel. Propag. Lett. **16**, 1618–1621 (2017)
25. Koziel, S., Sigurðsson, A.T.: Triangulation-based constrained surrogate modeling of antennas. IEEE Trans. Antennas Propag. **66**(8), 4170–4179 (2018)
26. Koziel, S., Pietrenko-Dabrowska, A.: Performance-based nested surrogate modeling of antenna input characteristics. IEEE Trans. Antennas Propag. **67**(5), 2904–2912 (2019)
27. Sim, C.Y.D., Chang, M.H., Chen, B.Y.: Microstrip-fed ring slot antenna design with wideband harmonic suppression. IEEE Trans. Antennas Propag. **62**(9), 4828–4832 (2014)
28. Koziel, S.: Fast simulation-driven antenna design using response-feature surrogates. Int. J. RF Microw. Comput. Aided Eng. **25**(5), 394–402 (2015)
29. Cheng, S.W., Dey, T., Shewchuk, J.: Delaunay Mesh Generation. Chapman and Hall, London (2013)
30. Chen, Y.C., Chen, S.Y., Hsu, P.: Dual-band slot dipole antenna fed by a coplanar waveguide. In: 2006 IEEE International Symposium Antennas and Propagation (ISAP) (2006)
31. Ai, M., Kong, X., Li, K.: A general theory for orthogonal array based Latin hypercube sampling. Statistica Sinica **26**(2), 761–777 (2016)
32. Koziel, S., Bekasiewicz, A.: Rapid simulation-driven multi-objective design optimization of decomposable compact microwave passives. IEEE Trans. Microw. Theory Tech. **64**(8), 2454–2461 (2016)

Aerodynamic Shape Optimization for Delaying Dynamic Stall of Airfoils by Regression Kriging

Vishal Raul[1], Leifur Leifsson[1(✉)], and Slawomir Koziel[2]

[1] Department of Aerospace Engineering, Iowa State University,
Ames, IA 5011, USA
{vvssraul,leifur}@iastate.edu
[2] Engineering Optimization and Modeling Center, Reykjavik University,
Reykjavik, Iceland
koziel@ru.is

Abstract. The phenomenon of dynamic stall produce adverse aerodynamic loading which can adversely affect the structural strength and life of aerodynamic systems. Aerodynamic shape optimization (ASO) provides an effective approach for delaying and mitigating dynamic stall characteristics without the addition of auxiliary system. ASO, however, requires multiple evaluations time-consuming computational fluid dynamics models. Metamodel-based optimization (MBO) provides an efficient approach to alleviate the computational burden. In this study, the MBO approach is utilized for the mitigation of dynamic stall characteristics while delaying dynamic stall angle of the flow past wind turbine airfoils. The regression Kriging metamodeling technique is used to approximate the objective and constrained functions. The airfoil shape design variables are described with six PARSEC parameters. A total of 60 initial samples are used to construct the metamodel, which is further refined with 20 infill points using expected improvement. The metamodel is validated with the normalized root mean square error based on 20 test data samples. The refined metamodel is used to search for the optimal design using a multi-start gradient-based method. The results show that an optimal design with a 3° delay in dynamic stall angle as well a reduction in the severity of pitching moment coefficients can be obtained.

Keywords: Dynamic stall · Unsteady CFD · Surrogate-based optimization · Regression Kriging · Expected improvement

1 Introduction

The dynamic stall phenomenon was first observed on retreating blades of helicopter rotor [6]. Horizontal and vertical axis wind turbines are prone to dynamic stall. Wind turbines are subjected to dynamic loading from multiple sources, such as wind shear, turbulence, yaw angles, upwind turbine wake, and tower shadow, that cause unsteady inflow to the turbine rotor which results in dynamic stall.

© Springer Nature Switzerland AG 2020
V. V. Krzhizhanovskaya et al. (Eds.): ICCS 2020, LNCS 12141, pp. 57–70, 2020.
https://doi.org/10.1007/978-3-030-50426-7_5

In vertical axis wind turbines (VAWT), dynamic stall arises from rapid changes in angle of attack on each blade in every rotation cycle [2,25]. The dynamic loading in wind turbines generates adverse loading conditions, significantly impacting the blade, hub, tower structure, performance and turbine life.

Significant research has been conducted to mitigate or control dynamic stall via active and passive control systems [10,15,27,28]. The addition of structures and control systems to the wind turbines increases their mass as well as their cost and complexity. Mitigating the adverse dynamic stall characteristics passively through aerodynamic shape optimization (ASO) has recently received interest from multiple researchers offering promising improvement in airfoil performance [12,16,23,24,26]. ASO studies for dynamic stall mitigation are typically done with adjoint-based computational fluid dynamics (CFD) simulations [4,12,16,26] and have shown promising results for multiple dynamic stall optimization cases. Adjoint-based CFD simulations is a modern approach to solve ASO problems using gradient-based optimization (GBO) algorithms [8]. The advantage of the adjoint method is the ability to estimate gradient information cheaply. The GBO approach, however, can get easily get stuck in local minima, especially if the CFD data is noisy. Wang et al. [23,24] used sequential quadratic programming (SQP) to alleviate aerodynamic loads during dynamic stall cycle on rotor airfoils.

Genetic algorithms have the ability to search the design space globally, but they require multiple design evaluations and can be impractical to use for high dimensional design problems. Ma et al. [11] used a multi-island genetic algorithm, which is a global search method, for VAWT performance improvement.

Metamodel-based optimization (MBO) (also called surrogate-based optimization) [22] is an approach to alleviate the computational burden of costly simulation-based design problems. In MBO, a metamodel (also called a surrogate) of the objective function is constructed using a limited number of the time-consuming simulations. The surrogate model is fast to evaluate and can be used within GBO or with genetica algorithms to search for the optimal design. To the best of our knowledge, MBO has not yet been utilized for ASO to mitigate dynamic stall characteristics of airfoils.

In this work, MBO is used for ASO of wind turbine airfoils to delay stall. The surrogate is constructed using regression Kriging [7] and is sequentially refinement using expected improvement infill criteria. The PARSEC airfoil parameterization technique [20] with six design variables is used for generating the airfoil shapes. The surrogate model is searched using a multi-start gradient-based optimizer.

The next section presents the problem statement for dynamic stall mitigation and the setup of the computational model. The following section describes the MBO approach. Results of numerical experiments are presented for the ASO. Conclusions and suggestions of future work are then described.

2   Problem Statement

This section describes the problem formulation and the airfoil parameterization method used for the current study, as well as the CFD modeling and validation.

2.1   Problem Formulation

The dynamic stall phenomenon is generally studied with sinusoidal oscillating airfoil in a uniform free-stream flow. The pitching motion of the airfoil is described using the angle of attack as a function of time t given as

$$\alpha(t) = \alpha_m + A\sin(\omega t),\tag{1}$$

where α_m, A and ω represent the mean angle of attack, amplitude of oscillation, and rotational rate, respectively. The reduced frequency, k, is another important parameter and is defined as

$$k = \frac{\omega c}{2U},\tag{2}$$

where c is the airfoil chord length, and U is the free-stream speed. In this work, a deep dynamic stall case from Lee et al. [9] is used. The parameters defining the case are: $\alpha_m = 10°$, $A = 15°$, $k = 0.05$, and a Reynolds number of $Re = 135,000$.

The objective of the study is to produce an optimum airfoil shape which mitigates the dynamic stall adverse loading by delaying the dynamic stall angle. This objective is achieved by delaying the formation of the dynamic stall vortex responsible for sudden divergence in the drag and pitching moment coefficients. The optimization problem is formulated as:

$$\min_{\mathbf{x}} \quad f(\mathbf{x}) = \left(\frac{\sum_{i=1}^{N} c_{d_i}}{F_{c_{d_0}}}\right) + \left(\frac{\sum_{i=1}^{N} |c_{m_i}|}{G_{c_{m_0}}}\right)\tag{3}$$

$$s.t. \quad g_1(\mathbf{x}) = \alpha_{ds0} + \Delta\alpha - \alpha_{ds} \le 0\tag{4}$$

$$\mathbf{x}_l \le \mathbf{x} \le \mathbf{x}_u\tag{5}$$

Here, $F_{c_{d_0}} = \sum_{i=1}^{N}(c_{d_0})_i$, $G_{c_{m_0}} = \sum_{i=1}^{N}|(c_{m_0})_i|$. $\mathbf{x}$ is the design variable vector. $\mathbf{x}_l$ and $\mathbf{x}_u$ are the lower and upper bounds of $\mathbf{x}$, respectively. The parameters c_{d_i}, c_{m_i}, α_{ds} represent the time variant drag coefficient, pitching moment coefficient at the ith timestep and dynamic stall angle of the airfoil. The subscript '0' represents the baseline airfoil shape, which is the NACA0012 airfoil. $\Delta\alpha$ denotes the minimum delay in the dynamic stall angle expected in the optimum design, which is set to $\Delta\alpha = 3°$ in this work. N denotes the number of time steps in each pitching cycle. For this study, we will only consider the upstroke part of the pitching cycle, which is predominantly affected by formation of dynamic stall vortex.

2.2   Design Variables

In this work, the PARSEC [20] parameterization technique is used for describing the airfoil shapes. In PARSEC, there are 12 parameters defining the airfoil shape of unit chord. The parameters affecting only the upper surface of the airfoil are considered in this study. The trailing edge offset and thickness are set to zero, which generates a sharp trailing edge airfoil. For this study, we have selected six parameters (see Table 1).

Table 1. Design variables and their bounds for upper airfoil surface

Description	x	x_u	x_l
Surface crest x coordinate	X	0.5011	0.2733
Surface crest z coordinate	Z	0.09	0.054
Second order surface derivative	Z_{xx}	−0.4036	−0.6726
Leading edge radius	R_{LE}	0.0222	0.0104
Trailing edge directional angle	θ_{TE}	−7.0294	−11.7156
TE wedge angle	β_{TE}	5.8803	3.52818

2.3 Computational Fluid Dynamics Modeling

The current study is performed with the Stanford University Unstructured (SU^2) unsteady compressible Navier-Stokes (URANS) solver [17]. The dynamic stall simulations are performed using dual time stepping strategy, rigid grid motion and Menter's shear stress transport (SST) turbulence model [14]. The convective fluxes calculated using second-order Jameson-Schmidt-Turkel (JST) scheme [17] and time discretization is done by the Euler implicit scheme [17] with maximum Courant-Friedrichs-Lewy (CFL) number selected as 4. The two-level multigrid W-cycle method [17] is also used for convergence acceleration. The Cauchy convergence criteria [1] is applied with Cauchy epsilon as 10^{-6} over last 100 iterations. No-slip boundary condition is used on airfoil surface with farfield condition on external boundary with Reynolds number of 135,000 and Mach number of 0.1. The c-grid mesh is set up an with outer boundary at 55c from airfoil is generated using blockmesh utility provided by OpenFoam [3]. The mesh is refined near the airfoil surface with first layer thickness to obtain $y^+ \leq 0.5$ and growth ratio of 1.05, which is necessary to accurately capture the onset of the dynamic stall vortex. Figure 1 show a coarse version of the mesh.

The grid and time independence study is done in two steps. Initially, the spatial resolution of the mesh is obtained by grid study. This mesh is then used to conduct time study to attain accurate physical time step. The flow and motion parameters are selected from study done by Lee et al. [9] as mentioned in Sect. 2.1. The grid study is done at Re=135,000, angle of attack $\alpha = 4°$ and turbulence intensity $TI = 0.08\%$. The details of grid study are shown in Table 2. Meshes 2, 3 and 4 show minimal change in lift coefficient $\Delta c_l \leq 0.003$ with the drag counts variation within 4 counts. Considering the simulation time requirement and accuracy of the results, mesh 2 with 387,000 cells is selected for the study.

After selecting the spatial resolution, a time independent study is conducted with multiple time steps of an airfoil in a sinusoidal pitching cycle in order to select the temporal resolution. This is done using the generalized Richardson extrapolation method (REM) [18] with the use of average drag coefficient per oscillation cycle $c_{d_{avg}}$ as a lower order value to an estimation parameter. The REM estimate $c_{d_{Est}}$ represents the average drag coefficient per cycle at a zero time step, which is calculated as $c_{d_{Est}} = 2,108$ counts. Table 3 summarizes the results. The simulation time and estimated error Est_{err} are then considered to select time step of 0.0015 for all further investigations.

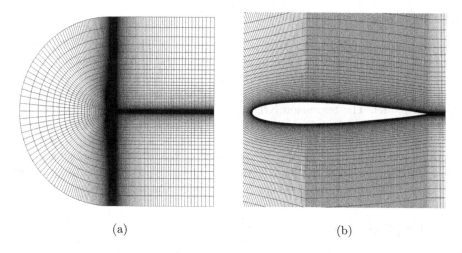

(a)   (b)

Fig. 1. Coarse mesh with NACA0012 airfoil (a) computational domain, (b) mesh around airfoil (zoom view)

Table 2. Grid convergence study at Re=135000, $\alpha = 4°$

Mesh	Number of cells $\times 10^3$	c_l	$c_d, counts$	*Simulation time (min)
1	259	0.395	174.3	75
2	387	0.414	180.4	146
3	540	0.416	184.7	220
4	720	0.417	184.2	298

*Computed on high-performance cluster with 64 processors

Table 3. Time study at $\alpha = 10° + 15° sin(\omega t)$ with $k = 0.05$ at $Re = 135,000$

dt [s]	$c_{d_{avg}}$ [counts/cycle]	**Simulation Time [hrs/cycle]	$Est_{err} = [c_{d_{avg}} - c_{d_{Est}}]$ [counts]
0.004	2,019	51	88.4
0.002	2,093	65	14.9
0.0015	2,103	69	4.8
0.0010	2,105	78	2.1
0.0005	2,107	99	0.52

**Computed on high-performance cluster with 112 processors

3   Methods

This section describes the MBO algorithm and the mathematical details of the metamodeling. In particular, the details of the workflow, sampling plan, regression Kriging, infill criteria, and validation are described.

3.1 Workflow

A flowchart of the MBO algorithm is shown in Fig. 2. The presented algorithm consist of an automated loop which sequentially improves the metamodel accuracy. The optimization algorithm starts with a sampling plan where the design space is sampled for initial samples. The initial samples are then evaluated with the CFD model. The regression Kriging metamodel is then constructed for the objective and constraint functions from the initial samples. The constructed metamodel is validated against a test data set. If the model does not pass termination criteria, then an infill strategy is used to refine the metamodel and the above steps are repeated until the metamodel accuracy satisfies the termination criteria. Finally, an optimum design is found by optimizing the metamodel.

3.2 Sampling Plan

The accurate construction of metamodel requires an appropriate sampling plan which captures the trend of objective function throughout design space. In this study, Latin hypercube sampling (LHS) [5,13] is used to generate initial and test data samples. For this study, an initial sample size is considered as ten times the number of design variables.

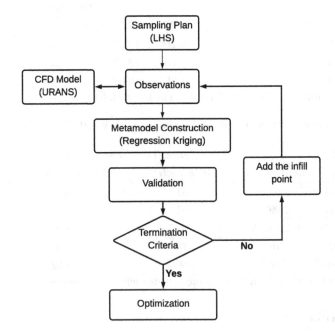

Fig. 2. Flowchart of the metamodel-based optimization algorithm

3.3 Regression Kriging

Kriging [19] is a Gaussian based interpolation method widely used in surrogate based optimization [19]. It mainly takes the training point as the realization of the unknown process and approximates as a combination of global trend function plus a localised departure as

$$y(\mathbf{x}) = G(\mathbf{x}) + Z(\mathbf{x}), \tag{6}$$

where $\mathbf{x}$ is any sample $\mathbf{x} = [x_1 \ x_2 \ ... \ x_P]^T \subset \mathbb{R}^P$, $y(\mathbf{x})$ is the unknown function, $G(\mathbf{x})$ is a known polynomial function and $Z(\mathbf{x})$ is a normally distributed Gaussian process with a zero mean, variance σ^2, and non-zero covariance [19] providing localised deviation to global trend function. The training samples $(\mathbf{x}^1, \mathbf{x}^2, ..., \mathbf{x}^{n_s})$ in design domain D are correlated with each other through covariance matrix of function $Z(\mathbf{x})$ given by

$$Cov[\ Z(\mathbf{x}^i), Z(\mathbf{x}^j)\ ] = \sigma^2 \ \mathbf{R}(\ [\ R(\mathbf{x}^i, \mathbf{x}^j)\ ]\ ), \tag{7}$$

where $\mathbf{R}$ is (n_s, n_s) symmetric correlation matrix with $\mathbf{R}_{ij} = R(\mathbf{x}^i, \mathbf{x}^j)$ a correlation function between any two sample points $\mathbf{x}^i$ and $\mathbf{x}^j$. In this work, we have used the Gaussian spatial correlation function

$$R(\mathbf{x}^i, \mathbf{x}^j) = exp\Big[-\sum_{p=1}^{P} \theta_p \, |x_p^i - x_p^j|^2 \Big], \tag{8}$$

where θ_p denotes pth component of vector $\theta = [\theta_1 \ \theta_2 ... \ \theta_P]^T$, a vector of unknown hyper-parameters to be tuned.

The Kriging predictor is given by [19]

$$\hat{y}(\mathbf{x}) = \hat{\beta} + \mathbf{r}^T(\mathbf{x}) \, \mathbf{R}^{-1}(\ \mathbf{y} - \mathbf{G}\hat{\beta}\ ), \tag{9}$$

where $\mathbf{y}$ is the column vector $(n_s, 1)$ containing response at sample points, $\mathbf{G}$ is a column vector $(n_s, 1)$ and filled with ones when $G(\mathbf{x})$ is considered constant. The vector $\mathbf{r}^T(\mathbf{x}) = [R(\mathbf{x}, \mathbf{x}^1), R(\mathbf{x}, \mathbf{x}^2), ..., R(\mathbf{x}, \mathbf{x}^{n_s})]$ is the correlation vector between known observed points $(\mathbf{x}^1, \mathbf{x}^2, ..., \mathbf{x}^{n_s})$ and the new sample points $\mathbf{x}$.

The vector $\hat{\beta}$ in (9) can be evaluated as

$$\hat{\beta} = (\mathbf{G}^T \mathbf{R}^{-1} \mathbf{G}) \, \mathbf{G}^T \mathbf{R}^{-1} \mathbf{y}. \tag{10}$$

The Kriging model is trained over sample data by tuning hyperparameters θ to maximize concentrated likelihood function [5] given by

$$l(\theta) = \frac{n_s}{2} \, ln(\hat{\sigma}^2) - \frac{1}{2} \, ln|\mathbf{R}|, \tag{11}$$

where estimated variance of Kriging model $\hat{\sigma}^2$ is computed as

$$\hat{\sigma}^2 = \frac{(\mathbf{y} - \mathbf{G}\hat{\beta})^T \, \mathbf{R}^{-1} \, (\mathbf{y} - \mathbf{G}\hat{\beta})}{n_s}. \tag{12}$$

The Kriging method assumes that the sampled responses are true and do not contain any errors. Typically, most of the engineering functions does have some inherent errors due to involved evaluation process. The objective function in this study could involve errors from the CFD simulation of separated flow region in the dynamic stall cycle. This would produce error in the Kriging approximation when more points are added in close proximity to each other during the optimization process. This problem can be alleviated by using regression Kriging [7], which allows the Kriging model to do a regression over the sampled data [7]. This is achieved by the addition of a regularization parameter λ to the diagonal terms of the Kriging correlation matrix $\mathbf{R}$, making it $\mathbf{R} + \lambda\mathbf{I}$ for regression Kriging method where $\mathbf{I}$ is an identity matrix. The regularization parameter λ is evaluated by maximizing likelihood function along with θ hyperparameters. The regression Kriging predictor is now given as [5]

$$\hat{y}_r = \hat{\beta}_r + \mathbf{r}^T(\mathbf{x})\,(\mathbf{R} + \lambda\mathbf{I})^{-1}(\,\mathbf{y} - \mathbf{G}\hat{\beta}_r\,), \tag{13}$$

where

$$\hat{\beta}_r = (\mathbf{G}^T(\mathbf{R} + \lambda\mathbf{I})^{-1}\mathbf{G})\,\mathbf{G}^T(\mathbf{R} + \lambda\mathbf{I})^{-1}\mathbf{y} \tag{14}$$

and variance $\hat{\sigma}_r{}^2$ of regression Kriging model is computed by

$$\hat{\sigma}_r{}^2 = \frac{(\mathbf{y} - \mathbf{G}\hat{\beta}_r)^T(\mathbf{R} + \lambda\mathbf{I})^{-1}(\mathbf{y} - \mathbf{G}\hat{\beta}_r)}{n_s}, \tag{15}$$

where the subscript r denotes regression.

3.4   Infill Criteria

The metamodel constructed with regression Kriging using the initial sample data is an approximation of true objective function. The search of the optimal design depends on the accuracy of the metamodel. Although, higher number of initial samples will improve model accuracy it is wise to add infill points strategically in the design space where further improvements in the metamodel are possible. For this study, we will use the expected improvement (EI) infill criteria to provide a balanced exploration and exploitation of the objective function. The EI for regression Kriging is an extension of the EI for Kriging, which uses a re-interpolation technique to make sure resampling of points are avoided. The infill points are obtained by the maximizing EI function, which is written as

$$E[I(\mathbf{x})] = \begin{cases} (y_{min} - \hat{y})\,\Phi\left(\frac{y_{min}-\hat{y}_r}{\hat{s}}\right) + \hat{s}\,\phi\left(\frac{y_{min}-\hat{y}_r}{\hat{s}}\right) & \text{when } \hat{s} > 0, \\ 0 & \text{when } \hat{s} = 0, \end{cases} \tag{16}$$

where y_{min} is the current minimum response, $\Phi()$ and $\phi()$ are normal cumulative distribution and probability density functions, respectively. The mean square error of the regression Kriging metamodel is given by

$$\hat{s}_{ri}(\mathbf{x}) = \hat{\sigma}_{ri}{}^2\left[1 - \mathbf{r}^T\mathbf{R}^{-1}\mathbf{r} + \frac{1 - \mathbf{G}^T\mathbf{R}^{-1}\mathbf{r}}{\mathbf{G}^T\mathbf{R}\,\mathbf{G}}\right], \tag{17}$$

where $\hat{\sigma}_{ri}^2$ is the variance of the metamodel with re-interpolation technique given as

$$\hat{\sigma}_{ri}^2 = \frac{(\mathbf{y} - \mathbf{G}\hat{\beta}_r)^T (\mathbf{R} + \lambda\mathbf{I})^{-1} \mathbf{R} (\mathbf{R} + \lambda\mathbf{I})^{-1} (\mathbf{y} - \mathbf{G}\hat{\beta}_r)}{n_s}. \tag{18}$$

The EI method for regression Kriging with the re-interpolation technique is described in detail by Forrester et al. [7].

3.5 Validation

In this work, the global accuracy of the metamodel is validated using the normalized root mean squared error $(NRMSE)$ defined as

$$NRMSE = \frac{\sqrt{\sum_{i=1}^{n_T} \frac{(y_{Test}^i - \hat{y}_{Test}^i)^2}{N}}}{(y_{max} - y_{min})_I}, \tag{19}$$

where y_{Test}^i and $\hat{y}_{Test}^i$ represent responses from the CFD evaluation and metamodel prediction at i^{th} test samples, respectively. The response value y could be an objective function $f(\mathbf{x})$ or constraint function $g_1(\mathbf{x})$ values for their respective error estimation. The n_T indicates the number of test data samples. The denominator of $(y_{max} - y_{min})_I$ represents maximum and minimum of response values of initial sample I data. In this work, $NRMSE \leq 10\%$ and a fixed budget of 20 infill samples are considered as acceptable criteria for accurate global metamodel.

3.6 Optimization

Once an accurate metamodel is obtained it is used by the optimizer to find an optimal design for given problem. For this study, we use a multi-start gradient-based search algorithm to find the optimal design. The sequential least squares programming (SLSQP) algorithm offered by Scipy [21] python package is utilized in this work. A total 240 starting points are used in this study. These start points are distributed over the design space by using the LHS technique. The best obtained result is reported as optimal design.

4 Results

This section presents the results of the metamodel generation and the validation study for the dynamic stall mitigation problem. The optimization results are discussed.

4.1 Metamodel Construction

As discussed earlier, the optimization algorithm generates the metamodel and sequentially refines it. Initially, the design space is sampled using LHS. A total of 60 design samples (10× number of design variables) are generated. Each design sample is then evaluated with the CFD module to generate the objective and constraint function values. Note that in this study we only simulated the upstroke of the pitching cycle where dynamic stall vortex formation occurs. The obtained observations are used to construct two separate metamodels, one for the objective and another for the constrained function. Both these metamodels are validated with 20 test data points (one third of initial samples). The test data points are also generated using LHS technique separately and evaluated with the CFD module. The global accuracy of the metamodel is tested using the $NRMSE$ metric. If the accuracy of the model satisfies the termination criteria then it is passed to the optimizer, else an infill point is evaluated and added to the initial sampling plan to construct a new metamodel. This process is iterated until the metamodel satisfies the termination criteria of $NRMSE \leq 10\%$ and fixed budget of 20 infill points.

Figure 3 shows a plot of the $NRMSE$ for the objective and constraints functions every 5 infill points. It can be seen that both the metamodels satisfy global accuracy error criteria well before infill points reach the fixed budget criteria. The constraint function metamodel shows a higher accuracy than the objective function metamodel reaching 2.4% and 8.8%, respectively, by total 80 sample points (60 initial samples plus 20 infill points).

4.2 Optimal Design

Figure 4 shows the baseline and optimum airfoil results. Table 4 gives the aerodynamic characteristics of the airfoils. There are major shape variations

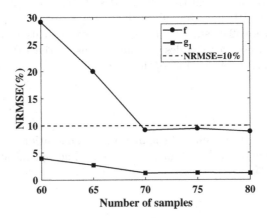

Fig. 3. Objective (**f**) and constraint (**g₁**) function metamodel validation

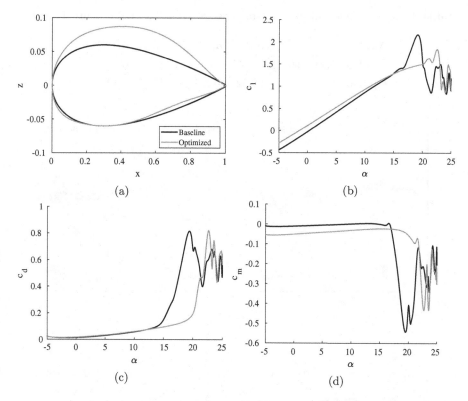

Fig. 4. Comparison between baseline and optimized designs (a) airfoil shapes (b) lift coefficient, (c) drag coefficient, (d) pitching moment coefficient. Time dependant aerodynamic coefficients results are with oscillation cycle parameters $\alpha = 10° + 15° \sin(\omega t)$ and $k = 0.05$

between the baseline (NACA0012) and the optimized airfoil. The optimized airfoil has a higher maximum thickness $(t/c_{max} = 0.146)$ with a maximum camber $(M) = 1.89\%$ located at $x/c = 0.62$. The optimum design is able to delay the dynamic stall angle (α_{ds}) by more than 3°, whereas the moment stall angle α_{ms} is delayed to 20.26°. The α_{ms} indicates formation of dynamic stall vortex which is responsible for sudden divergence in drag and pitching moment coefficients. The delay in dynamic stall vortex formation provides an increase in operational range without adverse loading on the airfoil. Moreover, optimum shape also shows the reduction in severity of pitching moment (Fig. 4d).

Figure 5 shows z-vorticity contour plots of baseline and optimum airfoil near moment stall and dynamic stall angles. It can be seen that near the moment stall and dynamic stall point of baseline airfoil, the optimal shape does not show any signs of dynamics stall vortex formation which verify details given in Table 4.

Table 4. Aerodynamic and shape characteristics of baseline and optimized airfoil

Airfoil	α_{ds}	α_{ms}	$(t/c)_{max}$	$M(\%)$
Baseline (NACA0012)	19.15°	16.55°	0.12	0
Optimized	22.52°	20.26°	0.146	1.89

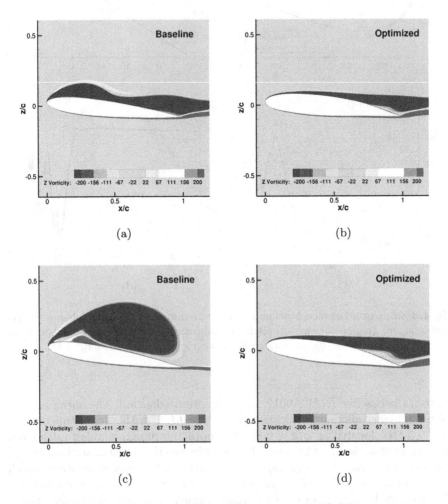

Fig. 5. Z-vorticity contour plot for (a) baseline at $\alpha = 16.55°$ (b) optimized at $\alpha = 16.55°$, (c) baseline at $\alpha = 18.9°$, (d) optimized at $\alpha = 18.9°$

5 Conclusion

In this work, efficient aerodynamic shape optimization using regression Kriging metamodeling is used for mitigating the adverse effects of dynamic stall on wind turbine airfoil shapes. The optimal airfoil shape shows a significant delay in the dynamic stall angle when compared to a baseline airfoil. It was found that the optimal shape has a higher maximum thickness and maximum camber compared to the baseline airfoil. Future work will consider global sensitivity analysis to provide the sensitivities of the individual variables with respect to objective and constraint functions, and to explore the interaction effects of variables. This will reveal how the airfoil aerodynamics affects dynamic stall response.

Acknowledgements. The second and third authors were supported in part by RAN-NIS grant number 174573.

References

1. Abbott, S.: Understanding Analysis. Springer, New York (2001). https://doi.org/10.1007/978-0-387-21506-8
2. Buchner, A., Lohry, M., Martinelli, L., Soria, J., Smits, A.: Dynamic stall in vertical axis wind turbines: comparing experiments and computations. J. Wind Eng. Ind. Aerodyn. **146**, 163–171 (2015)
3. Chen, G., Xiong, Q., Morris, P.J., Paterson, E.G., Sergeev, A., Wang, Y.: Openfoam for computational fluid dynamics. Not. AMS **61**(4), 354–363 (2014)
4. Economon, T., Palacios, F., Alonso, J.: Unsteady aerodynamic design on unstructured meshes with sliding interfaces. In: 51st AIAA Aerospace Sciences Meeting Including the New Horizons Forum and Aerospace Exposition, p. 632 (2013)
5. Forrester, A., Sobester, A., Keane, A.: Engineering Design via Surrogate Modelling: A Practical Guide. Wiley, Great Britain (2008)
6. Harris, F.D., Pruyn, R.R.: Blade stall half fact, half fiction. J. Am. Helicopter Soc. **13**(2), 27–48 (1968)
7. Forrester, A.I.J., Keane, A.J., Bressloff, N.W.: Design and analysis of "Noisy" computer experiments. AIAA J. **44**(10), 2331–2339 (2006)
8. Laurenceau, J., Meaux, M.: Comparison of gradient and response surface based optimization frameworks using adjoint method. In: 4th AIAA Multidisciplinary Design Optimization Specialists Conference, p. 1889 (2008)
9. Lee, T., Gerontakos, P.: Investigation of flow over an oscillating airfoil. J. Fluid Mech. **512**, 313–341 (2004)
10. Lee, T., Gerontakos, P.: Dynamic stall flow control via a trailing-edge flap. AIAA J. **44**(3), 469–480 (2006)
11. Ma, N., et al.: Airfoil optimization to improve power performance of a high-solidity vertical axis wind turbine at a moderate tip speed ratio. Energy **150**, 236–252 (2018)
12. Mani, K., Lockwood, B.A., Mavriplis, D.J.: Adjoint-based unsteady airfoil design optimization with application to dynamic stall. In: American Helicopter Society 68th Annual Forum Proceedings, vol. 68. American Helicopter Society Washington, DC (2012)
13. McKay, M.D., Beckman, R.J., Conover, W.J.: A comparison of three methods for selecting values of input variables in the analysis of output from a computer code. Technometrics **42**(1), 55–61 (2000)
14. Menter, F.R.: Two-equation eddy-viscosity turbulence models for engineering applications. AIAA J. **32**(8), 1598–1605 (1994)
15. Müller-Vahl, H.F., Nayeri, C.N., Paschereit, C.O., Greenblatt, D.: Dynamic stall control via adaptive blowing. Renew. Energy **97**, 47–64 (2016)

16. Nadarajah, S.K., Jameson, A.: Optimum shape design for unsteady flows with time-accurate continuous and discrete adjoint method. AIAA J. **45**(7), 1478–1491 (2007)
17. Palacios, F., et al.: Stanford university unstructured (SU 2): an open-source integrated computational environment for multi-physics simulation and design. In: 51st AIAA Aerospace Sciences Meeting including the New Horizons Forum and Aerospace Exposition, p. 287 (2013)
18. Roy, C.J.: Grid convergence error analysis for mixed-order numerical schemes. AIAA J. **41**(4), 595–604 (2003)
19. Simpson, T.W., Poplinski, J., Koch, P.N., Allen, J.K.: Metamodels for computer-based engineering design: survey and recommendations. Eng. Comput. **17**(2), 129–150 (2001)
20. Sobieczky, H.: Parametric airfoils and wings. In: Recent Development of Aerodynamic Design Methodologies, pp. 71–87. Vieweg+Teubner Verlag, Wiesbaden (1999)
21. Virtanen, P., et al.: SciPy 1.0-fundamental algorithms for scientific computing in python. arXiv preprint arXiv:1907.10121 (2019)
22. Wang, G.G., Shan, S.: Review of metamodeling techniques in support of engineering design optimization. J. Mech. Des. **129**(4), 370–380 (2006)
23. Wang, Q., Zhao, Q.: Rotor airfoil profile optimization for alleviating dynamic stall characteristics. Aerosp. Sci. Technol. **72**, 502–515 (2018)
24. Wang, Q., Zhao, Q., Wu, Q.: Aerodynamic shape optimization for alleviating dynamic stall characteristics of helicopter rotor airfoil. Chin. J. Aeronaut. **28**(2), 346–356 (2015)
25. Wang, S., Ingham, D.B., Ma, L., Pourkashanian, M., Tao, Z.: Numerical investigations on dynamic stall of low reynolds number flow around oscillating airfoils. Comput. Fluids **39**(9), 1529–1541 (2010)
26. Wong, T., O Malley, J., O Brien, D.: Investigation of effect of dynamic stall and its alleviation on helicopter performance and loads. In: Annual Forum Proceedings-American Helicopter Society, vol. 62, no. 3, p. 1749 (2006)
27. Yu, Y.H., Lee, S., McAlister, K.W., Tung, C., Wang, C.M.: Dynamic stall control for advanced rotorcraft application. AIAA J. **33**(2), 289–295 (1995)
28. Zhao, G., Zhao, Q.: Dynamic stall control optimization of rotor airfoil via variable droop leading-edge. Aerosp. Sci. Technol. **43**, 406–414 (2015)

Model-Based Sensitivity Analysis of Nondestructive Testing Systems Using Machine Learning Algorithms

Jethro Nagawkar[1], Leifur Leifsson[1(✉)], Roberto Miorelli[2], and Pierre Calmon[2]

[1] Iowa State University, Ames, IA 50011, USA
{jethro,leifur}@iastate.edu
[2] Département Imagerie Simulation pour le Contrôle, CEA, LIST,
91191 Gif-sur-Yvette, France
{roberto.miorelli,pierre.calmon}@cea.fr

Abstract. Model-based sensitivity analysis is crucial in quantifying which input variability parameter is important for nondestructive testing (NDT) systems. In this work, neural networks (NN) and convolutional NN (CNN) are shown to be computationally efficient at making model prediction for NDT systems, when compared to models such as polynomial chaos expansions, Kriging and polynomial chaos Kriging (PC-Kriging). Three different ultrasonic benchmark cases are considered. NN outperform these three models for all the cases, while CNN outperformed these three models for two of the three cases. For the third case, it performed as well as PC-Kriging. NN required 48, 56 and 35 high-fidelity model evaluations, respectively, for the three cases to reach within 1% accuracy of the physics model. CNN required 35, 56 and 56 high-fidelity model evaluations, respectively, for the same three cases.

Keywords: Nondestructive testing · Sensitivity analysis · Metamodeling · Neural networks · Convolutional neural networks

1 Introduction

The process of testing, inspecting or evaluating assemblies or components for discontinuities or damages without affecting the serviceability of the part is known as nondestructive testing (NDT) [1]. Various NDT methods, such as electromagnetic testing [2] and ultrasonic testing (UT) [3], have been developed and used in different engineering fields such as automobile manufacturing, in-service inspection of aircraft and wind turbines.

To quantify the effect of the different variability parameters on the model responses from NDT systems, sensitivity analysis (SA) [4] can be utilized. SA can be local [5] or global [6]. Local SA focuses on quantifying the effects of small perturbations near an input space value on the model response. Global SA is used to quantify the effects of the input variability on the output responses. For this study, global SA based on Sobol' indices [7] is used.

© Springer Nature Switzerland AG 2020
V. V. Krzhizhanovskaya et al. (Eds.): ICCS 2020, LNCS 12141, pp. 71–83, 2020.
https://doi.org/10.1007/978-3-030-50426-7_6

Traditional NDT measurements have relied heavily on experimental methods. These methods, however, are time-consuming and costly. To speed up this process, various physics-based NDT models, such as finite element methods [8] and boundary element methods [9], have been developed. Unfortunately, for SA, a large number of model evaluations are required in order to propagate the random input uncertainties to the model responses. This results in high computational cost, which renders SA for NDT systems challenging to complete within a required time frame.

To overcome the computational burden, metamodeling methods [10] can be used. These models replace the time-consuming, but accurate high-fidelity physics-based models with a computationally efficient one. Metamodeling methods can be broadly categorized in two classes: data-fit methods [11] and multi-fidelity methods [12]. In data-fit methods, a response surface is fit through the evaluated model responses at sampled high-fidelity data points. Multifidelity metamodeling reduces the computational burden by using information from two or more fidelities. Low-fidelity data can be used to provide the cost function trend to high-fidelity data.

In this work, two data-fit methods, namely neural networks (NN) [13] and convolutional NN (CNN) [14] are used as a part of model-based SA for three UT benchmark cases. The results from this case is compared to Kriging [15], least-angle regression (LARS) [16] based polynomial chaos expansions (PCE) [17] and polynomial chaos-based Kriging (PC-Kriging) [18]. Model-based SA can be used as a precursor to experimental testing, as it would provide important information on which input variability parameters are most crucial for NDT. This would result in a reduction of the total number of physical experiments that need to be performed, hence saving time and cost.

The paper is organised as follows. The next section introduces the method used to construct the NN and CNN as well perform SA. These algorithms are then applied to the three UT benchmark cases in the following section. The final section concludes the paper and provides suggestions for future work.

2   Methods

This section details the construction of the NN and CNN as well as the SA. The Keras [19] wrapper with Tensorflow [20] is used in this study to construct the NN and CNN. The following subsections describe the workflow, the sampling plan, the NN and CNN architectures, as well as SA using Sobol' indices.

2.1   Workflow

The model-based SA flowchart is shown in Fig. 1. The process starts by sampling for the training data. At these sample points, the high-fidelity physics-based model responses are observed from which the metamodel is constructed. This metamodel is then validated using a separate testing set. If the required accuracy is not met, resampling is done with a higher number of training points. Once

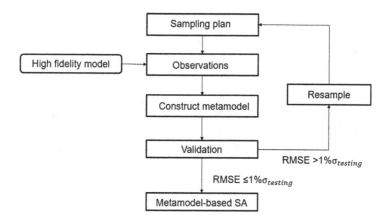

Fig. 1. Flowchart for the model-based sensitivity analysis

the required accuracy is met, the model-based SA is performed. The following subsections describe each step of the algorithm in Fig. 1.

2.2 Sampling Plan

The first step in the model-based SA involves sampling. In order to capture the trend of the model responses, sampling needs to be performed at a fixed combinations of the input variability parameters. The training data in this study is generated using Latin Hypercube sampling (LHS) [10], while the testing data is generated using Monte Carlo sampling (MCS) [21]. The training data is first generated using ten data points, which is increased during resampling until the required accuracy is met. The testing data is fixed to 1,000 points.

2.3 Neural Networks

Figure 2 depicts a NN [13] with an input layer with three inputs, one hidden layer with "n" neurons, and an output layer. A NN is constructed through linear combination of inputs followed by nonlinear transformations through activation functions. Each neuron has an activation function given by

$$z_j = a\left(\sum_{i=1}^{3} \omega_{ij} x_i\right), \tag{1}$$

where a is the activation function and ω_{ij} is the weight between the ith input layer and j^{th} neuron in the hidden layer. Here, the maximum value of i is three as there are three inputs to the NN. x and z are the input values and outputs of the neurons respectively. The NN output prediction is given by

$$\hat{y} = \sum_{i=1}^{n} \eta_i z_i, \tag{2}$$

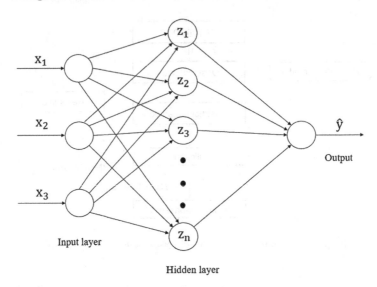

Fig. 2. Depiction of the neural network structure

where η_i is the weight between the ith neuron in the hidden layer and the output layer. To obtain the weights, a minimization problem is solved with a gradient-based algorithm. The cost function used is the mean squared error (MSE), given by

$$MSE = \sum_{i=1}^{N_t} (\hat{y}^{(i)} - y^{(i)})^2 / N_t, \qquad (3)$$

where the physics-based model observation are given by y. N_t is the total number of testing data points. For this study, the following hyperparameters are used: hyperbolic tangent activation function, 50 neurons in the hidden layer and total number of epochs (iterations) of 10,000. The Nesterov-accelerated Adaptive Moment Estimation [22] stochastic gradient descent algorithm with a batch size of 20, learning rate of 0.01 and momentum rate of 0.99 are also used. Details of these terms can be found in Goodfellow et al. [13]. In general, there is no rule-of-thumb on how to setup the NN architecture and its hyperparameters. This was done using trial and error in this work.

2.4   Convolutional Neural Networks

Figure 3 shows the CNN architecture used this in study. The input layer has been replaced by an input grid of size 3×1. CNN was originally developed to work on images [14], where each grid location contains the pixel value. For this study, the pixel values used are the variability parameters values, scaled to have values between zero and one. A kernel of size 1×1 passes over this grid from top to bottom. During this process the convolutional operation is preformed. The

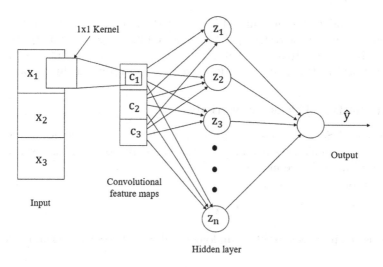

Fig. 3. Flowchart for the model-based sensitivity analysis

convolutional operation here is the product of the input and the value of the kernel. The resulting value in the convolutional feature map layer is given by

$$c_i = a\Big(\epsilon x_i\Big), \tag{4}$$

where c is the value in the feature map, a is the activation function, ϵ is the value of the kernel and x is the input value. This layer is then connected to the hidden layer, which is similar to that of the standard NN. The output of the hidden layer is given by

$$z_j = a\Big(\sum_{i=1}^{3} \omega_{ij} c_i\Big). \tag{5}$$

The weight between the ith grid location of the convolutional feature maps layer and j^{th} neuron in the hidden layer is ω_{ij}. The CNN output prediction $\hat{y}$ is same as (2). The hyperparameters used for this study, in the case of CNN, are the same as those of the NN, except for the batch size and the number of neurons in the hidden layer. These values are set to 10 and 100, respectively.

2.5 Validation

The root mean squared error (RMSE), given by

$$RMSE = \sqrt{MSE}, \tag{6}$$

and the normalized RMSE (NRMSE), given by

$$NRMSE = RMSE/(max(\mathbf{y}) - min(\mathbf{y})), \tag{7}$$

are used to validate the metamodel in this work. The maximum and minimum model observation of the testing points are given by $\max(\mathbf{y})$ and $\min(\mathbf{y})$, respectively. An RMSE less than or equal to $1\%\sigma_{testing}$ (standard deviation of testing points) is taken as the acceptable global accuracy criterion in this work.

2.6   Model-Based Sensitivity Analysis

Variance-based Sobol' indices [6] are used in this work. To determine by how much each variability parameter affects the model response. In this work, MCS is used to estimate these indices.

Given a black-box model,

$$M(\mathbf{X}) = f(\mathbf{X}), \tag{8}$$

where $\mathbf{X}$ is the input vector of m random variables. This equation can be decomposed as

$$M(\mathbf{X}) = f_0 + \sum_{i=1}^{m} f_i(X_i) + \sum_{i<j}^{m} f_{i,j}(X_i, X_j) + \dots + f_{1,2,\dots,m}(X_1, X_2, \dots, X_m), \tag{9}$$

where f_0 is a constant, and f_i is a function of X_i. The functional decomposition terms are orthogonal, which can then be decomposed in terms of conditional expected values

$$f_0 = \mathbb{E}(M(\mathbf{X})), \tag{10}$$

$$f_i(X_i) = \mathbb{E}(M(\mathbf{X})|X_i) - f_0, \tag{11}$$

$$f_{i,j}(X_i, X_j) = \mathbb{E}(M|X_i, X_j) - f_0 - f_i(X_i) - f_j(X_j), \tag{12}$$

and so on. The variance of (9) is then

$$\mathbb{V}ar(M(\mathbf{X})) = \sum_{i=1}^{m} V_i + \sum_{i<j}^{m} V_{i,j} + \dots + V_{1,2,\dots,m}, \tag{13}$$

where

$$V_i = \mathbb{V}ar_{X_i}(\mathbb{E}_{\mathbf{X}_{\sim i}}(M(\mathbf{X})|X_i)), \tag{14}$$

$$V_{i,j} = \mathbb{V}ar_{X_{i,j}}(\mathbb{E}_{\mathbf{X}_{\sim i,j}}(M(\mathbf{X})|X_i, X_j)) - V_i - V_j, \tag{15}$$

and so on, where $\mathbf{X}_{\sim i}$ notation denotes the set of all variables except X_i.

The main effect indices, given by the first-order Sobol' indices are

$$S_i = \frac{V_i}{\mathbb{V}ar(M(\mathbf{X}))}. \tag{16}$$

The total-order indices, given by the total-effect Sobol' indices are

$$S_{T_i} = \frac{\mathbb{E}_{\mathbf{X}_{\sim i}}(\mathbb{V}ar_{X_i}(M(\mathbf{X})|\mathbf{X}_{\sim i}))}{\mathbb{V}ar(M(\mathbf{X}))} = 1 - \frac{\mathbb{V}ar_{\mathbf{X}_{\sim i}}(\mathbb{E}_{X_i}(M(\mathbf{X})|\mathbf{X}_{\sim i}))}{\mathbb{V}ar(M(\mathbf{X}))}. \tag{17}$$

3   Numerical Examples

The model-based SA using NN and CNN used in this paper are demonstrated on three UT benchmark cases. These metamodels are compared to PCE, Kriging and PC-Kriging. The computational cost is computed as the total number of training samples required to reach the desired accuracy.

3.1   Problem Setup

Three benchmark cases developed by the World Federal Nondestructive Evaluation Center [23] are used in this work. The three cases are the spherically-void-defect case under focused transducer (Case 1), spherically-void-defect cases under planar transducer (Case 2), and the spherically-inclusion-defect case under focused transducer (Case 3).

The setup of the UT system is shown in Fig. 4. The variability parameters for Cases 1 and 3 are the probe angle (θ), the x location of the probe (x_p) and the F-number (F). The F-number is the focal length divided by the diameter of the transducer. For Case 2, the F-number is replaced with the y location of the probe (y_p). For all the cases θ and x_p have the normal distribution $N(0$ deg, 0.5^2 deg^2) and uniform distribution $U(0\,\text{mm}, 1\,\text{mm})$, respectively. F has a $U(13,15)$ and $U(8,10)$ for Case 1 and 3, respectively. y_p has a distribution $U(0\,\text{mm}, 1\text{mm})$. A summary of the variability parameter is given in Table 1.

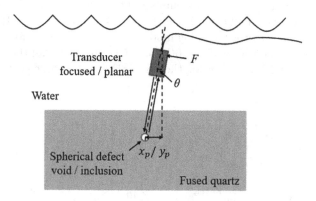

Fig. 4. Setup of the ultrasonic testing system for the benchmark cases

Table 1. The variability parameters used in the numerical examples

Parameters	Case 1	Case 2	Case 3
θ (deg)	$N(0, 0.5^2)$	$N(0, 0.5^2)$	$N(0, 0.5^2)$
x_p (mm)	$U(0, 1)$	$U(0, 1)$	$U(0, 1)$
y_p (mm)	N/A	$U(0, 1)$	N/A
F	$U(13, 15)$	N/A	$U(8, 10)$

The Thomspon Grey model [24] is used to predict the voltage wave forms at the receiver, while the multi-Gaussian beam model [25] evaluates the velocity diffraction coefficient. Separation of variables [26] is then used to calculate the scattering amplitude, which results in a closed-form expression. For this study, the transducer has a center frequency of 5 MHz. The fused quartz block has a density of 2,000 kg/m^3, a longitudinal wave speed of 5,969.4 m/s and a shear wave speed of 3,774.1 m/s. For more information about the models, refer to Du et al. [27].

3.2 Results

The NN and CNN model used in this study are compared to the PCE [17], Kriging [15] and PC-Kriging [18] metamodels. To measure the global accuracy, RMSE and NRMSE metrics are used. This is done at the defect size (a) of 0.5 mm. The number of training points used to generate the metamodel is increased until the desired accuracy of 1%$\sigma_{testing}$ is reached. Once this accuracy is reached, the model is retrained on data for different defect size and the accuracies measured again. Note that the same number of training points are used to measure accuracy at different defect sizes.

Figures 5(a), 6(a) and 7(a) show the RMSE for all the metamodels for an increasing number of training points, for Cases 1, 2 and 3, respectively. In Case 1, both NN and CNN outperform the remaining metamodels. CNN, however, required only 35 training points, compared to 48 for NN (Table 2). For Case 2, both CNN and NN require the same number of training data to reach the desired accuracy and outperform all the other metamodels (Table 3). Table 4, shows that NN outperforms all the other training models, however, CNN performs as well as the PC-Kriging model. Figures 5(b), 6(b) and 7(b) show that all the metamodels fall within the desired accuracy for the NRMSE for all the defect sizes and all the cases.

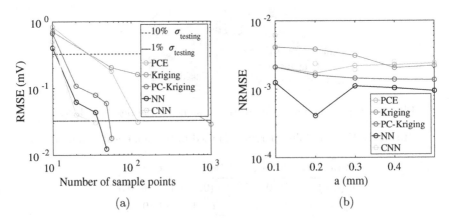

(a)   (b)

Fig. 5. Case 1 setup and model validation: (a) RMSE ($a = 0.5$ mm), (b) NRMSE with respect to defect size

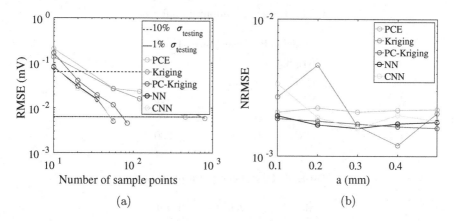

Fig. 6. Case 2 setup and model validation: (a) RMSE ($a = 0.5\,\mathrm{mm}$), (b) NRMSE with respect to defect size

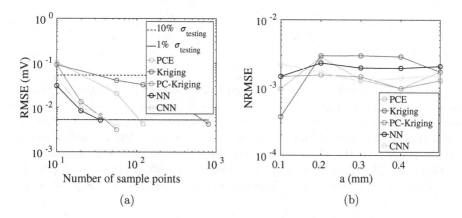

Fig. 7. Case 3 setup and model validation: (a) RMSE ($a = 0.5\,\mathrm{mm}$), (b) NRMSE with respect to defect size

Table 2. Case 1 computational cost

Model	Per defect size	Total
Kriging	1,000	5,000
PCE	120	600
PC-Kriging	56	280
NN	48	240
CNN	35	175

Table 3. Case 2 computational cost

Model	Per defect size	Total
Kriging	800	4,000
PCE	455	2,275
PCK	84	420
NN	56	280
CNN	56	280

Table 4. Case 3 computational cost

Model	Per defect size	Total
Kriging	800	4,000
PCE	120	600
PCK	56	280
NN	35	175
CNN	56	280

SA plots for the three cases are shown in Figs. 8, 9 and 10. 75,000 MCS were used to perform the physics-based model evaluations to obtain the sensitivity information for each of the three cases. This serves as the baseline to compare the metamodeling results. In the case of the PCE metamodel, its coefficients can be used to provide the 1st and total order Sobol' indices [27]. For the remaining metamodels, 75,000 MCS points were used to provide satisfactory results for the SA for each of the cases. Figures 8 and 10 show that the F-number has

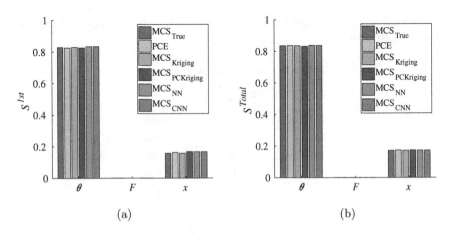

Fig. 8. Case 1 sensitivity analysis: (a) 1st-order Sobol' indices, (b) Total-order Sobol' indices

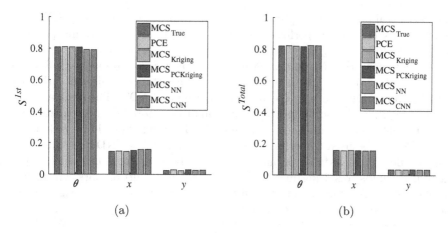

Fig. 9. Case 2 sensitivity analysis: (a) 1st-order Sobol' indices, (b) Total-order Sobol' indices

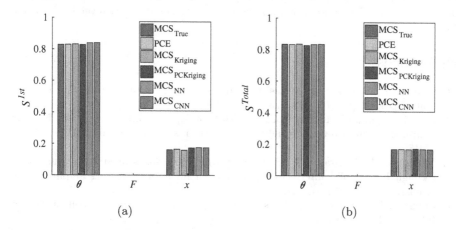

Fig. 10. Case 3 sensitivity analysis: (a) 1st-order Sobol' indices; (b) Total-order Sobol' indices

negligible effect on the model response and can be neglected while setting up the NDT experiments. For Case 2, y_p is small enough to be ignored, as shown in Fig. 9.

4   Conclusion

In this work, NN and CNN are used to perform model-based SA for three different UT benchmark cases. These metamodels are compared to other data-fit metamodels, namely, PCE, Kriging and PC-Kriging. NN is shown to outperform these three metamodeling methods for all the UT benchmark cases in terms of number of training points required to reach an accuracy of $1\%\sigma_{testing}$. CNN outperforms NN for Case 1, performs equally well for Case 2 as NN, and performs similarly to PC-Kriging for Case 3.

The SA shows that NN and CNN match the physics-based model results well. The F-number and y_p are shown to be of no importance while measuring the model response and can be neglected for experimental NDT. The remaining two parameters, x_p and θ, are important and cannot be ignored during experimental measurements. This study shows how machine learning algorithms, such as NN and CNN, can be used to perform accurate and fast SA for NDT systems. This help decide which variable parameters should be used for NDT using experimental methods, resulting in time and cost savings. Future work will include problems with high number of variability parameters, which require significantly more data to reach the desired accuracy.

Acknowledgements. The authors would like to thank Xiaosong Du for providing the UT data as well as the results from the PCE, Kriging and PC-Kriging metamodels. The first two authors are supported in part by NSF award number 1846862, and the Iowa State University Center for Nondestructive Evaluation Industry-University Research Program.

References

1. Crawley, P.: Non-destructive testing - current capabilities and future directions. J. Mater. Des. Appl. **215**, 213–223 (2001)
2. Gao, P., Wang, C., Li, Y., Cong, Z.: Electromagnetic and eddy current NDT in weld inspection: a review. Insight- Non-Destr. Test. Cond. Monit. **2015**, 337–345 (2015)
3. Thompson, R.B., Gray, T.A.: A model relating ultrasonic scattering measurements through liquid solid interfaces to unbounded medium scattering amplitudes. J. Acoust. Soc. Am. **74**(4), 1279–1290 (1983)
4. Lilburne, L., Tarantola, S.: Sensitivity analysis of spatial models. Int. J. Geogr. Inf. Sci. **23**, 151–168 (2009)
5. Castillos, E., Conejo, A., Minguez, R., Castillo, C.: A closed formula for local sensitivity analysis in mathematical programming. Eng. Optim. **38**, 93–112 (2007)
6. Sobol', I., Kuchereko, S.: Sensitivity estimates for nonlinear mathematical models. Math. Model. Comput. Exp. **1**, 407–414 (1993)
7. Sobol', I.: Global sensitivity indices for nonlinear mathematical models and their Monte Carlo estimates. Math. Comput. Simul. **55**, 271–280 (2001)
8. Zeng, Z., Udpa, L., Udpa, S.S.: Finite-element model for simulation of ferrite-core eddy-current probe. IEEE Trans. Magn. **46**, 905–909 (2009)
9. Zhang, C., Gross, D.: A 2D hyper singular time-domain traction BEM for transient elastodynamic crack analysis. Wave Motion **35**, 17–40 (2002)
10. Forrester, A.I.J., Sobester, A., Keane, A.J.: Engineering Design via Surrogate Modelling: A Practical Guide, 1st edn. Wiley, Hoboken (2008)
11. Queipo, N.V., Haftka, R.T., Shyy, W., Goel, T., Vaidyanathan, R., Tucker, P.K.: Surrogate-based analysis and optimization. Prog. Aerosp. Sci. **21**(1), 1–28 (2005)
12. Peherstorfer, B., Willcox, K., Gunzburger, M.: Survey of multifidelity methods in uncertainty propagation, inference, and optimization. Soc. Ind. Appl. Math. **60**(3), 550–591 (2018)

13. Goodfellow, I., Bengio, Y., Courville, A.: Deep Learning, 1st edn. MIT Press, Cambridge (2017)
14. LeCun, Y.: Generalization and network design strategies. Technical Report CRG-TR-89-4, University of Toronto
15. Krige, D.G.: Statistical approach to some basic mine valuation problems on the Witwatersrand. J. Chem. Metall. Min. Eng. Soc. South Africa **52**(6), 119–139 (1951)
16. Efron, B., Hatie, T., Johnstone, I., Tibshirani, R.: Least angle regression. Ann. Stat. **32**, 407–499 (2004)
17. Blatman, G.: Adaptive sparse polynomial chaos expansion for uncertainty propagation and sensitivity analysis. Ph.D. thesis, Blaise Pascal University - Clermont II. 3, 8, 9 (2009)
18. Schobi, R., Sudret, B., Wiart, J.: Polynomial-chaos-based kriging. Int. J. Uncertain. Quantif. **5**, 193–206 (2015)
19. Chollet, F.: Keras: deep learning library for theano and tensorflow (2016). https://keras.io/
20. Abadi, M., et al.: TensorFlow: a system for large-scale machine learning. In: 12th USENIX Symposium on Operating Systems Design and Implementation, pp. 265–283 (2016)
21. Shapiro, A.: Monte Carlo sampling methods. Handb. Oper. Res. Manag. Sci. **10**, 353–425 (2003)
22. Dozat, T.: Incorporating Nesterov momentum into Adam. In: CLR Workshop (2016)
23. Schmerr, L.W., Kim, H.J., Lopez, A.L., Sodov, A.: Simulating the experiments of the 2004 ultrasonic benchmark study. Rev. Progress Quant. Nondestr. Eval. **24**, 1880–1887 (2005)
24. Schmerr, L.W., Song, J.: Ultrasonic Nondestructive Evaluation Systems. Springer, Heidelberg (2007). https://doi.org/10.1007/978-0-387-49063-2
25. Wen, J.J., Breazeale, M.A.: A diffraction beam field expressed as the superposition of Gaussian beams. J. Acoust. Soc. Am. **83**, 1752–1756 (1988)
26. Schmerr, L.: Fundamentals of Ultrasonic Nondestructive Evaluation: A Modeling Approach. Springer, Heidelberg (2013)
27. Du, X., Leifsson, L., Meeker, W., Gurrala, P., Song, J., Roberts, R.: Efficient model-assisted probability of detection and sensitivity analysis for ultrasonic testing simulations using stochastic metamodeling. ASME J. Nondestr. Eval. **2**(4), 041002 (2019)

Application of Underdetermined Differential Algebraic Equations to Solving One Problem from Heat Mass Transfer

Viktor F. Chistyakov[1] , Elena V. Chistyakova[1]([⊠]) ,
and Anatoliy A. Levin[2]

[1] Institute for System Dynamics and Control Theory SB RAS, Irkutsk, Russia
chist@icc.ru, chistyak@gmail.com
[2] Energy Systems Institute SB RAS, Irkutsk, Russia
lirt@mail.ru

Abstract. This paper addresses a mathematical model of the boiling of subcooled liquid in an annular channel. The model is presented by a mixed system of ordinary differential equations, algebraic relations and a single partial differential equation, which, written together, can be viewed as an underdetermined differential algebraic equation with a partial differential equation attached. Using the tools of the differential algebraic equation theory, we reveal some important qualitative properties of this system, such as its existence domain, and propose a numerical method for its solution. The numerical experiments demonstrated that within the found existence domain the mathematical model adequately represents real-life boiling processes that occur in the experimental setup.

Keywords: Differential algebraic equations · Index · Boiling · Subcooled liquid · Heat mass transfer

1 Introduction

Many technical systems and processes can be described by mathematical models in the form of differential algebraic equations (DAEs). Generally speaking, a DAE is a generalization of ordinary differential equations and can be viewed as an incomplete system of differential equations closed with a set of algebraic relations. DAEs have received much attention due to a wide range applications in electric circuits simulation [1–3], mechanics of multibody systems [4–6], flow networks [7,8], etc. The complete coverage of the topic can be found, for example, in [9–12] and the references therein. In this paper, we focus on a particular case of DAEs that arises as a mathematical model of boiling of subcooled liquid in a steam generating channel of an experimental unit. This model appears to be

Supported by the Russian Foundation For Basic Research, Grant No. 18-29-10019.

an underdetermined DAE with a partial differential equation attached to it. We present a qualitative analysis of this particular system and reveal some important properties of underdetermined DAEs. We propose a numerical method that takes into consideration all important features of the model and discuss the compliance of the results of numerical experiments with the real-life process.

2   The Self-oscillatory Boiling Model

We consider a mathematical model of the coolant boiling-up under the conditions of unsteady heat generation on the wall of the flow-through duct, which detailed description can be found in [13], where the model was addressed by solving all equations individually. However, a closer look at a set of equations that comprise the model shows that it can be written as a system of the following form

$$A\dot{u}(t) + B(u(t)) = 0, \quad t \in [0, \vartheta], \vartheta \in R, \tag{1}$$

$$A = \begin{pmatrix} 1 & 0 & 0 & 0 & 0 & 0 & 0 & 0 \\ 0 & 1 & 0 & 0 & 0 & 0 & 0 & 0 \\ 0 & 0 & k_3 & 0 & 0 & 0 & 0 & 0 \\ 0 & 0 & 0 & k_3 & 0 & 0 & 0 & 0 \\ 0 & 0 & 0 & 0 & 1 & 0 & 0 & 0 \\ 0 & 0 & 0 & 0 & 0 & 1 & 0 & 0 \\ 0 & 0 & 0 & 0 & 0 & 0 & 0 & 0 \end{pmatrix}, \quad u(t) = \begin{pmatrix} u_1(t) \\ u_2(t) \\ u_3(t) \\ u_4(t) \\ u_5(t) \\ u_6(t) \\ u_7(t) \\ u_8(t) \end{pmatrix},$$

$$B(u) = \begin{pmatrix} B_1(u) \\ B_2(u) \\ B_3(u) \\ B_4(u) \\ B_5(u) \\ B_6(u) \\ B_7(u) \end{pmatrix} =$$

$$= - \begin{pmatrix} k_1 A_{kof}(u_6(t) - u_1(t)/u_2(t) \\ k_2 A_{kof}(u_6(t) - u_1(t)/u_2(t) \\ p_1 - u_7(t) - k_4 u_3(t)|u_3(t)| \\ u_7(t) - p_2 - k_4 u_4(t)|u_4(t)| \\ k_5(u_4(t) - u_3(t)) \\ \frac{R_v u_6(t)^2}{(L - R_v u_6(t)) u_5(t)} \left[k_5(u_4(t) - u_3(t)) k_7 A_{kof} \left(\frac{(u_1(t) - u_6(t))}{u_2} + u_8(t) \right) \right] \\ u_7(t) - p_s \exp \left(\frac{L}{R_v T_s} - \frac{L}{R_v u_6(t)} \right) \end{pmatrix},$$

where $u(t)$ is the desired vector-function with the following components: $u_1(t)$ is the temperature of the surface on which the unsteady heat transfer crisis develops; $u_2(t)$ is the thickness of the liquid microlayer, which evaporation dynamics determines heat transfer; $u_3(t)$ and $u_4(t)$ are the interfacial velocities in the upper and lower part of the channel, respectively; $u_5(t)$ is the vapor layer thickness which is determined by the interfacial velocity; $u_6(t)$ is the temperature inside the vapor layer; $u_7(t)$ is the vapor pressure inside the vapor layer; $u_8(t)$ is the liquid temperature gradient at the boundary with a cooling metal surface: $\langle \frac{\partial T}{\partial z} \rangle_{|z=0}$, where z is a coordinate axis perpendicular to the heating surface; k_j, $j = \overline{1,7}$, $A_{kof}, T_s, p_s, R_v, L, p_1, p_2$ are some given parameters (their detailed description and values can be found in [13]). All units of measurement are given in the SI metric system.

The initial conditions are also given

$$(u_1(0), u_2(0), u_3(0), u_4(0), u_5(0), u_6(0))^\top = (u_{1,0}, u_{2,0}, u_{3,0}, u_{4,0}, u_{5,0}, u_{1,6})^\top.$$
(2)

One of the issues in the problem statement is that the domain $[0, \vartheta]$ is unknown to us and should be determined on the basis of numerical calculations.

The DAE (1) is underdetermined and in [13] the authors of the model used a solution to a partial differential equation to obtain the necessary closure for the system

$$u_8(t) = 2(T_l - u_6(t))\sqrt{\frac{u_3(t) + u_4(t)}{2\pi A_{kof} b}}.$$

However, from a mathematical viewpoint, such underdetermined DAEs require special research. If we could not express the function $\langle \frac{\partial T}{\partial z} \rangle_{|z=0}$ via the other seven components of the vector-function $u(t)$, we would have to rely only on experimental observations. The practical meaning of studying the model (1) is ensued by the recently discovered heat transfer enhancement under conditions of the rapidly oscillating interface. It was shown in [13–16] that this enhancement is mainly associated with the short-term presence of a metastable liquid on a metal surface. The direct experimental observation and analysis of the dynamics of such processes are limited in view of the smallness of time periods and the geometric dimensions of the objects. Therefore, one of the most important tools in this area is the numerical integration of the equations that comprise the mathematical model.

3   Porperties of Underdetermined DAEs

The DAE (1) is underdetermined in the sense that the number of equations exceeds the dimension of the desired vector-function $u(t)$. Such systems can be generalized as

$$\Lambda_1(u) := A(u(t), t)\dot{u}(t) + B(u(t), t) = 0, \ t \in T = [\alpha, \beta], \alpha, \beta \in \mathbf{R}.$$
(3)

where $A(u, t)$ is an $(\nu \times n)$-matrix, $B(u, t)$ is an ν-dimensional vector-function, $\dot{u} = du(t)/dt$, $u \in \mathbf{U} \subseteq \mathbf{R}^n$, and the following condition holds

$$\operatorname{rank} A(u, t) < \min (n, \ \nu) \ \forall (u, t) \in \mathbf{U} \times T.$$
(4)

If $n = \nu$ in (3) (i.e. (3) is closed), the condition (4) is equivalent to

$$\det A(u,t) = 0 \quad \forall (u,t) \in \mathbf{U} \times T.$$

Usually, we also have a set of initial data

$$u(\alpha) = a \in \mathbf{U}. \tag{5}$$

We consider that $\mathbf{U} = \{u : \|a - u\| \leq \varrho \in \mathbf{R}^1, \varrho > 0\}$. Below we assume that all entries are sufficiently smooth. One of the most important characteristics of DAEs is index. An index is a way of measuring the distance from a DAE to its related ordinary differential equation (ODE). The concept of index for DAEs has several definitions, most of which are extensively covered in [12] and [18]. However, each new type of DAEs often requires a refinement of one of the popular notations.

Definition 1. *Let there exist an operator of the form*

$$\tilde{\Lambda}(u) := \sum_{j=0}^{l} L_j(\mathbf{u}) \left(\frac{d}{dt} \right)^j, \tag{6}$$

where $L_j(\mathbf{u}) = L_j(u^{(m)}, \cdots, \dot{u}, u, t)$ *are continuous in* $\mathbf{R}^{(m+1)n} \times T(\nu \times \nu)$-*matrices, and*

$$\tilde{\Lambda}(u)[\Lambda_1(u)] = A^l(u(t),t)\dot{u}(t) + B^l(u(t),t) \ \forall u(t) \in \mathbf{C}^{i+1}(T), \ i = \max\{l,m\} \tag{7}$$

where rank $A^l(a,\alpha) = \min(n, \nu)$. *Then, (6) it is said to be the left regularizing operator for the DAE (3) and the smallest possible* i *is said to be the index of the DAE (3).*

Now we will show one of the ways of building the LRO. Let there exist a $(\nu \times \nu)$-matrix $P(u,t) \in \mathbf{C}^1(\mathbf{U} \times T)$ with the properties: $\det P(u,t) \neq 0 \forall (u,t) \in \mathbf{C}^1(\mathbf{U} \times T)$,

$$P(u,t)\Lambda_1(u) = \begin{pmatrix} A_1(u,t) \\ 0 \end{pmatrix} \dot{u} + \begin{pmatrix} B_1(u,t) \\ B_2(u,t) \end{pmatrix} \ \forall u \equiv u(t) \in \mathbf{C}^1(T), \ u(t) \in \mathbf{U}, \tag{8}$$

where the zero block has the dimension $([\nu-r] \times n)$, $r = \max\{\mathrm{rank}A(u,t), (u,t) \in \mathbf{U} \times T\}$. Differentiate with respect to t the algebraic relations in (8). We obtain

$$\begin{pmatrix} E_r & 0 \\ 0 & (d/dt)E_{\nu-r} \end{pmatrix} P(u,t)\Lambda_1(u) = A^1(u,t)\dot{u} + B^1(u,t) =$$
$$= \begin{pmatrix} A_1(u,t) \\ \partial B_2(u,t)/\partial u \end{pmatrix} \dot{u} + \begin{pmatrix} B_1(u,t) \\ \partial B_2(u,t)/\partial t \end{pmatrix}. \tag{9}$$

Let rank $A^1(a,\alpha) = \min(n, \nu)$. Then the system

$$A^1(u(t),t)\dot{u}(t) + B^1(u(t),t) = 0$$

can be rewritten as

$$\left(A_{11}^1(u_1(t), u_2(t), t) \ A_{12}^1(u_1(t), u_2(t), t)\right) \begin{pmatrix} \dot{u}_1(t) \\ \dot{u}_2(t) \end{pmatrix} + B^1(u_1(t), u_2(t), t) = 0, \quad (10)$$

where the block $A_{11}^1(u_1, u_2, t)$ has the dimension $(\nu \times \nu)$, det $A_{11}^1(a, \alpha) \neq 0$, $u^\top = (u_1^\top, u_2^\top)^\top$. Set $u_2 = \psi(t)$, where $\psi(t)$ is an arbitrary smooth function, $\psi(\alpha) = a_2$, $(u_1^\top(\alpha), u_2^\top(\alpha))^\top) = (a_1, a_2)^\top = a^\top$. Then, taking into consideration that the matrix $A_{11}^1(u_1, u_2, t)$ is continuous, we can obtain a system of regular ODEs in the neighborhood of the point (a, α):

$$\dot{u}_1(t) = -\left(A_{11}^1(u_1(t), \psi(t), t)\right)^{-1} \times$$
$$\times \left[A_{12}^1(u_1(t), \psi(t), t)\dot{\psi}(t) - B^1(u_1(t), \psi(t), t)\right] \quad (11)$$

The reasoning presented above can be formulated as the following theorem.

Theorem 1. *If* rank $A(a, \alpha) = \text{rank}(A(a, \alpha)|B(a, \alpha))$, *then (8) satisfies the condition* $B_2(a, \alpha) = 0$, *and there exists a segment* $[\alpha, \ \alpha + \varepsilon] \subseteq T$, *on which the solution* $(u_1^\top(t), \psi^\top(t))^\top$ *to the initial problem (3), (5) can be found.*

The transition from (3) to (9) is equivalent to multiplication of (3) by the operator

$$\Omega_0 = R_0(u, t) + \frac{d}{dt} S_0(u, t) = \begin{pmatrix} P_1(u, t) \\ 0 \end{pmatrix} + \frac{d}{dt} \begin{pmatrix} 0 \\ P_2(u, t) \end{pmatrix},$$
$$P(u, t) = \begin{pmatrix} P_1(u, t) \\ P_2(u, t) \end{pmatrix}. \quad (12)$$

If in (9) rank $A^1(u, t) < \min(n, \nu) \ \forall (u, t) \in \mathbf{U} \times T$, then, similarly, we can build an operator

$$\Omega_1 = R_1(u, t) + \frac{d}{dt} S_1(u, t).$$

Suppose that after l steps of such a process we arrive at

$$\prod_{i=0}^{l-1} \Omega_i[\Lambda_1(u)] = A^l(u(t), t)\dot{u}(t) + B^l(u(t), t), \ u(t) \in \mathbf{U}, \quad (13)$$

where rank $A^l(a, \alpha) = \min(n, \nu)$ and

$$\Omega_i = R_i(u, t) + \frac{d}{dt} S_i(u, t).$$

If the conditions

$$\text{rank } A^i(a, \alpha) = \text{rank}(A^i(a, \alpha)|B^i(a, \alpha))$$

are satisfied, we can prove solvability of the initial problem (3), (5). The number l is said to be a rough index of the DAE (3).

If the matrices that define operators Ω_i in (13) are sufficiently smooth, then the product $\prod_{i=0}^{l-1} \Omega_i$ can be represented in the form of the operator $\tilde{\Lambda}_l(u)$.

If a DAE is closed, then the notion of index for underdetermined DAEs introduced in this paragraph fully coincides with the LRO based index from [17], which was further developed in [20] and [19].

Remark 1. The matrices $R_i(u,t)$, $S_i(u,t)$ from (13) can be built differently (see, for example, [12]).

Example 1. Consider the following DAE:

$$\Lambda_1(u) := \begin{pmatrix} u_2 & 0 & 0 \\ 0 & 0 & 0 \end{pmatrix} \dot{u} + Bu = 0, \ B = \begin{pmatrix} 1 & 0 & 1 \\ 0 & 1 & 0 \end{pmatrix},$$

where $u = \left(u_1,\, u_2,\, u_3\right)^{\top}$. The rough index for this DAE is not determined, because the matrix $A^1(u,t)$ in (9) is singular on the solutions of the system. However, the LRO exists and can be written as

$$\tilde{\Lambda}(u) := \begin{pmatrix} 0 & -\ddot{u}_1 \\ 0 & 0 \end{pmatrix} + \begin{pmatrix} 1 & -\dot{u}_1 \\ 0 & 1 \end{pmatrix} \frac{d}{dt},$$

and the index of the system is 1 in terms of Definition 1. Indeed,

$$\tilde{\Lambda}(u)[\Lambda_1(u)] = B\dot{u} \ \forall u \equiv u(t) \in \mathbf{C}^2(T).$$

Let the DAE have the form:

$$\Lambda_1(u) := \begin{pmatrix} 0 & u_2 & 0 \\ 0 & 0 & 0 \end{pmatrix} \dot{u} + Bu = 0,$$

In this case, the LRO is defined and

$$\tilde{\Lambda}(u) := \begin{pmatrix} 0 & -\ddot{u}_2 \\ 0 & 0 \end{pmatrix} + \begin{pmatrix} 1 & -\dot{u}_2 \\ 0 & 1 \end{pmatrix} \frac{d}{dt},$$

so the system has index 2 in terms of Definition 1. It can be readily seen that

$$\tilde{\Lambda}(u)[\Lambda_1(u)] = B\dot{u} \ \forall u \equiv u(t) \in \mathbf{C}^2(T).$$

It can be easily verified that this system has a rough index and it is equal to 2.

If in Example 1 we set u_3 as a given smooth function $\psi(t)$, we obtain closed systems with unique solutions. Technically, these systems have index 1: the LRO is of the first order, but if we take into account the derivatives that participate in the LRO's coefficients, the rough index coincides with the index in terms of Definition 1.

4 Analysis of the Physical Model

If we differentiate (1) with respect to t, we can see that its index is equal to 1 in terms of Definition 1. Therefore, (1) has a unique local solution in the neighborhood of the initial data. To find out if the system has a solution on a given segment, investigate its Lyapunov stability. Find the stationary solution of (1) by solving the following system of non-linear equations

$$B(u) = 0.$$

From the equations $B_1(u) = 0$, $B_3(u) = 0$, $B_7(u) = 0$ and the expression for u_8, we obtain

$$u_1^* = u_6^*, \quad u_4^* = u_3^*,$$
$$p_1 - p_2 = 2k_4 u_3^* |u_3^*|, \quad u_6^* = T_l,$$
$$u_7^* = p_s \exp\left(\frac{L}{R_v T_s} - \frac{L}{R_v T_l}\right).$$

Now multiply the first equation of the system by k_2/k_1 and subtract it from the second equation. We arrive at

$$\dot{u}_2(t) - \frac{k_2}{k_1}\dot{u}_1(t) = 0, \quad u_2(t) - \frac{k_2}{k_1}u_1(t) = c,$$

where c is an arbitrary constant.

For calculate c, we can use the stationary solution of the system. Taking into consideration the values for the initial data and coefficients from [13], we have

$$c = u_2(0) - \frac{k_2}{k_1}u_1(0) = u_{1,0} - \frac{k_2}{k_1}u_{1,0} = -0.01065,$$

$$u_2^* = c + \frac{k_2}{k_1}T_l = -0.003537.$$

The component $u_2(t)$ is present in the denominator of the right-hand part of the system: it participates in $B_1(u(t))$, $B_2(u(t))$, $B_7(u(t))$. At some point $t = \vartheta$, $u_2(t)$ turns into zero, since $u_2(0) = 0.0001 > 0$. Therefore, we can find the segment $[0, \vartheta]$, on which the solution to the initial problem (1), (2) exists. However, the system is quite stiff. We have the following values for the derivatives of the solution's components in the starting point:

$$\frac{du(t)}{dt}\Big|_{t=0} = \begin{pmatrix} -109063.32 \\ -2.47 \\ -1089.51 \\ 1107.58 \\ -0.4 \\ -428733.403 \end{pmatrix},$$

and the eigenvalues of the Jacobi matrix are of the same order. This fact preconditions the choice of the numerical method.

5 Numerical Method

Due to the fact that the right-hand part of (1) has only the first continuous derivative, we have to choose among the first order methods. Therefore, to solve (1) numerically, we applied the implicit Euler method:

$$\hat{A}\frac{w_{i+1} - w_i}{\tau} + B(w_{i+1}) = 0,$$
$$t \in (0, t^*), \ \tau = t^*/N, \ w_0 = u(0),$$

(14)

where N is a number of integration steps, $\hat{A}$ is a square matrix that was obtained from (1) by excluding a zero column from the matrix A, t^* is derived from the empirical data. We solve the nonlinear system (14) for each i by the Newton method taking only the first iteration. As the result, the calculation algorithm has the form:

$$[\hat{A} + \tau J(w_i)]w_{i+1} = \hat{A}w_i - \tau B(w_i) + \tau J(w_i)u_i,$$
$$J(w_i) = \frac{\partial B(u)}{\partial u}\Big|_{u=w_i}.$$

(15)

Since (1) has index 1, the matrix $[\hat{A} + \tau J(w_i)]$ is non-singular for a sufficiently small τ [17] and the method has the first convergence order. The integration step was selected taking into account the eigenvalues of the Jacobi matrix $J(w_0)$. The solution to the initial value problem (1), (2) in a mathematical sense exists only on the semi-interval $[0, \vartheta)$. However, numerical experiments showed that we can apply the difference scheme (15) beyond ϑ. Numerical calculations also revealed

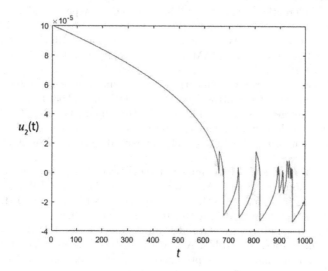

Fig. 1. Dynamics of the liquid layer thickness $u_2(t)$.

that when $u_2(t)$ reaches a zero-crossing, the solution surges up, which is fully consistent with the physical experiment (Fig. 1): at some moments, the liquid microlayer described by the component $u_2(t)$ completely evaporates and gets refilled.

In particular, the performed numerical calculations allowed us to define the semi-interval $[0, \vartheta) = [0, 0.02]$ millisecond. This result coincides with the empirical findings from [13].

6   Conclusions

We presented a qualitative and numerical study of the model for the boiling of subcooled liquid in an annular channel. We obtained the solvability conditions and proposed a numerical method of solution. The analysis of the mathematical model (1) was performed under the assumption that the elements of the matrix A and a number of other given parameters are constant. In fact, their values depend on the solution to the system and are located within a certain range determined by the physical meaning of the problem. Further research implies refinement of the model and investigation of its properties to determine the possible ranges for variable parameters.

References

1. Günther, M., Feldmann, U.: CAD-based electric-circuit modeling in industry I: mathematical structure and index of network equations. Surv. Math. Ind. **8**, 97–130 (1999)
2. Estévez Schwarz, D., Tischendorf, C.: Structural analysis of electric circuits and consequences for MNA. Int. J. Circ. Theory Appl. **28**(2), 131–162 (2000)
3. Riaza, R.: Differential-Algebraic Systems: Analytical Aspects and Circuit Applications. World Scientific, Singapore (2008). https://doi.org/10.1142/6746
4. Rabier, P.J., Rheinboldt, W.C.: Nonholonomic Motion of Rigid Mechanical Systems from a DAE Viewpoint. SIAM, Philadelphia (2000). https://doi.org/10.1137/1.9780898719536
5. Eich-Soellner, E., Führer, C.: Numerical Methods in Multibody Systems. Teubner Verlag, Wiesbaden (1998). https://doi.org/10.1007/978-3-663-09828-7
6. Arnold, M., Burgermeister, B., Führer, C., Hippmann, G., Rill, G.: Numerical methods in vehicle system dynamics: state of the art and current developments. Veh. Syst. Dyn. **49**, 1159–1207 (2011). https://doi.org/10.1080/00423114.2011.582953
7. Jansen, L., Tischendorf, C.: A unified (P)DAE modeling approach for flow networks. In: Schöps, S., Bartel, A., Günther, M., ter Maten, E.J.W., Müller, P.C. (eds.) Progress in Differential-Algebraic Equations. DEF, pp. 127–151. Springer, Heidelberg (2014). https://doi.org/10.1007/978-3-662-44926-4_7
8. Benner, P., Braukmüller, M., Grundel, S.: A direct index 1 DAE model of gas networks. In: Keiper, W., Milde, A., Volkwein, S. (eds.) Reduced-Order Modeling (ROM) for Simulation and Optimization, pp. 99–119. Springer, Cham (2018). https://doi.org/10.1007/978-3-319-75319-5_5

9. Brenan, K.E., Campbell, S.L., Petzold, L.R.: Numerical Solution of Initial-Value Problems in Differential-Algebraic Equations. SIAM, Philadelphia (1996). https://doi.org/10.1137/1.9781611971224

10. Hairer, E., Wanner, G.: Solving Ordinary Differential Equations II. Stiff and Differential-Algebraic Problems. Springer, Heidelberg (1996). https://doi.org/10.1007/978-3-642-05221-7

11. Kunkel, P., Mehrmann, V.: Differential-Algebraic Equations: Analysis and Numerical Solution. EMS Publishing House, Zürich (2006). https://doi.org/10.4171/017

12. Lamour, R., März, R., Tischendorf, C.: Differential-Algebraic Equations: A Projector Based Analysis. Springer, Berlin (2013). https://doi.org/10.1007/978-3-642-27555-5

13. Aktershev, S.P., Levin, A.A., Mesentsev, I.V., Mesentseva, N.N.: Self-oscillatory regime of boiling of a highly subcooled liquid in a flow-passage annular duct. Thermophys. Aeromech. **25**, 875–887 (2018). https://doi.org/10.1134/S0869864318060082

14. Levin, A.A., Tairov, E.A., Spiryaev, V.A.: Self-excited pressure pulsations in ethanol under heater subcooling. Thermophys. Aeromech. **24**, 61–71 (2017). https://doi.org/10.1134/S0869864317010073

15. Yagov, V.V., Zabirov, A.R., Kanin, P.K.: Heat transfer at cooling high-temperature bodies in subcooled liquids. Int. J. Heat Mass Transf. **126**, 823–830 (2018). https://doi.org/10.1016/j.ijheatmasstransfer.2018.05.018

16. Kang, J.Y., Lee, G.C., Kim, M.H., Moriyama, K., Park, H.S.: Subcooled water quenching on a super-hydrophilic surface under atmospheric pressure. Int. J. Heat Mass Transf. **117**, 538–547 (2018). https://doi.org/10.1016/j.ijheatmasstransfer.2017.09.006

17. Chistyakov, V.F.: Algebro-differentsial'nye operatory s konechnomernym yadrom (Algebraic Differential Operators with a Finite-Dimensional Kernel). Nauka, Novosibirsk (1996)

18. Mehrmann, V.: Index concepts for differential-algebraic equations. In: Engquist, B. (ed.) Encyclopedia of Applied and Computational Mathematics. Springer, Berlin (2015). https://doi.org/10.1007/978-3-540-70529-1120

19. Chistyakova, E.V., Chistyakov, V.F.: Solution of differential algebraic equations with the Fredholm operator by the least squares method. Appl. Numer. Math. **149**, 43–51 (2020). https://doi.org/10.1016/j.apnum.2019.04.013

20. Chistyakov, V.F., Chistyakova, E.V.: Evaluation of the index and singular points of linear differential-algebraic equations of higher order. J. Math. Sci. **231**(6), 827–845 (2018). https://doi.org/10.1007/s10958-018-3852-7

Fully-Asynchronous Fully-Implicit Variable-Order Variable-Timestep Simulation of Neural Networks

Bruno Magalhães[1], Michael Hines[2], Thomas Sterling[3], and Felix Schürmann[1(✉)]

[1] Blue Brain Project, École Polytechnique Fédérale de Lausanne, Biotech Campus, 1202 Genève, Switzerland
felix.schuermann@epfl.ch
[2] Department of Neuroscience, Yale University, New Haven, CT 06510, USA
[3] Department of Intelligent Systems Engineering, Indiana University, Bloomington, IN 47404, USA

Abstract. State-of-the-art simulations of detailed neurons follow the Bulk Synchronous Parallel execution model. Execution is divided in equidistant communication intervals, with parallel neurons interpolation and collective communication guiding synchronization. Such simulations, driven by stiff dynamics or wide range of time scales, struggle with fixed step interpolation methods, yielding excessive computation on intervals of quasi-constant activity and inaccurate interpolation of periods of high volatility in solution. Alternative adaptive timestepping methods are inefficient in parallel executions due to computational imbalance at the synchronization barriers. We introduce a distributed fully-asynchronous execution model that removes global synchronization, allowing for long variable timestep interpolations of neurons. Asynchronicity is provided by point-to-point communication notifying neurons' time advancement to synaptic connectivities. Timestepping is driven by scheduled neuron advancements based on interneuron synaptic delays, yielding an *exhaustive yet not speculative* execution. Benchmarks on 64 Cray XE6 compute nodes demonstrate reduced number of interpolation steps, higher numerical accuracy and lower runtime compared to state-of-the-art methods. Efficiency is shown to be activity-dependent, with scaling of the algorithm demonstrated on a simulation of a laboratory experiment.

Keywords: Simulation of neural networks · Asynchronous computing · Variable timestep · Parallel computing · Distributed computing

1 Introduction

Simulation of the electrical activity of large networks of biologically detailed neuron models is a major impact scientific problem, allowing for a better understanding of the brain. State-of-the-art neuron models follows from the Hodgkin-Huxley

© Springer Nature Switzerland AG 2020
V. V. Krzhizhanovskaya et al. (Eds.): ICCS 2020, LNCS 12141, pp. 94–108, 2020.
https://doi.org/10.1007/978-3-030-50426-7_8

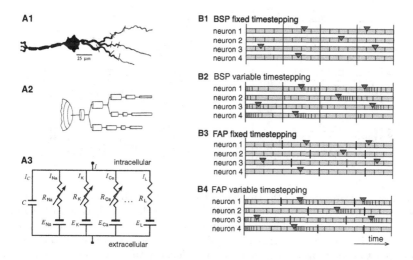

Fig. 1. Left: A1: a sample neuron morphology; **A2:** the spatially-discretized model of a compartmental tree of neuron dendrites; **A3:** The RC circuit representing the electrical activity of a compartment represented by the extended Hodgkin-Huxley model, with Sodium (Na), Potassium (K), Calcium (Ca) and leak (L) currents. **Right:** The four interpolation methods discussed, illustrated by left-to-right execution timelines of the simulation of four neurons. Gray cells represent the duration of interpolation steps. Inverted green triangles represent delivery of synaptic events. **B1:** Bulk Synchronous Parallel (BSP) model with fixed timestepping and collective synchronization. Vertical bars across all neurons represent collective communication; **B2:** BSP model with variable timestepping and collective synchronization implemented in NEURON [8]; **B3:** Fully-asynchronous parallel (FAP) fixed-step method, pioneered by our previous work [11]. Vertical bars across single neurons represent limit of stepping dictated by synaptic dependencies; **B4:** FAP variable-step method presented in this document.

(HH) formalism [5], modeling the electrical currents passing though connecting sections of neuron morphologies (spatially discretized as a tree of cylindrical leaky capacitors, henceforth referred to as **compartments**). Neurons are coupled via electro-chemical transductors, denominated **synapses**. When the voltage at a (**pre-synaptic**) neuron soma reaches a specific action potential threshold, it **spikes** (or **fires**), leading to a chain of biological reactions that changes the voltage at their **post-synaptic** counterparts. Due to the long simulation time required to express biological phenomena such as learning and synaptic plasticity, the acceleration of the simulation of neural networks is a relevant problem. Existing acceleration efforts follow the Bulk Synchronous Parallel (BSP) execution model, computing several neurons simultaneously via synchronized multithredead and distributed execution [4,7]. Execution time is divided in communication intervals equivalent to the time duration of the shortest **synaptic delay** across all neuron pairs in the network. Equidistant synchronous collec-

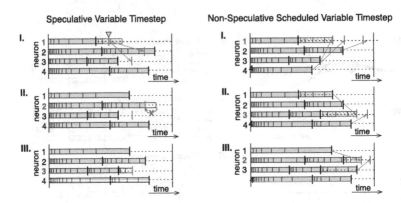

Fig. 2. Illustrative workflow of the two methods for variable timestep (vardt) interpolation described. **Left:** Speculative interpolation (state-of-the-art): **I.** Neuron 1 performs a step (blue area) and spikes during the step interval, with spike time marked with an inverted triangle, and delivery times marked with blue arrow heads; **II.** The current interpolation time of neuron 2 exceeds the spike delivery time. A back stepping to the delivery time follows (red arrow); **III.** Neuron 3 interpolates until the next (spike) event time. **Right:** Non-Speculative scheduled time-stepping. **I.** Neuron 1 advances to the earliest time instant allowed (green area), given by the time instant of pre-syn. neurons 2, 3 and 4 and the shortest synaptic delays 2→1, 3→1, and 4→1, respectively (green arrow heads); **II.** Neuron 3 is now the earliest neuron in time, and follows analogously based on the delays of neurons 1 and 4; **III.** Neuron 2 advances similarly. (Color figure online)

tive communication calls performs both synchronization of stepping and synaptic exchange. Interpolation of neurons is performed independently within the boundaries of each intervals, typically with fixed timestep methods, as illustrated in Fig. 1 B1). Variable timestep interpolation on the BSP execution model (Fig. 1 B2) has been presented on single compute nodes, with speculative interpolation of individual neurons [8]. Further acceleration can be achieved with finer-grained parallelism of individual neuron models via graph-parallelism of Ordinary Differential Equations (ODEs) [9] and branch-parallelism of neuron topology sections [10]. Cache-efficient acceleration has been demonstrated via a barrier-free fixed-timestep simulation on a fully-asynchronous parallel (FAP) execution model [11] (Fig. 1 B3), henceforth referred to as **our previous work**.

We introduce a method for the distributed fully-asynchronous variable-order variable-timestep implicit interpolation of detailed neuron models, benefiting from cache-efficient barrier-free synchronization and performing long variable timesteps on the FAP execution model, as illustrated in Fig. 1 B4). We study the numerical instability and performance dependency of our methods on the biological activity of the network. An implementation of our methods on the core kernel of the NEURON simulator [4] is detailed and benchmark on 64 Cray XE6 compute nodes. Distributed asynchrony and multicore executions on a global memory address space are provided by the HPX runtime system [13].

Benchmarks demonstrate lower time to solution, higher numerical accuracy, the removal of speculative computing and synchronization barriers, and good scaling properties, tested on a simulation of a laboratory experiment.

The methods presented provide insights for the exploration of modern asynchronous runtime systems on large networks of compute nodes, and for the redesign of future simulators across a wide range of scientific domains, driven by large systems of ODEs and highly-heterogeneous activity.

2   Methods

2.1   Mathematical Model

The RC circuit that models the electrical current passing through the membrane of a compartment n is modelled as:

$$C\frac{dV_n}{dt} = -\sum_i g_i x_i(V_n - E_i) + I(t) + \sum_{c:p(c)=n} \frac{V_c - V_n}{r_c} - \frac{V_n - V_{p(n)}}{r_{p(n)}} \quad (1)$$

where g_i and E_i describe the conductance and reversal potential of the ionic channels. x_i models the opening probability of the transmembrane ion channel currents, typically described by a voltage-gated ODE, and omitted for brevity. Synaptic currents or injected current stimuli, if any, are included in $I(t)$. The branching contributions are provided by Ohm's Law and the neuronal cable theory: the subscript $p(n)$ refers to the index of the parent compartment of n, and c to an iterator over the indices of its children in the compartmental tree. The variable r defines the axial resistance as a function of the diameter and the cytoplasmic resistivity. The RC circuit underlying the current passing through a compartment is illustrated in Fig. 1, layout A3. The solution of this system is solved numerically. The complexity of the spatio-temporal model of the neuron activity is reduced by performing a spatial discretization of the neuronal morphology, from biologically inspired to HH-based compartmental representation, and assume the spatial discretization to be small enough, so that the state across compartments' length is constant (Fig. 1, A1 and A2). Thus, interpolation of solution is performed for consecutive discrete time intervals only. The resolution follows a fixed step defined as *small enough* to capture the currents with fastest dynamics, set to 0.025 ms. The fastest synaptic delay across our network model has been measured as 0.1 ms or equivalently 4 computation steps, accounting for 0.13% of the total synapses.

Simple and Complex Neuron Models: A problem specific optimization allows for a speed-up on the resolution of **simple neuron models** such as the Hodgkin-Huxley, where state variables are described by linear ODEs and depend only on the voltage V, and vice-versa. An implicit resolution based on interleaved timestepping of voltage and states, by solving voltages at a given time t and states at time $t + \Delta t/2$, allows for the resolution of the system of ODEs as

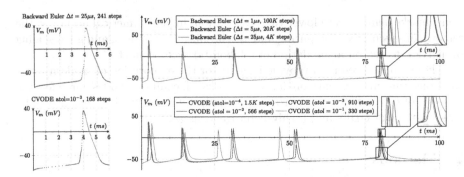

Fig. 3. Voltage potential at soma and interpolation steps for a sample neuron during 6 ms (left) and 100 ms simulation (right) of a 1.3 mA continuous current injection, interpolated with Backward Euler (top) and CVODE (bottom) methods. The reference implementations are the Backward Euler with $\Delta t = 1\,\mu s$ and CVODE with absolute tolerance 10^{-1}, presented in black and considered indistinguishable. The standard NEURON step size and tolerance are $\Delta t = 25\,\mu s$ and 10^{-3} and are presented in red. (Color figure online)

a system of linear equations. Resolution of **complex models**, including nonlinear and/or correlated state equations (such as synaptic plasticity presented by Graupner et al. [3]) cannot be resolved with the aforementioned method, and require a fully-implicit (non-staggered) resolution. Reliable resolutions rely on fixed-step iterative implicit methods such as Backward Euler. Alternatively, an implicit variable timestep method with variable order is possible. Its implementation to our use case is detailed next.

Variable Step Implementation: The CVODE (C Variable-step solver for ODEs, [2]) is an implementation of the Backward Differentiation Formula (BDF) for the variable-step multistep implicit method solving the Initial Value Problem (IVP, $\dot{y} = f(t,y)$, $y(t_0) = y_0$ where $y \in \mathbb{R}^N$) for ODEs as:

$$\sum_i^q \alpha_{n,i} y_{n-i} + \Delta t_n \beta_n \, \dot{y} = 0 \tag{2}$$

where $y = [..., V_{k-1}, V_k, V_{k+1}, .., x_{i-1}, x_i, x_{i+1}, ...]$ is a vector representing the state variables of a neuron or a set of neurons following the variable notation in Eq. 1, q is the order of the current iteration, and α and β are the q-dependent BDF-method coefficients. BDF-1 is the Backward Euler. In brief, CVODE returns the Δt_n and y_n that solve BDF-q for an user-provided tolerance **(atol)**. The computation is performed iteratively, with a suggested step size for each iteration based on the solution gradient and order q. Given and user-provided function that computes the ODE right-hand side and a Jacobian $j = \partial f / \partial y$ (or an approximation to it), the resolution relies on Newton iterations, with a stop condition based on the test $\|y_{n(m)} - y_{n(0)}\| \leq \epsilon$ for iteration m

in step n: If error is greater than threshold, a reiteration follows with a smaller $\Delta t_{n(m+1)}$; if error is smaller, proceeds to step $n+1$ with larger $\Delta t_{(n+1)(0)}$.

2.2 Asynchronous Timestepping of Neuron Networks

Control of neurons time advancement on synchronized distributed executions is a solved problem, by enforcing a BSP-like synchronization barrier [4], as in layouts B1 and B2 in Fig. 1. Barrier-free variable-timestep interpolation are possible with a speculative interpolator on a single compute node and small networks of neurons. Neurons are described by individual interpolators, and advance in time under the best assumption that no **discontinuity** of solution (synaptic current) will arrive with a delivery time earlier than the neuron current time. Discontinuities lead to a **reset of the IVP** problem and interpolator state history, and consequently to small steps in the following iterations. When a discontinuity is required to be delivered in a past instant in time, a **backstepping** operation must precede, in order to reset the recent step and interpolate neuron state back to a time instant of confidence (the time of the discontinuity). Simulations of small neuron networks on single compute nodes are possible and have previously shown a substantial runtime acceleration utilising this model [8]. Distributed executions and/or large neural network are infeasible with this method due large number of IVP resets, the complexity of backstepping cascades of events across several compute nodes, large amount of time spent on speculative stepping, and computational imbalance at synchronization barriers.

An alternative approach was implemented, based on the non-speculative asynchronous stepping methodology detailed in our previous work [11]. Neurons hold a map storing the time instant of their pre-synaptic connectivities. The map is updated by stepping notifications received actively at a certain frequency, throughout the stepping of its pre-synaptic dependencies. Neurons step to the maximum time allowed by their synaptic connectivities. This guarantees synapses to be delivered in future time instants, thus removing backstepping and reversion of sent synaptic spikes. The method is improved with an *earliest neuron steps first* scheduler at each compute node that keeps track of neurons advancement, and picks the earliest neuron in time as the next to interpolate. This guarantees the maximisation of the step length and provides a larger variable step interval, with the benefit of reduced communication and computation, the removal of solution resets and backstepping, and larger stepping intervals. For completion, both approaches described are pictured in Fig. 2.

2.3 Implementation Details

Our methods were implemented on the core kernel of the NEURON simulator [7]. Single Instruction Multiple Data (SIMD) capabilities—supported only by fixed-step method—were added to variable-step implementations. Communication, synchronisation control objects, memory allocation, threading, distributed memory space, distributed execution and parallelism were implemented with

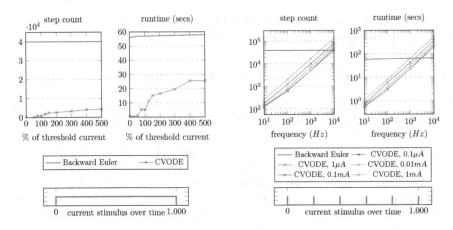

Fig. 4. Interpolation steps and runtime for 1000 ms simulation of a pyramidal cell on an Intel core i5 at 1.6 GHz. Left: injection of a continuous current as a percentage of the threshold current (0.206 mA). Right: injection of short 1 μs current pulses of different amplitudes I at different frequencies. Results presented for the Backward Euler with $\Delta t = 25\,\mu s$ and CVODE with atol $= 10^{-3}$.

HPX [13], the runtime system for the Parallex execution model [6]. The Implementation details have been covered in our previous manuscript [11], and are omitted for brevity. Efficient point-to-point communication and remote direct access memory is provided with specialized Infiniband network hardware.

3   Results

3.1   Numerical Accuracy

We compare the numerical accuracy of both fixed and variable step models by measuring the time difference of the main unit of interest in the activity of spiking neuron networks—the spiking time instants. Figure 3 presents the voltage trajectory and number of steps of a 1.3 mA current clamp experiment for a single (6 ms) and several (100 ms) spikes of a layer 5 pyramidal cell. Results on the single spike voltage trajectory (6 ms) display a reduced step count and better adaptation to trajectory change when comparing CVODE to Euler method. The rationale behind the better performance is adaptive stepping is gradient sensitive, thus better adapting to the trajectory of a neuron voltage. CVODE displays less steps during long periods of low gradient (e.g. 1–2.5 ms), and greater number of steps for steep trajectories (the spike trajectory). The 100 ms simulation displays a phase shift in solution (measured as the time difference between peak voltage values of reference and benchmark curves) that increases with the increase of the step size on the Euler methods. In practice, the timestep determines the fastest reaction time of the system, thus large timesteps will inevitably cause the system dynamics to be *slow*. The analysis show that a CVODE tolerance value of 10^{-2} approximates the resolution of the default Euler method (step

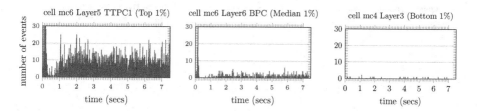

Fig. 5. Number of incoming spikes (bin size 0.25 ms) measured throughout 7.5 s of simulation, for a sample neuron collected from the top, median and bottom 1% on a network of 219K neurons.

size 25 μs), with a reduction of 7× in step count. At longer runs, the variable step demonstrated to be more precise, due to no accumulation of phase shift, with the maximum trajectory shift measured at approximately 1.1 ms. On the other hand, a tolerance value of 10^{-3} approximates closely the optimal solution with 40% less steps, and with a margin of error similar to its 5 μs Euler counterpart for the period of 100 ms, while yielding 22× less interpolations.

3.2   Performance Dependency on Stiffness and Discontinuities

We measured the response of both stepping methods to spiking frequency. Performance was measured in terms of steps count and time to solution on an Intel i5 at 1.6 GHz. Changes in trajectory were enforced by injecting a continuous current of a given amplitude on a neuron during 1000 ms. Current intensity is measured as a percentage of the **threshold current**, the minimum continuous current value that needs to be injected to force a neuron to spike. Results are presented in Fig. 4 (left), and demonstrate that high dynamics of the solution degrade the CVODE performance. This is due to CVODE requiring smaller steps on high trajectory variations in order to respect the absolute tolerance value. For the range of tested scenarios, the measured CVODE to Euler reductions were: (1) 434× in step count and 98× in runtime for injected currents below 50% of the threshold current; (2) 62× steps and 11.6× runtime for 100%; and (3) 9.4× and 2.5× runtime for 500%, a worst case scenario of little prob. of occurrence.

We measured the effect of discontinuities on both methods by injecting several current pulses at a fixed frequency on a neuron soma, mimicking synaptic events. The experiment results are displayed in Fig. 4 (right) and suggest that the CVODE performance depends on the trajectory change incurred by each discontinuity, i.e. the amplitude of the current injected: the larger the voltage increase, the larger the change in trajectory gradient, thus the more interpolation steps are required. As expected, results demonstrate that the number of discontinuities plays a major role in performance. CVODE is shown to deliver a reduction of steps in the order of 153–322× for a frequency of 10 discontinuities per second for current values of 1 mA to 0.1μA. The step count equilibrium between Euler and CVODE method lies in the interval of $10^{3.2}$–$10^{4.0}$Hz for similar currents interval. The runtime demonstrates a similar dependency on the

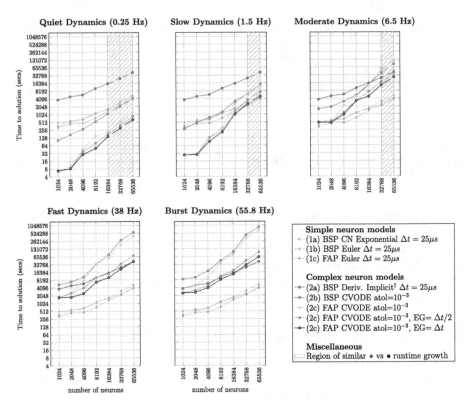

Fig. 6. Runtime for the simulation of one second of biological activity described by five spiking rate dynamics, measured for increasing input network sizes, on 64 Cray XE6 compute nodes. **Key:** BSP: Bulk Synchronous Parallel; FAP: Fully-Asynchronous Parallel; atol: absolute tolerance; EG: event grouping interval; †: able to solve non-linear ODEs implicitly, and unable to solve correlated mechanism states implicitly.

injected current, yielding a speed-up of $51\times$ for $10\,\mathrm{Hz}$ decreasing linearly up to the speed-up equilibrium value at $1000\,\mathrm{Hz}$ for the strongest current. For the lightest current injected, a speed-up of $100\times$ is visible for a $10\,\mathrm{Hz}$ discontinuity rate, decreasing to an Euler matching value at circa 1600 events/s ($10^{3.2}\,\mathrm{Hz}$).

3.3 Simulation of a Laboratory Experiment

We tested the suitability of variable step methods to our problem by measuring the spiking activity of a simulation of $7.5\,\mathrm{s}$ of electrical activity mimicking a laboratory experiment. The experimental set-up performs a fixed step simulation of the spontaneous activity of 219.247, detailed in section *Simulating Spontaneous Activity* in [12]. A representative distribution of discontinuity events for three groups of neurons—organized by highest 1%, median 1%, and lowest 1% number of discontinuities—is displayed in Fig. 5. The simulation incurred a total of circa 155 million events, with the following distribution: (a) top 1% of neurons, between 3040 and 6146 events in 7.5 s, or 405–820 Hz; (b) median 1%, from 541 to 558 events (72–74.4 Hz); and (c) bottom 1%: less than 100 events ($\leq$10 Hz). The average number of events was of 707 events for the 7.5 s of simulation, or equivalently, 94 Hz, significantly below the 1000 Hz threshold discussed in the previous section. Moreover, the results on the distributions of time interval between discontinuities, plotted in red on the right, display large periods of *silence* between events arrival in the median and bottom use cases, but not on the top, suggesting the suitability of adaptive stepping to most (but not all) neurons in the population.

3.4   Large-Scale Benchmark

We simulate one second of the electrical activity of a digitally reconstructed neural network extracted from the model of Markram et al. [12]. Execution times were collected on 64 Cray XE6 compute nodes, powered by an AMD Opteron 6380 with 16 cores at 2.5 GHz, 64 GB of RAM and 256-bit floating point units. CVODE was defined to utilise the default maximum BDF order value of 5.

On the set-up of the test bench, it is relevant to mention that neuronal activity is highly dependent on the mammal specie, brain region and momentary activity. Simulations must approximate real use cases, as spiking activity affects heavily the performance of variable step methods, as shown previously. Thus, our test bench benchmarks the efficiency of five different brain dynamics described in literature: **(1)** a model of *quiet dynamics* with a mean spiking rate of 0.25 Hz per neuron, representative of circa 90% of neurons in the human brain during regular activity; **(2)** *slow dynamics* at 1.5 Hz, representing the lower bound of active neurons; **(3)** *moderate dynamics* at 6.5 Hz, an approximation of the regime of slow oscillations regime from the Brunel network model [1] and an upper limit to the rat frontal cortex; **(4)** *fast dynamics* at 38 Hz, characterizing neuronal activity during periods of high vigilance; and the inhibition-dominated model of the Brunel Network [1]; and **(5)** *burst dynamics* at 55.8 Hz, typically a byproduct of strong current injections, similar to the first instants of simulation in Fig. 5; and the fast spiking regime of the Brunel Network [1]. Neurons activity is triggered by a constant current injection in all neurons throughout the whole duration of the simulation, strong enough to approximate the spiking rate to the regimes described. The input neural networks are retrieved from layers 4 and 5 of the rodent brain, where the longest dendritic trees and densest synaptic connectivity exist, thus representing a worst-case scenario for variable-step methods, and favourable to fixed-step methods. Thus, the results presented are a lower bound of possible acceleration. For complete coverage of the topic, we include the following state-of-the-art solvers for simple neuron models (labelled 1a to 1c, and restrained to linear ODEs with uncorrelated states) and complex models (2a–2c): **(1a)** the *cnexp* fixed step solver in NEURON, with added SIMD, providing an interleaved resolution of current and states as linear equations, with an analytical resolution of first-order ODEs describing state variables; **(1b)** the *Euler* solver in NEURON, with added SIMD, resolving the current-states dependency with an explicit Euler method with staggered timestepping, and as a linear

equations; (1c) the same *Euler* method on a FAP execution model, presented
in our previous work (Fig. 1 B3); (2a) the BSP fixed step *derivimplicit* solver
available in NEURON, with added SIMD, with interleaved-timestep resolution
of current as a linear equation, and implicit resolution of individual mechanism state ODEs; (2b) the BSP variable step method in NEURON with added
SIMD and a collective communication barrier (Fig. 1 B2); and (2c) the SIMD-
enabled FAP with variable timestepping introduced in this paper (Fig. 1 B4).
We tested our methods in neural networks ranging from 1024 to 65536 neurons,
a scale that approximates two columns in the rodent neocortex, and the maximum allowed due to memory requirements of the BDF order. The benchmark
results are presented in Fig. 6. The FAP variable step method (2c ○) is presented
alongside two variants—labelled 2c ● and 2c●—that group and deliver instantly
the discontinuity events within an interval equivalent to the timestep $\Delta t/2$ and
Δt of the interleaved- and fixed timestep methods, respectively. This approach
yields a level of reduced precision in the delivery of events—similar to fixed step
methods—while maintaining the same high variable-order variable-step accuracy
during continuous periods of activity and reducing significantly the number in
IVP resets. CVODE-based executions are displayed for an absolute tolerance
(atol) of 10^{-3}, the default value in the NEURON simulator. For brevity, we
omit CVODE executions with an absolute tolerance of 10^{-2}—tested and delivering a runtime reduction of 5%−8%—and the analysis comparing only simple
fixed-step solvers (1a–1c), covered in our previous work [11].

4   Discussion

4.1   Fixed- vs Variable-Timestep Interpolators

Fixed step methods do not yield significantly-different execution times across different spiking regimes. This is due to the homogeneous computation of neuron
state updates throughout time, and the light computation attached to synaptic
events and collective communication not yielding a substantial increase of runtime. As expected, variable step executions are penalized on regimes with high
discontinuity rates. Runtimes of fixed- and variable-step solvers approximate as
we increase the spiking rate, i.e. the increase of runtimes with the input size
is steeper for variable timestep (2b● and 2c●●●) compared to fixed timestep
methods (2a●). This is due to discontinuities in variable-step being delivered
throughout a continuous time line, compared to the discrete delivery instants of
the fixed-step methods,—therefore increase the number of interpolation steps;
and the iterative model of the variable timestep reinitializing the state computation with small step sizes on each IVP reset. A remarkable performance is
visible on the quiet dynamics use case, where our fully-implicit ODE solver of
complex models (with Newton iterations), still runs faster than the simple solver
resolving only a system of linear equations. The underlying rationale is that—
despite the inherent computation cost of Newton iterations in the variable step
methods—the low level of discontinuities allow for very long steps, that surpass
the simulation throughput of fixed step methods. The measured speed-up of our

reference method (2c•) compared to the reference fixed step method (2a•) was of 544-65× across input sizes for the quiet dynamics, down to 7.7-1.8× to the moderate dynamics. The fast dynamics presented a speed-up of twofold for the dataset of 1024 neurons, and a similar runtime for the 66K neurons. The burst dynamics, although of very unlikely probability of occurrence, demonstrated an acceleration of 1.5× for 1024 neurons and a deceleration of 1.5× for 66K neurons.

4.2   Variable Step Event Grouping

On the analysis of the performance of the CVODE with grouping of events within half fixed timestep (2c•), when applied to the largest dataset tested, the previous acceleration was reduced to 47× for quiet, 4.4× for slow, and 1.2× for moderate dynamics, with an inferior performance on the remaining regimes. A further reduction of speed-up to 33× for quiet and 1.9× for slow dynamics was noticeable on the CVODE implementation without events grouping (2c•), with lower performance for the remaining spike regimes. Although being more precise and solving correlated states implicitly, this method runs slower than the reference implicit fixed step method 2a• in the use cases characterized by a high number of neurons and/or strong network activity. This goes in line with the conclusions in Sect. 3.3, confirming that performance is activity dependent, and the performance depends on the network connectivity. The speed-up introduced across FAP CVODE variants (2c•••) increases with the amount of discontinuities in the system—correlated to high network activity or size—as the efficiency of the event grouping method is related to the amount of events in the same grouping interval that are delivered at once.

4.3   Fully-Asynchronous vs Bulk-Synchonous Execution Models

We study the performance difference between the BSP and FAP execution models. Results show that runtimes of both implementations approximate with an increase of input. This is visible by comparing the fixed step trajectories 1b♦ and 1c♦, and the variable step trajectories 2b• and 2c•. For small network sizes, the difference in runtime is few orders of magnitude higher than for larger network sizes. On large models, the runtimes are similar. This property was demonstrated in our previous work: in brief, an increase of network leads to a higher number of network connectivity, reducing the maximum stepping interval per neuron, and approximating it to the communication delay in BSP methods. On fixed step methods, it is noticeable a similar runtime on large (66K) networks of neurons, as timesteps are computationally homogeneous. On variable step methods, similar runtimes are only noticeable when significant network activity is present (moderate, fast and burst dynamics), as little network activity leads to few discontinuities and analogously large variable step intervals.

4.4   Runtime Dependency on Input Size and Spike Activity

It is known that, on simulations of small networks, variable-timestep methods yield a significant acceleration in time to solution compared to fixed timestep methods [8], due to little interneuron connectivity. However, the larger networks yield up to 10 thousand synapses per neuron, with the number of discontinuities in the system being related to the network activity. The question lies now on which conditions are required for similar computation complexity in both interpolators. To that extent, we measured the regions of similar runtime growth for the reference fixed step (2b●) and our variable step methods (2c●). The region is labelled as ▭ in Fig. 6. As expected, fixed step methods yield a quasi-linear runtime growth with the increase of the input size, and are independent of the spiking regime, due to almost ideal scaling of the algorithm in the BSP model. On the other hand, the runtime of variable timestep methods—dependent on the number of discontinuities—demonstrates a rapidly increasing growth with the input size outside the region of similar growth, and almost linearly inside. Moreover, as it depends on the network activity, the lower limit of the region increases with the spiking rate, and is delimited at 16.4K, 32.8K and 32.8K neurons or more for the quiet, slow and moderate dynamics, while not visible in the fast and burst dynamics. In the three spiking regimes where such region exists, the similar growth in both approaches provides a confidence of the scaling capabilities of our methods in larger network models. This is of high importance as it provides an estimation of runtime upper bound in simulations combining neurons with heterogeneous spiking rates, as discussed next.

4.5   Overall Runtime Speed-Up Estimation

To conclude our analysis, we computed an estimation of the performance acceleration on a simulation combining several spiking regimes. We measured the distribution of neuron spike rates and neurons per spiking regime, following the laboratory experiment simulation described in Sect. 3.3. Estimations were collected from 2–4 s of simulation time from the central minicolumn (31.3K neurons) of a 219K neurons network, to avoid boundary-effects from reduced connectivity and the initial artificial synaptic burst from the current injection. The measured percentage of neurons on each regime is 31.43%, 38.44%, 27.02%, 3.10% and 0.01%, relating to 68.9K, 84.3K, 59.2K, 6.8K and 22 neurons. Following the runtimes described in Sect. 4.1, the speed-up range for the interval of 1024-66K neurons when comparing our methods with the state-of-the-art solver for complex models (2a●) are estimated as: 224.5-11.9× for the variable step method with precise event delivery (2c●); 225.1-17.1× for the similar implementation with delivery of events within the next half timestep (2c●); and 228.5-24.6× for the use case with full-timestep event group delivery (2c●). Since the quiet, slow and moderate dynamics regimes weight over 95% in the runtime calculation, and as for datasets above 32.8K the reference vs benchmark runtimes have a similar runtime growth in those regimes, we believe the overall runtime and scaling properties are almost fully-preserved on larger networks.

5   Summary and Closing Remarks

This paper presented a distributed simulation of detailed neuron models with variable-order variable-timestep methods on a fully-asynchronous execution model, yielding asynchronous computation, communication and synchronisation. We detailed state-of-the-art approaches based on the Bulk Synchronous Parallel execution model (BSP), their limitations on the numerical resolution of complex neuron models, computation load imbalance, and speculative computing in variable-step simulations. We simulate five spiking regimes that characterize several dynamics of the mammal brain, on up to 65536 neurons on 64 Cray XE6 compute nodes, and compare our methods against five state-of-the-art numerical solvers. Results demonstrate higher numerical accuracy, with a speed-up of 544-65× for a quiet spiking regime of 0.25 Hz representing a majority of neurons in regular brain activity, down to 7.7-1.8× to a moderate regime of 6.5 Hz, and 2× to no acceleration for 38 Hz, a pattern of unlike occurrence or short duration. An analysis of performance achievable on the simulation of a laboratory experiment demonstrates a speed-up of 224.5-11.9× for an execution with precise delivery of events, increasing to 225.1-17.1× and 228.5-24.6× for two optimized alternatives that group events delivery in the next half and full timestep. With the scaling properties of our methods preserved on larger networks of neurons.

As a final remark, although being applied to a network of neurons, most methods presented are problem-independent and do not require intrinsic knowledge of the problem domain, therefore opening the prospectus for the acceleration of a wide domain of scientific problems modelled by complex systems of ODEs.

Acknowledgements. This study was supported by funding to the Blue Brain Project, a research center of the École polytechnique fédérale de Lausanne, from the Swiss government's ETH Board of the Swiss Federal Institutes of Technology. The computing infrastructures were provided by Indiana University. A portion of Michael Hines efforts was supported by NINDS grant R01NS11613.

References

1. Brunel, N.: Dynamics of sparsely connected networks of excitatory and inhibitory spiking neurons. J. Comput. Neurosci. **8**(3), 183–208 (2000). https://doi.org/10.1023/A:1008925309027
2. Cohen, S.D., Hindmarsh, A.C.: CVODE, a stiff/nonstiff ODE solver in C. Comput. Phys. **10**(2), 138–143 (1996)
3. Graupner, M., Brunel, N.: Calcium-based plasticity model explains sensitivity of synaptic changes to spike pattern, rate, and dendritic location. In: Proceedings of the National Academy of Sciences, pp. 3991–3996 (2012). https://www.pnas.org/content/pnas/109/10/3991.full.pdf
4. Hines, M.L., Carnevale, N.T.: The neuron simulation environment. Neural Comput. **9**(6), 1179–1209 (1997)
5. Hodgkin, A.L., Huxley, A.F.: A quantitative description of membrane current and its application to conduction and excitation in nerve. J. Physiol. **117**(4), 500–544 (1952)

6. Kaiser, H., Brodowicz, M., Sterling, T.: Parallex an advanced parallel execution model for scaling-impaired applications. In: International Conference on Parallel Processing Workshops, 2009 (ICPPW 2009), pp. 394–401. IEEE (2009)
7. Kumbhar, P., Hines, M., Fouriaux, J., Ovcharenko, A., King, J., Delalondre, F., Schürmann, F.: Coreneuron : an optimized compute engine for the neuron simulator. Front. Neuroinform. 13 (2019). https://doi.org/10.3389/fnif.2019.00063
8. Lytton, W.W., Hines, M.L.: Independent variable time-step integration of individual neurons for network simulations. Neural Comput. 17(4), 903–921 (2005)
9. Magalhaes, B.R.C., Sterling, T., Schürmann, F., Hines, M.: Exploiting flow graph of system of odes to accelerate the simulation of biologically-detailed neural networks. In: 2019 IEEE International Parallel and Distributed Processing Symposium (IPDPS), pp. 176–187. IEEE (2019). https://doi.org/10.1109/IPDPS.2019.00028
10. Magalhaes, B.R.C., Sterling, T., Hines, M., Schürmann, F.: Asynchronous branch-parallel simulation of detailed neuron models. Front. Neuroinform. 13 (2019). https://doi.org/10.3389/fninf.2019.00054
11. Magalhaes, B., Hines, M., Sterling, T., Schürmann, F.: Fully-asynchronous cache-efficient simulation of detailed neural networks. In: Proceedings of International Conference on Computational Science (ICCS), pp. 421–434 (2019). https://doi.org/10.1007/978-3-030-22744-9_33
12. Markram, H., et al.: Reconstruction and simulation of neocortical microcircuitry. Cell 163(2), 456–492 (2015)
13. Sterling, T., Anderson, M., Bohan, P.K., Brodowicz, M., Kulkarni, A., Zhang, B.: Towards exascale co-design in a runtime system. In: Exascale Applications and Software Conference, Stockholm, Sweden, April 2014

Deep Learning Assisted Memetic Algorithm for Shortest Route Problems

Ayad Turky[1(✉)], Mohammad Saiedur Rahaman[1], Wei Shao[1], Flora D. Salim[1], Doug Bradbrook[2], and Andy Song[1]

[1] School of Science, RMIT University, Melbourne, VIC 3000, Australia
{ayad.turky,saiedur.rahaman,wei.shao,flora.salim,andy.song}@rmit.edu.au
[2] Mornington Peninsula Shire Council, Rosebud, VIC 3939, Australia
doug.bradbrook@mornpen.vic.gov.au

Abstract. Finding the shortest route between a pair of origin and destination is known to be a crucial and challenging task in intelligent transportation systems. Current methods assume fixed travel time between any pairs, thus the efficiency of these approaches is limited because the travel time in reality can dynamically change due to factors including the weather conditions, the traffic conditions, the time of the day and the day of the week, etc. To address this dynamic situation, we propose a novel two-stage approach to find the shortest route. Firstly deep learning is utilised to predict the travel time between a pair of origin and destination. Weather conditions are added into the input data to increase the accuracy of travel time predicition. Secondly, a customised Memetic Algorithm is developed to find shortest route using the predicted travel time. The proposed memetic algorithm uses genetic algorithm for exploration and local search for exploiting the current search space around a given solution. The effectiveness of the proposed two-stage method is evaluated based on the New York City taxi benchmark dataset. The obtained results demonstrate that the proposed method is highly effective compared with state-of-the-art methods.

Keywords: Shortest route problems · Memetic algorithm · Deep learning · Travel times

1 Introduction

Finding shortest routes is crucial in intelligent transportation systems. Shortest route information can be utilised to enable route planners to compute and provide effective routing decisions [8,11,14,16,24]. However, shortest route computation is a challenging task partially due to dynamic environments [3]. For instance, the shortest path is impacted by various spatio-temporal factors, which are dynamic in nature, including weather, the time of the day, and the day of the week. That makes the current shortest route computation techniques ineffective [3,7]. Moreover, it is a challenging problem to incorporate these dynamic factors into shortest route computation.

© Springer Nature Switzerland AG 2020
V. V. Krzhizhanovskaya et al. (Eds.): ICCS 2020, LNCS 12141, pp. 109–121, 2020.
https://doi.org/10.1007/978-3-030-50426-7_9

In recent years, the proliferation of pervasive technologies has enabled the collection of spatio-temporal big data associated with user mobility and travel routes in a real-time manner [15]. Modern cars are equipped with telematics devices including in-car GPS (Global Positioning System) devices which can be used as a source of valuable information in traffic modelling [23]. The traces generated from GPS devices has been leveraged by many scenarios such as Spatio-temporal context recognition, taxi-passenger queue time prediction, study of city dynamics and transport demand estimation [3,12,13,17,23].

One important aspect of finding shortest routes in realistic environments, which are inherently dynamic, is travel time prediction [8,22]. Due to the dynamic nature of in the travel routes, traditional machine learning methods cannot be applied directly onto travel time prediction. One of the key challenge for traditional machine learning models is the unavailability of hand-crafted features which requires substantial involvement of domain experts. One relevant approach is the recent use of evolutionary algorithms in other domains to work along with deep learning models for effective feature extraction and selection [18–21]. In this study, we aim to identify relevant features for shortest route finding between an origin and destination, leveraging the auto-feature generation capability of deep learning. Thereby we propose a novel two-stage architecture for the travel time prediction and route finding task. In particular we design a customized memetic algorithm to find shortest route based on the predicted travel time from the earlier stage. The contributions of this research are summarised as follows:

- A novel two-stage architecture for the shortest route finding under dynamic environments.
- Development of a deep learning method to predict the travel time between a origin-destination pair.
- A customised memetic algorithm to find shortest route using the predicted travel time.

The rest of the paper is organized as follows. In Sect. 2, we present our proposed methodology for this study. Section 3 describes the experimental settings which is followed by the discussion of experimental results in Sect. 4. Finally, we conclude the paper in Sect. 5.

2   Proposed Methodology

In this paper, we propose a deep learning assisted memetic algorithm to solve the shortest route problems. The proposed method has two stages which are (1) prediction stage and (2) optimisation stage. The prediction stage is responsible to predict the travel times between a pair of origin and destination along the given route by using deep learning. The second stage uses memetic algorithm to actually find the shortest path to visit all locations along the given route. In the following subsections, we discuss the main steps of the proposed method and the components of each stage in detail. Figure 1 shows our proposed approach.

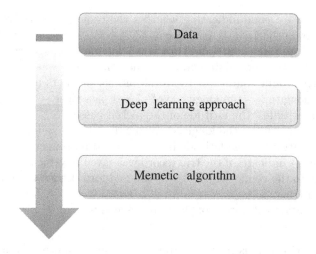

Fig. 1. Flowchart of the proposed two-stage approach

2.1   Prediction Stage

Conventional route finding methods assume fixed cost or travel time between any pairs of points. That is rarely the case in reality. One approach to the dynamic travel time issue is prediction. In this work, we incorporate the weather data along with the temporal-spatial data to develop a deep learning predictive approach. The goal of the proposed predictive approach is to predict future travel time between any points in the problem based on historical observations and weather condition. Specifically, given a group of historical travel time data, weather data and road network data, the aim is to predict travel time between source (s) and destination (d) $s_i, d_i \in R, i \in [1,2, ..., n]$, where n is the number of locations in the road network. Our predictive approach tries to predict the travel time at $t+1$ based on the given data at t. The proposed predictive approach has three parts: input data, data cleaning and aggregation, the prediction approach. Figure 2 shows the deep learning approach.

Input Data. In this work, we use data from three different sources. The data involves around 1.5 million trip records. These include the travel time data, weather data and road network data.

- **Travel time data.** The travel times between different locations were collected using 2016 NYC Yellow Cab trip record data.
- **Weather data.** We use the weather data in New York City - 2016. The data involves: date, maximum temperature, minimum temperature, average temperature, precipitation, snow fall and snow depth.
- **Road network data.** The road network data involves temporal and spatial information as follows:
  - Id - a trip identifier.

- Vendor_id - a code indicating whether the provider is involved with the trip record.
- Pickup_date-time - date and time when the meter was started.
- Drop-off_date-time - date and time when the meter was disconnected.
- Passenger_count - indicates the total number of riders in the vehicle.
- Pickup_longitude - the longitude of picked passenger.
- Pickup_latitude - the latitude of the picked passenger.
- Dropoff_longitude - the longitude of the dropped passenger.
- Dropoff_latitude - the latitude of the dropped passenger.
- Store_flag - indicates if the trip record was saved in vehicle memory before sending to the vendor where Y = store and forward; N = not a store and forward trip.
- Trip_duration - duration of the trip in seconds.

Data Preparation. This process involves removal of all error values, outliers, imputation of missing values and data aggregation. To facilitate the prediction we bound the data ranges between $(average + 2) \times standard\_deviation$ to $(average - 2) \times standard\_deviation$. Values outside of these ranges are considered as outliers and are removed. The missing values are imputed by the average values. Any overlapping pick-up and drop-off locations are also removed. In the aggregation step, we combine the travel time data, weather data and road network each time step so that it can be fed into our deep networks.

Prediction Approach. The main goal of this step is to provide high accuracy prediction of the travel times between different locations in the road network. The processed and aggregated data is provided as an input for the prediction approach. Once the prediction model is trained and retrieved, it is then ready to actually predict the travel times between given locations.

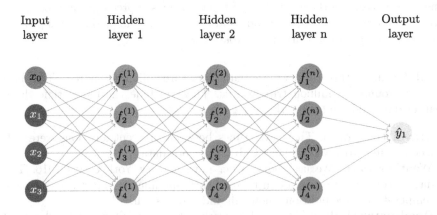

Fig. 2. Illustration of the deep network based prediction model

In this work, we propose a deep learning technique based on feedforward neural network to build our prediction approach. The deep neural network consists of one input layer, multiple hidden layers and one output layer. Each layer (input, hidden and output) involves a set of neurons. The total number of neurons in the input layer is same as the number of input variables in our input data. The output layer has one single neuron which represents the predicted value. In deep neural network, we have m number of hidden layers and each one has k number of neurons. The input layer takes the input data and then feed them into the hidden layers. The output of the hidden layers are used as an input for the output layer. Given the input data X ($X = x_1, .. x_n$) and the output value Y, the prediction approach aims to find the estimated value Y_{est} using a simple approach is as follows:

$$Y_{est} = x_1 w_1 + x_2 w_2 + x_3 w_3 + b \tag{1}$$

Where w is the weight and b is the bias. Using a four-layer (one input, two hidden and one output) neural network as example, the Y_{est} can be calculated as follows:

$$Y_{est} = x_1^{(4)} = f(w_{11}^{(2)} x_1^{(2)} + w_{12}^{(2)} x_2^{(2)} + w_{13}^{(2)} x_3^{(2)} + b_1^{(2)}) + f(w_{21}^{(3)} x_1^{(3)} + w_{22}^{(3)} x_2^{(3)} + w_{23}^{(3)} x_3^{(3)} + b_1^{(3)}) \tag{2}$$

Where is the $x_1^{(4)}$ is the output of the network and f is the activation function. In this work, Keras [1] based on TensorFlow [2] is used to develop our predication model.

2.2 Optimisation Stage

This subsection presents the proposed memetic algorithm (MA) for shortest route problems. MA is a population-based metaheuristic that combines the strengths of local search algorithm with population-based metaheuristic to improve the convergence process [9,10]. In this paper, we used genetic algorithm (GA) and local search (LS) algorithm to form our proposed MA. GA is responsible for exploring new areas in the search space of solutions. LS is used to accelerate the search convergence. The pseudocode of the proposed MA is presented in is shown in (1). The overview of the process is given below followed by detailed description of these steps.

Our proposed algorithm starts from setting parameters, creating a population of solutions, calculating the quality of each solution and identifying the best solution in the current population. Next, the main steps of MA will iterate over a number of generations until the stopping criterion is met. At each generation, good solutions are selected from the population by the selection procedure. Then the crossover operator is applied on the selected solutions to generate new solutions. After that the mutation operator is applied on the new solutions by randomly changing them. A repair procedure is applied to check the feasibility of the generated solutions and fix the infeasible solutions as some solutions are no longer feasible. Afterwards a local search algorithm is invoked to iteratively

---

Algorithm 1: The proposed memetic algorithm

---

1 **Input**: Population size, PS, crossover rate, CR, mutation rate, MR, the maximum number of generations, Max_G and consecutive non-improvement iterations;

2 Set P_Sol=Randomly generate a population of solutions (PS) ;

3 Evaluate the population of solutions;

4 Set $iter$=0;

5 **while** $iter < Max\_G$ **do**

6   /*Selection procedure*/ ;

7   FirstParent= Select_one_individual (P_Sol);

8   SecondParent= Select_one_individual (P_Sol);

9   /*Check the crossover probability*/;

10   **if** $Rand[0,1] < CR$ **then**

11   /*Apply the crossover operator*/;

12         $Offsprings_{cx}=$ Crossover(FirstParent, SecondParent);

13   **end**

14   /*Check the mutation probability*/;

15   **if** $Rand[0,1] < MR$ **then**

16   /*Apply the mutation operator*/;

17         $Offsprings_{mutation}=$Mutate($Offsprings_{cx}$);

18   **end**

19   /*Apply local search to Offsprings */;

20   Offsprings=LS($Offsprings_{mutation}$);

21   Update the population (P_Sol);

22     $iter = iter + 1$;

23 **end**

24 **Output** Best solution found ;

---

improve the current solutions. If one of the stopping criteria is satisfied, then the whole MA procedure will stop and the current best solution will be returned as the output. Otherwise, the fitness of the current pool of solutions will be calculated. Then the population is updated since new solutions have been generated by crossover, mutation, repair procedure and local search. After that a new iteration starts from the selection procedure again.

Set Parameters. The main parameters of the proposed MA are initialised in this step. The proposed MA has several parameters. These are: population size, the number of generations, crossover rate, mutation rate and the number of non improvement iterations for the local search.

Initial Population. The initial population is randomly generated. Each solution is represented as one chromosome, e.g. one-dimensional array. Each cell of the array contains an integer number which represent the location.

Fitness Function. In this step, the fitness value of each solution based on the objective function is calculated. The better the fitness value is, the higher chance the solution will be selected to reproduce the next generation of solutions. For shortest route problems, the fitness is the total travel time between the origin and destination locations. Therefore, solution with shortest travel time is the better.

Selection Procedure. This step is responsible for selecting two solutions for producing the next generation. In this paper, we adopted the traditional tournament selection mechanism [4–6]. The tournament size is set to 2, indicating that each tournament has two solutions competing with each other. At each call, two solutions are randomly selected from the current population and the one with highest fitness value will be added to the reproduction pool.

Crossover. This step is responsible to generate new solutions by taking the selected solutions and mixes their genetic materials to produce new offsprings. In this paper, single-point crossover method is used which only swap genetic materials at one point [5,6]. It first finds a common point between source node and destination node and then all points behind the common point are exchanged between the two solutions, thus resulting in two offspring's.

Mutation. Mutation operator helps explore a large search space by producing some random changes in various solutions. In this paper, we used a one-point mutation operator [5]. Crossover point is randomly selected and then all points behind the selected mutation point are changed with a random sequence.

Repair Procedure. The aim of this step is to turn infeasible solutions into feasible ones. After crossover and mutation operations, the resulting solutions may become infeasible [5,6]. In this paper, The MA in our experiments has repair procedure that ensure all infeasible solutions are repaired.

Local Search Algorithm. The main role of this step is to improve the convergence process of the search process in order to attain higher quality solutions [9,10]. In this paper, the utilised local search algorithm is the steepest descent algorithm. Steepest descent algorithm is a simple variation of the gradient descent algorithm. It starts with a given solution as an input and uses a neighbourhood structure to move the search process to other possibly better solutions. It uses an "accept only" improving acceptance criterion whereby only a better solution will be used as a new starting point. Given s_i, It applies a neighbourhood structure to create s_n. Replace s_n with s_i if s_n is better. The pseudocode of the steepest descent algorithm is shown in (2).

Algorithm 2: Steepest descent algorithm

---
1 Set $MaxIter$; $Iter = 0$;
2 $s_i \leftarrow GenerateInitialSolution$;
3 **while** $Iter < MaxIter$ **do**
4     $s_n \leftarrow$ apply neighbourhood structure to s_i;
5   **if** $f(s_n) <= f(s_i)$ **then**
6         $s_i \leftarrow s_n$;
7   **end**
8 **end**
9 Return the best solution;

Stopping Condition. If the stopping condition is met, terminate the search process and return the best found solution. For our proposed memetic algorithm, it will stop if the maximum number of generations is reached. Otherwise, go to step 24.

Table 1. The parameter settings of the deep learning approach

Parameter	Value
Number of input parameters	12
Number of output parameters	1
Number of hidden layers	2
Hidden units in each layer	45, 35
Activation function	ReLU

Table 2. The parameter settings of the memetic algorithm

Parameter	Tested range	Suggested value
Number of generations	5−100	40 fitness evaluations
Population size	5−30	20
Crossover rate	0.1−0.9	0.4
Mutation rate	0.1−0.9	0.2
Consecutive non-improvement iterations	5−20	10

3 Experimental Settings

In this section, the parameter settings of the deep learning and the proposed algorithm are provided. The values of parameters were selected empirically based on our preliminary experiments, where we tested the deep learning model and the proposed algorithm with different parameter combination using different values for each parameter. The values of these parameters are determined one by one through manually changing the value of one parameter, while fixing the others. Then, the best values for all parameters are recorded. The final parameter values of the deep learning and the proposed algorithm are presented in Tables 1 and 2.

4 Experimental Results

This section is divided into two subsections. The first examines the performance comparison between the deep learning approach and other machine learning models (Sect. 4.1). The second assesses the benefit of incorporating the proposed components on search performance (Sect. 4.2).

4.1 Deep Learning and Machine Learning Results

In this paper, we have implemented a number of machine learning models and the results of these models are compared with the deep learning model proposed in this work. We have tested the followings methods: XGBoost, Random forest, Artificial neural network, Multivariate regression.

The root-mean squared-error (RMSE) was used as an evaluation metric. Table 3 shows the results in term of RMSE on the NYC Taxi dataset.

Table 3. Comparing our deep prediction model with other machine learning models in term of RMSE

Model	RMSE
XGBoost	24.06
Random forest	21.34
Artificial neural network	70.21
Multivariate regression	27.19
Our deep learning model	**11.01**

In the table, the best obtained result is highlighted in bold. From Table 3, it can be seen that our deep prediction model is superior to the other machine learning models in term of RMSE. The best values with the lowest RMSE is 11.01 achieved by our approach, followed by 21.34 from random forest, 24.06 from XGBoost, 27.19 from multivariate regression and 70.21 from artificial neural network.

This good result can be attributed to the factor that deep learning consider all input features and then utilise best ones through the internal learning process. On the other hand, other machine learning methods require feature engineering step to identify the best subset of features which is a very time consuming and needs a human expert.

4.2 The Proposed Memetic Algorithm Results

This section evaluates the effectiveness of the machine learning models and the proposed memetic algorithm. To this end, genetic algorithm (GA) and memetic algorithm (MA) with different machine learning models are tested and compared against each other. These are: GA with XGBoost, GA with random forest, GA with artificial neural network, GA with multivariate regression, GA with deep prediction model, MA with XGBoost, MA with random forest, MA with artificial neural network, MA with multivariate regression and MA with deep prediction model. The main aim is to evaluate the benefit of using our deep prediction model and local search algorithm within MA.

Table 4. Results of the GA and MA with different prediction models (Part I)

Algorithm	Number of locations			
	500		1000	
	Best	std	Best	std
GA with XGBoost	5933.61	243.76	7671.84	136.72
GA with random forest	5844.78	194.14	7533.67	153.98
GA with artificial neural network	9401.86	215.61	10154.01	707.65
GA with multivariate regression	6217.77	227.02	7907.34	256.39
GA with deep prediction model	4602.03	153.69	6837.17	142.24
MA with XGBoost	5774.54	116.07	7405.12	120.02
MA with random forest	5637.81	96.16	7360.04	119.18
MA with artificial neural network	9293.44	196.01	10093.88	693.31
MA with multivariate regression	6171.63	138.09	7499.22	155.03
MA with deep prediction model	**3234.11**	**71.3**	**6411.72**	**127.69**

To ensure a fair comparison between the compared algorithms, the initial solution, number of runs, stopping condition and computer resources are the same for all instances. All algorithms were executed for 30 independent runs over all instances. We also used 4 instances with a different number of locations ranging between 500 and 2000 locations, which can be seen as small, medium, large and very large.

The computational comparisons of the above algorithms are presented in Tables 4 and 5. The comparison is in terms of the best cost (travel time) and standard deviation (std) for each number of locations, where the lower the better. The best results are highlighted in bold. A close scrutiny of Tables 4 and 5 reveals that, of all the instances, the proposed MA algorithm with deep learning approach outperforms the other algorithms in all instances. From Tables 4 and 5, we can make the following observations:

- GA with deep prediction model obtained better results when compared to GA with all other prediction models across all instances.

Table 5. Results of the GA and MA with different prediction models (Part II)

| Algorithm | Number of locations | | | |
| | 1500 | | 2000 | |
	Best	std	Best	std
GA with XGBoost	12908.11	998.91	13634.01	501.94
GA with random forest	12833.67	1076.14	13593.47	529.61
GA with artificial neural network	15751.01	1703.07	15882.33	992.84
GA with multivariate regression	13012.41	1047.41	13809.41	591.56
GA with deep prediction model	10571.09	607.99	12946.67	388.14
MA with XGBoost	12641.08	613.12	13436.42	481.08
MA with random forest	12607.91	684.74	13309.01	503.44
MA with artificial neural network	15603.17	980.01	15206.93	755.31
MA with multivariate regression	12988.14	721.36	13498.96	564.05
MA with deep prediction model	**9561.36**	**131.29**	**12721.45**	**129.15**

- GA with deep learning obtained better results when compared to MA with all other machine learning models (apart from MA with deep learning) across all instances.
- MA with deep learning obtained better results when compared to GA and MA with all other machine learning models across all instances.

This justifies the benefit of using deep learning approach to predict the travel time and the proposed memetic algorithm to exploit the current search space around the given solution.

5   Conclusion

In this study, we proposed a novel two-stage approach for finding the shortest route under dynamic environment where travel time changes. Firstly, we developed a deep learning method to predict the travel time between the origin and destination. We also added the weather conditions into the input to demonstrate that our approach can predict the travel time more accurately. Secondly, a customised memetic algorithm is developed to find shortest route using the predicted travel time. The effectiveness of the proposed method has been evaluated on New York City taxi dataset. The obtained results lead to our conclusion that the proposed two-stage shortest route is effective, compared with conventional methods. The proposed deep prediction model and memetic algorithm are beneficial.

Acknowledgements. This work is supported by the Smarter Cities and Suburbs Grant from the Australian Government and the Mornington Peninsula Shire Council.

References

1. Chollet, F.: Keras (2015). htttps://keras.io. Accessed 24 Dec 2019
2. Devin, M. et al.: TensorFlow: large-scale machine learning on heterogeneous systems (2015). https://www.tensorflow.org. Accessed 24 Dec 2019
3. Fu, T., Lee, W.: DeepIST: deep image-based spatio-temporal network for travel time estimation. In: Proceedings of the 28th ACM International Conference on Information and Knowledge Management, pp. 69–78. ACM (2019)
4. Goldberg, D.E., Deb, K.: A comparative analysis of selection schemes used in genetic algorithms. Found. Genet. Algorithms **1**, 69–93 (1991)
5. Goldberg, D.E., Holland, J.H.: Genetic algorithms and machine learning. Mach. Learn. **3**(2), 95–99 (1988)
6. Holland, J.H.: Genetic algorithms. Sci. Am. **267**(1), 66–73 (1992)
7. Lan, W., Xu, Y., Zhao, B.: Travel time estimation without road networks: an urban morphological layout representation approach. arXiv preprint arXiv:1907.03381 (2019)
8. Li, Y., Fu, K., Wang, Z., Shahabi, C., Ye, J., Liu, Y.: Multi-task representation learning for travel time estimation. In: Proceedings of the 24th ACM SIGKDD International Conference on Knowledge Discovery & Data Mining, pp. 1695–1704. ACM (2018)
9. Moscato, P., et al.: On evolution, search, optimization, genetic algorithms and martial arts: towards memetic algorithms. Caltech Concurrent Comput. Program, C3P Rep. **826**, 1989 (1989)
10. Neri, F., Cotta, C.: Memetic algorithms and memetic computing optimization: a literature review. Swarm Evol. Comput. **2**, 1–14 (2012)
11. Qin, K.K., Shao, W., Ren, Y., Chan, J., Salim, F.D.: Solving multiple travelling officers problem with population-based optimization algorithms. Neural Comput. Appl. 1–27 (2019). https://doi.org/10.1007/s00521-019-04237-2
12. Rahaman, M.S., Hamilton, M., Salim, F.D.: Predicting imbalanced taxi and passenger queue contexts in airport. In: PACIS, p. 172 (2017)
13. Rahaman, M.S., Hamilton, M., Salim, F.D.: Queue context prediction using taxi driver knowledge. In: Proceedings of the Knowledge Capture Conference, p. 35. ACM (2017)
14. Rahaman, M.S., Hamilton, M., Salim, F.D.: Coact: a framework for context-aware trip planning using active transport. In: 2018 IEEE International Conference on Pervasive Computing and Communications Workshops (PerCom Workshops), pp. 645–650. IEEE (2018)
15. Rahaman, M.S., Hamilton, M., Salim, F.D.: Using big spatial data for planning user mobility. In: Sakr, S., Zomaya, A. (eds.) Encyclopedia of Big Data Technologies, pp. 1–6. Springer, Cham (2018). https://doi.org/10.1007/978-3-319-77525-8
16. Rahaman, M.S., Mei, Y., Hamilton, M., Salim, F.D.: CAPRA: a contour-based accessible path routing algorithm. Inf. Sci. **385**, 157–173 (2017)
17. Rahaman, M.S., Ren, Y., Hamilton, M., Salim, F.D.: Wait time prediction for airport taxis using weighted nearest neighbor regression. IEEE Access **6**, 74660–74672 (2018)
18. Sabar, N.R., Turky, A., Song, A., Sattar, A.: Optimising deep belief networks by hyper-heuristic approach. In: 2017 IEEE Congress on Evolutionary Computation (CEC), pp. 2738–2745. IEEE (2017)
19. Sabar, N.R., Turky, A., Song, A., Sattar, A.: An evolutionary hyper-heuristic to optimise deep belief networks for image reconstruction. Appl. Soft Comput. 105510 (2019). https://doi.org/10.1016/j.asoc.2019.105510

20. Song, H., Qin, A.K., Salim, F.D.: Evolutionary model construction for electricity consumption prediction. Neural Comput. Appl. 1–18. https://doi.org/10.1007/s00521-019-04310-w

21. Song, H., Qin, A.K., Salim, F.D.: Multi-resolution selective ensemble extreme learning machine for electricity consumption prediction. In: Liu, D., Xie, S., Li, Y., Zhao, D., El-Alfy, E.-S.M. (eds.) ICONIP 2017. LNCS, vol. 10638, pp. 600–609. Springer, Cham (2017). https://doi.org/10.1007/978-3-319-70139-4_61

22. Wang, D., Zhang, J., Cao, W., Li, J., Zheng, Y.: When will you arrive? Estimating travel time based on deep neural networks. In: Thirty-Second AAAI Conference on Artificial Intelligence (2018)

23. Wang, H., Yang, H.: Ridesourcing systems: a framework and review. Transp. Res. Part B Methodol. **129**, 122–155 (2019)

24. Wang, Z., Fu, K., Ye, J.: Learning to estimate the travel time. In: Proceedings of the 24th ACM SIGKDD International Conference on Knowledge Discovery & Data Mining, pp. 858–866. ACM (2018)

A Relaxation Algorithm for Optimal Control Problems Governed by Two-Dimensional Conservation Laws

Michael Herty[1], Loubna Salhi[2(✉)], and Mohammed Seaid[3,4]

[1] RWTH Aachen University, Templegraben 55, 52056 Aachen, AG, Germany
herty@mathc.rwth-aachen.de
[2] Laboratory of Complex Systems Engineering & Human Systems,
University Mohammed VI Polytechnic, 43150 Benguerir, Morocco
loubna.salhi@um6p.ma
[3] Department of Engineering, University of Durham, South Road,
Durham DH1 3LE, UK
m.seaid@durham.ac.uk
[4] International Water Research Institute, University Mohammed VI Polytechnic,
43150 Benguerir, Morocco

Abstract. We develop a class of numerical methods for solving optimal control problems governed by nonlinear conservation laws in two space dimensions. The relaxation approximation is used to transform the nonlinear problem to a semi-linear diagonalizable system with source terms. The relaxing system is hyperbolic and it can be numerically solved without need to either Riemann solvers for space discretization or a non-linear system of algebraic equations solvers for time discretization. In the current study, the optimal control problem is formulated for the relaxation system and at the relaxed limit its solution converges to the relaxed equation of conservation laws. An upwind method is used for reconstruction of numerical fluxes and an implicit-explicit scheme is used for time stepping. Computational results are presented for a two-dimensional inviscid Burgers problem.

Keywords: Optimal control problems · Conservation laws · Relaxation approximation · Implicit-explicit schemes

1 Introduction

In many applications, optimal control problems consist of a class of differential equations whose evolution and the behavior of their solutions can be controlled by involving external control laws. In the current study, we are interested in optimal control problems subject to the following two-dimensional nonlinear conservation law

$$\partial_t u + \nabla \cdot \mathbf{F}(u) = 0, \qquad (x,y) \in \Omega, \quad t > 0, \tag{1a}$$

$$u(0,x,y) = u_0(x,y),$$

© Springer Nature Switzerland AG 2020
V. V. Krzhizhanovskaya et al. (Eds.): ICCS 2020, LNCS 12141, pp. 122–135, 2020.
https://doi.org/10.1007/978-3-030-50426-7_10

where Ω is an open bounded domain in $\mathbb{R}^2$, (x, y) the space coordinates, t the time, $u(t, x, y)$ is the control function, $u_0(x, y)$ the initial state and the flux $\mathbf{F}(u) = (f(u), g(u))^T$, with $f(u)$ and $g(u)$ are nonlinear functions. In practice, optimal control problems require minimizing a cost functional $J(u(T, x, y); u_d(x, y))$ based on the least-square method that associates a cost value to each possible behavior. Thus, the problem statement is

$$\min_{u_0} J(u(T, x, y); u_d(x, y)) := \min_{u_0} \frac{1}{2} \iint_{\Omega} (u(T, x, y) - u_d(x, y))^2 \, dx dy, \quad (1b)$$

subject to the conservation law (1a). In (1b), $u_d(x, y)$ is the desired state at the final time T. Optimal control problems of type (1b) have received growing attention in both theoretical and numerical studies over recent decades. In most of these studies, control problems governed by hyperbolic equations have been less extensively treated compared to elliptic and parabolic control problems. This is mainly due to the fact that the semi-group generated by the hyperbolic conservation law is non-differentiable in L^1 whereas its domain of definition is an L^1 closed subset of BV. In the case of nonlinear conservation laws in one space dimension, a differential structure on general BV solutions has been presented and discussed in [4,19] among others. The first-order optimality conditions for hyperbolic systems have been introduced in [5] based on the derived calculus. It turned out that the resulting adjoint equations are non-conservative which fail to recover stable solutions for problems with shocks. In [13,14,19], numerical results for one-dimensional scalar problems with distributed control have been presented. More results for the case of a one-dimensional linear hyperbolic systems can be found in [8,10,15]. In [11], a TVD Runge-Kutta method for the time discretization of such problems has been employed. It was shown that requiring high stability for both the discrete and adjoint states is too strong, limiting the method to first-order, regardless of the number of stages used in the method. Using the same discretization, authors in [11] have studied other conditions for the discrete adjoint such that the numerical approximation is of the best possible order. In [1], the emphasis was placed on high-order linear multistep schemes for the time discretization of adjoint equations arising within optimal control problems. The authors reported that the so-called Adams methods may reduce to the first-order accuracy and that only BDF schemes may be used as higher order discretization for the hyperbolic relaxation systems in combination with a Lagrangian scheme. Theoretical and numerical methods using finite difference schemes combined with an immersed boundary method have been developed in [9] for a special class of optimal control problems namely, problems involving the shallow water equations and a geometric parameter to be optimized in the terminal cost. More recently, theoretical studies including a posteriori error estimates have been carried out for numerical schemes to solve multi-dimensional problems, based on adjoint equations, see for instance [16,18].

In the present work, we are interested in developing numerical algorithms for control problems of two-dimensional nonlinear conservation laws to achieve numerical stability without need to inclusion of extra artificial diffusion in the problem under study. For this purpose we consider the relaxation approximation of nonlinear conservation laws in the same manner as introduced in [12]. This approach approximates the nonlinear problem to semi-linear system with linear characteristic speeds, while preserving the hyperbolic structure on the expense of an additional equation and stiff source terms. Thus, the resulting relaxation system is semi-linear which allows for a Riemann-solver free treatment. The relaxation methods have been investigated by many authors, see [3] among others. First studies of relaxation systems with respect to control problems have been reported in [2] in case of one-dimensional scalar conservation laws. Numerical results are still very limited in the multi-dimensional cases and we therefore restrict ourselves to a numerical study including a first-order relaxation approximation. For the space discretization, we consider an upwind reconstruction of the numerical fluxes and an implicit-explicit method is used for the time integration.

The remainder of this paper is structured as follows. In Sect. 2, the relaxation approximation for the coupled optimal control problem and the nonlinear conservation laws is formulated. The space and time discretizations along with the approximation procedure of the solution is presented in Sect. 3. In Sect. 4, numerical results are presented for a test example of inviscid Burgers equation. Section 5 contains concluding remarks.

2   Relaxation Approximations for Conservation Laws

Following [12], the relaxation approximation for (1a) allows to construct a corresponding linear hyperbolic system with a stiff source term that approximates the original problem with a small dissipative correction. Thus, the relaxation associated with (1a) reads

$$\partial_t u + \partial_x v + \partial_y w = 0,$$
$$\partial_t v + a^2 \partial_x u = -\frac{1}{\tau}\left(v - f(u)\right), \tag{2a}$$
$$\partial_t w + b^2 \partial_y u = -\frac{1}{\tau}\left(w - g(u)\right),$$

where τ is a small positive parameter that measures the relaxation rate, v and w are the relaxation variables, a^2 and b^2 are the characteristic speeds satisfying the sub-characteristic condition [12]

$$\frac{f'(u)^2}{a^2} + \frac{g'(u)^2}{b^2} \leq 1, \qquad \forall\, u. \tag{2b}$$

The initial conditions for the relaxation system (2a) are selected as

$$u(0, x, y) = u_0, \quad v(0, x, y) = f(u_0), \quad w(0, x, y) = g(u_0). \tag{2c}$$

It is clear that, when τ tends to 0, the relaxation system (2) converges to the system of conservation law (1a). Note that the main advantage of numerically solving the relaxation system (2) over the original conservation law (1a) lies in the special structure of the linear characteristic fields and localized lower order terms. Indeed, the linear hyperbolic nature of (2) allows to approximate its solution easily by underresolved stable numerical discretization that uses neither Riemann solvers spatially nor nonlinear system of algebraic equations solvers temporally. Hence, using the relaxation approximation, the optimal control problem (1b) becomes

$$\min_{u_0} \frac{1}{2} \iint_\Omega \left(\left(u(T,x,y) - u_d(x,y) \right)^2 + \left(v(T,x,y) - f(u_d(x,y)) \right)^2 + \right.$$

$$\left. \left(w(T,x,y) - g(u_d(x,y)) \right)^2 \right) dx dy, \qquad (3)$$

subject to the relaxation system (2a). Notice that a formal adjoint calculus leads to a first-order optimality conditions for the function u_0. The calculations are rigorous provided that the solutions have sufficient regularity which however in general is not the case. Hence, we formulate the adjoint equations for the system (2) as

$$-\partial_t p - a^2 \partial_x q - b^2 \partial_y r = \frac{1}{\tau} \left(qf'(u) + rg'(u) \right),$$

$$-\partial_t q - \partial_x p = -\frac{1}{\tau} q, \qquad (4a)$$

$$-\partial_t r - \partial_y p = -\frac{1}{\tau} r,$$

with terminal conditions given by

$$p(T,x,y) = u(T,x,y) - u_d(x,y), \quad q(T,x,y) = v(T,x,y) - f(u_d(x,y)),$$
$$r(T,x,y) = w(T,x,y) - g(u_d(x,y)). \qquad (4b)$$

It should be stressed that the adjoint equations (4) have to be solved backwards in time and the gradient of the reduced cost functional is defined as

$$p(0,x,y) + q(0,x,y)f'\left(u_0(x,y)\right) + r(0,x,y)g'\left(u_0(x,y)\right) = 0. \qquad (5)$$

Again, when τ tends to 0, the system (4) converges to the adjoint problem associated with the conservation law (1a). Then, from the second and third equations in (4a), an expansion in terms of τ gives

$$q = \tau \partial_x p + \mathcal{O}(\tau^2), \qquad r = \tau \partial_y p + \mathcal{O}(\tau^2).$$

Inserting these terms in the first equation of (4a) leads to

$$-\partial_t p - f'(u)\partial_x p - g'(u)\partial_y p = \tau \left(a^2 \partial_{xx} p + b^2 \partial_{yy} p \right), \qquad (6)$$

which is a viscous approximation to the formal adjoint of (4). Note that the gradient eventually vanishes at the minimum of the cost functional. Since u might develop discontinuities we will have to scope with discontinuous derivatives of the flux functions $f'(u)$ and $g'(u)$. However, since we use the relaxation approximation, the derivative functions $f'(u)$ and $g'(u)$ appear as source terms and not as a discontinuous transport coefficient as in (6). This problem has been investigated for one-dimensional problems in [19]. However, as pointed out in [2,19], the problem reappears in the small τ limit. In the one-dimensional case it can be shown that the first-order relaxation discretization converges to the reversible solution of a transport equation with discontinuous coefficient. Here, we focus on a numerical study of the optimality system (1a). Using the characteristic variables

$$v^\pm = v \pm au, \qquad w^\pm = w \pm bu,$$

an equivalent system associated with (2) can be reformulated as

$$\partial_t v^\pm \pm a^2 \partial_x v^\pm = -\frac{1}{\tau}\left(\frac{v^+ + v^-}{2} - f\left(\frac{v^+ - v^-}{2a}\right)\right),$$

$$\partial_t w^\pm \pm b^2 \partial_y w^\pm = -\frac{1}{\tau}\left(\frac{w^+ + w^-}{2} - g\left(\frac{w^+ - w^-}{2b}\right)\right). \tag{7}$$

The adjoint equations in characteristic form are therefore given by

$$-\partial_t s^\pm \mp a\partial_x s^\pm = -\frac{1}{\tau}\left(\frac{s^+ + s^-}{2} \mp \frac{s^+ + s^-}{2a} f'\left(\frac{(v + au) - (v - au)}{2a}\right)\right),$$

$$-\partial_t o^\pm \mp b\partial_y o^\pm_y = -\frac{1}{\tau}\left(\frac{o^+ + o^-}{2} \mp \frac{o^+ + o^-}{2b} g'\left(\frac{(v + bu) - (v - bu)}{2b}\right)\right). \tag{8}$$

This system is equivalent to a spatial splitting approximation of the adjoint equations (4). Introducing

$$q = s^+ + s^-, \qquad p = a\left(s^+ - s^-\right) \tag{9a}$$

we obtain from the equations in s that the solutions (p, q) satisfy

$$-\partial_t q - \partial_x p = -\frac{1}{\tau} q, \qquad -\partial_t p - a^2 \partial_x q = +\frac{1}{\tau}\left(q f'(u)\right). \tag{9b}$$

Similarly, for

$$r = o^+ + o^-, \qquad p = b(o^+ - o^-). \tag{9c}$$

we have

$$-\partial_t r - \partial_x p = -\frac{1}{\tau} r, \qquad -\partial_t p - b^2 \partial_x r = +\frac{1}{\tau}\left(r g'(u)\right). \tag{9d}$$

Hence, the formulation (9) is precisely the spatial splitting applied to (4). Therefore, the adjoints in characteristic form are the same as the adjoint of the characteristic form when applying a dimensional splitting in the spatial variable. For the optimize-then-discretize approach discussed below it is therefore sufficient to state the discretization of the forward equations in characteristic form. A rigorous discussion of the relation between discrete adjoints, the characteristic variables and higher-order schemes can be found in [2] in the case of one-dimensional scalar advection equations.

3   Numerical Solution of Optimal Control Problems

Relaxation schemes are in fact a combination of non-oscillatory upwind space discretization and an implicit-explicit time integration of the resulting semi-discrete system, see for instance [3, 12]. The fully discrete system of the equations (2a) is referred to as a *relaxing system*, while that of the limiting system as the relaxation rate τ tends to zero is called a *relaxed system*. In this section, we formulate the space and time discretizations used for the numerical solution of optimal control problems and also formulate the algorithm used for the discrete gradient.

3.1   Space and Time Discretizations

For the space discretization of the equations (2a), we cover the spatial domain with rectangular cells $C_{i,j} := [x_{i-\frac{1}{2}}, x_{i+\frac{1}{2}}] \times [y_{j-\frac{1}{2}}, y_{j+\frac{1}{2}}]$ of uniform sizes Δx and Δy for simplicity. The cells, $C_{i,j}$, are centered at $(x_i = i\Delta x, y_j = j\Delta y)$. We use the notations $\omega_{i\pm\frac{1}{2},j} := \omega(x_{i\pm\frac{1}{2}}, y_j, t)$, $\omega_{i,j\pm\frac{1}{2}} := \omega(x_i, y_{j\pm\frac{1}{2}}, t)$ and

$$\omega_{i,j} := \frac{1}{\Delta x}\frac{1}{\Delta y}\int_{x_{i-\frac{1}{2}}}^{x_{i+\frac{1}{2}}}\int_{y_{j-\frac{1}{2}}}^{y_{j+\frac{1}{2}}} \omega(x, y, t)dxdy,$$

to denote the point-values and the approximate cell-average of a generic function ω at $(x_{i\pm\frac{1}{2}}, y_j, t_n)$, $(x_i, y_{j\pm\frac{1}{2}}, t_n)$, and (x_i, y_j, t_n), respectively. We define the following finite differences

$$\mathcal{D}_x\omega_{i,j} := \frac{\omega_{i+\frac{1}{2},j} - \omega_{i-\frac{1}{2},j}}{\Delta x}, \qquad \mathcal{D}_y\omega_{i,j} := \frac{\omega_{i,j+\frac{1}{2}} - \omega_{i,j-\frac{1}{2}}}{\Delta y}. \tag{10}$$

Then, the semi-discrete approximation of (2a) is

$$\frac{du_{i,j}}{dt} + \mathcal{D}_x v_{i,j} + \mathcal{D}_y w_{i,j} = 0,$$

$$\frac{dv_{i,j}}{dt} + a^2\mathcal{D}_x u_{i,j} = -\frac{1}{\tau}\left(v_{i,j} - f\left(u_{i,j}\right)\right), \tag{11}$$

$$\frac{dw_{i,j}}{dt} + b^2\mathcal{D}_y u_{i,j} = -\frac{1}{\tau}\left(w_{i,j} - g\left(u_{i,j}\right)\right).$$

Similarly, the semi-discrete approximation of the adjoint equations (4a) is

$$-\frac{dp_{i,j}}{dt} - a^2\mathcal{D}_x q_{i,j} - b^2\mathcal{D}_y r_{i,j} = 0,$$

$$-\frac{dq_{i,j}}{dt} - \mathcal{D}_x p_{i,j} = -\frac{1}{\tau}q_{i,j}, \tag{12}$$

$$-\frac{dr_{i,j}}{dt} - \mathcal{D}_y p_{i,j} = -\frac{1}{\tau}r_{i,j}.$$

Most relaxation schemes can be described as fractional step methods, in which the relaxation step is just a projection of the system into the local equilibrium.

The fully-discrete formulation of systems (11) and (12) can be obtained by the well-established IMEX methods, see for instance [17]. Indeed, the special structure of the nonlinear terms in (11) and (12) makes it trivial to evolve the flux terms explicitly and the stiff source terms implicitly.

The semi-discrete formulations (11) or (12) can be rewritten in common ordinary differential equations notation as

$$\frac{d\mathcal{Y}}{dt} = \mathcal{F}(\mathcal{Y}) - \frac{1}{\tau}\mathcal{G}(\mathcal{Y}), \tag{13}$$

where the time-dependent vector functions are defined accordingly for the forward problem (11) or for the backward problem (12). Due to the presence of stiff terms in (13), one can not use fully explicit schemes to integrate the equations (13), particularly when τ tends to 0. On the other hand, integrating the equations (13) by fully implicit scheme, either linear or nonlinear algebraic equations have to be solved at every time step of the computational process. To find solutions of such systems is computationally very demanding. In this paper we consider an alternative approach based on the implicit-explicit (IMEX) Euler method. The non stiff stage of the splitting for $\mathcal{F}$ is straightforwardly treated by an explicit scheme, while the stiff stage for $\mathcal{G}$ is approximated by a diagonally implicit scheme.

Let $\Delta t = t_{n+1} - t_n$ be the time step and $\mathcal{Y}^n$ denotes the approximate solution at $t = n\Delta t$. We formulate the first-order IMEX scheme for the forward system (13) as

$$K_1 = \mathcal{Y}^n - \frac{\Delta t}{\tau}\mathcal{G}(K_1),$$

$$\mathcal{Y}^{n+1} = \mathcal{Y}^n + \Delta t\mathcal{F}(K_1) - \frac{\Delta t}{\tau}\mathcal{G}(K_1). \tag{14}$$

For the backward system (13), the IMEX scheme is implemented as

$$K_1 = \mathcal{Y}^{n+1} + \Delta t\mathcal{F}(K_1),$$

$$\mathcal{Y}^n = \mathcal{Y}^{n+1} + \Delta t\mathcal{F}(K_1) - \frac{\Delta t}{\tau}\mathcal{G}(K_1). \tag{15}$$

Note that, using the above relaxation scheme neither linear algebraic equation nor nonlinear source terms can arise. In addition the relaxation schemes are stable independently of τ, so that the choice of Δt is based only on the usual CFL condition

$$\text{CFL} = \max\left(\frac{\Delta t}{\delta}, a^2\frac{\Delta t}{\Delta x}, b^2\frac{\Delta t}{\Delta y}\right) \leq 1, \tag{16}$$

where δ denotes the maximum cell size, $\delta = \max(\Delta x, \Delta y)$. For the space discretization, a first-order upwind scheme is applied to the characteristic variables in (11) to obtain the numerical fluxes as

$$(v + au)_{i+\frac{1}{2},j} = (v + au)_{i,j}, \quad (v - au)_{i+\frac{1}{2},j} = (v - au)_{i+1,j},$$

$$(w + bu)_{i,j+\frac{1}{2}} = (w + bu)_{i,j}, \quad (w - bu)_{i,j+\frac{1}{2}} = (w - bu)_{i,j+1}. \tag{17}$$

Thus, a first-order reconstruction of the numerical fluxes in the forward problem (11) yields

$$
\begin{aligned}
u_{i+\frac{1}{2},j} &= \frac{u_{i,j} + u_{i+1,j}}{2} - \frac{v_{i+1,j} - v_{i,j}}{2a}, \\
u_{i,j+\frac{1}{2}} &= \frac{u_{i,j} + u_{i,j+1}}{2} - \frac{w_{i,j+1} - w_{i,j}}{2b}, \\
v_{i+\frac{1}{2},j} &= \frac{v_{i,j} + v_{i+1,j}}{2} - a\frac{u_{i+1,j} - u_{i,j}}{2}, \\
w_{i,j+\frac{1}{2}} &= \frac{w_{i,j} + w_{i,j+1}}{2} - b\frac{u_{i,j+1} - u_{i,j}}{2}.
\end{aligned}
\tag{18}
$$

The numerical fluxes in the backward problem (12) are obtained by applying first order upwind scheme to the characteristic variables

$$
\begin{aligned}
(p + aq)_{i+\frac{1}{2},j} &= (p + aq)_{i,j}, \quad (p - aq)_{i+\frac{1}{2},j} = (p - aq)_{i+1,j}, \\
(p + br)_{i,j+\frac{1}{2}} &= (p + br)_{i,j}, \quad (p - br)_{i,j+\frac{1}{2}} = (p - br)_{i,j+1}.
\end{aligned}
\tag{19}
$$

Thus, a first-order reconstruction of the numerical fluxes in the backward problem (12) yields

$$
\begin{aligned}
p_{i+\frac{1}{2},j} &= -\frac{p_{i,j} + p_{i+1,j}}{2} - a\frac{q_{i+1,j} - q_{i,j}}{2}, \\
p_{i,j+\frac{1}{2}} &= -\frac{p_{i,j} + p_{i,j+1}}{2} - b\frac{r_{i,j+1} - r_{i,j}}{2}, \\
q_{i+\frac{1}{2},j} &= -\frac{q_{i,j} + q_{i+1,j}}{2} - \frac{p_{i+1,j} - p_{i,j}}{2a}, \\
r_{i,j+\frac{1}{2}} &= -\frac{r_{i,j} + r_{i,j+1}}{2} - \frac{p_{i,j+1} - p_{i,j}}{2b}.
\end{aligned}
\tag{20}
$$

In this study, the characteristic speeds a and b in the relaxation systems (2) and (4) are calculated locally at every cell as

$$
a_{i+\frac{1}{2},j} = \max_{u \in \left\{ u_{i+\frac{1}{2},j}^{x,-}, u_{i+\frac{1}{2},j}^{x,+} \right\}} \left| f'(u) \right|, \qquad b_{i,j+\frac{1}{2}} = \max_{u \in \left\{ u_{i,j+\frac{1}{2}}^{y,-}, u_{i,j+\frac{1}{2}}^{y,+} \right\}} \left| g'(u) \right|.
\tag{21}
$$

It is worth saying that, larger a and b values usually add more numerical dissipation.

3.2   Discrete Gradient and Solution Procedure

The implementation of the iterative optimization along with the Eulerian-Lagrangian numerical approach used in the implementation are performed in the same way as detailed in [7]. Thus, starting from the basic optimal control problem formulated as follows: Given a terminal state $u_d(x, y)$, find an initial datum $u_0(x, y)$ which by time $t = T$ will either evolve into $u(T, x, y) = u_d(x, y)$ or will be as close as possible to u_d in the L^2-norm. To solve the problem iteratively, we implement the Algorithm 1 and generate a sequence of solutions $u_0^{(m)}(x, y)$, with $m = 0, 1, 2, \ldots$. It should also be pointed out that, the solution $u(t, x, y)$

Algorithm 1: Optimization procedure used in the present study.

$u_0^{(0)}(x,y)$: Chosen initial guess
$u_d(x,y)$: Desired solution
ε: Given tolerance
T: Final simulation time

- Solve the problem (2) subject to $u(0,x,y) = u_0^{(0)}(x,y)$, $v(0,x,y) = f\left(u_0^{(0)}(x,y)\right)$
  and $w(x,y,0) = g\left(u_0^{(0)}(x,y)\right)$ forward in time from $t=0$ to $t=T$ by using the
  relaxation method to obtain $u^{(0)}(T,x,y)$, $v^{(0)}(T,x,y) = f\left(u_0^{(0)}(T,x,y)\right)$ and
  $w^{(0)}(T,x,y) = g\left(u_0^{(0)}(T,x,y)\right)$.

for $m = 0,1,2,\ldots$ **do**

- Compute the cost function $J^{(m)} = \dfrac{1}{2}\iint\limits_{\Omega}\left(\left(u^{(m)}(x,y,T) - u_d(x,y)\right)^2 +\right.$

  $\left.\left(v^{(m)}(x,y,T) - f(u_d(x,y))\right)^2 + \left(w^{(m)}(x,y,T) - g(u_d(x,y))\right)^2\right)dxdy$

  while $J^{(m)} > \varepsilon$ *or* $\left|J^{(m)} - J^{(m-1)}\right| > \varepsilon$ **do**

  - Solve the linear system (4a) backward in time from $t=T$ to $t=0$ using
    the relaxation method to obtain $p^{(m)}(0,x,y)$, $q^{(m)}(0,x,y)$ and $r^{(m)}(0,x,y)$.
  - Update the control u_0, v_0 and w_0 using either a gradient descent or
    quasi-Newton method as described in [7].
  - Solve the problem (2) subject to $u(0,x,y) = u_0^{(m+1)}(x,y)$,
    $v(0,x,y) = f\left(u_0^{(m+1)}(x,y)\right)$ and $w(0,x,y) = g\left(u_0^{(m+1)}(x,y)\right)$ forward
    in time from $t=0$ to $t=T$ by using the relaxation method to obtain
    $u^{(m+1)}(T,x,y)$, $v^{(m+1)}(T,x,y) = f\left(u_0^{(m+1)}(T,x,y)\right)$ and
    $w^{(m+1)}(T,x,y) = g\left(u_0^{(m+1)}(T,x,y)\right)$.

  end

  - Set m:= m + 1.

end

---

does not have to be stored during the iterations by using the developed method. In addition, although Algorithm 1 is similar to the continuous approach used in [6], the focus is on the proposed numerical method to solve the problem (2) and thus, we do not need an approximation to the generalized tangent vectors to improve the gradient descent method.

4   Results for an Inviscid Burgers Problem

To examine the performance of the relaxation algorithm to solve optimal control we present numerical results for a two-dimensional inviscid Burgers problem. In

all the computational results presented in this section, the characteristic speeds
a and b are locally chosen as in (21), the CFL number is fixed to 0.5 and time
steps Δt are calculated according to the condition (16). Here, the flux functions
are defined by

$$f(u) = \frac{u^2}{2} \quad \text{and} \quad g(u) = \frac{u^2}{2}. \tag{22}$$

The optimal control problems are solved in the domain $[0,1] \times [0,1]$ subject to
period boundary conditions and equipped with the following initial data

$$u(0, x, y) = \sin^2(\pi x) \sin^2(\pi y).$$

We solve the optimization problem for terminal time $T = 0.2$ using a relaxation
rate $\tau = 10^{-6}$ on three different meshes with 100×100, 200×200 and 400×400
control volumes. For each of these runs, we display the initial data u_0, reference
solution and the optimized solution u_t along with the gradient of the reduced
cost functional defined in (5).

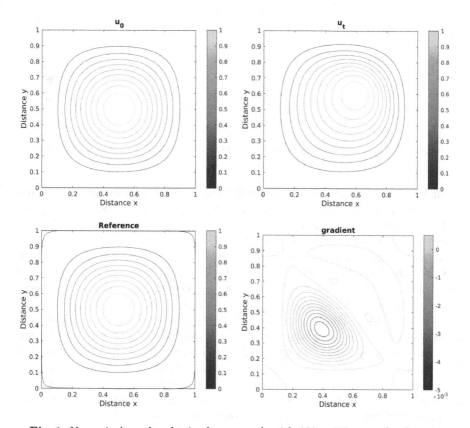

Fig. 1. Numerical results obtained on a mesh with 100×100 control volumes.

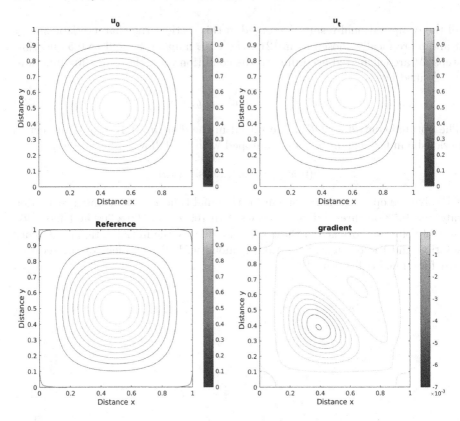

Fig. 2. Numerical results obtained on a mesh with 200×200 control volumes.

In Fig. 1 we present numerical results obtained on a mesh with 100×100 control volumes. Those results obtained on meshes with 200×200 and 400×400 control volumes are displayed in Fig. 2 and Fig. 3, respectively. It is clear that the proposed algorithm resolves the desired solution for this problem and it captures all small features appearing in computational domain. The reference solution and the initial condition appear to be similar confirming the convergence of the proposed numerical techniques. As can be seen in the presented results, a shock is formed in the solution u_t propagating along the main diagonal in the domain. The effect of mesh refinement on the computed solutions is noticeable in these figures. It is also clear that our relaxation methods accurately capture the shock and its propagation along the diagonal. However, due to the numerical dissipation, the resolved shock has been smeared out in the results obtained on a mesh with 100×100 control volumes. As expected, the numerical results obtained on this mesh are more diffusive than those computed using meshes with 200×200 and 400×400 control volumes. To further visualize this effect we display in Fig. 4 the cross-sections along the main diagonal $y = x$ for the results on the considered meshes.

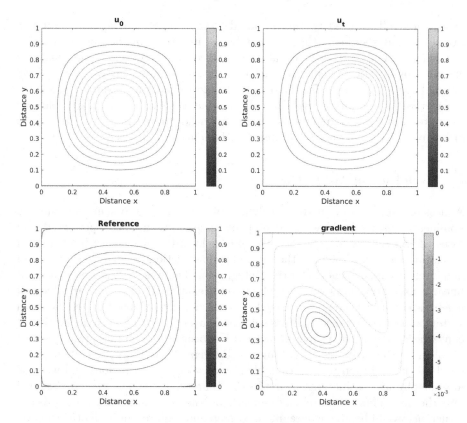

Fig. 3. Numerical results obtained on a mesh with 400×400 control volumes.

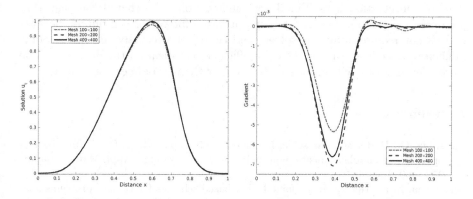

Fig. 4. Cross-sectional results at the main diagonal $y = x$ on different meshes.

It is apparent that the gradient resolution is deteriorated with the excessive dissipation included by the coarse mesh with 100×100 control volumes. On the other hand, the solutions are completely free of spurious oscillations and the shocks are well resolved by the proposed method without nonlinear computational tools. It should be that the number of iterations in the optimal control problem does not overpass 23 iterations for all considered meshes. These features clearly demonstrate the efficiency achieved by the proposed method for solving optimal control problems for the inviscid Burgers equation. The performance of the method is very attractive since the computed solution remains stable and accurate even when coarse meshes are used without requiring Riemann solvers or complicated techniques to reconstruct the numerical fluxes.

5   Concluding Remarks

A class of numerical methods for solving optimal control problems governed by nonlinear conservation laws in two space dimensions has been presented and assessed. As solvers for the forward and backward problems we implement a relaxation method combining the upwind reconstruction for space discretization and implicit-explicit scheme for time integration. These techniques solve the nonlinear conservation laws without relying on Riemann solvers or linear solvers of algebraic equations. The optimal control problem is formulated for the relaxation system and at the relaxed limit its solution converges to the relaxed equation of conservation laws. The proposed method has been tested on an optimal control problem for the two-dimensional inviscid Burgers. The obtained results indicate good shock resolution with reasonable accuracy in smooth regions and without any nonphysical oscillations near the shock areas. Although, we have studied only the case of first-order relaxation methods, the extension to high-order reconstructions would be an encouraging next step and requires an in-depth study on optimal control problems to deal with the nonlinear structure of hyperbolic systems of conservation laws. Finally, we should point out that d the algorithm presented in this paper can be highly optimized for vector computers, because it does not require nonlinear solvers and contain no recursive elements. Some difficulties arise from the fact that for efficient vectorization the data should be stored contiguously within long vectors rather than two-dimensional arrays.

References

1. Albia, G., Herty, M., Pareschic, L.: Linear multistep methods for optimal control problems and applications to hyperbolic relaxation systems. Appl. Math. Comput. **354**, 460–477 (2019)
2. Banda, M.K., Herty, M.: Adjoint IMEX-based schemes for control problems governed by hyperbolic conservation laws. Comput. Optim. Appl. **51**(2), 909–930 (2012). https://doi.org/10.1007/s10589-010-9362-2
3. Banda, M., Seaid, M.: Higher-order relaxation schemes for hyperbolic systems of conservation laws. J. Numer. Math. **13**(3), 171–196 (2005)
4. Bianchini, S.: On the shift differentiability of the flow generated by a hyperbolic system of conservation laws. Discrete Continuous Dyn. Syst. **6**(2), 329–350 (2000)

5. Bressan, A., Shen, W.: Optimality conditions for solutions to hyperbolic balance. In: Control Methods in PDE-Dynamical Systems, vol. 426, p. 129. American Mathematical Society(2007)
6. Castro, C., Palacios, F., Zuazua, E.: An alternating descent method for the optimal control of the inviscid burgers equation in the presence of shocks. Math. Models Methods Appl. Sci. **18**(03), 369–416 (2008)
7. Chertock, A., Herty, M., Kurganov, A.: An Eulerian-Lagrangian method for optimization problems governed by multidimensional nonlinear hyperbolic PDEs. Comput. Optim. Appl. **59**(3), 689–724 (2014)
8. Coron, J., Nguyen, H.: Optimal time for the controllability of linear hyperbolic systems in one-dimensional space. SIAM J. Control Optim. **57**(2), 1127–1156 (2019)
9. Court, S., Kunisch, K., Pfeiffer, L.: Optimal control problem for viscous systems of conservation laws, with geometric parameter, and application to the Shallow-Water equations. Interfaces Free Boundaries **21**, 273–311 (2019)
10. Gugat, M., Leugering, G.: Solutions of lp-norm-minimal control problems for the wave equation. Comput. Appl. Math. **21**(1), 227–244 (2002)
11. Hajian, S., Hintermüller, M., Ulbrich, S.: Total variation diminishing schemes in optimal control of scalar conservation laws. IMA J. Numer. Anal. **39**(1), 105–140 (2019)
12. Jin, S., Xin, Z.: The relaxation schemes for systems of conservation laws in arbitrary space dimensions. Commun. Pure Appl. Math. **48**(3), 235–276 (1995)
13. Käppeli, R., Balsara, D., Chandrashekar, P., Hazra, A.: Optimal, globally constraint-preserving, DG (TD) 2 schemes for computational electrodynamics based on two-derivative Runge-Kutta timestepping and multidimensional generalized riemann problem solvers-a von Neumann stability analysis. J. Comput. Phys. **408**, 109238 (2020)
14. Liu, Z., Sandu, A.: On the properties of discrete adjoints of numerical methods for the advection equation. Int. J. Numer. Methods Fluids **56**(7), 769–803 (2008)
15. Mordukhovich, B., Raymond, J.: Optimal boundary control of hyperbolic equations with pointwise state constraints. Nonlinear Anal. Theory Methods Appl. **63**(5–7), 823–830 (2005)
16. Nordström, J., Ghasemi, F.: The relation between primal and dual boundary conditions for hyperbolic systems of equations. J. Comput. Phys. **401**, 109032 (2020)
17. Pareschi, L., Russo, G.: Implicit-explicit Runge-Kutta schemes for stiff systems of differential equations. Recent Trends Numer. Anal. **3**, 269–289 (2000)
18. Polat, G., Özer, T.: The group-theoretical analysis of nonlinear optimal control problems with hamiltonian formalism. J. Nonlinear Math. Phys. **27**(1), 106–129 (2020)
19. Ulbrich, S.: A sensitivity and adjoint calculus for discontinuous solutions of hyperbolic conservation laws with source terms. SIAM J. Control Optim. **41**(3), 740–797 (2002)

Genetic Learning Particle Swarm Optimization with Interlaced Ring Topology

Bożena Borowska[✉]

Institute of Information Technology, Lodz University of Technology,
Wólczańska 215, 90-924 Lodz, Poland
bozena.borowska@p.lodz.pl

Abstract. Genetic learning particle swarm optimization (GL-PSO) is a hybrid optimization method based on particle swarm optimization (PSO) and genetic algorithm (GA). The GL-PSO method improves the performance of PSO by constructing superior exemplars from which individuals of the population learn to move in the search space. However, in case of complex optimization problems, GL-PSO exhibits problems to maintain appropriate diversity, which leads to weakening an exploration and premature convergence. This makes the results of this method not satisfactory. In order to enhance the diversity and adaptability of GL-PSO, and as an effect of its performance, in this paper, a new modified genetic learning method with interlaced ring topology and flexible local search operator has been proposed. To assess the impact of the introduced modifications on performance of the proposed method, an interlaced ring topology has been integrated with GL-PSO only (referred to as GL-PSOI) as well as with a flexible local search operator (referred to as GL-PSOIF). The new strategy was tested on a set of benchmark problems and a CEC2014 test suite. The results were compared with five different variants of PSO, including GL-PSO, GGL-PSOD, PSO, CLPSO and HCLPSO to demonstrate the efficiency of the proposed approach.

Keywords: Genetic learning particle swarm optimization · Enhanced diversity · Particle swarm optimization · Optimization

1 Introduction

Developed by Kennedy and Eberhart [1, 2] particle swarm optimization (PS0) is a stochastic optimization method modeled on social behavior and intelligence of animal such as flocks of birds and fish schooling. Similar to other evolutionary methods, it is based on the population. The mechanism of the PSO method relies on particles following their best personal particle and globally the best particle in the swarm towards the most promising areas of the search space. Because of its easy implementation and high convergence rate, it is widely used in solving various optimization problems, including energetic [3], mechanics [4], scheduling problem [5], antenna design [6, 7], control systems [8], image classification [9] and many others. However likewise other evolutionary algorithms, PSO encounters some troubles including stagnation in local optima, excessive loss of diversity and premature convergence [10]. A variety of

© Springer Nature Switzerland AG 2020
V. V. Krzhizhanovskaya et al. (Eds.): ICCS 2020, LNCS 12141, pp. 136–148, 2020.
https://doi.org/10.1007/978-3-030-50426-7_11

different variants of PSO have been introduced to counteract these disadvantages and enhance the efficiency of PSO. Among them, the following improvements can be distinguished:

- Adjustment of basic coefficients. According to Shi and Eberhart [11], a key to the improvement of the PSO performance is inertia weight, which should be linearly decreased from 0.9 to 0.4. Clerc [12] recommended to use fixed factors, and indicates that inertia weight of 0.729 with fixed acceleration coefficients of 1.494 can enhance convergence speed. Five years later Trelea [13] proved that PSO with inertia weight of 0.6 and constant acceleration coefficients of 1.7 allowed to get faster convergence than that achieved by Eberhart [11] and Clerc [12]. The PSO method with nonlinear factors were proposed by Borowska [14, 15]. Furthermore, the efficiency of changing factors was examined by Ratnawera et al. [16]. The cited authors concluded that time-varying acceleration coefficients (TVAC) helped to control local and global searching process more efficiently.
- Modification of the update equations. To improve searching process the researches propose to use a new update equation [17, 18] or add a new component to existing velocity equation [19]. Another approach is to introduce, for ineffective particles, a repair procedure [10] with other velocity updating equations that helps more precisely determine swarm motion and stimulate particles when their efficiency decreases.
- Topology structure. According to Kennedy [20] topology structure affects the way information exchange and the swarm diversity. Many different topological structures have been proposed including: square, four clusters, ring, pyramid and the von Neumann topology [20–23]. Another approach is a multi-swarm structure recommended by Liang and Suganthan [24] and Chen et al. [25]. In contrast, Gong et al. [22] have introduced a two-cascading-layer structure. In turn, Wang et al. [26] developed PSO based on multiple layers.
- Learning strategy. It is used to improve performance of algorithm by breading high quality exemplars from which other swarm particles can acquire knowledge and learn to search space. A multi-swarm PSO based on dynamic learning strategy has been presented by Ye et al.[27]. Likewise, Liang et al.[28] has proposed a comprehensive learning strategy (CLPSO) according to which, particle velocity is updated based on historical best information of all other particles. To greater improve the performance and adaptability of CLPSO, Lin et al. [29] recommend to use an adaptive comprehensive learning strategy with dynamically adjusting learning probability level according to the performance of the particles during the optimization process. Another approach is based on social learning PSO as described by Cheng et al. [30].
- Hybrid methods combine beneficial features of two or more approaches. They are used to strength PSO efficiency and achieve faster convergence as well as better accuracy of the resultant solution. Holden et al. [31] have proposed to join PSO with an ant colony optimization method. Li et al.[32] have combined PSO with jumping mechanism of SA (simulated annealing). A modified version based on PSO and SA has been developed by Shieh et al. [33]. In turn, PSO with chaos has been presented by Tian and Shi[34], whereas Chen et al. [35] have proposed

learning PSO based on biogeography. Furthermore, a hybrid approach based on improved PSO, cuckoo search and clustering method has been developed by Bouer and Hatamlou [36].

In order to enhance the PSO performance, Gong et al. [22] have merged the latter two categories and proposed genetic learning particle swarm optimization (GL-PSO). In GL-PSO, except PSO and genetic operators, a two layer structure have been applied in which the former is used to generate exemplars whereas the latter to update particles through the PSO algorithm.

The GL-PSO method improves the performance of PSO by constructing superior exemplars from which individuals of the population learn to move in the search space. Unfortunately, this approach is not free from disadvantages. In fact, the algorithm can achieve high convergence rate but in case of complex problems, due to global topology, the particle diversity quickly decreases and, as a result, impairs the exploration capability.

In order to enhance the diversity and adaptability of GL-PSO as well as to improve its performance in solving complex optimization problems, in this paper, a new modified genetic learning method, referred to as GL-PSOIF, has been demonstrated. The proposed GL-PSOIF method is based on GL-PSO in which two modifications have been introduced. Specifically, instead of global topology, an interlaced ring topology has been introduced. The second modification relies on introducing a flexible local search operator. The task of the interlaced ring topology is to increase the population diversity and improve effectiveness of the method by generating better quality exemplars. In turn, a flexible local search operator has been introduced to enrich searching and improve the exploration and the exploitation ability. To evaluate the impact of the proposed modifications on performance of the proposed method, the interlaced ring topology has been first integrated with GL-PSO only (referred to as GL-PSOI) and then together with a flexible local search operator (referred to as GL-PSOIF). Both methods were tested on a set of benchmark problems and a CEC2014 test suite [38]. The results were compared with five different variants of PSO, including the genetic learning particle swarm optimization (GL-PSO) [22], the comprehensive particle swarm optimizer (CLPSO) [28], the standard particle swarm optimization (PSO), the global genetic learning particle swarm optimization (GGL-PSOD) [23], and the heterogeneous comprehensive learning particle swarm optimization (HCLPSO) [39].

2   The PSO Method

The PSO method was inspired by the social behavior of flocks of organisms (bird flocking, fish schooling, bees swarm) living in their natural environment [2, 3]. Likewise other evolutionary method, PSO is based on a population. Individuals of the population are called particles, and the population itself is called a swarm. In the PSO, the optimisation process is achieved by migration the particles towards the most promising area of the search space. Assuming that migration occurs in the D-dimensional search space, we can imagine particle swarm as a set of points each of which possess knowledge about: its actual position described by the position vector $x_j = (x_{j1}, x_{j2}, ..., x_{jD})$, its current speed of movement described by velocity vector $v_j = (v_{j1}, v_{j2}, ..., v_{jD})$, its best position encountered by itself described by $pbest_j = (pbest_{j1}, pbest_{j2}, ..., pbest_{jD})$, and the best position encountered in all swarm described as $gbest = (gbest_1, gbest_2, ..., gbest_D)$. In the first iteration, the position vector value and the velocity vector value are randomly generated. In subsequent iterations, values of the vectors are updated based on the knowledge and acquired experience of the particles. Changing of the particles velocity is achieved based on the Eq. (1).

$$v_j(l+1) = w \cdot v_j(l) + c_1 \cdot r_1 (pbest_j - x_j(l)) + c_2 \cdot r_2 (gbest - x_j(l)) \tag{1}$$

Changing the particle position is realized by adding its actual velocity to its previous position (2)

$$x_j(l+1) = x_j(l) + \cdot v_j(l+1) \tag{2}$$

where: w - inertia weight, $pbest_j$ -the best j particle position., $gbest$ - the best position. in a swarm, r_1, r_2 .-random numbers generated from (0, 1), c_1, c_2- acceleration coefficients.

3 Genetic Learning Particle Swarm Optimization

In contrast to PSO, the GL-PSO algorithm possess a two-cascading-layer structure. One layer is used to generate exemplars, the other to update particles position and velocity through the PSO algorithm. To generate exemplars, three operators (crossover, mutation and selection) of the GA algorithm [37] are applied.

Exemplars e_j are selected from offspring. To generate offspring o_j for each dimension of particle j, a crossover operator is applied according to the formula:

$$o_j = \begin{cases} r \cdot pbest_j, + (1 - r) \cdot gbest, & if\, f(pbest_j) < f(pbest_k) \\ pbest_k & otherwise \end{cases} \tag{3}$$

where k is random selected particle, r –random number from (0, 1).

Next, for each dimension, a random number $r[0, 1]$ is generated and then if $r < p_m$, (where p_m probability mutation) the offspring is mutated. Then the offspring undergoes the selection operation according to the formula:

$$e_j \leftarrow \begin{cases} o_j, & if\, f(o_j) < f(e_j) \\ e_j, & otherwise \end{cases} \tag{4}$$

The particle velocity is updated based on the following equation:

$$v_j(l+1) = w \cdot v_j(l) + c \cdot r(e_j - x_j(l)) \tag{5}$$

where e_j is the exemplar of the j particle.

4 The Proposed Method

In order to improve the performance of global genetic learning particle swarm optimization (GL-PSO), in this article two modifications have been proposed: interlaced ring topology and flexible local search operator.

4.1 Interlaced Ring Topology

One of the main reason for inability to obtain and pursue satisfactory performance of the GL-PSO is the lack or weakennes ability to maintain diversity of the population (swarm). This leads to a loss of balance between exploration and exploitation and consequently to premature convergence and unsatisfactory results. To avoid this, it is necessary to develop tools that could help increase adaptability of the algorithm, which, in turn, should give satisfactory results.

Lin et al. [22] have introduced ring and a global learning component with linearly adjusted control parameters to enhance a GL-PSO diversity. This improves the adaptability of the method but is not sufficient. Hence, the problem remains open and other solutions should be sought. To improve the adaptability of the GL-PSO, in this paper, instead of global learning, the interlaced ring topology has been proposed. This approach uses two neighbour particles, like in the ring topology, but in every next iteration (except the first one), the order of the particles is changed as follows. The particle collection is divided into two parts (sets) and particles of the second part take up spaces between the particles of the first part alternately (one particle from the first set, another particle from the second set, and next one from the first set, another from the second set etc.) according to Eq. 6 and 7.

$$n_j = \frac{j+1}{2} \quad for \ odd \ j \tag{6}$$

$$n_j = \frac{j+N}{2} \quad for \ even \ j \tag{7}$$

where n_j is the position of the particle to be moved to the j place in the ring, $j = 1 \dots N$, N is a swarm size (for example $n_2 = 5$ means that the second position in the ring is occupied by a particle from 5th place in the swarm).

Then, the position of exemplars are generated according to Eqs. 8, 9 and 10.

$$o_j = r \cdot pbest_{n_{j1}} + (1 - r) \cdot pbest_{n_{j2}} \tag{8}$$

$$n_{j1} = \begin{cases} N, & j = 1 \\ j - 1, & j > 1 \end{cases} \tag{9}$$

$$n_{j2} = \begin{cases} 1, & j = N \\ j + 1, & j < N \end{cases} \tag{10}$$

where according to the ring topology n_{j1} and n_{j2} are the indexes of the adjacent particles from the left and right side of the particle j, respectively.

4.2   Flexible Local Search Operator

To improve the searching behavior of PSO and improve the exploitation capacity of the swarm, a flexible local search operator is introduced. The particle positions are updated according to the formula:

$$x_j^{k+1} = \begin{cases} pbest_j \cdot (1 + N(0,1)), & otherwise \\ x_j^k + v_j^{k+1}, & if\ p < s \end{cases} \qquad (11)$$

where p is a a randomly selected number in the range $[0,1]$, s is a real number linearly increasing from 0.6 to 0.8. This means that each particle has a 40 to 20% possibility to perform search in the vicinity of its personal best position. This means that, according to [16], in the early stage of the optimization process, the exploration is enhanced, and the local exploitation in the latter stage is facilitated.

Table 1. Optimization test functions.

Function	Formula	f_{min}	Range	Accept
Sphere	$f_1 = \sum_{i=1}^{n} x_i^2$	0	$[-100, 100]^n$	10^{-5}
Schwefel	$f_2 = \sum_{i=1}^{n} (\sum_{j=1}^{i} x_j)^2$	0	$[-100, 100]^n$	100
Rosenbrock	$f_3 = \sum_{i=1}^{n-1} [100(x_{i+1} - x_i^2)^2 + (x_i - 1)^2$	0	$[-5, 5]^n$	10^{-5}
Rastrigin	$f_4 = \sum_{i=1}^{n} (x_i^2 - 10\cos(2\pi x_i) + 10)$	0	$[-32, 32]^n$	10^{-5}
Ackley	$f_5 = -20 \exp\left(-0.2 \sqrt{\frac{1}{n}\sum_{i=1}^{n} x_i^2}\right) - \exp\left(\frac{1}{n}\sum_{i=1}^{n} \cos(2\pi x_i)\right) + 20 + e$	0	$[-600,600]^n$	10^{-5}
Penalized	$f_6 = 0.1\{\sin^2(3\pi x_1) + \sum_{i=1}^{n-1} (x_i - 1)^2 [1 + \sin^2(3\pi x_{i+1})]$ $+ (x_n - 1)^2 [1 + \sin^2(2\pi x_n)]\} + \sum_{i=1}^{n} u(x_i, 5, 100, 4),$ $u(z, a, k, m) = \begin{cases} k(z-a)^m, & z > a, \\ 0, & -a \le z \le a, \\ k(-z-a)^m, & z < -a. \end{cases}$	0	$[-50, 50]^n$	10^{-5}

5   Test Results

In order to investigate the efficiency of the proposed modifications, the GL-PSOI (in which only the interlaced ring topology was adopted) and GL-PSOIF (with interlaced ring topology and flexible local search operator) were evaluated, separately. Both strategies were tested on a set of classical benchmark problems, and on the CEC2014 test suite. Twelve of them (6 selected benchmark function and 6 CEC2014 functions) are described in this article and depicted in Tables 1 and 2.

Table 2. Selected CEC2014 test suite.

	Functions Name	Range	$F(x*)$
F7	Rotated Bent Cigar Function	$[-100,100]^n$	100
F8	Shifted and Rotated Rosenbrock's Function	$[-100,100]^n$	400
F9	Shifted and Rotated Ackley's Function	$[-100,100]^n$	500
F10	Shifted Rastrigin's Function	$[-100,100]^n$	800
F11	Shifted and Rotated Rastrigin's Function	$[-100,100]^n$	900

The results of the tests were compared with performances of CLPSO, HCLPSO, PSO, GL-PSO and GGL-PSOD. The parameter settings of this algorithms are listed in Table 3.

Table 3. Parameters settings.

Algorithm	Parameter settings
CLPSO	$w = 0.9$-0.4, $c = 1.496$
HCLPSO	$w = 0.99$-0.2, $c_l = 2.5$-0.5, $c_2 = 0.5$-2.5, $c = 3$-1.5
PSO	$w = 0.9$-0.4, $c_1 = 2.0$, $c_2 = 2.0$
GL-PSO	$w = 0.7298$, $c = 1.49618$, $p_m = 0.01$, $s_g = 7$
GL-PSOD	$w = 0.7298$, $c = 1.49618$, $p_m = 0.01$, $s_g = 7$

Both in the GL-PSOI and GL-PSOIF, the inertia weight $w = 0.6$ [13]. The acceleration coefficients used in the computations were equal $c_1 = c_2 = 1.7$. In case of the set of benchmark functions, the population consisted of 20 particles, the dimension of the search space was 30, the maximum number of function evaluations was 300000. The search range depends on the function used as shown in Table 1. For each problem, the simulations were run 30 times. For CEC2014 functions, the population consisted of 50 particles, the dimension of the search space was $D = 30$, and the maximum number of function evaluations was $D \times 10^4$. The search range was $[-100,100]^n$. For CEC2014 functions, the algorithms were run 31 times independently.

The exemplary results of the tests are summarized in Tables 4 and 5.

Table 4. The comparison test results of the PSO algorithms on the benchmark functions.

Functions	Criteria	CLPSO	HCLPSO	GL-PSO	PSO	GGL-PSOD	GL-PSOI	GL-PSOIF
F1	Mean	0.00E+00(=)	0.00E+00(=)	0.00E+00(=)	3.48E−25(+)	0.00E+00(=)	0.00E+00	0.00E+00
	Std	0.00E+00	0.00E+00	0.00E+00	2.08E−24	0.00E+00	0.00E+00	0.00E+00
F2	Mean	6.88E+01(+)	5.57E+00(+)	2.43E−20(+)	2.71E−11(+)	6.74E−20(+)	**3.15E−22**	4.52E−21
	Std	3.24E+01	4.03E+00	3.16E−20	4.29E−11	4.82E−20	2.67E−21	3.84E−20
F3	Mean	2.34E+01(+)	2.16E+00(+)	6.48E−01(+)	4.16E+01(+)	6.53E−01(+)	**5.02E−01**	5.16E−01
	Std	1.58E+01	4.24E+00	2.54E−01	3.92E+01	6.07E−01	5.48E−01	2.58E−01
F4	Mean	1.02E−11(+)	6.32E−12(+)	7.14E−14(+)	3.89E+01(+)	4.32E−14(+)	6.44E−15	**3.50E−16**
	Std	3.21E−12	8.40E−12	3.62E−14	9.22E+00	5.36E−14	5.37E−14	3.68E−15
F5	Mean	2.05E−14(+)	1.41E−12(+)	7.86E−15(+)	3.59E−13(+)	6.29E−15(+)	5.85E−15	**5.32E−16**
	Std	3.41E−15	4.07E−13	3.92E−15	7.91E−14	2.23E−15	2.73E−15	1.98E−15
F6	Mean	1.82E−32(+)	1.65E−32(+)	1.73E−31(+)	3.47E−02(+)	2.11E−31(+)	1.62E−32	**1.57E−32**
	Std	5.56E−48	5.56E−48	1.94E−32	5.89E−02	3.73E−32	5.04E−36	4.86E−34

Table 5. The comparison test results of the PSO algorithms on the CEC2014 test suite.

Functions	Criteria	CLPSO	HCLPSO	GL-PSO	PSO	GGL-PSOD	GL-PSOI	GL-PSOIF
F7	Mean	**3.24E+02(-)**	4.15E +02(-)	5.96E+02(+)	8.09E+02(+)	7.12E+02(+)	4.58E+02	4.41E+02
	Std	4.85E+02	6.73E+02	3.63E+02	3.34E+02	7.29E+02	6.73E+02	1.18E+02
F8	Mean	6.93E+01(+)	3.82E+01(-)	**2.76E+01(-)**	1.62E+02(+)	6.27E+01(+)	5.75E+01	4.64E+01
	Std	3.15E+01	3.36E+01	6.59E+01	5.16E+01	3.49E+01	5.18E+01	2.37E+01
F9	Mean	2.08E+01(=)	**2.00E+01(=)**	2.05E+01(=)	2.32E+01(+)	**2.00E+01(=)**	2.00E+01	2.00E+01
	Std	5.37E−02	6.24E−03	3.42E−02	8.89E−02	3.27E−02	2.83E−02	2.12E−02
F10	Mean	4.07E−02(+)	2.38E−01(+)	1.95E−10(+)	2.66E+01(+)	2.43E−12(+)	2.35E−13	**1.57E−13**
	Std	2.19E−02	5.40E−01	7.23E−11	8.19E+00	7.68E−13	6.48E−13	1.88E−13
F11	Mean	4.20E+01(+)	4.43E+01(+)	5.84E+01(+)	7.81E+01(+)	3.57E+01(+)	2.97E+01	**2.35E+01**
	Std	7.17E+00	1.26E+01	2.13E+01	2.69E+01	1.49E+01	1.56E+01	1.06E+01

Table 6. The comparison test results of the PSO algorithms.

Signature	CLPSO	HCLPSO	GL-PSO	PSO	GGL-PSOD
+	8	7	8	11	8
−	1	2	1	0	1
=	2	2	2	0	2

The exemplary charts showing the mean fitness selected functions in the following iterations for GL-PSO, GGL-PSO, CLPSO, HCLPSO, PSO, GL-PSOI and GL-PSOIF algorithms, are depicted in Figs. 1, 2 and 3.

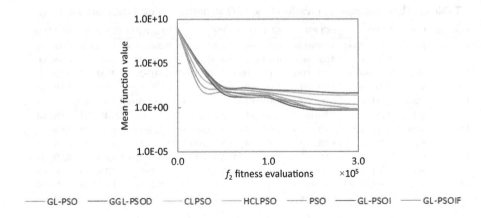

Fig. 1. Convergence performance for f_2 function.

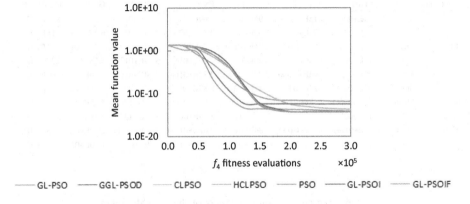

Fig. 2. Convergence performance for f_4 function.

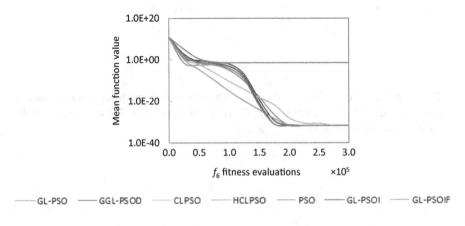

Fig. 3. Convergence performance for f_6 function.

The results of the tests confirmed that both GL-PSOI and GL-PSOIF are more effective and can achieve superior performance over the remaining tested methods. In case of unimodal functions, the GL-PSOI with interlaced ring topology obtained superior results over the ones for GL-PSOIF. For multimodal functions superior results were achieved by GL-PSOIF.

In case f2 function, GL-PSO achieved worse results than GL-PSOI and GL-PSOIF but better than those obtained by the CLPSO, HCLPSO and PSO. For f3 function, GL-PSOI achieved the best result. The performance of GL-PSO was worse than that obtained by GL-PSOI but superior then performance of GL-PSOIF. For unimodal f7 function the best results were obtained by CLPSO. The outcomes achieved by GL-PSOI and GL-PSOIF were worse than results obtained by CLPSO but better than the results achieved by the remaining tested methods. For multimodal functions, the results show that (almost in all cases) GL-PSOIF exhibit the best performance.

The convergence curves presented in Figs. 1, 2, and 3 indicate that both GL-PSOI and GL-PSOIF converge slower in the early stage of the optimization process than most of the compared methods. At this stage, each algorithm, except PSO, is faster. Then both algorithms accelerate and converge faster than the others.

In case of the unimodal f2 function, both algorithms initially revealed slower convergence, which was followed by a further rapid acceleration after about $5x10^4$ iterations showing superiority over the rest evaluated methods. For the unimodal f2 function, GL-PSOIF performed a bit slower than GL-PSOI, which could be due to the introduction of flexible search operator, which did not improved the GL-PSOIF run. In case multimodal functions (Figs. 2 and 3), GL-PSOIF converges slowly (other methods are faster) but after about $1.3x10^5$ iterations accelerates and after $2x10^5$ iterations becomes the fastest.

6 Statistical Test

In order to evaluate the differences between algorithms, a statistical t-test was used. A confidence level of 0, 05 was selected for all statistical comparisons. Tables 4 and 5 shows the results of the t-test performed on the test functions. The signature '+' indicates that GL-PSOIF is significantly better than the other algorithms, '−' worse to the other algorithms, and '=' equal to the other algorithms. The rows in Table 6 named '+', '−'and '=' mean the number of times that the GL-PSOIF is better than, worse than or equal to the other algorithms. The results of the t-test indicate that proposed algorithm is significantly better than other methods with 95% confidence level in a statistically meaningful way.

7 Conclusion

In this study, a new genetic learning particle swarm optimization with interlaced ring topology and flexible local search operator (GL-PSOIF) has been proposed. To assess the impact of introduced modifications on performance of the evaluated method, first the interlaced ring topology was integrated with GL-PSO only (referred to as GL-

PSOI) and then with the flexible local search operator (GL-PSOIF). The efficiency of the new strategy was tested on a set of benchmark problems and the CEC2014 test suite. The results were compared with five different variants of PSO, including GL-PSO, GGL-PSOD, PSO, CLPSO and HCLPSO. The results of the experimental trials indicated that the genetic learning particle swarm optimization with interlaced ring topology is effective for unimodal function. In case of the multimodal function, GL-PSOIF showed superior performance over the remaining tested methods.

References

1. Kennedy, J., Eberhart, R.C.: Particle Swarm Optimization. In: IEEE International Conference on Neural Networks, pp. 1942–1948. Perth, Australia (1995)
2. Kennedy, J., Eberhart, R.C., Shi, Y.: Swarm Intelligence. Morgan Kaufmann Publishers, San Francisco (2001)
3. Ignat, A., Lazar, E., Petreus, D.: Energy management for an islanded microgrid based on Particle Swarm Optimization. In: IEEE 24th International Symposium for Design and Technology in Electronic Packaging (SIITME 2018), Romania, pp. 213–216 (2018)
4. Wu, D., Gao, H.: An adaptive particle swarm optimization for engine parameter optimization. Proc. Natl. Acad. Sci. India Sect. A: Phys. Sci. **88**, 121–128 (2018). https://doi.org/10.1007/s40010-016-0320-y
5. Hu, Z., Chang, J., Zhou, Z.: PSO scheduling strategy for task load in cloud computing. Hunan Daxue Xuebao/J. Hunan Univ. Nat. Sci. **46**(8), 117–123 (2019)
6. Zhang, X., Lu, D., Zhang, X. et al.: Antenna array design by a contraction adaptive particle swarm optimization algorithm. J Wireless Commun. Netw. **2019**, p. 57 (2019). https://doi.org/10.1186/s13638-019-1379-3
7. Yu, M., Liang, J., Qu, B., Yue, C.: Optimization of UWB antenna based on particle swarm optimization algorithm. In: Li, K., Li, W., Chen, Z., Liu, Y. (eds.) ISICA 2017. CCIS, vol. 874, pp. 86–97. Springer, Singapore (2018). https://doi.org/10.1007/978-981-13-1651-7_7
8. You, Z., Lu, C.: A heuristic fault diagnosis approach for electro-hydraulic control system based on hybrid particle swarm optimization and Levenberg–Marquardt algorithm. J. Ambient Intell. Humanized Comput. 1–10 (2018). https://doi.org/10.1007/s12652-018-0962-5
9. Junior, F.E.F., Yen, G.G.: Particle swarm optimization of deep neural networks architectures for image classification. Swarm Evol. Comput. **49**, 62–74 (2019)
10. Borowska, B.: An improved CPSO algorithm. In: International Scientific and Technical Conference Computer Sciences and Information Technologies (CSIT), pp. 1–3, IEEE, Lviv (2016). https://doi.org/10.1109/stc-csit.2016.7589854
11. Shi, Y., Eberhart, R.C.: Empirical study of particle swarm optimization. In: Congress on evolutionary computation, Washington D.C., USA, pp. 1945–1949 (1999)
12. Clerc, M.: The swarm and the queen: towards a deterministic and adaptive particle swarm optimization. In: Proceedings of the ICEC, Washington, DC, pp. 1951–1957 (1999)
13. Trelea, I.C.: The particle swarm optimization algorithm: convergence analysis and parameter selection. Inf. Process. Lett. **85**, 317–325 (2003)
14. Borowska, B.: Nonlinear inertia weight. in particle swarm optimization. In: International Scientific and Technical Conference, Computer Science and Information Technologies (CSIT 2017), Lviv, Ukraine, pp. 296–299 (2017)

15. Borowska, B.: Influence of social coefficient on swarm motion. In: Rutkowski, L., Scherer, R., Korytkowski, M., Pedrycz, W., Tadeusiewicz, R., Zurada, J.M. (eds.) ICAISC 2019. LNCS (LNAI), vol. 11508, pp. 412–420. Springer, Cham (2019). https://doi.org/10.1007/978-3-030-20912-4_38
16. Ratnaveera, A., Halgamuge, S.K., Watson, H.C.: Self-organizing hierarchical particle swarm optimizer with time-varying acceleration coefficients. IEEE Trans. Evol. Comput. **8**(3), 240–255 (2004)
17. Lu, H., Chen, W.: Self-adaptive velocity particle swarm optimization for solving constrained optimization problems. J. Glob. Optim. **41**, 427–445 (2008)
18. Borowska, B.: Novel algorithms of particle swarm optimisation with decision criteria. J. Exp. Theor. Artif. Intell. **30**(5), 615–635 (2018). https://doi.org/10.1080/0952813X.2018.1467491
19. Mahmoud, K.R., El-Adawy, M., Ibrahem, S.M.M.: A comparison between circular and hexagonal array geometries for smart antenna systems using particle swarm optimization algorithm. Prog. Electromagnet. Res. **72**, 75–90 (2007)
20. Kennedy, J., Mendes, R.: Population structure and particle swarm performance. In: Proceedings of the IEEE Congress Evolutionary Computations, Honolulu, HI, USA, vol. 2, pp. 1671–1676 (2002)
21. Mendes, R., Kennedy, J., Neves, J.: The fully informed particle swarm: simpler, maybe better. IEEE Trans. Evol. Comput. **8**, 204–210 (2004)
22. Gong, Y.J., et al.: Genetic learning particle swarm optimization. IEEE Trans. Cybern. **46**(10), 2277–2290 (2016)
23. Lin, A., Sun, W., Yu, H., Wu, G., Tang, H.: Global genetic learning particle swarm optimization with diversity enhanced by ring topology. Swarm Evol. Comput. **44**, 571–583 (2019)
24. Liang, J.J., Suganthan, P.N.: Dynamic multi-swarm particle swarm optimizer. In: Proceedings of the Swarm Intelligence Symposium, pp. 124–129 (2005)
25. Chen, Y., Li, L., Peng, H., Xiao, J., Wu, Q.T.: Dynamic multi-swarm differential learning particle swarm optimizer. Swarm Evol. Comput. **39**, 209–221 (2018)
26. Wang, L., Yang, B., Chen, Y.H.: Improving particle swarm optimization using multilayer searching strategy. Inf. Sci. **274**, 70–94 (2014)
27. Ye, W., Feng, W., Fan, S.: A novel multi-swarm particle swarm optimization with dynamic learning strategy. Appl. Soft Comput. **61**, 832–843 (2017)
28. Liang, J.J., Qin, A.K., Suganthan, P.N., Baskar, S.: Comprehensive learning particle swarm optimizer for global optimization of multimodal functions. IEEE Trans. Evol. Comput. **10**(3), 281–295 (2006)
29. Lin, A., Sun, W., Yu, H., Wu, G., Tang, H.: Adaptive comprehensive learning particle swarm optimization with cooperative archive. Appl. Soft Comput. J. **77**, 533–546 (2019)
30. Cheng, R., Jin, Y.: A social learning particle swarm optimization algorithm for scalable optimization. Inf. Sci. **291**, 43–60 (2015)
31. Holden, N., Freitas, A.A.: A hybrid particle swarm/ant colony algorithm for the classification of hierarchical biological data. In: Proceedings of the IEEE SIS, pp. 100–107 (2005)
32. Li, L., Wang, L., Liu, L.: An effective hybrid PSOSA strategy for optimization and its application to parameter estimation. Appl. Math. Comput. **179**, 135–146 (2006)
33. Shieh, H.L., Kuo, C.C., Chiang, C.M.: Modified particle swarm optimization algorithm with simulated annealing behavior and its numerical verification. Appl. Math. Comput. **218**, 4365–4383 (2011)
34. Tian, D., Shi, Z.: MPSO: modified particle swarm optimization and its applications. Swarm Evol. Comput. **41**, 49–68 (2018)

35. Chen, X., Tianfield, H., Mei, C., et al.: Biogeography-based learning particle swarm optimization. Soft. Comput. **21**, 7519–7541 (2017). https://doi.org/10.1007/s00500-016-2307-7
36. Bouyer, A., Hatamlou, A.: An efficient hybrid clustering method based on improved cuckoo optimization and modified particle swarm optimization algorithms. Appl. Soft Comput. **67**, 172–182 (2018)
37. Duraj, A., Chomatek, L.: Outlier detection using the multiobjective genetic algorithm. J. Appl. Comput. Sci. **25**(2), 29–42 (2017)
38. Liang, J.J., Qu, B.Y., Suganthan, P.N.: Problem definitions and evaluation criteria for the CEC 2014 special session and competition on single objective real-parameter numerical optimization, Computational Intelligence Laboratory, Zhengzhou University, Zhengzhou China. Technical report, Nanyang Technological University, Singapore (2013)
39. Lynn, N., Suganthan, P.N.: Heterogeneous comprehensive learning particle swarm optimization with enhanced exploration and exploitation. Swarm Evol. Comput. **24**, 11–24 (2015)

Low Reynolds Number Swimming with Slip Boundary Conditions

Hashim Alshehri, Nesreen Althobaiti, and Jian Du$^{(\boxtimes)}$

Florida Institute of Technology, Melbourne, FL 32901, USA
jdu@fit.edu

Abstract. We investigate the classical Taylor's swimming sheet problem in a viscoelastic fluid, as well as in a mixture of a viscous fluid and a viscoelastic fluid. Extensions of the standard Immersed Boundary (IB) Method are proposed so that the fluid media may satisfy partial slip or free-slip conditions on the moving boundary. Our numerical results indicate that slip may lead to substantial speed enhancement for swimmers in a viscoelastic fluid and in a viscoelastic two-fluid mixture. Under the slip conditions, the speed of locomotion is dependent in a nontrivial way on both the viscosity and elasticity of the fluid media. In a two-fluid mixture with free-slip network, the swimming speed is also significantly affected by the drag coefficient and the network volume fraction.

Keywords: Swimming sheet · Viscoelastic fluid · Slip condition · Immersed boundary method

1 Introduction

How micro-organisms move in their surrounding fluid environment is of significant biological and clinical importance. Examples include the locomotion of E.coli in intestinal fluid [1], and the swimming of mammalian spermatozoa within cervical mucus in the process of reproduction [2]. Such problems involve the dynamical interactions between elastic boundaries and a complex fluid medium, which often exhibits complicated Non-Newtonian responses. Recent theoretical, experimental and computational investigations are characterized by the complexity of different ways in which biological locomotion may depend on fluid properties. Analysis of the infinite undulatory sheet with small amplitude found that fluid elasticity always reduces the swimming speed [3]. Further analytical work indicated that swimming can be boosted by elasticity under specified gaits [4]. Numerical simulations of finite swimmers with large amplitude of motion showed that swimming speed may be enhanced by elasticity [5]. Experimentally, the self-propulsion of C. elegans was observed to be hindered significantly in viscoelastic fluid [6]. However, the artificial swimmers in [7] exhibited systematic elastic speed-ups. In [8] and [9], it was shown that favorable stroke asymmetry, swimmer body dynamics and fluid elasticity may work together to cause increases in speed.

© Springer Nature Switzerland AG 2020
V. V. Krzhizhanovskaya et al. (Eds.): ICCS 2020, LNCS 12141, pp. 149–162, 2020.
https://doi.org/10.1007/978-3-030-50426-7_12

In most of the analytical and numerical works to date, the fluid environment is treated as a single continuous medium. No-slip boundary condition is assumed on the swimmer's surface so that the fluid medium always moves together with the swimmer. Such models and assumptions may not be appropriate for many applications. First, biological fluids such as mucus are mixtures of a solvent and a polymer network. There may be significant relative motions between different components within the mixture so that it can not be adequately described by a single phase continuum medium [10]. Furthermore, it has been long known that slip may occur for polymer solutions near a solid boundary. This can be caused by the phase separation over the solvent-rich boundary region where the polymer phase is driven away [11]. Recent studies highlight the importance of boundary conditions and fluid models in locomotion problems. The analysis in [12] examined swimming in a medium consisting of a mixture of a Newtonian fluid and an elastic solid. Both elastic speed-up and slow-down can be obtained, depending on the type of boundary conditions imposed. In [13], it was shown analytically that the introduction of apparent slip or the reduction of fluid viscosity near the swimmer in Newtonian fluids may lead to faster swimming. In [14] and [15], different variations of the Immersed Boundary Method were proposed to simulate interactions between elastic boundaries and a two-phase medium. Despite these advances, a comprehensive analysis for the role of slip on swimmers in viscoelastic media is lacking.

In this paper, we present the first computational investigation of the role of slip for Taylor's classical swimming sheet in a single phase viscoelastic fluid, as well as in a mixture of a viscous fluid and a viscoelastic fluid. Our computational method is based on extensions of the classical Immersed Boundary Method [16] so that elastic boundaries are allowed to slip through the surrounding fluid media. In Sect. 2 and 3, the model equations and numerical methods are presented first, followed by simulation results which highlight the influence of slip on locomotion in complex fluids. The concluding remarks are given in Sect. 4.

2   Swimming in a Single Phase Viscous/Viscoelastic Fluid

2.1   Model Equations

Consider an infinite 2D sheet immersed in a incompressible, viscoelastic Oldroyd-B fluid. In its own frame, the movement of the sheet is described by $y = \epsilon \sin(kx - \omega t)$. The fluid equations are given by:

$$\nabla \cdot \boldsymbol{\sigma} - \nabla p = 0, \tag{1}$$

$$\nabla \cdot \mathbf{u} = 0, \tag{2}$$

where $\mathbf{u}$ is the fluid velocity, and p is the pressure. The total stress tensor is composed of viscous and polymeric contributions: $\boldsymbol{\sigma} = \mu_s(\nabla \mathbf{u} + \nabla \mathbf{u}^T) + \boldsymbol{\sigma}_p$, with

μ_s be the shear viscosity of the fluid. The polymer stress $\boldsymbol{\sigma}_p$ evolves according to constitutive equation:

$$\boldsymbol{\sigma}_p + \lambda \left(\frac{\partial \boldsymbol{\sigma}_p}{\partial t} + \mathbf{u} \cdot \nabla \boldsymbol{\sigma}_p - \nabla \mathbf{u}^T \cdot \boldsymbol{\sigma}_p - \boldsymbol{\sigma}_p \cdot \nabla \mathbf{u} \right) = \mu_p (\nabla \mathbf{u} + \nabla \mathbf{u}^T). \quad (3)$$

Here μ_p is the polymer viscosity and λ is the polymer relaxation time. On the sheet surface Γ, the fluid velocity $\mathbf{u}$ satisfies the following boundary conditions:

$$[\mathbf{u} \cdot \mathbf{n}]|_\Gamma = 0. \quad (4)$$

$$[\mathbf{u} \cdot \boldsymbol{\tau}]|_\Gamma = 2\Xi (\boldsymbol{\tau} \cdot \boldsymbol{\sigma} \cdot \mathbf{n})|_\Gamma. \quad (5)$$

$\mathbf{n}$ and $\boldsymbol{\tau}$ are unit vectors normal and tangential to the surface, respectively. The square bracket terms represent the components of the fluid velocity relative to the surface of the sheet (slip velocity). Ξ is the slip coefficient. Condition (4) states that the fluid and the sheet move together in the direction normal to the sheet surface. According to (5), the fluid is allowed to slip relative to the sheet in its tangential direction. The extent of slip is proportional to the local shear stress, as well as the slip constant Ξ. This is the well known Navier Slip Condition [17]. Note that the boundary conditions (4) and (5) apply to both the upper and lower surfaces of the sheet. Since Taylor's classical work [18], there have been many analytical and computational studies on different versions of the swimming sheet problem. See [19] for a complete review.

2.2   IB Method with Partial Slip Condition

The "classical" Immersed Boundary (IB) Method [16] is a powerful computational method capable of handling dynamic fluid-structure interactions. An Eulerian description is used for the fluid variables such as velocity and pressure, while a Lagrangian coordinate is used for each immersed elastic object. The simplicity and robustness of the IB method have led to its successful applications to many biological problems. Let $\mathbf{x}$ denote the fixed Eulerian coordinates and $\mathbf{X}(q, t)$ be the physical location of material points on the immersed object, which is parameterized by q. Let Ω be the fluid domain and Γ denote the Lagrangian domain. The equations for the coupled fluid-structure system are given by:

$$\nabla \cdot \boldsymbol{\sigma} - \nabla p + \mathbf{f} = 0, \quad (6)$$

$$\mathbf{f}(\mathbf{x}, t) = \int_\Gamma \mathbf{F}(q, t) \delta (\mathbf{x} - \mathbf{X}(q, t)) dq = S\mathbf{F}, \quad (7)$$

$$\frac{\partial \mathbf{X}(q, t)}{\partial t} = \int_\Omega \mathbf{u}(\mathbf{x}, t) \delta (\mathbf{x} - \mathbf{X}(q, t)) d\mathbf{x} = S^* \mathbf{u}. \quad (8)$$

Here δ denotes the Dirac delta function. (7) describes how the Lagrangian force density $\mathbf{F}$ is spread to the fluid and S represents the force spreading operator. (8) is based on the assumption that the immersed object moves with local fluid velocity (no-slip condition). S* is the velocity interpolation operator which is the adjoint of the spreading operator S.

IB method described above needs to be modified to handle slip conditions such as (5). This involves the evaluation of the interfacial fluid stresses on the irregular boundary, which can be computationally challenging [20]. On a Stokes swimmer, the elastic force $\mathbf{F}$ is balanced by the hydrodynamics forces (both viscous and viscoelastic), which can be calculated from the jump in fluid stress across the swimmer. For Taylor's sheet within an infinite domain, the tangential hydrodynamics forces on the two surfaces (Γ_+ and Γ_-) of the sheet are equal because of symmetry. So we have $\tau \cdot \sigma_+ \cdot \mathbf{n} = -\tau \cdot \sigma_- \cdot \mathbf{n}$. Thus the force balance on the sheet in the tangential direction gives $\mathbf{F} \cdot \tau = -\tau \cdot [\sigma] \cdot \mathbf{n} = -2\tau \cdot \sigma_+ \cdot \mathbf{n}$, where $[\sigma] = \sigma_+ - \sigma_-$ is the stress jump across the sheet. Therefore, the tangential component of the elastic force (which is straightforward to compute in IB method) can be directly used to enforce the slip boundary condition. Denote the boundary fluid velocity obtained from right hand side of (8) by $\mathbf{U}(\mathbf{X}(q,t))$, the sheet velocity $\mathbf{U}_\Gamma$ can then be computed by:

$$\mathbf{U}_\Gamma(\mathbf{X}) \cdot \mathbf{n} = \mathbf{U}(\mathbf{X}) \cdot \mathbf{n}, \tag{9}$$

$$\mathbf{U}_\Gamma(\mathbf{X}) \cdot \tau = \mathbf{U}(\mathbf{X}) \cdot \tau + \Xi \mathbf{F} \cdot \tau. \tag{10}$$

2.3   Discretization and Numerical Solutions

All fluid variables are discretized using a Cartesian grid, with constant grid space h. A MAC-type staggered computational grid is used for spatial discretization. Scalars are located at the grid centers and vectors are located at the grid edges. All components of the viscoelastic stress tensor σ_p are placed at the cell centers. The sheet is represented by a set of discrete IB points. Using centered difference for all spatial derivatives, the discretized equations from time t^k to $t^{k+1} = t^k + \Delta t$ are:

$$\mu_s \Delta_h \mathbf{u}^{k+1} + \nabla_h \cdot \sigma_p^{k+1} - \nabla_h p^{k+1} + S_h^k \mathbf{F}(\mathbf{X}^k) = 0, \tag{11}$$

$$\nabla_h \cdot \mathbf{u}^{k+1} = 0, \tag{12}$$

$$\mathbf{X}^{k+1} = \mathbf{X}^k + \Delta t \left((S_h^*)^k \mathbf{u}^{k+1} + \Xi \left(\mathbf{F}(\mathbf{X}^k) \cdot \tau^k \right) \tau^k \right). \tag{13}$$

Here Δ_h and ∇_h are discretized Laplacian and gradient operators, respectively. S_h^k and S_h^* are discretized version of the spreading and interpolation operators as defined in (7) and (8). The time iteration for the proposed scheme can be summarized as following:

1. Compute the elastic forces $\mathbf{F}(\mathbf{X}^k)$ on the sheet from its geometric configuration at t^k. Spread the Lagrangian force to the fluid grid.
2. Update the viscoelastic stress tensor σ_p^{k+1} from the discretization of (3) using extrapolated velocity at time level $t^{k+1/2}$ from values at t^k and t^{k-1}.
3. Solve (11) and (12) to get the values of $\mathbf{u}$ and p at t^{k+1}.
4. Update the positions of the IB points on the sheet according to (13).

Each IB point is connected by linear springs to its two neighboring points. It is also connected by a stiff spring to a corresponding "tether" point whose role is to impose the desired motion of the sheet. The unit tangent vector τ_j at the jth IB point $\mathbf{X}_j$ is approximated by $\tau_j = \frac{\tau_{j+1/2}+\tau_{j-1/2}}{2}$, where $\tau_{j+1/2} = \frac{\mathbf{X}_{j+1}-\mathbf{X}_j}{\|\mathbf{X}_{j+1}-\mathbf{X}_j\|}$. Surface normal $\mathbf{n}_j$ is obtained by a $\pi/2$ rotation of τ_j. The discretized operators S_h^k and S_h^* are constructed with the four-point cosine-based discrete delta function proposed by Peskin [16]. A multigrid solver with the box-type smoother is used to solve the coupled linear system from (11) and (12) [21]. Finally, to solve the stress Eq. (3), a high-resolution unsplit Godunov scheme is used to approximate the advection term explicitly. Crank-Nicolson approximation is used for the remaining terms. For each Eulerian grid cell, a 3×3 linear system is solved to update all components of $\boldsymbol{\sigma}_p$. See [22] for the detailed algorithm.

Our simulations are carried out in the domain $[0, 1] \times [-1, 1]$. The boundary condition in the x direction is periodic and that at $y = \pm 2$ is no-slip. The grid size is 128×256 and a constant time step $\Delta t = 10^{-4}$ is used for all simulations. For all results presented in this paper, we use $\epsilon = 0.012$, $k = \omega = 2\pi$, and $\mu_s = 1$. The swimming speed of the sheet is calculated by averaging the x velocity over all the IB points and over one wave period until a steady state value is obtained. To verify the proposed method, we first set $\boldsymbol{\sigma}_p$ to zero and compare the numerical results with the analytical solution given by [13]:

$$\frac{U}{U_0} = 1 + 4k\mu_s \Xi, \tag{14}$$

where U and U_0 are the second order swimming speeds of the sheet with and without slip, respectively. The no-slip swimming speed is given by $U_0 = -\frac{1}{2}k\omega\epsilon^2$. Note that the slip velocity in [13] is proportional to the shear rate, instead of the shear stress. So the slip length Λ as defined in [13] is related to our slip coefficient by $\Lambda = 2\mu_s \Xi$. From Fig. 1, it is clear that the numerical swimming speed increases linearly with the slip coefficient. And our simulation results agree well with the analytical solution. Next, we study the effect of slip on the swimmer in

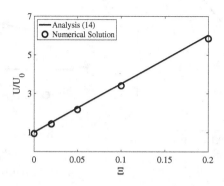

Fig. 1. Scaled swimming speed as a function of the slip coefficient: Taylor's sheet in a viscous fluid.

a viscoelastic medium. We carry out simulations with different slip coefficients under three fixed values of the relaxation time $\lambda = 2$, $\lambda = 0.2$, and $\lambda = 0.05$, respectively. The polymer viscosity is fixed at $\mu_p = 2$. The scaled swimming speed $\frac{U}{U_0}$ is plotted as the function of the slip coefficient in Fig. 2(a). Here the Deborah Number defined as $De = \lambda\omega$ is used to quantify the fluid elasticity. Note that in the plot, the analytical solution is plotted from (14), with μ_s replaced by the total viscosity of the fluid $\mu_s + \mu_p$. The numerical results indicate that apparent slip always enhances the swimming speed in a viscoelastic fluid. It seems that for a fixed Deborah Number, the swimming speed increases linearly with the slip coefficient Ξ, which is similar to the swimmer in a viscous fluid. For the same slip coefficient, the swimming speed decreases with the increase of the fluid elasticity. As the Deborah Number $De \to 0$, the numerical solutions approach asymptotically to the analytical solution for the viscous fluid. Next, we fix the relaxation time $\lambda = 0.2$ and study the influence of polymer viscosity on swimming under different slip coefficients. As shown in Fig. 2(b), when $\Xi = 0$, the swimming speed decreases monotonically with the increase of μ_p. The result matches well with the analytical solution given by (15) [3]. When the slip coefficient is moderately increased to 0.02, the swimming speed is not significantly impacted by the change of μ_p. And the variation is no longer monotone. For larger Ξ values of 0.05 and 0.1, greater values of μ_p always lead to a faster swimmer, whose speed changes more dramatically with μ_p than the one with smaller Ξ. Overall, the simulation results indicate that there exists a slip threshold beyond which the polymer viscosity can benefit swimming.

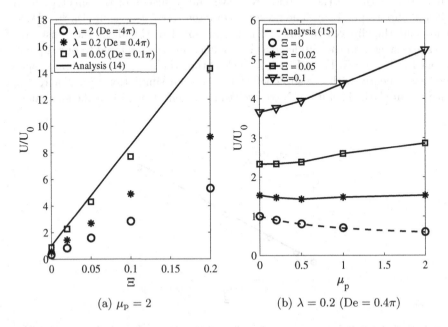

(a) $\mu_p = 2$ (b) $\lambda = 0.2$ ($De = 0.4\pi$)

Fig. 2. Scaled swimming speed as a function of the slip coefficient (a) and polymer viscosity (b): Taylor's sheet in an Oldroyd-B fluid.

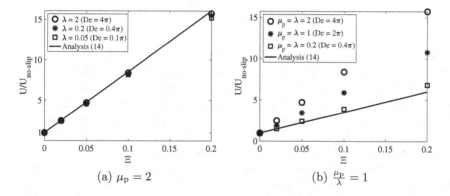

Fig. 3. Relative boost in swimming speed as a function of the slip coefficient: Taylor's sheet in an Oldroyd-B fluid. Note that $U_{no-slip}$ has different values for curves with different Deborah Numbers.

In Fig. 3(a), a different scaling is used for the same swimming speed U shown in Fig. 2(a). Here $U_{no-slip}$ is the analytical second order swimming speed for an infinite sheet in an Oldroyd-B fluid (without slip) [3]:

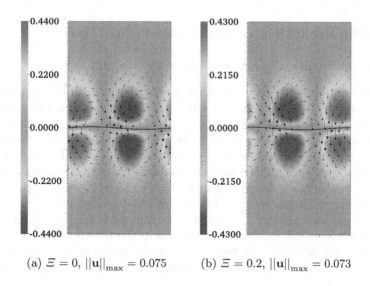

(a) $\Xi = 0$, $||\mathbf{u}||_{max} = 0.075$ (b) $\Xi = 0.2$, $||\mathbf{u}||_{max} = 0.073$

Fig. 4. Distribution of $\mathbf{u}$ and σ_p^{12} at $t = 8$ for different Ξ.

$$U_{\text{no-slip}} = \frac{1 + \frac{\mu_s}{\mu_s + \mu_p}\text{De}^2}{1 + \text{De}^2}U_0. \tag{15}$$

Therefore, $\frac{U}{U_{\text{no-slip}}}$ measures the relative slip boost for a swimmer in the same medium. Interestingly, for a fixed μ_p, the numerical results with different Deborah Numbers all match well with the analytical solution with De $= 0$. Therefore, for the range of parameters tested in this work, our results indicate that the relative slip boost for the infinite waving sheet is similar for a single phase viscous and a single phase viscoelastic fluid (with fixed fluid viscosity). In Fig. 3(b), the scaled speed is plotted for a fixed ratio of polymer viscosity to relaxation time $\frac{\mu_p}{\lambda} = 1$. Here the analytical solution is plotted from (14) without viscosity contribution from the polymer ($\mu_p = 0$). For fixed μ_s and ω, the ratio $\frac{\mu_p}{\lambda}$ measures the relative contribution of the polymeric stress to the force balance in fluid [5]. It is clear that for the same slip coefficient, the relative speed boost increases with the increase of Deborah Number. As the values of μ_p and λ decrease, the fluid behaves more like a viscous fluid with viscosity μ_s. In Fig. 4, the distributions of fluid velocity $\mathbf{u}$ and stress component σ_p^{12} at $t = 8$ are plotted for two simulations both with $\mu_p = 2$ and $\lambda = 0.2$. The one on the left has no-slip condition while the one on the right has slip coefficient $\Xi = 0.2$. Compared with the no-slip case, the magnitude of σ_p^{12} and $\mathbf{u}$ is slightly lower for the simulation with slip.

3   Swimming in Viscoelastic Two-Fluid Mixture

3.1   Model Equations and Two-Phase IB Method

In this section, we study the swimming sheet problem in a two-fluid mixture, which is modeled as a mixture of a viscous solvent phase (denote by s) and a viscoelastic network phase (denoted by n). The viscous solvent fluid satisfies the standard no-slip condition on the swimmer while the viscoelastic network fluid can slip freely in the direction tangential to the swimmer. Two-fluid models of this kind have been widely used to describe dynamics of biofluids such as blood clot, biofilm and cytoplasm [10, 23]. At any spatial location $\mathbf{x}$, the relative amounts of the two fluids are given by their volume fraction, $\theta_s(\mathbf{x}, t)$ and $\theta_n(\mathbf{x}, t)$ for the solvent and network, respectively. In this work, we treat θ_s and θ_n as model parameters with spatially uniform values constant in time. The solvent and network fluids move with their own velocity fields, $\mathbf{u}_s(\mathbf{x}, t)$ and $\mathbf{u}_n(\mathbf{x}, t)$. Mass conservation gives the incompressibility condition on the volume-averaged velocity:

$$\nabla \cdot (\theta_s \mathbf{u}_s + \theta_n \mathbf{u}_n) = 0. \tag{16}$$

For a small Reynolds number, the force balance equations for the two fluids are given by:

$$\nabla \cdot (\theta_s \boldsymbol{\sigma}_s) - \theta_s \nabla p + \xi \theta_n \theta_s (\mathbf{u}_n - \mathbf{u}_s) + \mathbf{f}_s = 0, \tag{17}$$

$$\nabla \cdot (\theta_n \boldsymbol{\sigma}_n) + \nabla \cdot (\theta_n \boldsymbol{\sigma}_p) - \theta_n \nabla p + \xi \theta_n \theta_s (\mathbf{u}_s - \mathbf{u}_n) + \mathbf{f}_n = 0. \tag{18}$$

Here, $\boldsymbol{\sigma}_s$ and $\boldsymbol{\sigma}_n$ are the viscous stress tensors for the solvent and network, respectively. $\boldsymbol{\sigma}_p$ is the viscoelastic stress tensor for the network fluid. $\xi\theta_n\theta_s(\mathbf{u}_n - \mathbf{u}_s)$ represents the frictional drag force between the two fluids due to relative motions where ξ is the drag coefficient. $\mathbf{f}_s$ and $\mathbf{f}_n$ are force densities generated by immersed elastic structures on the two fluids. $\boldsymbol{\sigma}_s$ and $\boldsymbol{\sigma}_n$ are taken to be those of Newtonian fluids:

$$\boldsymbol{\sigma}_s = \mu_s(\nabla \mathbf{u}_s + \nabla \mathbf{u}_s{}^T) + (\lambda_s \nabla \cdot \mathbf{u}_s)I, \tag{19}$$

$$\boldsymbol{\sigma}_n = \mu_n(\nabla \mathbf{u}_n + \nabla \mathbf{u}_n{}^T) + (\lambda_n \nabla \cdot \mathbf{u}_n)I. \tag{20}$$

Here I is the identity tensor, μ_s and μ_n are the shear viscosities and $\lambda_{s,n}+2\mu_{s,n}/d$ are the bulk viscosities of the solvent and network (d is the dimension). We choose $\lambda_{s,n} = -\mu_{s,n}$ so that the bulk viscosities in both phases are zero. The network fluid is treated as an Oldroyd-B fluid with constitutive equation given by (3), where $\mathbf{u}$ is replaced by network velocity $\mathbf{u}_n$. In [14], an Immersed Boundary Method was proposed to simulate interactions between elastic structures and mixtures of two fluids. A penalty method was used to enforce the no-slip condition for both fluids on the elastic boundaries. In this work, we propose an extension to the method which allows the elastic structure to slip through one of the materials in the mixture. As shown in Fig. 5, the infinite sheet is represented by the immersed boundary Γ^s, where the associated IB points $\mathbf{X}_s(q,t)$ move with local solvent velocity $\mathbf{u}_s$ (no-slip condition). A "virtual IB" Γ^n is introduced to enforce the no-penetration condition for the network fluid on the boundary. Material points on the virtual IB are denoted by $\mathbf{X}_n(q,t)$, which move with local network velocity $\mathbf{u}_n$. As indicated in the figure, each $\mathbf{X}_n$ is connected by a stiff spring (with zero rest length) to a corresponding "anchor point" $\mathbf{X}_n^a$ located on Γ^s. Similarly, each $\mathbf{X}_s$ is connected to an anchor point $\mathbf{X}_s^a$ located on Γ^n. The resulting force penalizes separation between Γ^s and Γ^n in the normal direction, without penalizing the relative motion tangential to the sheet surface. Using an analog of (7), Lagrangian force on Γ^s is distributed only to the solvent fluid and Lagrangian force on Γ^n is distributed only to the network fluid. In (17) and (18), the Eulerian force densities have the form $\mathbf{f}_s = \theta_s S\mathbf{F}_s^o + \theta_s\theta_n S\mathbf{F}_s^p$ and $\mathbf{f}_n = \theta_n S\mathbf{F}_n^o + \theta_s\theta_n S\mathbf{F}_n^p$. Here S is the force spreading operator defined before.

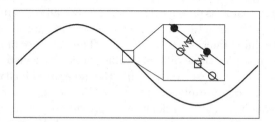

Fig. 5. Dual IB representation of an infinite swimmer $\bullet - \mathbf{X}_s$, $\circ - \mathbf{X}_n$, $\square - \mathbf{X}_s^a$, $\triangle - \mathbf{X}_n^a$. $\mathbf{X}_n$ and $\mathbf{X}_s$ are material points on the boundary. $\mathbf{X}_s^a$ and $\mathbf{X}_n^a$ are anchor points to enforce the no-penetration boundary condition for the network fluid.

$\mathbf{F}_s^p$ and $\mathbf{F}_n^p$ are penalty forces on Γ^s and Γ^n, respectively. At each IB point and the associated anchor point, we have $\mathbf{F}_s^p = -\mathbf{F}_n^p$. The spread contributions from the penalty forces are scaled by the product of the volume-fractions $\theta_s \theta_n$ so that no penalty force is applied if either of the volume fractions goes to zero. This also ensures that the total net penalty forces applied to the two fluids approximately add up to zero, provided that an IB point and its anchor point are always close (small normal separation between Γ^n and Γ^s). Other Lagrangian forces $\mathbf{F}_s^o$ and $\mathbf{F}_n^o$ are scaled by the fluid's volume fraction after they are spread to that fluid. These include forces from the springs connecting an IB point to its two neighbors. Additionally, $\mathbf{F}_s^o$ also include forces from the springs connecting Γ^s to tether points with prescribed waving motion.

3.2   Numerical Solutions

To solve the model equations presented in the previous section, we use the same space-time discretization as described in Sect. 2.3. The time iteration scheme is given by:

1. From the boundary configurations $\mathbf{X}_s(q, t^k)$ and $\mathbf{X}_n(q, t^k)$, identify the anchor points $\mathbf{X}_s^a$ and $\mathbf{X}_n^a$ for all IB points on the two boundaries. Compute boundary forces $\mathbf{F}_s^o$, $\mathbf{F}_n^o$, $\mathbf{F}_s^p$, and $\mathbf{F}_n^p$ at t^k. Use the values to calculate the Eulerian force densities $\mathbf{f}_s$ and $\mathbf{f}_n$ on the two fluids.
2. Update the viscoelastic stress tensor $\boldsymbol{\sigma}_p^{k+1}$ using extrapolated network velocity at time level $t^{k+1/2}$.
3. Solve discrete versions of (16), (17) and (18) to get the values of $\mathbf{u}_s$, $\mathbf{u}_n$ and p at t^{k+1}.
4. Update the positions of all IB points by $\mathbf{X}_j(q, t^{k+1}) = \mathbf{X}_j(q, t^k) + \Delta t (S_h^*)^k \mathbf{u}_j^{k+1}$ for $j = s, n$.

In step 1, $\mathbf{F}_s^p$ and $\mathbf{F}_n^p$ are computed at and spread from all IB and anchor points. For a specific $\mathbf{X}_s$, the associated anchor point $\mathbf{X}_s^a$ is defined as the point on the piece-wise linear boundary Γ^n such that $\|\mathbf{X}_s - \mathbf{X}_s^a\|$ is the shortest distance between $\mathbf{X}_s$ and Γ^n. The anchor point $\mathbf{X}_n^a$ on Γ^s for $\mathbf{X}_n$ is identified similarly. In step 3, a multigrid preconditioned GMRES solver is used to solve the linear system [22]. All simulation parameters such as computational domain, grid size and time step are the same as ones used in Sect. 2.3. For all simulations, we set the viscosity values to $\mu_s = \mu_n = 1.0$. In the first set of test, we fix the drag coefficient $\xi = 1.0$ and fluid volume fractions $\theta_s = \theta_n = 0.5$. The influence of relaxation time on swimming is studied for three different values of polymer viscosities $\mu_p = 0.5$, $\mu_p = 2$ and $\mu_p = 4$. In Fig. 6(a) and (b), the relative velocity $\mathbf{u}_n - \mathbf{u}_s$ and the stress component σ_p^{12} are plotted for $\mu_p = 0.5$ and $\mu_p = 2$, respectively, at $t = 12$. In both plots, the relative velocity is approximately tangent to the sheet, indicating the boundary condition is properly enforced. With a larger polymer viscosity, the stress component has larger magnitude while the motion of the network relative to the solvent fluid is less significant. The scaled swimming speed is shown in Fig. 7(a) as the function of λ for different values of μ_p. The

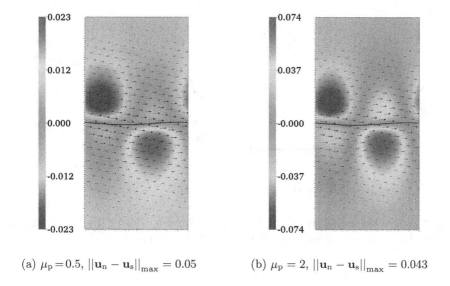

(a) $\mu_{\mathrm{p}} = 0.5$, $\|\mathbf{u}_{\mathrm{n}} - \mathbf{u}_{\mathrm{s}}\|_{\max} = 0.05$ (b) $\mu_{\mathrm{p}} = 2$, $\|\mathbf{u}_{\mathrm{n}} - \mathbf{u}_{\mathrm{s}}\|_{\max} = 0.043$

Fig. 6. Distribution of $\mathbf{u}_{\mathrm{n}} - \mathbf{u}_{\mathrm{s}}$ and σ_{p}^{12} for $\lambda = 2$ at t = 12.

plots indicate that the sheet always swims much faster in the mixture than in a viscous fluid, even when the mixture contains a highly elastic network. For fixed polymer viscosity, the increase of the network elasticity monotonically hinders the swimming speed. For a fixed λ, the sheet moves faster in mixtures with larger values of μ_{p}. The speed enhancement due to polymer viscosity is more significant for less elastic mixture. Next, we carry out simulations in which both μ_{p} and λ are varied while the values of their ratio $\frac{\mu_{\mathrm{p}}}{\lambda}$ remain fixed. As seen from Fig. 7(b), with fixed $\frac{\mu_{\mathrm{p}}}{\lambda}$, the swimming speed is always moderately enhanced when μ_{p} and λ increase with the same rate. Together with the data shown in Fig. 2(b), our simulation results suggest that for a swimmer that is allowed to slip through a viscoelastic material (or mixture of materials), the speed of locomotion is dependent in a nontrivial way on both the viscosity and elasticity of the material. In Fig. 8(a), the swimming speed is plotted as the function of the drag coefficient ξ for $\theta_{\mathrm{n}} = 0.5$. The sheet moves slower with the increase of drag. For a drag coefficient of $\xi = 10^4$, the swimming speed is about 60% of that in a viscous fluid. In Fig. 8(b), $\frac{U}{U_0}$ is plotted for different network volume fraction θ_{n} with ξ fixed at 1. The increase of the network volume fraction in the mixture leads to significant swimming speed-ups. In a separate test with no-slip network, we observe smaller swimming speed with the increase of θ_{n} (result not shown).

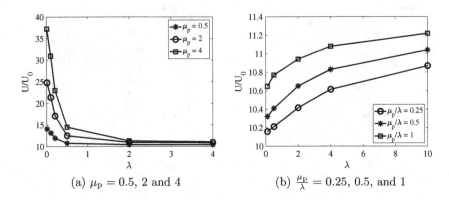

(a) $\mu_p = 0.5$, 2 and 4   (b) $\frac{\mu_p}{\lambda} = 0.25$, 0.5, and 1

Fig. 7. Scaled swimming speed as a function of relaxation time λ: Taylor's sheet in a two-fluid mixture with no-slip solvent fluid and free-slip network fluid.

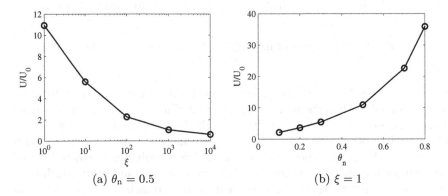

(a) $\theta_n = 0.5$ (b) $\xi = 1$

Fig. 8. Scaled swimming speed as a function of the drag coefficient (a) and the network volume fraction (b): Taylor's sheet in a two-fluid mixture. $\mu_p = 2$, $\lambda = 2$.

4   Conclusion

We simulate the infinite swimming sheet problem in complex fluids under slip boundary conditions with extensions of the classical IB method. For swimmers in a viscoelastic fluid, interpolated fluid velocities are modified using tangential components of the Lagrange force to account for the partial slip condition. This can be thought as the single-phase version of the force calculation strategy proposed in [15]. In a viscoelastic two-fluid mixture, a dual IB representation of the immersed structure is used where the free-slip condition is enforced through a penalty method. Instead of the projection-based fractional step methods as used in [15], we solve the momentum equations and the incompressibility constraint simultaneously. This makes it more straightforward to enforce the velocity boundary conditions. Furthermore, our method can be directly applied to problems where fluid volume fractions are spatially variable. For such problems, methods for Stokes equations that decouple the velocity and the pressure, such as the pressure-Poisson formulation, can not be used. Our numerical results show that: (1) Slip may lead to substantial speed enhancement for the swimmer in a viscoelastic fluid or two-fluid mixture relative to the swimmer in a no-slip viscous fluid. (2) For a viscoelastic fluid with fixed viscosity and relaxation time, the swimming speed increases linearly with the slip coefficient. With fixed viscosity and slip coefficient, the swimming speed decreases with the increase of relaxation time (fluid elasticity). (3) While polymer viscosity always hinders swimming for a no-slip viscoelastic fluid, it can benefit the swimmer in a viscoelastic fluid if the slip coefficient is large enough. (4) In a two-fluid mixture where the swimmer is allowed to slip freely through the viscoelastic network, speed enhancement can be obtained by reducing the drag coefficient, increasing the polymer viscosity, and increasing the network volume fraction.

References

1. Berg, H.: E. Coli in Motion. Springer, New York (2004). https://doi.org/10.1007/b97370
2. Fauci, L., Dillon, R.: Biofluidmechanics of reproduction. Ann. Rev. Fluid Mech. **38**, 371–394 (2006)
3. Lauga, E.: Propulsion in a viscoelastic fluid. Phys. Fluids **19**(8), 083104 (2007)
4. Riley, E., Lauga, E.: Small-amplitude swimmers can self-propel faster in viscoelastic fluids. J. Theor. Biol. **382**, 345–355 (2015)
5. Teran, J., Fauci, L., Shelley, M.: Viscoelastic fluid response can increase the speed and efficiency of a free swimmer. Phys. Rev. Lett. **104**(3), 038101 (2010)
6. Shen, X., Arratia, P.: Undulatory Swimming in viscoelastic fluids. Phys. Rev. Lett. **106**, 208101 (2011)
7. Espinosa-Garcia, J., Lauga, E., Zenit, R.: Fluid elasticity increases the locomotion of flexible swimmers. Phys. Fluids **25**, 031701 (2013)
8. Thomases, B., Guy, R.: Mechanisms of elastic enhancement and hindrance for finite length undulatory swimmers in viscoelastic fluids. Phys. Rev. Lett. **113**(9), 098102 (2014)
9. Thomases, B., Guy, R.: The role of body flexibility in stroke enhancements for finite-length undulatory swimmers in viscoelastic fluids. J. Fluid Mech. **825**, 109–132 (2017)
10. Cogan, N., Guy, R.: Multiphase flow models of biogels from crawling cells to bacterial biofilms. HFSP J. **4**(1), 11–25 (2010)
11. Barnes, H.: A Review of the slip (Wall Depletion) of polymer solutions, emulsions and particle suspensions in viscometers. J. Non-Newton. Fluid Mech. **56**(3), 221–251 (1995)
12. Fu, H., Shenoy, V., Powers, T.: Low-Reynolds-number swimming in gels. Europhys. Lett. **91**(2), 24002 (2010)
13. Man, Y., Lauga, E.: Phase-separation models for swimming enhancement in complex fluids. Phys. Rev. E **92**, 023004 (2015)
14. Du, J., Guy, R., Fogelson, A.: An immersed boundary method for two-fluid mixtures. J. Comput. Phys. **262**, 231–243 (2014)
15. Lee, P., Wolgemuth, C.: An immersed boundary method for two-phase fluids and gels and the swimming of C. elegans through viscoelastic fluids. Phys. Fluids **28**(1), 011901 (2016)

16. Peskin, C.: The immersed boundary method. Acta Numerica **11**, 479–517 (2002)
17. Sochi, T.: Slip at fluid-solid interface. Polym. Rev. **51**(4), 309–340 (2011)
18. Taylor, G.: Analysis of the swimming of microscopic organisms. Proc. Roy. Soc. A **209**, 447–461 (1951)
19. Lauga, E., Powers, T.: The hydrodynamics of swimming microorganisms. Rep. Prog. Phys. **72**(19), 096601 (2009)
20. Williams, H., Fauci, L., Gaver III, D.: Evaluation of interfacial fluid dynamical stresses using the immersed boundary method. Disc. Continuous Dyn. Syst. Ser. B **11**(2), 519–540 (2009)
21. Vanka, S.: Block-implicit multigrid solution of Navier-Stokes equations in primitive variables. J. Computat. Phys. **65**(1), 138–158 (1986)
22. Wright, G., Guy, R., Du, J., Fogelson, A.: A high-resolution finite-difference method for simulating two-fluid, viscoelastic gel dynamics. J. Non-Newton. Fluid Mech. **166**, 1137–1157 (2011)
23. Du, J., Fogelson, A.: A two-phase mixture model of platelet aggregation. Math. Med. Biol. **35**(2), 225–256 (2018)

Trilateration-Based Multilevel Method for Minimizing the Lennard-Jones Potential

Jithin George[1] and Zichao (Wendy) Di[2(✉)]

[1] Department of Engineering Sciences and Applied Mathematics,
Northwestern University, Evanston, IL, USA
[2] Mathematics and Computer Science Division, Argonne National Laboratory,
Lemont, IL, USA
wendydi@anl.gov

Abstract. Simulating atomic evolution for the mechanics and structure of materials presents an ever-growing challenge due to the huge number of degrees of freedom borne from the high-dimensional spaces in which increasingly high-fidelity material models are defined. To efficiently exploit the domain-, data-, and approximation-based hierarchies hidden in many such problems, we propose a trilateration-based multilevel method to initialize the underlying optimization and benchmark its application on the simple yet practical Lennard-Jones potential. We show that by taking advantage of a known hierarchy present in this problem, not only a faster convergence, but also a better local minimum can be achieved comparing to random initial guess.

Keywords: Multilevel optimization · Lennard-Jones potential · Nonlinear optimization

1 Introduction

Simulating interactions between atoms/molecules is particularly essential for understanding materials properties in materials science, chemistry, and biology. However, such problem presents an ever-growing challenge related to the huge number of degrees of freedom borne from the high-dimensional spaces in which increasingly high-fidelity material models are defined. One way is to simulate the arrangement of atoms by minimizing the potential energy. Among various mathematical models describing the interaction, the Lennard-Jones (LJ) potential [9] has attracted a lot of theoretical and computational attention due to its mathematical simplicity yet practical importance of discovering low energy configurations of clusters of atoms. There are two main difficulties in solving this problem. First, the non-convexity and highly non-linearity of many variables lead to a large number of local minima [3,13], and second, the intensive computational burden of many variables leads to slow convergence [3,6]. Therefore, an acceleration of LJ potential with potentially better local minimum is desired.

© Springer Nature Switzerland AG 2020
V. V. Krzhizhanovskaya et al. (Eds.): ICCS 2020, LNCS 12141, pp. 163–175, 2020.
https://doi.org/10.1007/978-3-030-50426-7_13

Given the domain-, data-, and approximation-based hierarchies present in this problem, it is natural to exploit them to speedup the convergence of LJ potential. One known property is that the Mackay icosahedron [4,10,19,20] dominates the structures of LJ clusters at least in the size range of 10–150 atoms. In this work, we observe similar property as icosahedronal packing by examining the local minimum obtained by truncated-Newton method [12] using random initial guess. This observation inspires us to adapt the successive refinement [2,15] developed in traditional multigrid method to accelerate the LJ convergence. Therefore, we propose a novel interpolation method as part of a successive refinement framework so that a better initial guess can be obtained for a larger system from the solution of a smaller system. We benchmark our method on various sizes of systems and show that the proposed method can lead to a much faster convergence. Furthermore, in our examples, we show that the proposed hierarchical approach can potentially lead to an overall better local minimum with a lower energy than using random initial guess.

2 Mathematical Model

Atoms arrange themselves spatially to minimize a potential that is an accumulation of all the different kinds of attraction and repulsion at that scale. The Lennard-Jones potential is a simplistic attempt to model such interaction.

Given a configuration of N atoms in a cluster $\mathbf{V} = \{\mathbf{v}_1, \ldots, \mathbf{v}_N\}$, its potential is given by

$$E_N(\mathbf{V}) = 4\varepsilon \sum_{i<j} \left(\left(\frac{\sigma}{d(\mathbf{v}_i, \mathbf{v}_j)} \right)^{12} - \left(\frac{\sigma}{d(\mathbf{v}_i, \mathbf{v}_j)} \right)^6 \right), \tag{1}$$

where $\mathbf{v}_i$ and $\mathbf{v}_j$ are the locations of the ith and jth atoms, and $d(\mathbf{v}_i, \mathbf{v}_j)$ is the Euclidean distance between atoms $\mathbf{v}_i, \mathbf{v}_j \in \mathbb{R}^3$. The physical constants ε and σ are the depth of the potential well and inter-atom reaction limit, respectively, both depending on the type of atom. Because of its importance in computational chemistry, finding optimal configurations that locally minimize the potential energy (1) remains an active research area. For example, Maranas et al. [11] proposed an exotic optimization algorithm to find many stationary points, and Asenjo et al. [1] studied the mapping of the basins of attraction for various optimization algorithms. In this work, we follow the longstanding convention in [8] and consider the reduced-unit optimization problem

$$\min_{\mathbf{V}} E_N = \sum_{i<j} \left(\frac{1}{||\mathbf{v}_i - \mathbf{v}_j||^{12}} - \frac{1}{||\mathbf{v}_i - \mathbf{v}_j||^6} \right). \tag{2}$$

The corresponding gradient is given by

$$\nabla E_N(\mathbf{v}_i) = -\sum_{i<j} \left(\frac{12}{||\mathbf{v}_i - \mathbf{v}_j||^{14}} - \frac{6}{||\mathbf{v}_i - \mathbf{v}_j||^8} \right)(\mathbf{v}_i - \mathbf{v}_j). \tag{3}$$

To avoid unnecessary degrees of freedom, we fix atom 1 at the origin, atom 2 on the x-axis and atom 3 on the x-y plane. Figure 1 illustrates the classical Leonard-Jones potential E_2. We can see that even minimizing the smallest system can be challenging due to either the singularity as $\|v_i - v_j\| \to 0$ or stationary but not local minimum as $\|v_i - v_j\| \to \infty$. We can further extend this visualization to E_3, where we fix v_1 at the origin and v_2 on the y-axis at the optimal distance of $r^* = 2^{\frac{1}{6}}$ (discussed more in Sect. 3) from the origin, then we examine the optimization performance by freeing v_3 on x-y plane. Again, through the paper, we use truncated-Newton method [12] to perform all optimization procedure due to its robust performance. Figure 2a shows the number of iterations (indicated by color intensity) needed for convergence at different initial locations of v_3 on x-y plane. Figure 2b shows the corresponding energy (indicated by color intensity) obtained while optimization converges. Again, even for the three atoms case, the basins of attractions are so rich in its complexity that a good initial guess is desired for a fast optimization process.

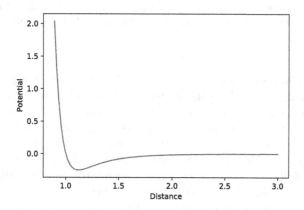

Fig. 1. The LJ potential E_2 by varying only v_2.

3   Method

One standard practice for initialization of the minimization of E_N is by using random initial guess. The random locations of the atoms are obtained by uniform sampling in a box of size $2\sqrt{3}r^*$ [1]. As a side note, we compared it to another practice where the box sizes vary as $\left(\frac{N}{0.8442}\right)^{1/3}$ [21] (0.8442 is a specific density value). We observe that the fixed box has comparable performance for the range of number of atoms we study, therefore we use it through the paper.

There is often visual modularity associated with the LJ minima. This fact inspired many researchers to leverage the geometric principles of LJ minima to generate good initial guesses (e.g., [17,22]). For example, for minimizing E_N, instead of random initial guess, we initialize the iterate by simply adding an

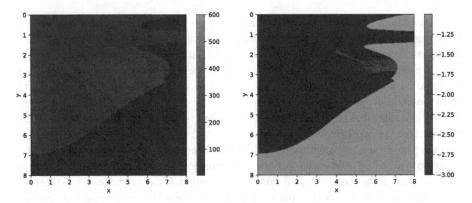

Fig. 2. Left: for E_3 with only $\mathbf{v}_3$ can freely move on the x-y plane, the number of iterations (indicated by color) needed for convergence for different initial location of $\mathbf{v}_3$. The global minimum exists at approximately (0.56, 0.97), which can only be achieved if the initial guess is placed near it. Right: the energy landscape (indicated by color) of the final iterate when optimization converges. Notice that starting at the points in yellow or green regions converges to the local but not global minimum. (Color figure online)

atom randomly to a local minimum of E_{N-1}. As we can see in Fig. 3, compared to a random initial guess, using the initial guess built upon the local minimum of E_{N-1} requires much fewer number of iterations to converge, without sacrificing the quality of the local minimum. This observation prompts us to search for a multilevel approach to the LJ optimization, where one could generate good initial guesses by exploiting the hierarchical structure embedded in different systems, i.e., by interpolating solutions from small systems to serve as initial guess for large system, we could expect a more robust convergence.

Let us first consider the pairwise interaction between atoms $\mathbf{v}_i$ and $\mathbf{v}_j$, which is expressed by the following:

$$\frac{1}{||\mathbf{v}_i - \mathbf{v}_j||^{12}} - \frac{1}{||\mathbf{v}_i - \mathbf{v}_j||^6}. \tag{4}$$

It is easy to derive that the minimum of Eq. 4 is obtained when $||\mathbf{v}_i - \mathbf{v}_j|| = r^* = 2^{1/6}$, where r^* is the ideal distance of separation between two atoms. Therefore, the global minima of E_3 and E_4 in 3D are equilateral triangle and tetrahedron, respectively, with pairwise distance r^*.

For larger systems, we first define the neighboring atoms of the atom $\mathbf{v}_i$ as the ones connected with $\mathbf{v}_i$ by edges of Delaunay triangulation [7]. Notice that the optimal pairwise distance of r^* is not maintained between all neighboring atoms for $N > 3$ in 2D and $N > 4$ in 3D. Then an intuitive expectation is that an arrangement of the atoms, where all the pairwise neighboring distances are approximating r^*, can be a good initial guess. This motivates the main idea we propose in this paper, which is to add new atoms to the boundary of a coarser atomic configuration (i.e., a smaller system with fewer atoms) at its local

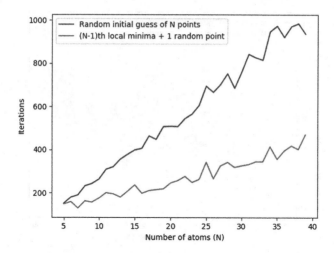

Fig. 3. The number of iterations needed to find a local minimum for E_N in 3D greatly reduces when a local minimum of E_{n-1} in addition to a random point is used as the initial guess

minimum so that the distance between the new atom and its nearest neighbours from amongst the existing atoms is r^*.

We employ the method generally known as trilateration to perform this action. Simply speaking, in 2D, we choose two neighboring atoms (i.e., connected by an edge in the Delaunay triangulation) and draw circles around them with radius r^*, then the two intersection points of the circles provide two candidates as the new atoms to be added. For example, given two atoms at $(0,0)$ and $(d,0)$ where $d \le 2r^*$, respectively, then the intersection points of trilateration are $\left(\dfrac{d}{2}, \pm\sqrt{r^{*2} - \dfrac{d^2}{4}}\right)$. This idea is illustrated in Fig. 5 and can be extended to 3D. This trilateration technique of circle-circle intersection in 2D or sphere-sphere-sphere intersection in 3D has been used in various applications such as surveying and GPS systems [5,14,23].

After performing trilateration on every edge on the boundary of the atomic configuration, a list of candidate locations for the new atoms are obtained as shown in Fig. 6. Each location is evaluated by adding an atom there and seeing if the resulted system has smaller energy, since the energies of stable systems should decrease as the number of atoms increase. If adding an atom at a location increases the energy of the system, this location is discarded. For simplicity, we summarize the detailed procedure (denoted as trilateration-based multilevel method) in 2D as follows (Fig. 4),

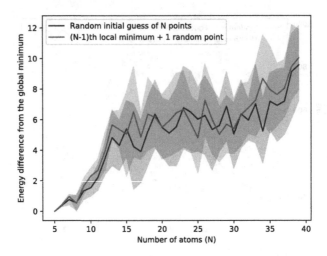

Fig. 4. Comparable potential energies obtained by the two different ways of initialization. The shaded area denotes the standard deviation. (Color figure online)

- Given a local minimum $\mathbf{V}_N$ of E_N (use the global minimum if it is available), apply Delaunay triangulation on it to generate triangles (simplices in 3D).
- Generate the smallest convex hull which encloses all the N atoms, and identify its corresponding edges.
- Selection of edges:

  - Select edges that only have two atoms lying on it.
  - If an edge on the convex hull has more than 2 atoms on it, divide that edge into multiple edges comprising only of 2 atoms and select all the edges thus formed.
  - If the length of an edge is longer than $2r^*$, identify the triangular simplex that the edge belongs to. Select the other two edges in the triangular simplex instead of the original edge. If any of the edges formed this way have more than 2 atoms, divide it up into multiple edges with 2 atoms and select them all.

- On each edge among the selected edges, use the trilateration method to find candidate locations to place the new atoms.
- Take all the candidate locations found this way and add them one at a time to the original system of atoms. If the corresponding potential energy decreases, the new atom becomes part of the system. Otherwise, it is rejected.
- Once the new configuration is obtained with m atoms where $m > N$, it is used as an initial guess for minimizing E_m. One could also perform Delaunay triangulation and the trilateration procedure again on the new system to get a larger system of atoms.
- This process can be repeated until we reach a desired system size.
- Perform optimization on the final iterate giving us the generated initial guess to find a minimum.

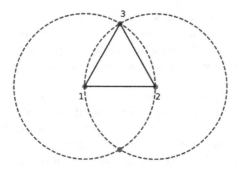

Fig. 5. Using trilateration on the bottom edge to find two new locations at a distance r^* from both atoms on the bottom edge. One of the locations is already occupied by an existing atom and thus only one new position is obtained (in green). (Color figure online)

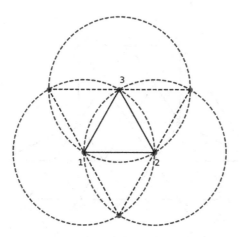

Fig. 6. Given existing atoms v_1, v_2, and v_3, trilateration on all the edges is used to find locations to place the new atoms (in green). The resulting structure, where the dashed lines are the new edges, serves as an ideal initial guess for the minimum of E_6. (Color figure online)

If the user desires a minimum for a particular number of atoms, this method can be applied to find a minimum for a system of size close to the desired number. Then, the remaining few atoms can be added either randomly or using trilateration as a heuristic. The resulted configuration can still serve as a good initial guess as suggested by Fig. 3.

The selection of the edge for trilateration is a critical step. For example, trilateration does not provide any points of intersection if the edge on the boundary (obtained from the convex hull) is longer than $2r^*$. If this fact is not taken into consideration, eventually the interpolation will stagnate as shown in Fig. 8. Therefore, in the proposed algorithm, we provide a threshold to carefully choose the right boundary for trilateration, i.e., if the boundary edge is too long (e.g., $>2r^*$), the algorithm opts into the concave boundary as suggested in Fig. 7. This guarantees our method to be scalable and avoids problems as shown in Fig. 8 where the boundaries are too long for the algorithm to perform trilateration anywhere.

In terms of the computational complexity added by the proposed multilevel method, since the pairwise distance between all atoms is already available by calculating the potential, it is trivial to perform the threshold criteria for edge selection. The trilateration for each selected edge has a relatively simple analytical formula, therefore, this part of calculation is negligible as well. The main cost then is on the generation of the convex hull and the Delaunay triangulation which are $O(N \log N)$. Since these operations are one-time upfront cost, the overall added complexity by the proposed trilateration-based multilevel method for initialization is negligible comparing to one iteration cost of the underlying optimization.

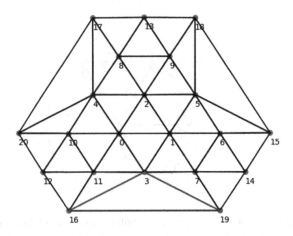

Fig. 7. The blue edge created by convex hull is too long for trilateration. Hence, our algorithm chooses the concave green boundary obtained by Delaunay triangulation and performs trilateration on the two edges that make it. (Color figure online)

4   Numerical Results

In this section, we demonstrate the performance of our proposed method for various sizes of systems primarily in 2D, with a preliminary exploration on 3D. Notice that in all experiments, we perform the optimization at the final step when the number of atoms reaches a desirable number. For the optimization step, we emphasize that any general-purpose optimization methods can be used in conjunction with the multilevel approach proposed in this paper. For our illustration purposes, we use the Truncated-Newton algorithm in the Scipy Python

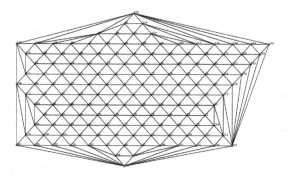

Fig. 8. Not selecting the edges carefully (i.e., simply choosing the boundary of the convex hull) results in long edges of the convex hull enclosing all the atoms, and therefore the stagnation of the proposed trilateration method due to no intersection points.

package [18]. Again, as for comparison, we evaluate the performance of our proposed method against randomly initialized guesses which is a standard practice for initialization.

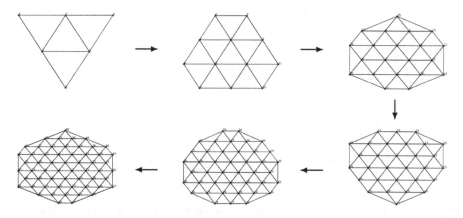

Fig. 9. The multilevel process to extend a small system to a large system based on the proposed trilateration approach.

4.1   The 2 Dimensional Case

Figure 9 demonstrates the iterates obtained at every step of the trilateration process with the initial system as the global minimum of E_3. In Fig. 10, we compare the number of iterations needed for the convergence of the optimization process for E_N when the initial guess is chosen randomly versus if the initial guess is obtained using the trilateration procedure. We see dramatic savings in the number of iterations required.

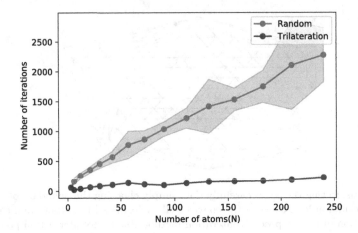

Fig. 10. The number of iterations needed for the convergence of the optimization algorithm. Using an initial guess obtained by the trilateration method results in far fewer iterations. The shaded area signifies the standard deviation in the random trials.

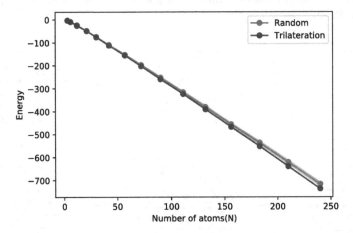

Fig. 11. The energy of the local minimum obtained by using the initial guess from the trilateration process as opposed to taking a random initial guess.

The question that remains is if the minimum found using the initial guess from trilateration is better than a random initial guess. Figure 11 shows that in average, the minimum obtained by the proposed trilateration has a lower energy than the one from averaged trials of random initial guesses.

4.2   The 3 Dimensional Case

We also demonstrate a preliminary extension of the proposed method to 3D. As briefly shown in Fig. 12, the reduced number of iterations needed for convergence by trilateration is maintained in the 3D case. In Fig. 13, we use the energy

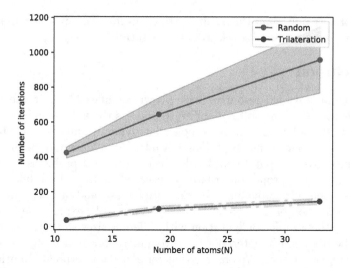

Fig. 12. The number of iterations needed for the convergence of the optimization algorithm in 3D. The proposed method shows promise for speeding up the optimization process with reduced number of iterations.

of the global minimum extracted from the online database [20] as a reference, and compare it against the energies of the minima found by random initial guesses and those found through the trilateration method, respectively. Due to the increased geometric complexity in terms of locating the concave regions arising in 3D, the edge selection criteria in 2D can not be trivially extended to

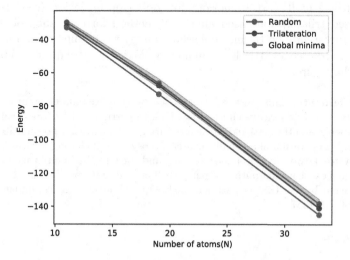

Fig. 13. The energy of the local minimum obtained by using the initial guess from the trilateration process as opposed to taking a random initial guess in 3D.

3D, therefore, we only show the results up to 35 atoms. We leave the development of more optimal way for edge selection to the future.

5   Conclusion

Although computational resources have grown significantly over the last few years, their growth alone will not suffice to address the volume and complexity encountered in applications such as molecular dynamic simulations. Multilevel methods presents one avenue for taking advantage of a known hierarchy to overcome the intensive computational burden bearing in such problems.

In this work, we propose a trilateration-based multilevel method to speed up the convergence by providing a better initial guess, and prototype its benefit on the minimization of the Lennard-Jones potential. We exploit the hierarchies embedded in the atomic configuration with different sizes so that a good initial guess can be obtained by interpolating through trilateration from a local minimum of a smaller system. We observe that for 2D, the interpolated initial guess can save the number of iteration greatly comparing to random initial guess, without sacrificing the quality of the minimum. Although only preliminary results for 3D are shown for limited sizes of system, it is promising that the proposed method can be scaled to practical use. However, a more careful design of the framework is needed, for example, how to more optimally choose the edge to perform trilateration is critical for the scalability of the proposed method.

Although the current work can not immediately address the challenge of finding global minimum, one simple way to take advantage of our trilateration-based multilevel method is to plug in to a more complicated framework for global optimization such as the one proposed in [3]. For example, the random initial guess can be simply replaced by the proposed method in this work. Alternatively, one can utilize multistart framework for global optimization (e.g, [16]) with the proposed trilateration-based multilevel method for each subproblem. One could also investigate better interpolation strategy to combine trilateration with physics to guarantee a better local minimum. We leave these questions to be the focus for the future.

Acknowledgments. This work was supported by the Exascale Computing Project (17-SC-20-SC), a collaborative effort of two U.S. Department of Energy organizations (Office of Science and the National Nuclear Security Administration) responsible for the planning and preparation of a capable exascale ecosystem, including software, applications, hardware, advanced system engineering, and early testbed platforms, in support of the nation's exascale computing imperative. The material was also based in part on work supported by the U.S. Department of Energy, Office of Science, under contract DE-AC02-06CH11357.

References

1. Asenjo, D., Stevenson, J.D., Wales, D.J., Frenkel, D.: Visualizing basins of attraction for different minimization algorithms. J. Phys. Chem. B **117**(42), 12717–12723 (2013)

2. Bornemann, F.A., Deuflhard, P.: The cascadic multigrid method for elliptic problems. Numerische Mathematik **75**(2), 135–152 (1996)
3. Cai, W., Feng, Y., Shao, X., Pan, Z.: Optimization of Lennard-Jones atomic clusters. J. Mol. Struct.: THEOCHEM **579**(1–3), 229–234 (2002)
4. Daven, D., Tit, N., Morris, J., Ho, K.: Structural optimization of Lennard-Jones clusters by a genetic algorithm. Chem. Phys. Lett. **256**(1–2), 195–200 (1996)
5. Fang, B.T.: Trilateration and extension to global positioning system navigation. J. Guidance Control Dyn. **9**(6), 715–717 (1986)
6. Frantsuzov, P.A., Mandelshtam, V.A.: Size-temperature phase diagram for small Lennard-Jones clusters. Phys. Rev. E **72**(3), 037102 (2005)
7. Geršgorin, S.: Bulletin de l'Académie des sciences de l'URSS. classe des sciences mathématiques et naturelles (1931)
8. Hoare, M., Pal, P.: Physical cluster mechanics: statics and energy surfaces for monatomic systems. Adv. Phys. **20**(84), 161–196 (1971)
9. Jones, J.E.: On the determination of molecular fields, II: from the equation of state of a gas. Proc. Roy. Soc. Lond. A **106**(738), 463–477 (1924)
10. Mackay, A.: A dense non-crystallographic packing of equal spheres. Acta Crystallographica **15**(9), 916–918 (1962)
11. Maranas, C.D., Floudas, C.A.: A global optimization approach for Lennard-Jones microclusters. J. Chem. Phys. **97**(10), 7667–7678 (1992)
12. Nash, S.G.: A survey of truncated-Newton methods. J. Comput. Appl. Math. **124**(1–2), 45–59 (2000)
13. Retzlaff, M., Munson, T., Di, Z.W.: Data compression for optimization of a molecular dynamics system: preserving basins of attraction. In: Rodrigues, J.M.F., et al. (eds.) ICCS 2019. LNCS, vol. 11538, pp. 457–470. Springer, Cham (2019). https://doi.org/10.1007/978-3-030-22744-9_36
14. Thomas, F., Ros, L.: Revisiting trilateration for robot localization. IEEE Trans. Robot. **21**(1), 93–101 (2005)
15. Trottenberg, U., Oosterlee, C.W., Schuller, A.: Multigrid. Elsevier, London (2000)
16. Ugray, Z., Lasdon, L., Plummer, J., Glover, F., Kelly, J., Martí, R.: Scatter search and local NLP solvers: a multistart framework for global optimization. INFORMS J. Comput. **19**(3), 328–340 (2007)
17. Uppenbrink, J., Wales, D.J.: Packing schemes for Lennard-Jones clusters of 13 to 150 atoms: minima, transition states and rearrangement mechanisms. J. Chem. Soc. Faraday Trans. **87**(2), 215–222 (1991)
18. Virtanen, P., et al.: SciPy 1.0-Fundamental Algorithms for Scientific Computing in Python. arXiv e-prints arXiv:1907.10121, July 2019
19. Wales, D.J., Doye, J.P.: Global optimization by basin-hopping and the lowest energy structures of Lennard-Jones clusters containing up to 110 atoms. J. Phys. Chem. A **101**(28), 5111–5116 (1997)
20. Wales, D.J., Scheraga, H.A.: Global optimization of clusters, crystals, and biomolecules. Science **285**(5432), 1368–1372 (1999)
21. Weik, F., et al.: ESPResSo 4.0-an extensible software package for simulating soft matter systems. Eur. Phys. J. Spec. Top. **227**(14), 1789–1816 (2019)
22. Xiang, Y., Cheng, L., Cai, W., Shao, X.: Structural distribution of Lennard-Jones clusters containing 562 to 1000 atoms. J. Phys. Chem. A **108**(44), 9516–9520 (2004)
23. Yang, Z., Liu, Y., Li, X.Y.: Beyond trilateration: on the localizability of wireless ad-hoc networks. In: IEEE INFOCOM 2009, pp. 2392–2400. IEEE (2009)

A Stochastic Birth-Death Model of Information Propagation Within Human Networks

Prasidh Chhabria$^{(\boxtimes)}$ and Winnie Lu

Harvard University, Cambridge, MA 02138, USA
chhabria@college.harvard.edu

Abstract. The fixation probability of a mutation in a population network is a widely-studied phenomenon in evolutionary dynamics. This mutation model following a Moran process finds a compelling application in modeling information propagation through human networks. Here we present a stochastic model for a two-state human population in which each of N individual nodes subscribes to one of two contrasting messages, or pieces of information. We use a mutation model to describe the spread of one of the two messages labeled the mutant, regulated by stochastic parameters such as talkativity and belief probability for an arbitrary fitness r of the mutant message. The fixation of mutant information is analyzed for an unstructured well-mixed population and simulated on a Barabási-Albert graph to mirror a human social network of $N = 100$ individuals. Chiefly, we introduce the possibility of a single node speaking to multiple information recipients or listeners, each independent of one another, per a binomial distribution. We find that while in mixed populations, the fixation probability of the mutant message is strongly correlated with the talkativity (sample correlation $\rho = 0.96$) and belief probability ($\rho = -0.74$) of the initial mutant, these correlations with respect to talkativity ($\rho = 0.61$) and belief probability ($\rho = -0.49$) are weaker on BA graph simulations. This indicates the likely effect of added stochastic noise associated with the inherent construction of graphs and human networks.

Keywords: Evolutionary dynamics · Human networks · Birth-death process

1 Introduction

The spread of information within human networks remains a particularly complex phenomenon, of particular interest in a time when information channels are varied and numerous [2]. Today, information propagates through more media than ever before; as a result, modeling the spread of information is difficult

Supported by Harvard University.

V. V. Krzhizhanovskaya et al. (Eds.): ICCS 2020, LNCS 12141, pp. 176–188, 2020.
https://doi.org/10.1007/978-3-030-50426-7_14

and must account for complexities arising from heterogeneity in communication across different channels. The phenomenon of information propagation on a network can be seen at play in politics, public health information, and social networks, among other common scenarios. Thus, understanding the dynamics of its spread is crucial to developing systems that encourage the spread of beneficial, factual information while deterring the spread of misleading or malicious information. The evolutionary dynamics of two competing, and often contrasting, pieces of information can be meaningfully characterized in both deterministic and stochastic models of human social networks.

In this paper, we model the spread of a mutant piece of information throughout an unstructured, well-mixed population of size N. We consider the case of one-to-one communication in which a speaker is able to speak to only one individual at a time. We then consider the case of one-to-many communication in which a speaker is able to simultaneously speak to multiple receivers. We then replicate this analysis by simulating on a Barabási-Albert graph, as an approximation of a classical human network.

1.1   Observed Information Propagation Models

Several mathematical models have been constructed to explain information spread. The majority of established studies have used deterministic models [4] such as the classic Spreader-Ignorant (SI) [5] and Spreader-Ignorant-Stifler (SIS) [6] rumor models. A *spreader* is an infected, or mutated, node that can spread the rumor to its neighbors. An *ignorant* is a susceptible node, which has not been exposed to the rumor. Finally, *stiflers* are nodes which have been exposed to the rumor but do not spread it. The present study attempts to establish a continuous spectrum across these three classes of nodes, by introducing probabilistic parameters such as talkativity and belief probability, to quantify the likelihood of each node believing and or sharing a piece of information. This allows for greater flexibility in the model rather than classifying each node into one of three categories. Several existing studies using deterministic models encounter problems while fitting suitable parameter values to a system of differential equations. These are unable to account for stochastic noise or probabilistic barriers. Since the spread of information depends on probabilities at each stage, the present study presents it as a stochastic process. We use a mutation model following a birth-death process to illustrate the transmission of information through a population, in a manner similar to a mutation albeit with altered parameters.

1.2   The Moran Birth-Death Process

The Moran process is a birth-death process [7] used to study the fixation of a mutant allele in an other wild-type population of size N. There are two absorbing states, at $i = 0$: where there are no mutants in the population, and $i = N$: where the mutant has fixated, i.e. spread to every node in the population [1]. In each time step of the process, there is always a single birth and a single death, so the number of mutants in the population can change by at most one. The present study will extend this to a one-to-many process enabling group communication.

1.3   Barabási-Albert Graph [8]

The Barabási-Albert network is a widely used graph architecture for reproducing the structure of a real human social network [9]. It is a stochastic scale-free network generated by a distribution specified by the number of nodes and the average number of neighbors for each node. The present study relies on the BA graph, following its widespread use in literature for understanding the dynamic nature of human activities in social networks [10] (Fig. 1).

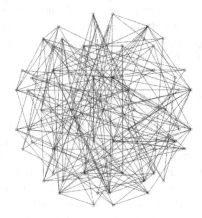

Fig. 1. A simple Barabási-Albert Graph computer-generated by a population of $N = 60$ nodes and an average number of $k = 10$ neighbors for each node.

2   Results

We outline a model for both one-to-one and one-to-many communication, beginning in an unstructured, well-mixed population. In both cases, we assume that the population is comprised of individuals (also known as 'nodes'). Within this population, we consider two competing and often contrasting pieces of information: the 'mutated' message and the 'non-mutated' message. The former is the variant of information whose propagation and fixation are of interest to our study. Following a mutation model, it begins as a rare event in the population and can either fixate, die out, or coexist with the non-mutated message. We consider only two pieces of information; each node believes one or the other, but not both. We say that the "mutant" confers a fitness of r to all believers (mutant nodes); $r < 1$ if knowledge and belief of the mutated message make an individual less likely to share the information, while $r > 1$ would suggest a heightened need to spread the mutant message.

A node that has been exposed to the mutated message and believes it is a mutant. A non-mutant is a node that has either not been exposed to or does not believe the mutated message. Past exposure to the mutant message is independent of a node's future belief probability if/when it is re-exposed to the message. A speaker is a node who communicates a piece of information, regardless of whether or not it is mutated. The speaker acts as the reproducing individual in a birth-death process. A receiver is an individual who listens to a message, representing the 'death' node, regardless of whether it is mutated or not. A node at position k has a talkativity τ_k where $0 \leq \tau_k \leq 1$. A higher τ_k here indicates that the node is more likely to be a speaker and spread the message to which it subscribes. A node at position k has belief probability β_k where $0 \leq \beta_k \leq 1$. A larger β_k here indicates that the node is more likely to believe the information it receives from a speaker.

In a well-mixed population of N individuals with i mutants, index positions 1 through i are mutants and index positions $i + 1$ through N are non-mutants. When a node mutates, they are re-indexed into the ith position. We disregard order within the population of mutants and within the population of non-mutants, as only the number of mutants and non-mutants is pertinent to the model.

2.1   One-to-One Communication Model

We begin modeling an unstructured, well-mixed population in which one speaker node communicates to only one recipient node at a time. The population begins with one mutant (allowing for fixation to be studied) but can have any number $0 \leq i \leq N$ of mutants at any given stage.

Population Shifts from i to $i-1$ Mutants: To facilitate this state change, a non-mutant must speak to a mutant who receives and believes the non-mutated message. The probability of this state change (see **Model Equations and Analytic Results** for derivation) is:

$$p_{i,i-1} = \frac{\sum_{k=i+1}^{N} \tau_k}{r \sum_{k=1}^{i} \tau_k + \sum_{k=i+1}^{N} \tau_k} \frac{\sum_{k=1}^{i} \beta_k}{N} \tag{1}$$

Population Shifts from i to $i+1$ Mutants: In order for this state change to occur, a mutant must speak to a non-mutant who listens and believes the mutated information. The probability of this transition is (see **Model Equations and Analytic Results** for derivation):

$$p_{i,i+1} = \frac{r \sum_{k=1}^{i} \tau_k}{r \sum_{k=1}^{i} \tau_k + \sum_{k=i+1}^{N} \tau_k} \frac{\sum_{k=i+1}^{N} \beta_k}{N} \tag{2}$$

2.2   One-to-Many Communication Model

The following explains the transition in population states in an unstructured, well-mixed population, in which one speaker node communicates to more than one recipient node simultaneously. Similar to the one-to-one model, the population begins with 1 mutant for the sake of this paper, but the population can have $0 \leq i \leq N$ mutants at any given stage.

The probability $p_{m,n}$ that a population goes from m mutants to n mutants where $m, n \in \mathbb{R}$ and $0 \leq m, n \leq N$ is given by an $(N+1)$ by $(N+1)$ transition matrix P. Let P consist of elements $p_{m,n}$:

$$
P_{(N+1 \times N+1)} =
\begin{bmatrix}
1 & 0 & 0 & 0 & \cdots & 0 \\
p_{1,0} & p_{1,1} & p_{1,2} & p_{1,3} & \cdots & p_{1,N} \\
0 & \cdots & p_{2,2} & \cdots & \cdots & \vdots \\
\vdots & & \ddots & \ddots & \ddots & \vdots \\
\vdots & \vdots & \vdots & \ddots & \ddots & \vdots \\
p_{N-1,0} & \cdots & \cdots & \cdots & \cdots & p_{N-1,N} \\
0 & 0 & 0 & \cdots & \cdots & 1
\end{bmatrix}
\tag{3}
$$

The transition matrix displays two clear absorbing states in its first and last rows. The probabilities within the transition matrix P are given by the ones below.

Population Shifts from i to $i+j$ $(j \geq 0)$ Mutants: In the case in which the number of mutants in the population is increasing by more than one in one time step, the probability of a population of N individuals going from a state of i mutants to $i + j$ mutants is:

$$
p_{i,i+j} = \frac{r \sum\limits_{k=1}^{i} \tau_k}{r \sum\limits_{k=1}^{i} \tau_k + \sum\limits_{i+1}^{N} \tau_k} \binom{N}{j} \left(\frac{p \sum\limits_{k=i+1}^{N} \beta_k}{N} \right)^j \left(\frac{N - p \sum\limits_{k=i+1}^{N} \beta_k}{N} \right)^{N-j}
\tag{4}
$$

where p is the probability of an individual being chosen to speak. This is constant for all N individuals in the population but is appropriately weighted by talkativity τ.

Population Shifts from i to $i - j$ $(j \geq 0)$ Mutants: In the case in which the number of mutants in the population is decreasing by more than one in one time step, the probability of a population of N individuals going from i mutants to ij mutants, where $j \geq 0$, is:

$$
p_{i,i-j} = \frac{\sum\limits_{k=i+1}^{N} \tau_k}{r \sum\limits_{k=1}^{i} \tau_k + \sum\limits_{i+1}^{N} \tau_k} \binom{N}{j} \left(\frac{p \sum\limits_{k=1}^{i} \beta_k}{N} \right)^j \left(\frac{N - p \sum\limits_{k=1}^{i} \beta_k}{N} \right)^{N-j}
\tag{5}
$$

Population State Does Not Shift: We consider the final case in which the number of mutants remains unchanged. The initial and final state of the population is i mutants. The general form of $p_{i,i}$ is given by

$$p_{i,i} = 1 - \sum_{j=1}^{N-i} p_{i,i+j} - \sum_{j=1}^{i} p_{i,i-j} \tag{6}$$

A few key elements of the transition matrix P (3) include:

$$p_{1,0} = \frac{\sum_{k=2}^{N} \tau_k}{r\tau_1 + \sum_{k=2}^{N} \tau_k} (p\beta_1)\left(\frac{N - p\beta_1}{N}\right)^{N-1} \tag{7}$$

$$p_{1,1} = 1 - \sum_{j=1}^{N-1} p_{1,1+j} - p_{1,0} \tag{8}$$

$$p_{1,N} = \frac{r\tau_1}{r\tau_1 + \sum_{k=2}^{N} \tau_k} \left(\frac{p\sum_{k=2}^{N}\beta_k}{N}\right)^{N-1} \left(N - p\sum_{k=2}^{N}\beta_k\right) \tag{9}$$

$$p_{N-1,N} = \frac{r\sum_{k=1}^{N-1} \tau_k}{r\sum_{k=1}^{N-1} \tau_k + \tau_N} (p\beta_N)\left(\frac{N - p\beta_N}{N}\right)^{N-1} \tag{10}$$

P is a transition matrix for an absorbing Markov chain and can be written in canonical form where the transient states precede the absorbing states [12]. We rearrange the horizontal and vertical indices to $\{N-1, 1, ..., N-2, 0, N\}$. The canonical form of P is

$$S = \left(\begin{array}{c|c} Q & R \\ \hline 0 & I \end{array}\right)$$

$$S_{(N+1)\times(N+1)} = \begin{bmatrix} p_{N-1,N-1} & p_{N-1,1} & \cdots & \cdots & \cdots & \cdots & p_{N-1,0} & p_{N-1,N} \\ p_{1,N-1} & p_{1,1} & p_{1,2} & p_{1,3} & \cdots & p_{1,N-2} & p_{1,0} & p_{1,N} \\ \vdots & \ddots & \ddots & \ddots & \ddots & \ddots & & \vdots \\ \vdots & \ddots & \ddots & \ddots & \ddots & \ddots & & \vdots \\ p_{N-2,N-1} & \cdots & \cdots & \cdots & \cdots & p_{N-2,N-2} & p_{N-2,0} & p_{N-2,N} \\ 0 & 0 & \cdots & \cdots & \cdots & 0 & 1 & 0 \\ 0 & 0 & \cdots & \cdots & \cdots & 0 & 0 & 1 \end{bmatrix}$$

Then, the matrices Q and R are:

$$Q_{(N-1)\times(N-1)} = \begin{bmatrix} p_{N-1,N-1} & p_{N-1,1} & \cdots & \cdots & \cdots & \cdots \\ p_{1,N-1} & p_{1,1} & p_{1,2} & p_{1,3} & \cdots & p_{1,N-2} \\ \vdots & \ddots & \ddots & \ddots & \ddots & \vdots \\ \vdots & \ddots & \ddots & \ddots & \ddots & \vdots \\ p_{N-2,N-1} & \cdots & \cdots & \cdots & \cdots & p_{N-2,N-2} \end{bmatrix}$$

$$R_{(N-1)\times 2} = \begin{bmatrix} p_{N-1,0} & p_{N-1,N} \\ p_{1,0} & p_{1,N} \\ \ddots & \vdots \\ p_{N-2,0} & p_{N-2,N} \end{bmatrix}$$

To obtain the fundamental matrix $M_{(N-1)\times(N-1)}$, we must invert the $I_{(N-1)} - Q_{(N-1)\times(N-1)}$ matrix:

$$I_{(N-1)} - Q_{(N-1)\times(N-1)} = \begin{bmatrix} 1-p_{N-1,N-1} & -p_{N-1,1} & \cdots & \cdots & \cdots & \cdots \\ -p_{1,N-1} & 1-p_{1,1} & -p_{1,2} & -p_{1,3} & \cdots & -p_{1,N-2} \\ \vdots & \ddots & \ddots & \ddots & \ddots & \vdots \\ \vdots & \ddots & \ddots & \ddots & \ddots & \vdots \\ -p_{N-2,N-1} & \cdots & \cdots & \cdots & \cdots & 1-p_{N-2,N-2} \end{bmatrix}$$

$$M_{(N-1)\times(N-1)} = \left[I_{(N-1)} - Q_{(N-1)\times(N-1)} \right]^{-1}$$
$$= \begin{bmatrix} 1-p_{N-1,N-1} & -p_{N-1,1} & \cdots & \cdots & \cdots & \cdots \\ -p_{1,N-1} & 1-p_{1,1} & -p_{1,2} & -p_{1,3} & \cdots & -p_{1,N-2} \\ \vdots & \ddots & \ddots & \ddots & \ddots & \vdots \\ \vdots & \ddots & \ddots & \ddots & \ddots & \vdots \\ -p_{N-2,N-1} & \cdots & \cdots & \cdots & \cdots & 1-p_{N-2,N-2} \end{bmatrix}^{-1}$$

B is the absorption probability matrix, a t by q matrix with entries b_{ij} where t is the number of transient states $(N-1)$ and q is the number of absorbing states (two; absorbing states are $i = 0$ and $i = N$).

$$B_{(N-1)\times 2} = M_{(N-1)\times(N-1)} R_{(N-1)\times 2}$$
$$= \begin{bmatrix} 1-p_{N-1,N-1} & -p_{N-1,1} & \cdots & \cdots & \cdots & \cdots \\ -p_{1,N-1} & 1-p_{1,1} & -p_{1,2} & -p_{1,3} & \cdots & -p_{1,N-2} \\ \vdots & \ddots & \ddots & \ddots & \ddots & \vdots \\ \vdots & \ddots & \ddots & \ddots & \ddots & \vdots \\ -p_{N-2,N-1} & \cdots & \cdots & \cdots & \cdots & 1-p_{N-2,N-2} \end{bmatrix}^{-1} \begin{bmatrix} p_{N-1,0} & p_{N-1,N} \\ p_{1,0} & p_{1,N} \\ \vdots & \vdots \\ \vdots & \vdots \\ p_{N-2,0} & p_{N-2,N} \end{bmatrix}$$

We examine fixation probability as the probability of beginning with one mutant and resulting in N mutants. Thus, the absorption probability that represents fixation, the probability of interest, is $b_{1,N}$.

For a well-mixed population, having set up the transition matrix Q in canonical form, we find the fixation probability $b_{1,N}$ as the entry in the second row and second column of the matrix MR. We go on to simulate this for well-mixed populations and extend the result to birth-death communication as per the constraints of our model on Barabási-Albert graphs.

We simulate the birth-death process with one-to-many communication on a Barabási-Albert graph of size $N = 100$ nodes and an average of $k = 16$ connections for each node. This choice of illustrative parameters was informed by a recent study revealing that on average, Americans have sixteen friends, each with whom they would willingly communicate in some capacity [13].

2.3   Simulations on a Well-Mixed Population

For simulations on a well-mixed population, we studied fixation probability of the mutated message given that the population begins with only one mutant. All talkativities and belief probabilities were generated as random number vectors. We examine the trend between talkativity or belief probability of the original mutant and the likelihood of its fixation. The percentage of simulations that reached fixation provides an indication of the fixation probability with respect to the talkativity and the belief probability of the initial mutant.

2.4   Simulations on a Barabási-Albert Graph

For simulations on a Barabási-Albert (BA) graph, we study the effect of talkativity and belief probability on the percentage of simulations in which the mutated message spread to at least one additional node. Fixation is fairly unlikely on a BA graph given the added stochastic noise and the probabilistic barriers created by the existence or lack of connection between two nodes. Thus, we chose to study the reduced case of simulations in which the mutated message is spread to at least one other node. The percentage of simulations that involved the mutation spreading to at least one other individual provides an indication of the fixation probability with respect to the talkativity and belief probability of the initial mutant.

3   Discussion

The model offers an understanding of how word-of-mouth information propagation, the most common mechanism of information spread [14], works in human networks. Specifically, two significant factors, talkativity and belief probability, that affect the likelihood of information spread are incorporated to understand the stochastic barriers a message must overcome to fixate in a population. The fixation probability of mutated messages in different population structures with respect to the talkativity and the belief probability of the initial mutant illuminates several important trends. In a well-mixed population, levels of fixation increase near-linearly with the level of talkativity of the initial mutant (correlation $\rho = 0.96$) (Fig. 2a). This is an intuitive result, as an initial mutant who is more talkative is more likely to spread the mutated information, contributing to a greater likelihood of fixation. Decreasing levels of belief probability of the initial mutant ($\rho = -0.74$) correspond to decreasing fixation probabilities (Fig. 2b). One explanation for this trend is that if the initial mutant is more likely to believe information, particularly the non-mutated message, its mutated information has a reduced probability of spreading or fixating. Nevertheless, this explanation may not manifest in reality, as belief probability likely depends on the content of the information, while β in our case is independent of the message. Thus, another parameter might be necessary to explain this trend.

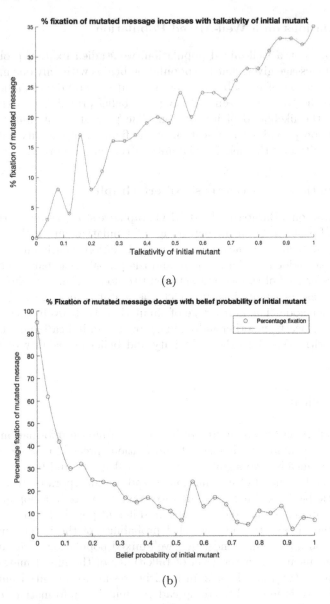

(a)

(b)

Fig. 2. Simulations of fixation of the mutated message in an unstructured, well-mixed population of $N = 100$ individuals at incremental values of talkativity and belief probability of the initial mutant. In (a), the percentage of fixation of the mutated message is measured against incremental τ values of talkativity. For each τ, 100 simulations were run, each with 10^5 time steps. In (b), the percentage of fixation of the mutated message is measured against incremental β values of the belief probability of the initial mutant. As with talkativity, 100 simulations were run for each x value of belief probability, each with 10^5 time steps. Smooth scatter lines simply connect consecutive points and do not imply any curvature or functional form of the relationship between outcome and parameters.

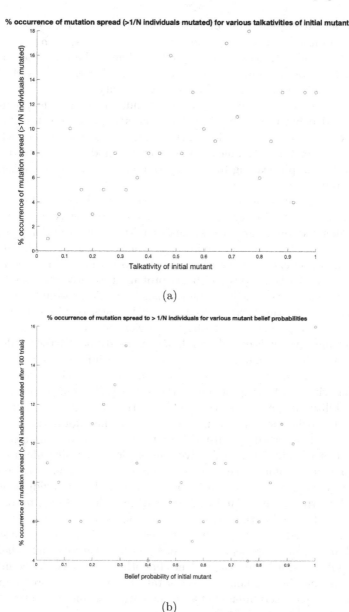

Fig. 3. Simulations of mutation spread on a BA graph of $N = 100$ individuals at incremental values of talkativity and belief probability of the initial mutant. In (a), the percentage of mutation spread is measured against incremental τ values of talkativity. At each τ, 100 simulations were run, each with 10^5 time steps. In (b), the percentage of mutation spread is measured against incremental x values of the belief probability of the initial mutant. As with talkativity, 100 simulations were run for each β value of belief probability, each with 10^5 time steps. Smooth scatter lines simply connect consecutive points and do not imply any curvature or functional form of the relationship between outcome and parameters.

Given the convoluted nature of the BA graph generated for Fig. 3a and 3b, fixation was unlikely because of added stochastic barriers, such as the probability that two nodes are connected, thereby reducing the available number of possible channels of communication. However, a weaker trend was observed in the percentage of mutation spread against talkativity ($\rho = 0.61$) and against belief probability ($\rho = -0.49$) of the initial mutant. Although, theoretically, talkativity and belief probability still have the same effect on the graph, the effects are diluted by other random factors intrinsic to the graph such as the likelihood of successful communication between two nodes. This introduces randomness on the graph, leading to no clear trends with respect to talkativity or belief probability.

An interesting social implication of the mathematical results of our model is the sheer difficulty of information becoming common knowledge or "fixating." Even in smaller subpopulations, the stochastic nature of information propagation renders it difficult for information to not only reach every individual, but also for every individual to believe the information. Due to stochastic barriers presented by the individuals' talkativities and belief probabilities, even the most valuable of information may never reach the individuals it should or would most benefit, as it cannot surpass the various probabilistic barriers required for successful transmission. An argument for stochastic barriers proving useful against fake news and misinformation is equally valid, but can be studied further to determine the probabilistic factors that enable the spread of misinformation.

A limitation of the present model is that the parameters considered are limited to talkativity and belief probability. Particularly, β in our model is a node's belief probability independent of the content of the message, but in human networks, a node's belief probability likely depends on the information itself.

An additional factor worth consideration is the relationship strength between two individuals or nodes. While talkativity and belief probability affect the likelihood of two nodes communicating, the strength of their relationship also impacts how likely any piece of information is communicated. Two individuals with low talkativity but with a high relationship strength may have a high probability of fixation despite having low talkativity. This may necessitate the use of a weighted network in the model. Along these lines, the effect of sub-networks is also significant. Within smaller networks of nodes, one node's talkativity and belief probability may not necessarily be independent of another node's. Nodes connected in a sub-network may be of a similar level of talkativity or belief probability. Our model also does not discriminate between the nature of communication through word-of-mouth and online platforms.

Understanding how information travels through the massive network that is humans is valuable for understanding how human beliefs evolve. Insightful contributions made to the existing pool of human knowledge may not necessarily manifest in beneficial ways due to various stochastic barriers; hence, understanding the degree to which said barriers impede communication can be helpful in ensuring information achieves its intended targets, particularly on technological platforms.

4 Appendix: Assumptions and Derivations

4.1 Model Assumptions:

Communication between nodes is restricted to at most one speaker speaking at a time. Only one speaker communicates at each time step, regardless of the number of recipients. Information spread from one node to another is assumed to have a binary trait of being mutated or non-mutated information. The model assumes the default state that a population begins with all the nodes believing the non-mutated information. Any pieces of information that is not the non-mutated information is considered a mutation. In reality, there are multiple speakers communicating at once and there may be more nuances as to the classification of different information.

We model mutation state as a binary state space: there are only two states a node can be in, either it is mutated or it is non-mutated. A mutated node in this model is a node that believes the mutated information and non-mutated node is one that either does not believe the mutated information and/or believes the non-mutated information.

The two forms of communication we model are one-to-one and one-to-many communication. In the one-to-many case, we assume that the number of receivers of a piece of information follows a binomial distribution $\text{Bin}(N, p)$. N here is the total population size in a well-mixed population. On a graph, N reduces to the number of neighbors of the speaker node. p is the probability of speaking to any one individual, independently.

4.2 Model Equations and Analytic Results

Analytic Results for One-to-One Communication:

- *Population shifts from i to $i - 1$ mutants:*
  (1) is the product of the following probabilities:

$$\frac{\sum_{k=i+1}^{N} \tau_k}{r \sum_{k=1}^{i} \tau_k + \sum_{k=i+1}^{N} \tau_k} \tag{11}$$

$$\frac{\sum_{k=1}^{i} \beta_k}{N} \tag{12}$$

(11) is the probability that a non-mutant speaks, weighted by the ratio of its talkativity to the sum of all individuals' talkativity and (12) is the probability that a mutant is the receiver and believes the non-mutated information.
- *Population shifts from i to $i + 1$ mutants:*
  (2) is the product of the following probabilities:

$$\frac{r \sum_{k=1}^{i} \tau_k}{r \sum_{k=1}^{i} \tau_k + \sum_{k=i+1}^{N} \tau_k} \tag{13}$$

$$\frac{\sum_{k=i+1}^{N} \beta_k}{N} \tag{14}$$

(13) is the probability that a mutant speaks, weighted by talkativity, and (14) is the probability that a non-mutant is the receiver and believes the mutated information.

Acknowledgments. We thank Martin A. Nowak and Alexander Heyde of the Program for Evolutionary Dynamics (PED) at Harvard University for their guidance and support.

References

1. Nowak, M.A.: Evolutionary Dynamics: Exploring the Equations of Life, 1st edn. Harvard University Press, Cambridge (2006)
2. Yan, Q.: Information propagation in online social network based on human dynamics. Abstr. Appl. Anal. (2013). https://doi.org/10.1155/2013/9534068
3. Frias-Martinez, E., Williamson, G., Frias-Martinez, V.: An agent-based model of epidemic spread using human mobility and social network information. In: Proceedings of the 2011 IEEE Third International Conference on Privacy, Security, Risk and Trust and 2011 IEEE Third International Conference on Social Computing (2011)
4. Haeupler, B.: Simple, fast and deterministic gossip and rumor spreading. J. ACM (2015). https://doi.org/10.1145/2767126
5. Li, D., Ma, J.: How the government's punishment and individual's sensitivity affect the rumor spreading in online social networks. Phys. A (2017). https://doi.org/10.1016/j.physa.2016.11.033
6. Rabajante, J.F.: Investigating the propagation and death of information in human subpopulation networks. Int. J. Appl. Math. Res. **1**(4) (2012). https://doi.org/10.14419/ijamr.v1i4.336
7. Moran, P.A.P.: Random processes in genetics. Math. Proc. Cambridge Philos. Soc. (1958). https://doi.org/10.1017/S0305004100033193
8. Barabasi, A.L.: Network Science The Barabasi-Albert Model (2014)
9. Mazzoli, M., Re, T., Bertilone, R., Maggiora, M., Pellegrino, J.: Agent Based Rumor Spreading in a scale-free network (2018)
10. Rezvan, A., Meybodi, M.R.: Stochastic graph as a model for social networks. Comput. Hum. Behav. (2016). https://doi.org/10.1016/j.chb.2016.07.032
11. Souza, E.P., Ferreira, E.M., Neves, A.G.M.: Fixation probabilities for the Moran process in evolutionary games with two strategies: graph shapes and large population asymptotics (2018)
12. Ermon, S., Gomes C.P., Sabharwal A., Selman B.: Designing fast absorbing Markov chains. In: Proceedings of the Twenty-Eighth AAAI Conference on Artificial Intelligence (2014)
13. The reasons American adults find it hard to make new friends. https://www.independent.co.uk/life-style/friends-adults-american-how-to-friendship-difficulty-a8906861.html. Accessed 7 Nov 2019
14. Aramendia-Muneta, M.E.: Spread the word - the effect of word of mouth in e-marketing. Comm. Commun. Digital Age (2017). https://doi.org/10.1515/9783110416794-013

A Random Line-Search Optimization Method via Modified Cholesky Decomposition for Non-linear Data Assimilation

Elias D. Nino-Ruiz[✉][ID]

Applied Math and Computer Science Lab, Department of Computer Science,
Universidad del Norte, Barranquilla 0800001, Colombia
enino@uninorte.edu.co
https://sites.google.com/a/vt.edu/eliasnino/

Abstract. This paper proposes a line-search optimization method for non-linear data assimilation via random descent directions. The iterative method works as follows: at each iteration, quadratic approximations of the Three-Dimensional-Variational (3D-Var) cost function are built about current solutions. These approximations are employed to build sub-spaces onto which analysis increments can be estimated. We sample search-directions from those sub-spaces, and for each direction, a line-search optimization method is employed to estimate its optimal step length. Current solutions are updated based on directions along which the 3D-Var cost function decreases faster. We theoretically prove the global convergence of our proposed iterative method. Experimental tests are performed by using the Lorenz-96 model, and for reference, we employ a Maximum-Likelihood-Ensemble-Filter (MLEF) whose ensemble size doubles that of our implementation. The results reveal that, as the degree of observational operators increases, the use of additional directions can improve the accuracy of results in terms of ℓ_2-norm of errors, and even more, our numerical results outperform those of the employed MLEF implementation.

Keywords: Ensemble Kalman filter · Line-search optimization · Modified Cholesky decomposition

1 Introduction

Data Assimilation is the process by which imperfect numerical forecasts are adjusted according to real observations [1]. In sequential methods, a numerical forecast $\mathbf{x}^b \in \mathbb{R}^{n \times 1}$ is adjusted according to an array of observations $\mathbf{y} \in \mathbb{R}^{m \times 1}$ where n and m are the number of model components and the number of observations, respectively. When Gaussian assumptions are made in prior and observational errors, the posterior mode $\mathbf{x}^a \in \mathbb{R}^{n \times 1}$ can be estimated via the minimization of the Three Dimensional Variational (3D-Var) cost function:

$$\mathcal{J}(\mathbf{x}) = \frac{1}{2} \cdot \left\| \mathbf{x} - \mathbf{x}^b \right\|_{\mathbf{B}^{-1}}^2 + \frac{1}{2} \cdot \left\| \mathbf{y} - \mathcal{H}(\mathbf{x}) \right\|_{\mathbf{R}^{-1}}^2, \tag{1}$$

© Springer Nature Switzerland AG 2020
V. V. Krzhizhanovskaya et al. (Eds.): ICCS 2020, LNCS 12141, pp. 189–202, 2020.
https://doi.org/10.1007/978-3-030-50426-7_15

where $\mathbf{B} \in \mathbb{R}^{n \times n}$ and $\mathbf{R} \in \mathbb{R}^{m \times m}$ are the background error and the data error covariance matrices, respectively. Likewise, $\mathcal{H}(\mathbf{x}) : \mathbb{R}^{n \times 1} \to \mathbb{R}^{m \times 1}$ is a (non-) linear observation operator which maps vector states to observation spaces. The solution to the optimization problem

$$\mathbf{x}^a = \arg \min_{\mathbf{x}} \mathcal{J}(\mathbf{x}), \tag{2}$$

is immediate when $\mathcal{H}(\mathbf{x})$ is linear (i.e., closed-form expressions can be obtained to compute $\mathbf{x}^a$) but, for non-linear observation operators, numerical optimization methods such as Newton's one must be employed [2]. However, since Newton's step is derived from a second-order Taylor polynomial, it can be too large with regard to the actual step size. Thus, line search methods can be employed to estimate optimal step lengths among Newton's method iterations. A DA method based on this idea is the Maximum-Likelihood-Ensemble-Filter (MLEF), which performs the assimilation step onto the ensemble space. However, the convergence of this method is not guaranteed (i.e., as the mismatch of gradients cannot be bounded), and even more, analysis increments can be impacted by sampling noise. We think that there is an opportunity to enhance line-search methods in the non-linear DA context by employing random descent directions onto which analysis increments can be estimated. Moreover, the analysis increments can be computed onto the model space to ensure global convergence.

This paper is organized as follows: in Sect. 2, we discuss topics related to linear and non-linear data assimilation as well as line-search optimization methods. Section 3 proposes an ensemble Kalman filter implementation via random descent directions. In Sect. 4, experimental tests are performed to assess the accuracy of our proposed filter implementation by using the Lorenz 96 model. Conclusions of this research are stated in Sect. 5.

2   Preliminaries

2.1   The Ensemble Kalman Filter

The Ensemble Kalman Filter (EnKF) is a sequential Monte-Carlo method for parameter and state estimation in highly non-linear models [3]. The popularity of the EnKF obeys to his simple formulation and relatively ease implementation. In the EnKF, an ensemble of model realizations is employed to estimate moments of the background error distribution [4]:

$$\mathbf{X}_k^b = \left[\mathbf{x}^{b[1]}, \mathbf{x}^{b[2]}, \dots, \mathbf{x}^{b[N]} \right] \in \mathbb{R}^{n \times N} \tag{3}$$

where $\mathbf{x}^{b[e]} \in \mathbb{R}^{n \times 1}$ stands for the e-th ensemble member, for $1 \leq e \leq N$, at time k, for $0 \leq k \leq M$. Then, the ensemble mean:

$$\overline{\mathbf{x}}^b = \frac{1}{N} \cdot \sum_{e=1}^{N} \mathbf{x}^{b[e]} \in \mathbb{R}^{n \times 1}, \tag{4}$$

and the ensemble covariance matrix:

$$\mathbf{P}^b = \frac{1}{N-1} \cdot \boldsymbol{\Delta}\mathbf{X}^b \cdot \left[\boldsymbol{\Delta}\mathbf{X}^b\right]^T \in \mathbb{R}^{n \times n}, \tag{5}$$

act as estimates of the background state $\mathbf{x}^b$ and the background error covariance matrix $\mathbf{B}$, respectively, where the matrix of member deviations reads:

$$\boldsymbol{\Delta}\mathbf{X}^b = \mathbf{X}^b - \bar{\mathbf{x}}^b \cdot \mathbf{1}^T \in \mathbb{R}^{n \times N}. \tag{6}$$

Posterior members can be computed via the use synthetic observations:

$$\mathbf{X}^a = \mathbf{X}^b + \boldsymbol{\Delta}\mathbf{X}^a, \tag{7}$$

where the analysis increments can be obtained via the solution of the next linear system:

$$\left[\left[\mathbf{P}^b\right]^{-1} + \mathbf{H}^T \cdot \mathbf{R}^{-1} \cdot \mathbf{H}\right] \cdot \boldsymbol{\Delta}\mathbf{X}^a = \mathbf{H}^T \cdot \mathbf{R}^{-1} \cdot \mathbf{D}^s \in \mathbb{R}^{n \times N}, \tag{8}$$

and $\mathbf{D}^s \in \mathbb{R}^{m \times N}$ is the innovation matrix on the synthetic observations whose e-th column reads $\mathbf{y} - \mathbf{H} \cdot \mathbf{x}^{b[e]} + \boldsymbol{\varepsilon}^{[e]} \in \mathbb{R}^{m \times 1}$ with $\boldsymbol{\varepsilon}^{[e]} \sim \mathcal{N}(\mathbf{0}_m, \mathbf{R})$. In practice, model dimensions range in the order of millions while ensemble sizes are constrained by the hundreds and as a direct consequence, sampling errors impact the quality of analysis increments. To counteract the effects of sampling noise, localizations methods are commonly employed [5], in practice. In the EnKF based on a modified Cholesky decomposition (EnKF-MC) [6] the following estimator is employed to approximate the precision covariance matrix of the background error distribution [7]:

$$\widehat{\mathbf{B}}^{-1} = \widehat{\mathbf{L}}^T \cdot \widehat{\mathbf{D}}^{-1} \cdot \widehat{\mathbf{L}} \in \mathbb{R}^{n \times n}, \tag{9}$$

where the Cholesky factor $\mathbf{L} \in \mathbb{R}^{n \times n}$ is a lower triangular matrix,

$$\left\{\widehat{\mathbf{L}}\right\}_{i,v} = \begin{cases} -\beta_{i,v} & , v \in P(i,r) \\ 1 & , i = v \\ 0 & , otherwise \end{cases}, \tag{10}$$

whose non-zero sub-diagonal elements $\beta_{i,v}$ are obtained by fitting models of the form,

$$\mathbf{x}_{[i]}^T = \sum_{v \in P(i,r)} \beta_{i,v} \cdot \mathbf{x}_{[v]}^T + \boldsymbol{\gamma}_i \in \mathbb{R}^{N \times 1}, 1 \le i \le n, \tag{11}$$

where $\mathbf{x}_{[i]}^T \in \mathbb{R}^{N \times 1}$ denotes the i-th row (model component) of the ensemble (3), components of vector $\boldsymbol{\gamma}_i \in \mathbb{R}^{N \times 1}$ are samples from a zero-mean Normal distribution with unknown variance σ^2, and $\mathbf{D} \in \mathbb{R}^{n \times n}$ is a diagonal matrix whose diagonal elements read,

$$\{\mathbf{D}\}_{i,i} = \widehat{\text{var}} \left(\mathbf{x}_{[i]}^T - \sum_{v \in P(i,r)} \beta_{i,v} \cdot \mathbf{x}_{[j]}^T \right)^{-1} \tag{12}$$

$$\approx \text{var} \left(\boldsymbol{\gamma}_i \right)^{-1} = \frac{1}{\sigma^2} > 0, \text{ with } \{\mathbf{D}\}_{1,1} = \widehat{\text{var}} \left(\mathbf{x}_{[1]}^T \right)^{-1}, \tag{13}$$

where $\text{var}(\bullet)$ and $\widehat{\text{var}}(\bullet)$ denote the actual and the empirical variances, respectively. The analysis equations can then be written as follows:

$$\mathbf{X}^a = \mathbf{X}^b + \left[\widetilde{\mathbf{L}}^T \cdot \widetilde{\mathbf{D}}^{-1/2} \right]^{-1} \cdot \mathbf{E} \in \mathbb{R}^{n \times N}, \tag{14}$$

where

$$\widehat{\mathbf{A}}^{-1} = \widetilde{\mathbf{L}}^T \cdot \widetilde{\mathbf{D}}^{-1} \cdot \widetilde{\mathbf{L}} = \widehat{\mathbf{B}}^{-1} + \mathbf{H}^T \cdot \mathbf{R}^{-1} \cdot \mathbf{H} \tag{15}$$

$$= \widehat{\mathbf{L}}^T \cdot \widehat{\mathbf{D}}^{-1} \cdot \widehat{\mathbf{L}} + \mathbf{H}^T \cdot \mathbf{R}^{-1} \cdot \mathbf{H} \in \mathbb{R}^{n \times n},$$

is an estimate of the posterior precision covariance matrix while the columns of matrix $\mathbf{E} \in \mathbb{R}^{n \times N}$ are formed by samples from a standard Normal distribution, $\widetilde{\mathbf{L}}^T \in \mathbb{R}^{n \times n}$ is a lower triangular matrix (with the same structure as $\widehat{\mathbf{L}}$), and $\widetilde{\mathbf{D}}^{-1} \in \mathbb{R}^{n \times n}$ is a diagonal matrix. Given the special structure of the left-hand side in (14), the direct inversion of the matrix $\widetilde{\mathbf{L}} \cdot \widetilde{\mathbf{D}}^{-1/2} \in \mathbb{R}^{n \times n}$ can be avoided [8, Algorithm 1].

2.2   Maximum Likelihood Ensemble Filter (MLEF)

To handle non-linear observation operators during assimilation steps, optimization based methods can be employed to estimate analysis increments. A well-known method in this context is the Maximum-Likelihood-Ensemble-Filter (MLEF) [9,10]. This square-root filter employs the ensemble space to compute analysis increments, this is:

$$\overline{\mathbf{x}}^a - \overline{\mathbf{x}}^b \in \text{range} \left\{ \boldsymbol{\Delta} \mathbf{X} \right\},$$

which is nothing but a pseudo square-root approximation of $\mathbf{B}^{1/2}$. Thus, vector states can be written as follows:

$$\mathbf{x} = \overline{\mathbf{x}}^b + \boldsymbol{\Delta} \mathbf{X} \cdot \mathbf{w}, \tag{16}$$

where $\mathbf{w} \in \mathbb{R}^{N \times 1}$ is a vector in redundant coordinates to be computed later. By replacing (16) in (1) one obtains:

$$\mathcal{J}(\mathbf{x}) = \mathcal{J} \left(\overline{\mathbf{x}}^b + \boldsymbol{\Delta} \mathbf{X} \cdot \mathbf{w} \right) = \frac{N-1}{2} \cdot \|\mathbf{w}\|^2 + \frac{1}{2} \cdot \left\| \mathbf{y} - \mathcal{H} \left(\overline{\mathbf{x}}^b + \boldsymbol{\Delta} \mathbf{X} \cdot \mathbf{w} \right) \right\|_{\mathbf{R}^{-1}}^2. \tag{17}$$

The optimization problem to solve reads:

$$\mathbf{w}^* = \arg \min_{\mathbf{w}} \mathcal{J} \left(\overline{\mathbf{x}}^b + \boldsymbol{\Delta} \mathbf{X} \cdot \mathbf{w} \right). \tag{18}$$

This problem can be numerically solved via Line-Search (LS) and/or Trust-Region methods. However, convergence is not ensured since gradient approximations are performed onto a reduce space whose dimension is much smaller than that of the model one.

2.3   Line Search Optimization Methods

The solution of optimization problems of the form (2) can be approximated via Numerical Optimization. In this context, solutions are obtained via iterations:

$$\mathbf{x}_{k+1} = \mathbf{x}_k + \boldsymbol{\Delta}\mathbf{s}_k, \tag{19}$$

wherein k denotes iteration index, and $\boldsymbol{\Delta}\mathbf{s}_k \in \mathbb{R}^{n \times 1}$ is a descent direction, for instance, the gradient descent direction [11]

$$\boldsymbol{\Delta}\mathbf{s}_k = -\nabla\mathcal{J}\left(\mathbf{x}_k\right), \tag{20a}$$

the Newton's step [12],

$$\nabla^2\mathcal{J}\left(\mathbf{x}_k\right) \cdot \boldsymbol{\Delta}\mathbf{s}_k = -\nabla\mathcal{J}\left(\mathbf{x}_k\right), \tag{20b}$$

or a quasi-Newton based method [13],

$$\mathbf{P}_k \cdot \boldsymbol{\Delta}\mathbf{s}_k = -\nabla\mathcal{J}\left(\mathbf{x}_k\right), \tag{20c}$$

where $\mathbf{P}_k \in \mathbb{R}^{n \times n}$ is a positive definite matrix. A concise survey of Newton based methods can be consulted in [14]. Since step lengths in (20) are based on first or second order Taylor polynomials, the step size can be chosen via line search [15] and/or trust region [16] methods. Thus, we can ensure global convergence of optimization methods to stationary points of the cost function (1). This holds as long as some assumptions over functions, gradients, and (potentially) Hessians are preserved [17]. In the context of line search, the following assumptions are commonly done:

C1 A lower bound of $\mathcal{J}(\mathbf{x})$ exists on $\Omega_0 = \{\mathbf{x} \in \mathbb{R}^{n \times 1}, \mathcal{J}(\mathbf{x}) \leq \mathcal{J}\left(\mathbf{x}^\dagger\right)\}$, where $\mathbf{x}^\dagger \in \mathbb{R}^{n \times 1}$ is available.
C2 There is a constant $\mathbf{L}$ such as:

$$\|\nabla\mathcal{J}(\mathbf{x}) - \nabla\mathcal{J}(\mathbf{z})\| \leq L \cdot \|\mathbf{x} - \mathbf{z}\|, \text{ for } \mathbf{x}, \mathbf{z} \in B, \text{ and } L > 0,$$

where B is an open convex set which contains Ω_0. These conditions together with iterates of the form,

$$\mathbf{x}_{k+1} = \mathbf{x}_k + \alpha \cdot \boldsymbol{\Delta}\mathbf{s}_k, \tag{21}$$

ensure global convergence [18] as long as α is chosen as an (approximated) minimizer of

$$\alpha^* = \arg\min_{\alpha \geq 0} \mathcal{J}\left(\mathbf{x}_k + \alpha \cdot \boldsymbol{\Delta}\mathbf{s}_k\right). \tag{22}$$

In practice, rules for choosing step-size such as the Goldstein rule [19], the Strong Wolfe rule [20], and the Halving method [21] are employed to partially solve (22). Moreover, soft computing methods can be employed for solving (22) [22].

3 Proposed Method: An Ensemble Kalman Filter Implementation via Line-Search Optimization and Random Descent Directions

In this section, we propose an iterative method to estimate the solution of the optimization problem (2). We detail our filter derivation, and subsequently, we theoretically prove the convergence of our method.

3.1 Filter Derivation

Starting with the forecast ensemble (3), we compute an estimate $\widehat{\mathbf{B}}^{-1}$ of the precision covariance $\mathbf{B}^{-1}$ via modified Cholesky decomposition. Then, we perform an iterative process as follows: let $\mathbf{x}_0 = \overline{\mathbf{x}}^b$, at iteration k, for $0 \leq k \leq K$, where K is the maximum number of iterations, we build a quadratic approximation of $\mathcal{J}(\mathbf{x})$ about $\mathbf{x}_k$

$$\mathcal{J}_k(\mathbf{x}) = \frac{1}{2} \cdot \|\mathbf{x} - \mathbf{x}_k\|^2_{\widehat{\mathbf{B}}^{-1}} + \frac{1}{2} \cdot \left\|\mathbf{y} - \widehat{\mathcal{H}}_k(\mathbf{x})\right\|^2_{\mathbf{R}^{-1}}, \tag{23a}$$

where

$$\widehat{\mathcal{H}}_k(\mathbf{x}) = \mathcal{H}(\mathbf{x}_k) + \mathbf{H}_k \cdot [\mathbf{x} - \mathbf{x}_k],$$

and $\mathbf{H}_k$ is the Jacobian of $\mathcal{H}(\mathbf{x})$ at $\mathbf{x}_k$. The gradient of (23a) reads:

$$\nabla \mathcal{J}_k(\mathbf{x}) = \widehat{\mathbf{B}}^{-1} \cdot [\mathbf{x} - \mathbf{x}_k] - \mathbf{H}_k^T \cdot \mathbf{R}^{-1} \cdot [\mathbf{d}_k - \mathbf{H}_k \cdot \mathbf{x}]$$
$$= \left[\widehat{\mathbf{B}}^{-1} + \mathbf{H}_k^T \cdot \mathbf{R}^{-1} \cdot \mathbf{H}_k\right] \cdot \mathbf{x} - \mathbf{H}_k^T \cdot \mathbf{R} \cdot \mathbf{d}_k \in \mathbb{R}^{n \times 1},$$

where $\mathbf{d}_k = \mathbf{y} - \mathcal{H}(\mathbf{x}_k) + \mathbf{H}_k \cdot \mathbf{x}_k \in \mathbb{R}^{m \times 1}$. Readily, the Hessian of (23a) is

$$\nabla^2 \mathcal{J}_k(\mathbf{x}) = \widehat{\mathbf{B}}^{-1} + \mathbf{H}_k^T \cdot \mathbf{R}^{-1} \cdot \mathbf{H}_k \in \mathbb{R}^{n \times n}, \tag{23b}$$

and therefore, the Newton's step can be written as follows:

$$\mathbf{p}_k(\mathbf{x}) = -\left[\widehat{\mathbf{B}}^{-1} + \mathbf{H}_k^T \cdot \mathbf{R}^{-1} \cdot \mathbf{H}_k\right]^{-1}$$
$$\cdot \left[\left[\widehat{\mathbf{B}}^{-1} + \mathbf{H}_k^T \cdot \mathbf{R}^{-1} \cdot \mathbf{H}_k\right] \cdot \mathbf{x} - \mathbf{H}_k^T \cdot \mathbf{R} \cdot \mathbf{d}_k\right],$$
$$= -\mathbf{x} + \left[\widehat{\mathbf{B}}^{-1} + \mathbf{H}_k^T \cdot \mathbf{R}^{-1} \cdot \mathbf{H}_k\right]^{-1} \cdot \mathbf{H}_k^T \cdot \mathbf{R} \cdot \mathbf{d}_k. \tag{23c}$$

As we mentioned before, the step size (23c) is based on a quadratic approximation of $\mathcal{J}(\mathbf{x})$ and depending how highly non-linear is $\mathcal{H}(\mathbf{x})$, the direction (23c) can poorly estimate the analysis increments. Thus, we compute U random directions based on the Newton's one as follows:

$$\mathbf{q}_{u,k} = \Pi_u \cdot \mathbf{p}_k(\mathbf{x}_k) \in \mathbb{R}^{n \times 1}, \text{ for } 1 \leq u \leq U, \tag{23d}$$

where the matrices $\Pi_u \in \mathbb{R}^{n \times n}$ are symmetric positive definite and these are randomly formed with $\|\Pi_u\| = 1$. We constraint the increments to the space spanned by the vectors (23d), this is

$$\mathbf{x}_{k+1} - \mathbf{x}_k = \mathbf{range}\,\{\mathbf{Q}_k\},$$

where the u-th column of $\mathbf{Q}_k \in \mathbb{R}^{n \times U}$ reads $\mathbf{q}_{u,k}$. Thus,

$$\mathbf{x}_{k+1} = \mathbf{x}_k + \mathbf{Q}_k \cdot \boldsymbol{\gamma}^*, \tag{23e}$$

where $\boldsymbol{\gamma}^* \in \mathbb{R}^{U \times 1}$ is estimated by solving the following optimization problem

$$\boldsymbol{\gamma}^* = \arg\min_{\boldsymbol{\gamma}} \mathcal{J}\left(\mathbf{x}_k + \mathbf{Q}_k \cdot \boldsymbol{\gamma}\right). \tag{23f}$$

To solve (23f), we proceed as follows: generate Z random vectors $\boldsymbol{\gamma}_z \in \mathbb{R}^{U \times 1}$, for $1 \leq z \leq Z$, with $\|\boldsymbol{\gamma}_z\| = 1$. We then, for each direction $\mathbf{Q}_k \cdot \boldsymbol{\gamma}_z \in \mathbb{R}^{n \times 1}$, we solve the following one-dimensional optimization problem

$$\alpha_z^* = \arg\min_{\alpha_z} \mathcal{J}\left(\mathbf{x}_k + \alpha_z \cdot [\mathbf{Q}_k \cdot \boldsymbol{\gamma}_z]\right), \tag{23g}$$

and therefore, an estimate of the next iterate (23e) reads:

$$\mathbf{x}_{k+1} = \mathbf{x}_k + \mathbf{Q}_k \cdot [\alpha_k^* \cdot \boldsymbol{\gamma}_k], \tag{23h}$$

where the pair $(\alpha_k^*, \boldsymbol{\gamma}_k)$ is chosen as the duple $(\alpha_z^*, \boldsymbol{\gamma}_z)$ which provide the best profit (minimum value) in (23g), for $1 \leq z \leq Z$. The overall process detailed in equations (23) is repeated until some stopping criterion is satisfied (i.e., we let a maximum number of iterations K).

Based on the iterations (23h), we estimate the analysis state as follows:

$$\overline{\mathbf{x}}^a = \overline{\mathbf{x}}^b + \sum_{k=1}^{K} \mathbf{Q}_k \cdot [\alpha_k^* \cdot \boldsymbol{\gamma}_k] = \mathbf{x}_K. \tag{24}$$

The inverse of the Hessian (23b) provides an estimate of the posterior error covariance matrix. Thus, posterior members (analysis ensemble) can be sampled as follows:

$$\mathbf{x}^{a[e]} \sim \mathcal{N}\left(\overline{\mathbf{x}}^a, \left[\nabla^2 \mathcal{J}_K\left(\overline{\mathbf{x}}^a\right)\right]^{-1}\right). \tag{25}$$

To efficiently perform the sampling process (25) the reader can consult [23]. Afterwards, the analysis members are propagated in time until a new observation is available. We name this formulation the Random Ensemble Kalman Filter (RAN-EnKF).

3.2   Convergence of the Analysis Step in the RAN-EnKF

For proving the convergence of our method, we consider the assumptions C1, C2, and

$$\nabla \mathcal{J}\left(\mathbf{x}_k\right)^T \cdot \mathbf{q}_{u,k} < 0, \ \text{for } 1 \leq u \leq U. \tag{26}$$

The next Theorem states the necessary conditions in order to ensure global convergence of the analysis step in the RAN-EnKF.

Theorem 1. *If* (2.3)*,* (2.3)*, and* (26) *hold, then the RSLS-RD with exact line search generates an infinite sequence* $\{\mathbf{x}_k\}_{u=0}^{\infty}$*, then*

$$\lim_{k \to \infty} \left[\frac{-\nabla J(\mathbf{x}_k)^T \cdot \mathbf{Q}_k \cdot \boldsymbol{\gamma}^*}{\|\mathbf{Q}_k \cdot \boldsymbol{\gamma}^*\|} \right]^2 = 0 \tag{27}$$

holds.

Proof. By Taylor series and the Mean Value Theorem we know that,

$$J\left(\mathbf{x}^{(u)} + \alpha^* \cdot \mathbf{Q}_k \cdot \boldsymbol{\gamma}^*\right) = J(\mathbf{x}_k)$$
$$+ \alpha^* \cdot \int_0^1 \nabla J(\mathbf{x}_k + \alpha^* \cdot t \cdot \mathbf{Q}_k \cdot \boldsymbol{\gamma}^*)^T$$
$$\cdot \mathbf{Q}_k \cdot \boldsymbol{\gamma}^* \cdot dt,$$

and therefore,

$$J(\mathbf{x}_k) - J(\mathbf{x}_{k+1}) \geq -\alpha^* \cdot \int_0^1 \nabla J(\mathbf{x}_k + \alpha^* \cdot t \cdot \mathbf{Q}_k \cdot \boldsymbol{\gamma}^*)^T$$
$$\cdot \mathbf{Q}_k \cdot \boldsymbol{\gamma}^* \cdot dt$$

for any $\mathbf{x}_{k+1}$ on the ray $\mathbf{x}_k + \alpha \cdot \mathbf{Q}_k \cdot \boldsymbol{\gamma}^*$, with $\alpha \in [0, 1]$, we have

$$J(\mathbf{x}_k) - J(\mathbf{x}_{k+1}) \geq J(\mathbf{x}_k) - J(\mathbf{x}_k + \alpha^* \cdot \mathbf{Q}_k \cdot \boldsymbol{\gamma}^*),$$

hence:

$$J(\mathbf{x}_k) - J(\mathbf{x}_{k+1}) \geq -\alpha^* \cdot \nabla J(\mathbf{x}_k)^T \cdot \mathbf{Q}_k \cdot \boldsymbol{\gamma}^*$$
$$- \alpha^* \cdot \int_0^1 [\nabla J(\mathbf{x}_k + \alpha^* \cdot t \cdot \mathbf{Q}_k \cdot \boldsymbol{\gamma}^*) - \nabla J(\mathbf{x}_k)]^T$$
$$\cdot \mathbf{Q}_k \cdot \boldsymbol{\gamma}^* \cdot dt,$$

by the Cauchy Schwarz inequality we have

$$J(\mathbf{x}_k) - J(\mathbf{x}_{k+1}) \geq -\alpha^* \cdot \nabla J(\mathbf{x}_k)^T \cdot \mathbf{Q}_k \cdot \boldsymbol{\gamma}^*$$
$$- \alpha^* \cdot \int_0^1 \|\nabla J(\mathbf{x}_k + \alpha^* \cdot t \cdot \mathbf{Q}_k \cdot \boldsymbol{\gamma}^*) - \nabla J(\mathbf{x}_k)\|$$
$$\cdot \|\mathbf{Q}_k \cdot \boldsymbol{\gamma}^*\| \cdot dt$$
$$\geq -\alpha^* \cdot \nabla J(\mathbf{x}_k)^T \cdot \mathbf{Q}_k \cdot \boldsymbol{\gamma}^*$$
$$- \alpha^* \cdot \int_0^1 L \cdot \|\alpha^* \cdot t \cdot \mathbf{Q}_k \cdot \boldsymbol{\gamma}^*\| \cdot \|\mathbf{Q}_k \cdot \boldsymbol{\gamma}^*\| \cdot dt$$
$$= -\alpha^* \cdot \nabla J(\mathbf{x}_k)^T \cdot \mathbf{Q}_k \cdot \boldsymbol{\gamma}^*$$
$$- \alpha^* \cdot L \cdot \|\mathbf{Q}_k \cdot \boldsymbol{\gamma}^*\| \cdot \int_0^1 \|t \cdot \alpha^* \cdot \mathbf{Q}_k \cdot \boldsymbol{\gamma}^*\| \cdot dt$$
$$= -\alpha^* \cdot \nabla J(\mathbf{x}_k)^T \cdot \mathbf{Q}_k \cdot \boldsymbol{\gamma}^* - \frac{1}{2} \cdot \alpha^{*2} \cdot L \cdot \|\mathbf{Q}_k \cdot \boldsymbol{\gamma}^*\|^2,$$

choose

$$\alpha^* = -\frac{\nabla \mathcal{J}\left(\mathbf{x}_k\right)^T \cdot \mathbf{Q}_k \cdot \boldsymbol{\gamma}^*}{L \cdot \|\mathbf{Q}_k \cdot \boldsymbol{\gamma}^*\|^2},$$

therefore,

$$
\mathcal{J}\left(\mathbf{x}_k\right) - \mathcal{J}\left(\mathbf{x}_{k+1}\right) \geq \frac{\left[\nabla \mathcal{J}\left(\mathbf{x}_k\right)^T \cdot \mathbf{Q}_k \cdot \boldsymbol{\gamma}^*\right]^2}{L \cdot \|\mathbf{Q}_k \cdot \boldsymbol{\gamma}^*\|^2}
$$

$$
-\frac{1}{2} \cdot \frac{\left[-\nabla \mathcal{J}\left(\mathbf{x}_k\right)^T \cdot \mathbf{Q}_k \cdot \boldsymbol{\gamma}^*\right]^2}{L \cdot \|\mathbf{Q}_k \cdot \boldsymbol{\gamma}^*\|^2}
$$

$$
= \frac{1}{2 \cdot L} \cdot \left[-\frac{\nabla \mathcal{J}\left(\mathbf{x}_k\right)^T \cdot \mathbf{Q}_k \cdot \boldsymbol{\gamma}^*}{\|\mathbf{Q}_k \cdot \boldsymbol{\gamma}^*\|}\right]^2.
$$

By (2.3), and (26), it follows that $\{\mathcal{J}\left(\mathbf{x}_k\right)\}_{k=0}^{\infty}$ is a monotone decreasing number sequence and it has a bound below, therefore $\{\mathcal{J}\left(\mathbf{x}_k\right)\}_{k=0}^{\infty}$ has a limit, and consequently (27) holds.

We are now ready to test our proposed method numerically.

4   Experimental Results

For the experiments, we consider non-linear observation operators, a current challenge in the context of DA [6,24]. We make use of the Lorenz-96 model [25] as our surrogate model during the experiments. The Lorenz-96 model is described by the following set of ordinary differential equations [26]:

$$
\frac{dx_j}{dt} = \begin{cases} (x_2 - x_{n-1}) \cdot x_n - x_1 + F & \text{for } j = 1, \\ (x_{j+1} - x_{j-2}) \cdot x_{j-1} - x_j + F & \text{for } 2 \leq j \leq n-1, \\ (x_1 - x_{n-2}) \cdot x_{n-1} - x_n + F & \text{for } j = n, \end{cases} \tag{28}
$$

where F is external force and $n = 40$ is the number of model components. Periodic boundary conditions are assumed. When $F = 8$ units the model exhibits chaotic behavior, which makes it a relevant surrogate problem for atmospheric dynamics [27,28]. A time unit in the Lorenz-96 represents 7 days in the atmosphere. We create the initial pool $\widehat{\mathbf{X}^b}_0$ of $\widehat{N} = 10^4$ members. The error statistics of observations are as follows:

$$
\mathbf{y}_k \sim \mathcal{N}\left(\mathcal{H}_k\left(\mathbf{x}_k^*\right), [\epsilon^o]^2 \cdot \mathbf{I}\right), \text{ for } 0 \leq k \leq M,
$$

where the standard deviations of observational errors $\epsilon^o = 10^{-2}$. The components are randomly chosen at the different assimilation cycles. We use the non-smooth and non-linear observation operator [29]:

$$\{\mathcal{H}(\mathbf{x})\}_j = \frac{\{\mathbf{x}\}_j}{2} \cdot \left[\left(\frac{|\{\mathbf{x}\}_j|}{2} \right)^{\beta-1} + 1 \right], \tag{29}$$

where j denotes the j-th observed component from the model state. Likewise, $\beta \in \{1, 3, 5, 7, 9\}$. Since the observation operator (29) is non-smooth, gradients of (1) are approximated by using the ℓ_2-norm. A full observational network is available at assimilation steps. The ensemble size for the benchmarks is $N = 20$. These members are randomly chosen from the pool $\widehat{\mathbf{X}^b}_0$ for the different experiments in order to form the initial ensemble $\mathbf{X}_0^b$ for the assimilation window. Evidently, $\mathbf{X}_0^b \subset \widehat{\mathbf{X}^b}_0$. The ℓ_2-norm of errors are utilized as a measure of accuracy at the assimilation step k,

$$\mathcal{E}(\mathbf{x}_k, \mathbf{x}^*) = \sqrt{[\mathbf{x}^* - \mathbf{x}_k]^T \cdot [\mathbf{x}^* - \mathbf{x}_k]}, \tag{30}$$

where $\mathbf{x}^*$ and $\mathbf{x}_k$ are the reference and current solution at iteration k, respectively. The initial background error, in average, reads $\epsilon^b \approx 31.73$. By convenience, this value is expressed in the log scale: $\log(\epsilon^b) = 3.45$. We consider a single assimilation cycle for the experiments. We try sub-spaces of dimensions $U \in \{10, 20, 30\}$ and number of samples from those spaces of $Z \in \{10, 30, 50\}$. We set a maximum number of iterations of 40. We compare our results with those obtained by the MLEF with $N = 40$, note that, the ensemble size in the MLEF doubles the ones employed by our method.

We group the results in Figs. 1 and 2 by sub-space size and sample size (sub-space dimension), respectively. As can be seen, the RAN-EnKF outperforms the MLEF in terms of ℓ_2-norm of errors, for all cases. Note that the error differences between the compared filter implementations are given by order of magnitudes. This can be explained as follows: the MLEF method performs the assimilation step onto a space given by the ensemble size; this is equivalent to perform an assimilation process by using the sample covariance matrix (5) whose quality is impacted by sampling errors. Contrarily, in our formulation, we employ sub-spaces whose basis vectors rely on the precision covariance (9) and, therefore, the impact of sampling errors is mitigated during optimization steps. As the degree β of the observation operator increases, the accuracy of the MLEF degrades, and consequently, this method diverges for the largest β value. On the other hand, convergence is always achieved in the RAN-EnKF method; this should be expected based on the theoretical results of Theorem 1. It should be noted that, as the β value increases, the 3D-Var cost function becomes highly non-linear, and as a consequence, more iterations are needed to decrease errors (as in any iterative optimization method). In general, it can be seen that as the number of samples Z increases, the results can be improved regardless of the sub-space dimension U (i.e., for $Z = 10$). However, it is clear that, for highly non-linear observation operators, it is better to have small sub-spaces and a large number of samples.

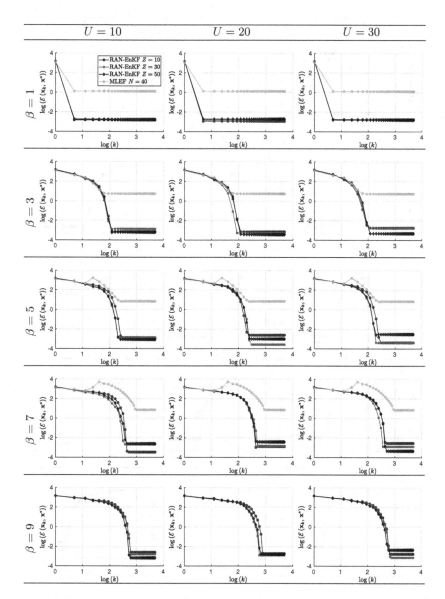

Fig. 1. ℓ_2-norm of errors in the log-scale for the 3D-Var Optimization Problem with different degrees β of the observation operator and dimension of sub-spaces U. For the largest β value, the MLEF diverges and therefore, its results are not reported.

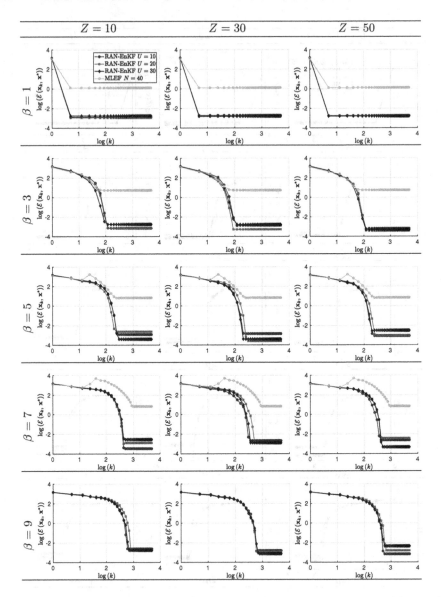

Fig. 2. ℓ_2-norm of errors in the log-scale for the 3D-Var Optimization Problem with different degrees β of the observation operator and number of samples Z. For the largest β value, the MLEF diverges and therefore, its results are not reported.

5   Conclusions

In this paper, we propose an ensemble Kalman filter implementation via line-search optimization; we name it a Random Ensemble Kalman Filter (RAN-EnKF). The proposed method proceeds as follows: an ensemble of model realization is employed to estimate background moments, and then, quadratic approximations of the 3D-Var cost function are obtained among iterations via the linearization of the observation operator about current solutions. These approximations serve to estimate descent directions of the 3D-Var cost function, which are perturbed to obtain additional directions onto which analysis increments can be computed. We theoretically prove the global convergence of our optimization method. Experimental tests are performed by using the Lorenz 96 model and the Maximum-Likelihood-Ensemble-Filter formulation. The results reveal that the RAN-EnKF outperforms the MLEF in terms of ℓ_2-norm of errors, and even more, it is able to achieve convergence in cases wherein the MLEF diverges.

Acknowledgment. This work was supported by the Applied Math and Computer Science Lab at Universidad del Norte in Barranquilla, Colombia.

References

1. Nino-Ruiz, E.D., Guzman-Reyes, L.G., Beltran-Arrieta, R.: An adjoint-free four-dimensional variational data assimilation method via a modified Cholesky decomposition and an iterative Woodbury matrix formula. Nonlinear Dyn. **99**(3), 2441–2457 (2020)
2. Nino-Ruiz, E.D.: Non-linear data assimilation via trust region optimization. Comput. Appl. Math. **38**(3), 1–26 (2019). https://doi.org/10.1007/s40314-019-0901-x
3. Evensen, G.: The ensemble Kalman filter: theoretical formulation and practical implementation. Ocean Dyn. **53**(4), 343–367 (2003)
4. Stroud, J.R., Katzfuss, M., Wikle, C.K.: A Bayesian adaptive ensemble Kalman filter for sequential state and parameter estimation. Mon. Weather Rev. **146**(1), 373–386 (2018)
5. Greybush, S.J., Kalnay, E., Miyoshi, T., Ide, K., Hunt, B.R.: Balance and ensemble Kalman filter localization techniques. Mon. Weather Rev. **139**(2), 511–522 (2011)
6. Nino-Ruiz, E.D., Sandu, A., Deng, X.: An ensemble Kalman filter implementation based on modified Cholesky decomposition for inverse covariance matrix estimation. SIAM J. Sci. Comput. **40**(2), A867–A886 (2018)
7. Bickel, P.J., Levina, E., et al.: Regularized estimation of large covariance matrices. Ann. Statist. **36**(1), 199–227 (2008)
8. Nino-Ruiz, E.: A matrix-free posterior ensemble Kalman filter implementation based on a modified Cholesky decomposition. Atmosphere **8**(7), 125 (2017)
9. Zupanski, M.: Maximum likelihood ensemble filter: theoretical aspects. Mon. Weather Rev. **133**(6), 1710–1726 (2005)
10. Zupanski, D., Zupanski, M.: Model error estimation employing an ensemble data assimilation approach. Mon. Weather Rev. **134**(5), 1337–1354 (2006)
11. Savard, G., Gauvin, J.: The steepest descent direction for the nonlinear bilevel programming problem. Oper. Res. Lett. **15**(5), 265–272 (1994)
12. Pan, V.Y., Branham, S., Rosholt, R.E., Zheng, A.-L.: Newton's iteration for structured matrices. In: Fast Reliable Algorithms for Matrices with Structure, pp. 189–210. SIAM (1999)
13. Nocedal, J.: Updating Quasi-Newton matrices with limited storage. Math. Comput. **35**(151), 773–782 (1980)

14. Knoll, D.A., Keyes, D.E.: Jacobian-free Newton-Krylov methods: a survey of approaches and applications. J. Comput. Phys. **193**(2), 357–397 (2004)
15. Hosseini, S., Huang, W., Yousefpour, R.: Line search algorithms for locally Lipschitz functions on Riemannian manifolds. SIAM J. Optim. **28**(1), 596–619 (2018)
16. Conn, A.R., Gould, N.I.M., Toint, Ph.L.: Trust region methods, vol. 1. SIAM (2000)
17. Shi, Z.-J.: Convergence of line search methods for unconstrained optimization. Appl. Math. Comput. **157**(2), 393–405 (2004)
18. Zhou, W., Akrotirianakis, I.G., Yektamaram, S., Griffin, J.D.: A matrix-free linesearch algorithm for nonconvex optimization. Optim. Methods Softw. **34**, 1–24 (2017)
19. Dunn, J.C.: Newton's method and the Goldstein step-length rule for constrained minimization problems. SIAM J. Control Optim. **18**(6), 659–674 (1980)
20. Dai, Y.-H., Yuan, Y.: A nonlinear conjugate gradient method with a strong global convergence property. SIAM J. Optim. **10**(1), 177–182 (1999)
21. Ravindran, A., Reklaitis, G.V., Ragsdell, K.M.: Engineering Optimization: Methods and Applications. Wiley, Hoboken (2006)
22. Nino-Ruiz, E.D., Yang, X.-S.: Improved Tabu Search and Simulated Annealing methods for nonlinear data assimilation. Appl. Soft Comput. **83**, 105624 (2019)
23. Nino-Ruiz, E.D., Beltran-Arrieta, R., Mancilla Herrera, A.M.: Efficient matrix-free ensemble Kalman filter implementations: accounting for localization. In: Kalman Filters - Theory for Advanced Applications. InTech, February 2018
24. Nino-Ruiz, E.D., Cheng, H., Beltran, R.: A robust non-Gaussian data assimilation method for highly non-linear models. Atmosphere **9**(4), 126 (2018)
25. Gottwald, G.A., Melbourne, I.: Testing for chaos in deterministic systems with noise. Phys. D Nonlinear Phenom. **212**(1–2), 100–110 (2005)
26. Karimi, A., Paul, M.R.: Extensive chaos in the Lorenz-96 model. Chaos Interdiscip. J. Nonlinear Sci. **20**(4), 043105 (2010)
27. Wilks, D.S.: Comparison of ensemble-MOS methods in the Lorenz'96 setting. Meteorol. Appl. **13**(3), 243–256 (2006)
28. Fertig, E.J., Harlim, J., Hunt, B.R.: A comparative study of 4D-VAR and a 4D ensemble Kalman filter: perfect model simulations with Lorenz-96. Tellus A **59**(1), 96–100 (2007)
29. van Leeuwen, P.J., Cheng, Y., Reich, S.: Nonlinear Data Assimilation. FADSRT, vol. 2. Springer, Cham (2015). https://doi.org/10.1007/978-3-319-18347-3

A Current Task-Based Programming Paradigms Analysis

Jérôme Gurhem[1,2]($\boxtimes$) and Serge G. Petiton[1,2]

[1] Univ. Lille, CNRS, UMR 9189 - CRIStAL - Centre de Recherche en Informatique
Signal et Automatique de Lille, 59000 Lille, France
{jerome.gurhem,serge.petiton}@univ-lille.fr
[2] Université Paris-Saclay, UVSQ, CNRS, CEA, Maison de la Simulation,
91191 Gif-sur-Yvette, France

Abstract. Task-based paradigm models can be an alternative to MPI. The user defines atomic tasks with a defined input and output with the dependencies between them. Then, the runtime can schedule the tasks and data migrations efficiently over all the available cores while reducing the waiting time between tasks. This paper focus on comparing several task-based programming models between themselves using the LU factorization as benchmark.

HPX, PaRSEC, Legion and YML+XMP are task-based programming models which schedule data movement and computational tasks on distributed resources allocated to the application. YML+XMP supports parallel and distributed tasks with XscalableMP, a PGAS language. We compared their performances and scalability are compared to ScaLAPACK, an highly optimized library which uses MPI to perform communications between the processes on up to 64 nodes. We performed a block-based LU factorization with the task-based programming model on up to a matrix of size 49512×49512. HPX is performing better than PaRSEC, Legion and YML+XMP but not better than ScaLAPACK. YML+XMP has a better scalability than HPX, Legion and PaRSEC. Regent has trouble scaling from 32 nodes to 64 nodes with our algorithm.

Keywords: Parallel and distributed programming paradigms · Task-based programming models · Supercomputers

1 Introduction

Task-based programming models are an interesting alternative to the Message Passing Interface (MPI) [9]. MPI is widely used and very efficient on the current architectures. However, MPI may not be a solution efficient enough on exascale machines, especially in terms of fault tolerance and check-pointing [16]. Task-based approach can help in managing fault tolerance and check-pointing since the tasks could be restarted on another location and data from tasks saved at

© Springer Nature Switzerland AG 2020
V. V. Krzhizhanovskaya et al. (Eds.): ICCS 2020, LNCS 12141, pp. 203–216, 2020.
https://doi.org/10.1007/978-3-030-50426-7_16

any moment. The goal of this paper is to analyze and compare the task-based programming models and languages between themselves.

While MPI focus on exchanging data between processes, other programming models may be efficient to parallelize applications with good scalability without being held back by global synchronizations. For instance, Partitioned Global Address Space Languages (PGAS) programming models which let the users see the distributed memory as a global memory address space that is partitioned across each processing element [5] is an alternative to MPI. Task-based programming models which allows to define fine-grain tasks (computations) on a specific set of data (input and output) are also an alternative. Runtime systems which can optimize execution are another one. Moreover, task-based programming models (first level of parallelism) combined with coarse-grain tasks implemented in a PGAS language (second level of parallelism) can also be one of them. Usually, fine-grain tasks use one process (eventually multi-threaded) whereas coarse-grain tasks perform on several processes (eventually with distributed memory).

In Sect. 2, we will describe the languages we are using to implement the application and how we did it. We focus on a PGAS language, task-based programming models and a combination of the two. Furthermore, we will present the performed experimentations and the results we obtained in Sect. 3.

2   Programming Paradigms

In this section, we present and use several languages to implement a LU factorization. MPI, a message passing library, and XcalableMP, a PGAS language, are used to implement a regular and distributed LU factorization (non blocked). Legion, PaRSEC, HPX and YML+XMP are used to implement the block-based version in which tasks make matrix operations on the sub-blocks of the matrix with a subset of the processes allocated to the application. Legion, PaRSEC, HPX are programming models based on a graph of task and YML+XMP is based on a graph of parallel and distributed tasks. They will be succinctly described afterwards.

2.1   MPI

The Message Passing Interface (MPI) [9] is a standardized norm designed to work on a wide variety of parallel computers. It defines the syntax of the core library routines to write message-passing applications. The application uses several processes to make computation at the same time on different cores and uses MPI to send data from one process to another one. MPI has several implementations (OpenMPI, MPICH2, IntelMPI, ...) which consist of a set of routines that can be called from C, C++ and Fortran.

The MPI library interface includes point-to-point send/receive operations, aggregate functions involving communication between all processes, synchronous nodes (barrier operations), one-way communication, dynamic process management and I/O operations. MPI provides synchronous and asynchronous routines as well as blocking and non-blocking operations. Collective routines involve communications between all processes in a group, for instance MPI_Bcast (broadcast that sends an array to all of the other processes).

MPI can be used with shared memory programming models like OpenMP or with libraries to send data and make computations on CPUs like CUDA. This hybrid model is called MPI+X.

2.2   XcalableMP

XcalableMP (XMP) [13] is a directive-based language extension for C and Fortran, which allows users to develop parallel programs for distributed memory systems easily and to tune the performance by having minimal and simple notations. XMP supports (1) typical parallelization methods based on the data-/task-parallel paradigm under the "global-view" model and (2) the co-array feature imported from Fortran 2008 for "local-view" programming.

The Omni XMP compiler translates an XMP-C or XMP-Fortran source code into a C or Fortran source code with XMP runtime library calls, which uses MPI and other communication libraries as its communication layer.

2.3   Legion (Regent)

Legion [1] is a data-centric parallel programming model. It aims to make the programming system aware of the structure of the data in the program. Legion provides explicit declaration of data properties (organization, partitioning, privileges, and coherence) and their implementation via the logical regions. They are the fundamental abstraction used to describe data in Legion applications. Logical regions can be partitioned into sub-regions and data structures can be encoded in logical regions to express locality describing data independence.

Regent [15] is a programming model which simplifies Legion. Regent compiler translates Regent programs into efficient implementations for Legion. It results in programs that are written with fewer lines of codes and at a higher level.

2.4   PaRSEC

PaRSEC [2,3] (Parallel Runtime Scheduling and Execution Controller) is an engine for scheduling tasks on distributed hybrid environments.

It offers a flexible API to develop domain specific languages. It aims to shift the focus of developers from repetitive architectural details toward meaningful algorithmic improvements. Two domain specific languages are supported by Parsec, the Parameterized Task Graph [6] (PTG) and Dynamic Task Discovery [11] (DTD).

2.5 YML+XMP

YML [8] is a development and execution environment for scientific workflow applications over various platforms, such as HPC, Cloud, P2P and Grid with multilevel of parallelism. YML defines an abstraction over the different middlewares, so the user can develop an application that can be executed on different middlewares without making changes related to the middleware used. YML can be adapted to different middlewares by changing its back-end. Currently, the proposed back-end [17] uses OmniRPC-MPI [14], a grid RPC which supports master-worker parallel and distributed programs based on multi SPMD programming paradigms. This back-end is developed for large scale clusters such as Japanese K-Computer [17]. A back-end for peer to peer networks is also available.

For the experiments, we use XMP to develop the YML components as introduced in [17]. This allows two levels programming. The higher level is the graph (YML) and the second level is the PGAS component (XMP). In the components, YML needs complementary information to manage the computational resources and the data at best: the number of XMP processes for a component and the distribution of the data in the processes (template). With this information, the scheduler can anticipate the resource allocation and the data movements. The scheduler creates the processes that the XMP components need to run the component. Then each process will get the piece of data which will be used in the process from the data repository.

2.6 HPX

High Performance ParalleX (HPX) [12] is a C++ Standard Library for Concurrency and Parallelism. HPX API implements the interfaces defined by the C++11/14/17/20 ISO standard and respects the programming guidelines used by the Boost collection of C++ libraries. It also extends the C++ Standard APIs to the distributed case. It aims to improve the scalability of current applications. It also tries to expose new levels of parallelism which are necessary to take advantage of the future systems.

HPX is an open-source implementation of the ParalleX execution model. This model focuses on overcoming the four main barriers to achieve scalability (Starvation, Latencies, Overhead, Waiting for contention resolution).

3 Performance Experiences

3.1 Cluster Description

The tests were performed on Poincare, the cluster of *La Maison de la Simulation* in France. It is an IBM cluster mainly composed of iDataPlex dx360 M4 servers, hosted at IDRIS, the CNRS supercomputer centre in Saclay, France. There is

77 compute nodes with 2 Sandy Bridge E5-2670 processors (8 cores each, so 16 cores per nodes) and 32 GB of RAM. The file system is constituted of two parts: a replicated file system with the homes of the users and a scratch file system with a faster access from the nodes. The network is based on QLogic QDR InfiniBand.

3.2 Experiments Details

We performed performance tests on up to 64 nodes of Poincare with the LU factorization implemented via MPI, ScaLAPACK, XMP, YML+XMP, HPX, PaRSEC and Regent. We used several sizes of matrices: 16384×16384, 32768×32768 and 49512×49512. 16384×16384 is the largest size we can use to perform the tests on one node since YML+XMP cannot perform the LU factorization with greater sizes of matrices on one node.

In HPX, PaRSEC and Regent, the performances depends on the number of blocks in each dimension (thus, the size of the blocks). We used several values for the number of blocks. Table 1, Table 2 and Table 3 show the block parameters which obtained the fastest execution time for each size of matrix. The execution times shown here are the case in which we obtained the fastest time for each number of nodes. We performed those test several times and computed the execution times mean of the same case. We will compare the results of the task-based programming languages to those obtained with ScaLAPACK. We will also compare them to our MPI and XMP implementations. Tests were run on several number of nodes in order to extract strong scaling information which will be discussed in Sect. 3.4. We used 1, 2, 4, 8, 16, 32 and 64 nodes to factorize the 16384×16384 values matrix. Then, we used 4, 8, 16, 32 and 64 nodes for the 32768×32768 values matrix. And finally, we used 8, 16, 32 and 64 nodes for the 49512×49512 values matrix.

Table 1. Number of blocks for the fastest case on a 16384×16384 matrix with number of processes per tasks between parenthesis

	1	2	4	8	16	32	64
HPX	$90^2(1)$	$45^2(1)$	$80^2(1)$	$45^2(1)$	$45^2(1)$	$55^2(1)$	$55^2(1)$
PaRSEC	$150^2(1)$	$200^2(1)$	$70^2(1)$	$120^2(1)$	$210^2(1)$	$240^2(1)$	$250^2(1)$
Regent	$50^2(1)$	$50^2(1)$	$50^2(1)$	$35^2(1)$	$40^2(1)$	$35^2(1)$	$30^2(1)$
YML+XMP	$4^2(8)$	$8^2(8)$	$8^2(16)$	$8^2(32)$	$4^2(128)$	$4^2(128)$	$4^2(128)$

Table 2. Number of blocks for the fastest case on a 32768 × 32768 matrix with number of processes per tasks between parenthesis

	4	8	16	32	64
HPX	$90^2(1)$	$90^2(1)$	$90^2(1)$	$75^2(1)$	$81^2(1)$
PaRSEC	$70^2(1)$	$120^2(1)$	$270^2(1)$	$380^2(1)$	$420^2(1)$
Regent	$70^2(1)$	$70^2(1)$	$60^2(1)$	$50^2(1)$	$50^2(1)$
YML+XMP	$1^2(64)$	$4^2(64)$	$8^2(32)$	$8^2(128)$	$8^2(128)$

Table 3. Number of blocks for the fastest case on a 49512 × 49512 matrix with number of processes per tasks between parenthesis

	8	16	32	64
HPX	$148^2(1)$	$148^2(1)$	$148^2(1)$	$145^2(1)$
PaRSEC	$250^2(1)$	$250^2(1)$	$400^2(1)$	$420^2(1)$
Regent	$70^2(1)$	$70^2(1)$	$70^2(1)$	$70^2(1)$
YML+XMP	$1^2(128)$	$2^2(128)$	$4^2(128)$	$8^2(128)$

Our MPI application is MPI-only so we used MPI support for shared memory and used one MPI rank per core i.e. 16 processes per core.

ScaLAPACK has a MPI only distributed implementation so it is run with one MPI rank process per core.

Our XMP implementation only uses pure XMP directives which are converted to MPI calls. It is launched as a MPI only application with one MPI rank process per core.

Regent is a compiler that translates a Lua based code into Legion. Regent applications are launched by passing the MPI command to Regent launcher which will compile and run the application. It creates a Legion worker on each node. Each one of them spawns a process to manage the local tasks, a process to manage data and a process to execute the tasks by default. Then, the user has to specify the number of processes on which the tasks will be executed by passing specific arguments to Legion runtime. We used 14 processes on each node to execute the tasks.

To launch our HPX application, we used the *mpirun* command to execute one instance of HPX runtime on each node as one would use MPI to execute one process per node. Then, HPX is able to infer the node configuration. HPX runtime spawns a worker process on each core of the node and tasks are run as light-weight threads on those processes. HPX is able to detect that there is two sockets on the node and manages them internally.

PaRSEC runtime depends on MPI and is used in the applications. Therefore, PaRSEC applications has to be run with the *mpirun* command. We created one MPI rank per core i.e. 16 MPI processes per node.

YML scheduler is launched with MPI on one core (the first one in the machine file) which launch XMP tasks with *MPI_Comm_spawn* routine on the leftover cores available.

3.3 Performances

Figure 1 shows the performances obtained for the LU factorization with HPX, MPI, PaRSEC, Regent, ScaLAPACK, XMP and YML+XMP on three sizes of matrices 16384×16384 (top), 32768×32768 (middle) and 49512×49512 (bottom).

On a 16384×16384 matrix, MPI is close to XMP on a small amount of nodes. When the number of node increases, MPI becomes significantly faster than XMP. Indeed, Fig. 1 middle and bottom charts show that MPI is significantly faster than XMP for each number of node. MPI and XMP applications share the same algorithm and a similar implementation but expressed with two different models. This may be due to an overhead from the PGAS description and access of the data in XMP compared to MPI.

Regent, HPX and PaRSEC are relatively close to one another on a small number of cores. However, we can outline tendencies. PaRSEC is faster than HPX on the lower number of node then HPX becomes faster when the number of node increases. It also seems that when the size of the matrix increases, HPX and PaRSEC performances are becoming closer and that HPX becomes faster than PaRSEC on the larger number of nodes. Indeed, HPX becomes faster than PaRSEC after 4 nodes for a matrix of size 16384×16384, after 16 nodes for a matrix of size 32768×32768 and after 64 nodes 49512×49512. For the later value, the difference between the two is very small (330s vs 331) so we expect HPX to become significantly faster for this size of matrix with a greater number of nodes.

Regent is a little bit behind HPX and PaRSEC on each number of nodes and size of matrices except for 2 and 4 nodes on a 16384×16384 matrix where Regent is very efficient. We can also notice that Regent is taking more time on 64 nodes than on 32. This may be related to the fact that Regent does not seem to be able to manage a large number of tasks on a large number of nodes since the number of sub-matrices is decreasing when the number of cores is increasing as Table 1, Table 2 and Table 3 are showing. However, other task based languages obtain better results when the number of sub-matrices they process increase with the number of cores. It creates more task and parallelism so that the runtime can use the resources most efficiently.

The YML+XMP applications are the slowest compared to the applications implemented with the other models. However, YML+XMP is the only model where tasks are also parallel and distributed. Moreover, it also uses the file system to perform the communications between the tasks so the communications between tasks are not efficient.

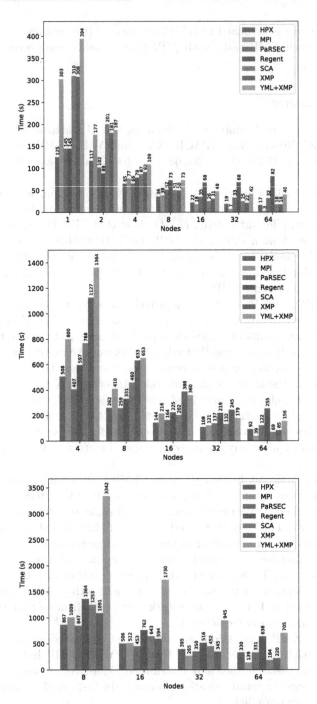

Fig. 1. Execution times obtained with the block-based LU factorization implemented with several task-based programming models on a 16384 × 16384 matrix (top), a 32768 × 32768 matrix (middle) and a 49512 × 49512 matrix (bottom)

Our last application uses the ScaLAPACK library to compute the LU factorization. It performs very well on large number of nodes but HPX, PaRSEC and Regent are faster on lower number of cores for each size of matrix. They are not using the same block based algorithm but ScaLAPACK is using a tiled algorithm that makes computations on rows and columns of the matrix [4]. Therefore, it is an interesting comparison to our block-based algorithms where the operations on the blocks are implemented with tasks. For a 16384 × 16384 matrix ScaLA-PACK and HPX are close on 64 nodes but ScaLAPACK is faster for greater size of matrices. This may be due to the cyclic distribution of data in ScaLAPACK which induces a different communication pattern very efficient on this kind of machine and algorithm.

3.4 Strong Scaling

Figure 2 shows the speed-up extracted from the performances values from Fig. 1 for HPX, MPI, PaRSEC, Regent, ScaLAPACK, XMP and YML+XMP on three sizes of matrices 16384 × 16384 (top), 32768 × 32768 (middle) and 49512 × 49512 (bottom). The speed-up corresponds to the ratio t_S/t_N where t_N is the execution time for N nodes and t_F is the execution time of the first number of nodes considered in the test. In the top chart of Fig. 2, t_F is t_1 since the experiments start with 1 node. In the middle (bottom) chart, t_F corresponds to 4 (8). It translates how efficiently we are managing the addition of more resources to solve the same problem.

Our MPI regular LU factorization is scaling very well as we can see on the charts. It even exceeds the ideal speed-up with matrices of size 16384 × 16384 (Fig. 2 top chart) and 32768 × 32768 (Fig. 2 middle chart). We think that it may be due to processes not having enough computations to do on 32 and 64 nodes matrices of size 16384 × 16384. Indeed, when increasing the size of the matrix to 32768 × 32768, the strong scalability for our MPI application seems more reasonable. The same situation occurs for 64 nodes when increasing the size of the matrix from 32768 × 32768 to 49512 × 49512.

Our task based applications obtain better scalability with the increase of the data size and the number of tasks processed by the applications. Table 1, Table 2 and Table 3 show that the number of tasks for a given number of nodes increases with the size of the matrix for each task based programming model. It produces more parallelism and opportunities to optimize the scheduling of the tasks and improve the use of the computing resources.

Regent strong scalability decreases from 32 to 64 nodes for each size of matrix. We expect its strong scalability to decrease even more with the increase of the number of cores.

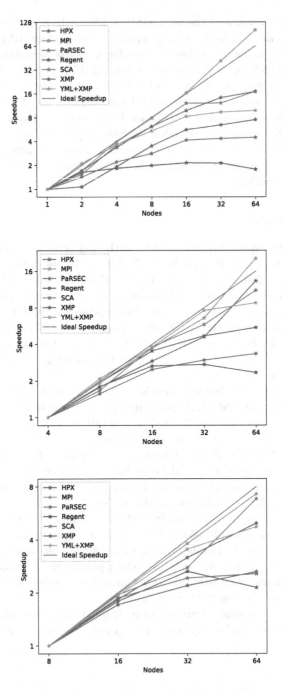

Fig. 2. Speed-ups obtained with the block-based LU factorization implemented with several task-based programming models on a 16384 × 16384 matrix (top), a 32768 × 32768 matrix (middle) and a 49512 × 49512 matrix (bottom) - log_2 scale for the y-axis

Our HPX application is scaling better than our PaRSEC application with matrices of size 16384 × 16384 and 32768 × 32768. It corresponds to the results we obtained in the previous section. We can also see that PaRSEC and HPX are very close with matrices of size 49512 × 49512 and that HPX is exceeding PaRSEC after 32 nodes. It seems that HPX may have a better scalability than PaRSEC on more than 64 nodes with matrices of size 49512 × 49512 if more nodes were available.

Finally, our YML+XMP application has the best strong scalability compared to the other task-based programming models. Therefore, we think that this programming model will be well adapted to larger machines with a distributed system and integrated schedulers.

3.5   Results Summary

As expected on a relatively small cluster, MPI has the best results and scalability on 64 nodes but the application does not use partial pivoting so it is not comparable to ScaLAPACK. It is also faster than XMP since MPI routines are highly optimized. Even though, XMP translates its directives into MPI code, the PGAS model used in XMP is not as efficient as using directly MPI.

ScaLAPACK is faster than the applications implemented with the task based programming models but it uses very efficient kernel routines to perform computations and communications internally whereas we are using unoptimized routines.

In term of task based programming models where we implemented everything (the description of the tasks and the computations executed by the tasks) with the language of the programming model, HPX is the most efficient on 64 nodes. However, PaRSEC also shows interesting performances in specific circumstances. Furthermore, Regent applications performances are not improving while increasing the number of nodes from 32 to 64. We think that the difference of performances between those programming models comes from their ability to manage the number of tasks, the dependencies between the tasks, tasks workload and the data migrations between nodes. Indeed, Regent performs best with a smaller amount of tasks than HPX and PaRSEC but its performances are behind them (see Table 1, Table 2 and Table 3 where we can see the number of tasks executed with the programming models to obtain their best performances). HPX and PaRSEC are performing better than Regent and YML+XMP with a larger number of smaller tasks since the number of tasks increase but not the data global size. HPX and PaRSEC seem to distribute very efficiently the tasks on the resources and optimize the data migrations between the nodes whereas Regent does not seem to be able to do so since the user has to reserve resources in Legion to manage the data and the computations. We would expect Legion to be able to reserve those resources by itself. Moreover, we used the default mapper for data and tasks provided by Regent and Legion. The default mapper may not be efficient enough to be used on production environment. Implementing a new mapper more adapted to our use may improve the performances of our Regent application. However, this means that the responsibility of implementing a good data and computation mapper is pushed to the user. Therefore, the optimization of the scheduling of the computations, the data positioning and data migrations is relegated to the user and not feature of the task based programming model anymore.

Finally, YML+XMP is the only programming model using tasks which are also distributed but the data migrations between the tasks are performed by reading/writing them on the file system which decrease its efficiency compared to node to node communications. YML+XMP performances may not be impressive on this number of nodes but with the increase of the number of nodes and its strong scalability higher than the other task-based programming languages, YML+XMP could be able to perform better than the other programming models on a very large scale as already experienced on the K computer [10]. Moreover, changing the use of the file system to make the communications between the tasks to in-memory communications could improve the performances even more.

4   Conclusion and Perspectives

We experimented several programming models on a cluster composed of 77 nodes. Indeed, we performed strong scalability tests on up to 64 nodes (1024 cores) with our implementation of the LU factorization in several programming paradigms. We implemented it with XMP, a PGAS language, with Regent (Legion), HPX and PaRSEC, three fine-grain task-based programming models and with YML+XMP, a coarse-grain task-based programming model combined with a PGAS language. We compared the performances we obtained with the different task-based programming models. We also compared to the ScaLAPACK library and our MPI implementation.

Our study has shown that ScaLAPACK performed better than task-based languages with a problem large enough. ScaLAPACK is expected to run better since it uses high performance libraries (e.g. BLACS and LAPACK) to perform the inner computations and the data migrations so it may also explain why it is faster. Moreover, we also showed that HPX is performing better than the other task-based languages on a large number of cores and that PaRSEC is more efficient than HPX on smaller number of cores. Unfortunately, Regent performances are close to HPX and PaRSEC but we encountered difficulties to make it scale from 32 to 64 nodes. However, we expect fine-grain task-based programming models them to get better performances with the increase of compute nodes and the use of optimized routines to implement tasks. Finally YML+XMP, is the less efficient one due to the communications between the tasks being held by the file system. Furthermore, coarse-grain task-based programming models with two levels of parallelism (the graph of tasks and the tasks implemented in a PGAS language) are not adapted to this kind of machine and number of nodes. They could possibly obtain better results with an adapted scheduler and a greater number of nodes as shown in [10].

As new perspectives, since, we only used a relatively small amount of nodes on an already old cluster, the number of cores could be increased as well as the size of the problem. We think that task-based programming models may get better performances than MPI+X when the size of the problem, the number of computing resources and the communication network involved in its solution will greatly increase. The task approach allows to describe the computations, the data migrations and the dependencies between them more precisely and at a finer grain. Therefore, the scheduler will be able to predict and anticipate the location where the data will be required. The scheduler could also optimize load balancing in the processes available as well as run different type of task at the same time compared to MPI where each process does almost the same thing at the same time. Moreover, the scheduler could be able to launch computations on resources where the data are stored and place the data in a way that reduces their movement during the execution. Other graph of tasks based frameworks like Pegasus [7] could also be studied.

Finally, our applications could get even better results by using existing and efficient libraries to perform the operations on the sub-matrices. Another improvement is to manage data sizes which are not divisible by the number of blocks and introduce pivoting to improve numerical stability.

Acknowledgment. Thanks to George Bosilca for his help with PaRSEC. Thanks to Harmut Kaiser and Mickael Simberg for their answers to my questions about HPX. Thanks to Elliott Slaughter for his help in installing and in using Regent properly. We would like to thank TOTAL SA which supported and allowed this work.

References

1. Bauer, M., Treichler, S., Slaughter, E., Aiken, A.: Legion: expressing locality and independence with logical regions. In: Proceedings of the International Conference on High Performance Computing, Networking, Storage and Analysis, SC 2012, pp. 66:1–66:11. IEEE Computer Society Press, Los Alamitos (2012). http://dl.acm.org/citation.cfm?id=2388996.2389086
2. Bosilca, G., Bouteiller, A., Danalis, A., Faverge, M., Herault, T., Dongarra, J.J.: Parsec: exploiting heterogeneity to enhance scalability. Comput. Sci. Eng. **15**(6), 36–45 (2013). https://doi.org/10.1109/MCSE.2013.98
3. Bosilca, G., Bouteiller, A., Danalis, A., Herault, T., Lemarinier, P., Dongarra, J.: Dague: a generic distributed dag engine for high performance computing. In: 2011 IEEE International Symposium on Parallel and Distributed Processing Workshops and Ph.D. Forum, pp. 1151–1158, May 2011. https://doi.org/10.1109/IPDPS.2011.281
4. Choi, J., Dongarra, J.J., Pozo, R., Walker, D.W.: Scalapack: a scalable linear algebra library for distributed memory concurrent computers. In: [Proceedings 1992] The Fourth Symposium on the Frontiers of Massively Parallel Computation, pp. 120–127, October 1992. https://doi.org/10.1109/FMPC.1992.234898
5. Coarfa, C., et al.: An evaluation of global address space languages: co-array fortran and unified parallel c, pp. 36–47, January 2005. https://doi.org/10.1145/1065944.1065950

6. Danalis, A., Bosilca, G., Bouteiller, A., Herault, T., Dongarra, J.: PTG: an abstraction for unhindered parallelism. In: 2014 Fourth International Workshop on Domain-Specific Languages and High-Level Frameworks for High Performance Computing, pp. 21–30, November 2014. https://doi.org/10.1109/WOLFHPC.2014.8

7. Deelman, E., et al.: Pegasus: a framework for mapping complex scientific workflows onto distributed systems. Sci. Program. J. **13**(3), 219–237 (2005). http://pegasus.isi.edu/publications/Sci.pdf

8. Delannoy, O., Emad, F., Petiton, S.: Workflow global computing with YML. In: 2006 7th IEEE/ACM International Conference on Grid Computing, pp. 25–32, September 2006. https://doi.org/10.1109/ICGRID.2006.310994

9. Forum, M.P.: MPI: A message-passing interface standard. Technical report, Knoxville, TN, USA (1994)

10. Gurhem, J., Tsuji, M., Petiton, S.G., Sato, M.: Distributed and parallel programming paradigms on the k computer and a cluster. In: Proceedings of the International Conference on High Performance Computing in Asia-Pacific Region, HPC Asia 2019, pp. 9–17. ACM, New York (2019). https://doi.org/10.1145/3293320.3293330

11. Hoque, R., Herault, T., Bosilca, G., Dongarra, J.: Dynamic task discovery in parsec: a data-flow task-based runtime. In: Proceedings of the 8th Workshop on Latest Advances in Scalable Algorithms for Large-Scale Systems, ScalA 2017, pp. 6:1–6:8. ACM, New York (2017). https://doi.org/10.1145/3148226.3148233

12. Kaiser, H., Heller, T., Adelstein-Lelbach, B., Serio, A., Fey, D.: HPX: a task based programming model in a global address space. In: Proceedings of the 8th International Conference on Partitioned Global Address Space Programming Models, PGAS 2014, pp. 6:1–6:11. ACM, New York (2014). https://doi.org/10.1145/2676870.2676883

13. Lee, J., Sato, M.: Implementation and performance evaluation of xcalablemp: a parallel programming language for distributed memory systems. In: 2010 39th International Conference on Parallel Processing Workshops, pp. 413–420, September 2010. https://doi.org/10.1109/ICPPW.2010.62

14. Sato, M., Hirano, M., Tanaka, Y., Sekiguchi, S.: OmniRPC: a grid RPC facility for cluster and global computing in OpenMP. In: Eigenmann, R., Voss, M.J. (eds.) WOMPAT 2001. LNCS, vol. 2104, pp. 130–136. Springer, Heidelberg (2001). https://doi.org/10.1007/3-540-44587-0_12

15. Slaughter, E., Lee, W., Treichler, S., Bauer, M., Aiken, A.: Regent: a high-productivity programming language for HPC with logical regions. In: SC 2015: Proceedings of the International Conference for High Performance Computing, Networking, Storage and Analysis, pp. 1–12, November 2015. https://doi.org/10.1145/2807591.2807629

16. Snir, M., et al.: Addressing failures in exascale computing. Int. J. High Perform. Comput. Appl. **28**(2), 129–173 (2014). https://doi.org/10.1177/1094342014522573

17. Tsuji, M., Sato, M., Hugues, M., Petiton, S.: Multiple-SPMD programming environment based on PGAS and workflow toward post-petascale computing. In: 2013 42nd International Conference on Parallel Processing, pp. 480–485, October 2013. https://doi.org/10.1109/ICPP.2013.58

Radial Basis Functions Based Algorithms for Non-Gaussian Delay Propagation in Very Large Circuits

Dmytro Mishagli[✉][iD] and Elena Blokhina[iD]

University College Dublin, Belfield, Dublin 4, Ireland
dmytro.mishagli@gmail.com, elena.blokhina@ucd.ie

Abstract. In this paper, we discuss methods for determining delay distributions in modern Very Large Scale Integration design. The delays have a non-Gaussian nature, which is a challenging task to solve and is a stumbling block for many approaches. The problem of finding delays in VLSI circuits is equivalent to a graph optimisation problem. We propose algorithms that aim at fast and very accurate calculations of statistical delay distributions. The speed of execution is achieved by utilising previously obtained analytical results for delay propagation through one logic gate. The accuracy is achieved by preserving the shapes of non-Gaussian delay distribution while traversing the graph of a circuit. The discussion on the methodology to handle non-Gaussian delay distributions is the core of the present study. The proposed algorithms are tested and compared with delay distributions obtained through Monte Carlo simulations, which is the standard verification procedure for this class of problems.

Keywords: Timing analysis · Statistical static timing analysis · Delay propagation · Uncertainty · Non-Gaussian · Graph optimisation

1 Introduction

The decrease of the feature size of modern Very Large Scale Integration (VLSI) design and the increase of the transistor count on a single chip inevitably approaches us to the end of Moore's law. Recently, Integrated Circuit (IC) technology has already reached the 5 nm technological node. At such scales, the role of uncertainty during the manufacturing process arises naturally due to physical fluctuations of various parameters such as the transistor channel width, its, length, etc. The design verification for a VLSI circuit has now become even more important since the complexity of design increases, which inevitably increase their cost. The standard way of circuit verification is to perform the timing analysis of a design [1,7,11,13]. The most reliable analysis is done by running Monte Carlo (MC) simulations that consider all possible variations of every parameter in a system. However, such computations can last for weeks for a single design.

© Springer Nature Switzerland AG 2020
V. V. Krzhizhanovskaya et al. (Eds.): ICCS 2020, LNCS 12141, pp. 217–229, 2020.
https://doi.org/10.1007/978-3-030-50426-7_17

Thus, semi-analytical methods based on delay models have been developed and are used in addition of MC simulations.

Deterministic Static Timing Analysis (or simply STA) can effectively take into account systematic process variations, and it was a dominant tool for several decades. However, STA gives too pessimistic predictions of delays, which significantly increases the cost of a chip in attempt to mitigate predicted delays. Therefore, the need for handling random correlated processes have arisen, and statistical approaches, known as Statistical Static Timing Analysis (SSTA), have been developed [2,5]. Within SSTA, all the delays in a system are treated as random variables (RVs) with some distributions. The goal of such an approach is to determine the mean value of the delay across selected paths and critical delays that may jeopardise the logic operation of the whole circuit. In addition, it is required to determine the standard deviation of the delay, its distribution, probability density function (PDF) and/or cumulative distribution function (CDF).

It is accepted now that the distribution of the delay generated by an individual logic gate is generally non-Gaussian [8]. A number of methods have been proposed to address this issue. However, treating non-Gaussian distributions of delays in a very large graph it still a challenge. Traditionally, approximations to the actual forms of distributions are used, as, for example, in studies [4,15,16], where the so-called *canonical model* of a delay has been proposed. Within this method, the delay is described as a linear function of parameter variations. Such approaches include (i) numerical approximations to the max-operator (see Sect. 3 for the details on this issue) and/or linearisation of nonlinear functions, and (ii) approximations to the actual distributions with Gaussians. For example, paper [12] discusses another modification of the canonical form based on adding a quadratic term and using skew-normal distributions.

In this study, we propose a fast and accurate algorithm for determining delay distributions taking into account their non-Gaussian nature. The problem of finding delays in VLSI circuits is formulated as that of a graph optimisation. The speed of the algorithm execution is achieved by utilising previously obtained analytical results for delay propagation through one logic gate, which was proposed in [6]. The accuracy is achieved by preserving the shapes of non-Gaussian delay distribution while traversing the graph of a circuit. This requires presenting a PDF of a gate's delay as a mixture of radial basis functions (RBFs) and solving the corresponding optimisation problem. The discussion on the methodology to handle non-Gaussian delay distributions is the core of the present study. The proposed optimisation strategies are incorporated in a traversal algorithm and tested and compared with delay distributions obtained through Monte Carlo simulations. The latter is the standard verification procedure for this class of problems.

The paper is organised as follows. In Sect. 2, we give a brief introduction to SSTA and discuss general statement of the problem. Section 3 discusses the model of delay propagation through a logic gate. The model allows us to built an algorithm for an accurate and fast calculation of the critical delay through a graph, which is summarised in Sect. 4. The key steps of the algorithm and the optimisation problem are discussed in Sect. 5. Section 6 concludes this paper with the verification of the algorithm via simulations and overall discussion.

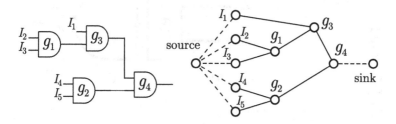

Fig. 1. Example logic circuit and its timing graph.

2   Statement of the Problem

2.1   Definitions

Throughout the paper, we will use the following commonly accepted terminology [2]. A logic circuit can be represented as a timing graph $G(E, N)$ as illustrated in Fig. 1 where the *graph* and its *paths* are defined as follows.

Definition 1. *A timing graph $G(E, N)$ is an acyclic directed graph, where E and N are the sets of edges and nodes correspondingly. Nodes reflect pins (logic gates) of a circuit. The timing graph always has only one source and one sink. Edges are characterised by weights d_i that describe delays.*

The timing graph is called a *statistical timing graph* within SSTA when the edges of the graph are described by RVs. The task then is to determine the critical (longest) path.

Definition 2. *Let $p_i(i = 1, \ldots, N)$ be a path of ordered edges from the source to the sink in a timing graph G and let D_i be the path length of p_i. Then $D_{max} = \max(D_1, \ldots, D_n)$ is referred to as the SSTA problem of a circuit.*

Therefore, SSTA is aimed at determining the distribution of the circuit delay, which is equivalent to calculating the longest path in the graph formalism. Since SSTA operates with random variables, we will also use the following notation for the probability density function of the Gaussian distribution:

$$g(x|\mu, \sigma) \stackrel{\text{def}}{=} \frac{1}{\sqrt{2\pi}\sigma} \varphi\left(\frac{x-\mu}{\sigma}\right), \quad \varphi(x) \stackrel{\text{def}}{=} e^{-x^2/2} \tag{1}$$

where μ and σ^2 are the mean value and variance respectively, and $\varphi(x)$ is the Gaussian kernel function. The cumulative density function will be denoted as follows:

$$\Phi(x|\mu, \sigma) \stackrel{\text{def}}{=} \frac{1}{2}\left[1 + \text{erf}\left(\frac{1}{\sqrt{2}}\frac{x-\mu}{\sigma}\right)\right], \tag{2}$$

where $\text{erf}(x)$ is the error function.

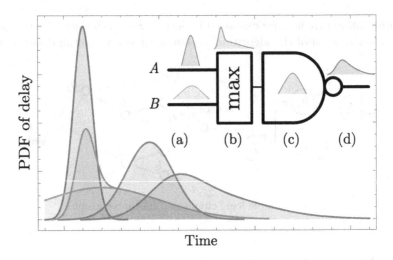

Fig. 2. Illustration of delay propagation through a logic gate. At stage (a), two signals arrive at the input of the gate. At stage (b), the max-operation is performed, which gives a skewed PDF. At the same time, the gate has its own operation time described by some distribution (c). Thus, the distribution of the gate delay (d) requires the convolution of the obtained distribution (b) and given (c). This convolution results in a new RV, which clearly has a non-Gaussian form.

2.2 Gate Level Analysis

We will start by explaining briefly how the overall delay is generated in a single logic gate. At the gate level, delay propagation is described by two operations: computing the maximum (max) of two delays entering a gate and the summation of the latter with the delay of the gate. From the statistical point of view, when these operations are applied to RVs, the delay of a gate with two inputs reads:

$$\max(X_1, X_2) + X_0, \tag{3}$$

where X_1 and X_2 are the RVs that describe the arrival times of input signals, and X_0 is the RV that gives the gate operation time. The operation of a logic gate in terms of the arrival and operation time distributions is presented in Fig. 2. Therefore, Eq. (3) is the *convolution* of the $\max(X_1, X_2,)$ and X_0 probability density functions. The analytical solution to combination (3) exists in a limited number of cases. We discuss these cases in the next section.

3   Model for a Logic Gate Delay

Consider a logic gate with two inputs, A and B, and suppose that the gate operation time is distributed according to the normal law with a mean μ_0 and a variance σ_0^2, i.e., it is a Gaussian RV. Assume now that the arrival times of both signals are also Gaussian RVs with means and variances μ_1, σ_1^2 and μ_2,

σ_2^2 respectively. Even if the individual distribution in the example of Fig. 2 are Gaussian, the application of formula (3) leads to a non-Gaussian distribution of the delay at the gate output as shown in that figure.

In our previous work [6], the exact expression for the PDF of such an RV was presented:

$$f(x) = \frac{1}{\sqrt{2\pi}} \sum_{\substack{i,j=1,2 \\ i \neq j}} \frac{1}{\tilde{\sigma}_i} \varphi\left(\frac{x - \mu_0 - \mu_i}{\tilde{\sigma}_i}\right) \Phi\left[\frac{1}{\sqrt{1 + \kappa_{ij}^2}} y(\ldots)\right], \qquad (4)$$

where

$$\kappa_{ij} = \frac{\sigma_0 \sigma_i}{\sigma_j \tilde{\sigma}_i}, \quad \tilde{\sigma}_i = \sqrt{\sigma_0^2 + \sigma_i^2}, \quad y(x) = \frac{\sigma_i^2(x - \mu_0) + \sigma_0^2 \mu_i}{\tilde{\sigma}_i^2 \sigma_j} - \frac{\mu_j}{\sigma_j}. \qquad (5)$$

Expression (4) does not take into account possible correlations between the arrival signals. This issue will not be addressed in this study. Instead, we are interested in demonstrating how this exact solution can speed up SSTA for a given graph keeping precision high. Formula (4) assumes all initial delays (arrival and gate itself) to have Gaussian distributions. In principle, both the arrival signal and gate delay do not have to be Gaussian. If they can be decomposed into a linear superposition of Gaussian kernel functions, the PDF of the gate output delay can be presented as a linear combination of expressions (4) due to the linearity of the integration operation. This idea constitutes the core of a delay propagation algorithm which we discuss in the next sections.

4   Delay Propagation Algorithm

The algorithm for the calculation of the delay propagation through a timing graph is outlined below. The high-level description of the algorithm is shown in Fig. 3.

Algorithm 1. Returns a list with the parameters of Gaussian mixtures for each node of the graph G.

Input: graph G; distributions, means and variances for graph nodes

1. Perform preprocessing: decompose each non-Gaussian PDF for input nodes and gates and get corresponding Gaussian mixtures
2. Do forward propagation in G.
   for node in G:
   – calculate PDF $f(x)$, mean μ_{gate} and standard deviation σ_{gate}
   – represent via Gaussian mixture
   – append to a list
   Output: a list with PDFs for all nodes in the RBF representation

This algorithm relies on the decomposition procedure. This procedure aims to represent a skewed (non-Gaussian) distribution with a mixture of RBFs that have Gaussian form, which allows one to use the result (4). In the next section, we discuss how such a decomposition can be performed.

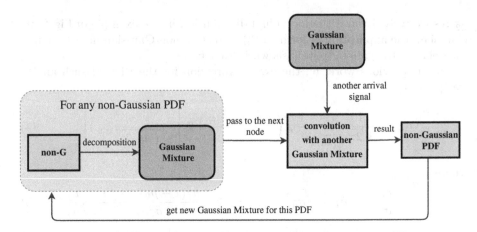

Fig. 3. High-level diagram of the Algorithm flow.

5   Optimisation

The exact function (4) can be written as $f_{\text{rbf}}(x)$, a sum of RBFs; each of these RBFs has a Gaussian-like shape. In other words, it can be decomposed to a Gaussian mixture [9,14]. The decomposition procedure is equivalent to fitting the actual PDF with a sum of RBFs, which brings us to an optimisation problem. In this study, we discuss the minimisation of the sum of squares of the residuals

$$\min \sum_i |f_{\text{rbf}}(x_i) - y_i|^2, \tag{6}$$

subject to constraints (specified below). Here y_i are the data points corresponding to the actual function $f(x)$ that we want to fit.

Depending on the form of the RBFs, the minimisation of (6) can vary significantly, *e.g.*, an approximate function $f_{\text{rbf}}(x)$ can be either linearly or non-linearly dependent on the fitting parameters. We discuss two alternative approaches below.

5.1   Choice of the Cost Function

Let us consider the RBFs $f_{\text{rbf}}(x)$ in the following form:

$$f_{\text{rbf}}(x) = \sum_{i=1}^{m} w_i \varphi \left(\frac{x - b_i}{c_i} \right), \quad \forall w_i, c_i > 0, \tag{7}$$

where w_i are the weight coefficients, b_i determine positions of the corresponding RBFs and c_i are the shape parameters. The constraints on w_i and c_i are chosen so that the resulting mixture of RBFs has the meaning of a PDF. In the most general case, for m RBFs there are $3\,m$ parameters to be determined from the solution to the least-squares problem (6).

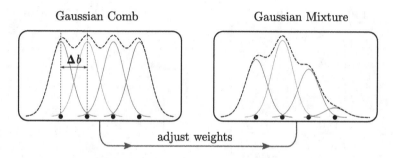

Fig. 4. A sketch of a Gaussian comb. For a given shape parameter c, the Gaussian kernels $\varphi(x)$, being equally separated by Δb, form a 1D grid, which we call a Gaussian comb. By adjusting the weights w_i, the Gaussian comb can give various shapes. The black dashed line shows the resulting curve from a mixture of Gaussian kernels in a comb.

Allowing all three parameters to vary, the fit function $f_{\mathrm{rbf}}(x)$ gives the desired shape using a small number m of RBFs. At the same time, obtaining a minimum of (6) in such a case is not a trivial problem even for a few RBFs: the optimisation solver may stuck in one of the local minima, which makes this approach less reliable.

One can simplify the problem by setting a 1D grid of RBFs with given parameters b_i and c_i, therefore, only the weights w_i remain unknown. Note that we get only *linear* to the Gaussian kernels $\varphi(x)$ unknown parameters in such a case. To make the problem even simpler, let us fix the shape parameters c_i for all kernels, $c_i = \mathrm{const}$, and let us distribute the kernels with equal separation Δb. We shall call such mixtures of RBFs a *Gaussian comb* (see Fig. 4). At the same time, such a simplification will require more terms in a mixture of RBFs to get the desired shape.

Thus, to construct the Gaussian comb, one needs to set up the number m of RBFs in the comb, the Gaussian comb step Δb and the shape parameter c. The weight coefficients w_i then to be determined from (6). In this study, we assume m and c are known *a priori*, and the step Δb is determined as

$$\Delta b = \frac{\Delta y}{m}, \qquad (8)$$

where Δy is the effective interval of values over which the function (PDF) must be fit (we call it *bandwidth*). The choice of the bandwidth is discussed below. In principle, the optimisation problem can be formulated in such a way that m and c are determined simultaneously with the weights w_i, making the solution self-consistent. This will be reported elsewhere.

The distinguishable feature of such RBF decompositions is that one obtains simple functions that allow fast and simple computations of a mean and a vari-

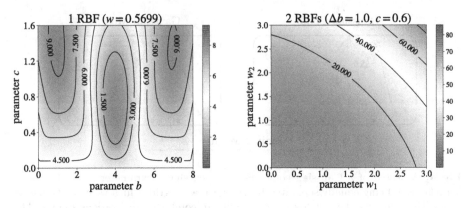

Fig. 5. Cost functions for the least square problem of fitting a Gaussian PDF with $\mu = 4$ and $\sigma = 0.7$. *Left:* only one RBF is used; for $w = 0.5699$, the minimum is located at $b = 4$ and $c = 0.7$. *Right:* two RBFs are used; for $b_1 = 3.5$, $b_2 = 4.5$, $\Delta = 1.0$ and $c = 0.6$, the minimum is located at $w_1 = w_2 = 0.5299$.

ance for the PDF:

$$\mu_{\text{rbf}} = \sqrt{2\pi} \sum_{i=1}^{m} w_i b_i c_i, \quad \sigma_{\text{rbf}}^2 = \sqrt{2\pi} \sum_{i=1}^{m} w_i c_i (b_i^2 + c_i^2) - \mu_{\text{rbf}}^2; \qquad (9)$$

for a case of Gaussian comb, $\forall c_i = c$.

5.2 Comparison

For the sake of illustration, consider a problem of fitting the Gaussian PDF, $g(x|\mu = 4, \sigma = 0.7)$, with only one RBF. If we pre-set the weight w, only two parameters, b and c, are left to find. Choosing $w = 1/(\sqrt{2\pi}0.7) \approx 0.6599$, the parameters should be then exactly $b = 4$ and $c = 0.7$. A contour plot of the cost function (6) for such a case is shown in Fig. 5 on the left. The cost function has only one minimum, which lies in a valley (the point with $b = 4$ and $c = 0.7$). However, if we allow the weight w to vary, there will be 3 parameters to find and local minima appear. Increasing the number m of RBFs inevitably brings us to a problem of omitting such minima.

The Gaussian comb consist of two Gaussian kernels with $c = 0.6$ and $\Delta b = 1$ ($b_1 = 3.5$ and $b_2 = 4.5$) leads to the cost function shown in Fig. 5 on the right. One can see that the minimum is located in the bottom of a steep well that rises dramatically as one goes away from the minimum. Thus, there will be no difficulties in finding the minimum in higher dimensions, moreover, this problem allows exact solution.

Consider another example. The PDF $f_{\text{LN}}(x)$ of the lognormal distribution reads

$$f_{\text{LN}}(x) = \frac{1}{x} \cdot \frac{1}{\sigma\sqrt{2\pi}} \exp\left(-\frac{(\ln x - \mu)^2}{2\sigma^2}\right), \qquad (10)$$

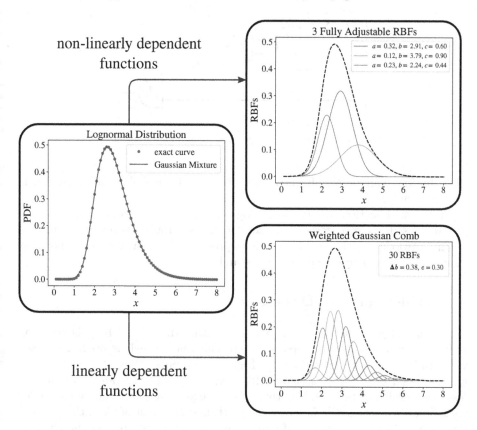

Fig. 6. Case study: representation of the lognormal distribution with a Gaussian mixture. Two strategies are realised: (i) 3 RBFs with nonlinear to kernels fitting parameters and (ii) 30 RBFs forming the Gaussian comb with linear to kernels parameters. Both strategies give the perfect fit and desired values of $\mu_{LN} = 3$ and $\sigma_{LN} = 0.9$ via the approximation (9).

where μ and σ^2 are the mean and variance of the corresponding normal distribution. The reciprocal relation between these μ and σ^2 and the mean and variance of the lognormal distribution, μ_{LN} and σ_{LN}^2, is as follows:

$$\mu_{LN} = \exp\left(\mu + \frac{\sigma^2}{2}\right), \quad \sigma_{LN}^2 = \left[\exp\left(\sigma^2\right) - 1\right] \cdot \exp\left(2\mu + \sigma^2\right). \quad (11)$$

Two different decompositions of (10) with $\mu_{LN} = 3$ and $\sigma_{LN} = 0.9$ are shown in Fig. 6. One can see that both approaches give perfect result, while require different number of RBFs: (i) 3 RBFs with 3 fitting parameters each (w_i, b_i and c_i) and (ii) 30 RBFs in the Gaussian comb with 30 fitting parameters (weights w_i). The bandwidth is chosen as $\Delta y = [\mu_{LN} - 4\sigma_{LN}, \mu_{LN} + 4\sigma_{LN}]$. We shall keep this choice for the bandwidth in this study although it is not optimal

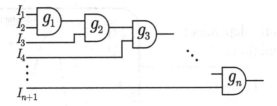

Fig. 7. A sequence of logic gates used in simulations.

(distributions can be significantly skewed). This issue will be addressed in a separate study.

In the next section, the proposed Algorithm with the decomposition strategies discussed above is tested and compared with the numerical experiments (Monte Carlo simulations). The goal is to proof the concept, determine possible issues and define future steps in the research.

6   Verification and Discussion

Let us investigate whether an error in computing the delays' PDF is accumulating or not when the decomposition into Gaussian kernels is used. To do so, we consider a model circuit shown in Fig. 7. The inputs' delays I_i $(i = 1, \ldots, n)$ are described by the corresponding RVs, X_i. For simplicity (and without any loss of generality), the operation time of the gates is considered to be the same, thus, described by an RV X_0. From the mathematical point of view, the forward traversing of such a sequence of gates is equivalent to computing the chained expressions of type (3). Thus, for the nth gate we have

$$\max\{\underbrace{\ldots \max[\max(X_1, X_2) + X_0, X_3] + X_0 \ldots}_{n-1 \text{ times}}, X_{n+1}\} + X_0. \qquad (12)$$

We have conducted a series of runs of the Algorithm and MC simulations for the sequence of $n = 20$ (the source code is available from [10]). We have chosen the initial delays X_i to be distributed as $X_i \sim \mathcal{N}(\mu_i, \sigma_i)$. The values for the inputs' means and standard deviations, μ_i and σ_i, were randomly drawn from $\sim \mathcal{U}(2, 7)$ and $\sim \mathcal{U}(0.2, 1.3)$ respectively for each run. One of the realisations of the experiments is shown in Fig. 8. Since the absolute values of delays are not important in the present study, the performance of the algorithm is measured by *relative errors* in the mean values and standard deviations of delays with respect to the Monte Carlo simulations.

For the Gaussian comb, $m = 55$ kernels have used with the shape parameter $c = 0.15$. The relative error is less than 0.01% and remains at that level until it starts to grow dramatically (see Fig. 8). This occurs when the chosen topology of the comb becomes non-optimal, as it is shown for node 19. Also note that the relative error in the standard deviation increased faster than that for the mean, which is expected.

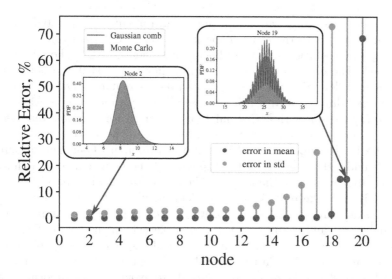

Fig. 8. Results for the Gaussian comb approach ($m = 55$ kernels with $c = 0.15$): the relative error in determination of the mean and standard deviation with respect to the Monte Carlo values versus number of gates passed in the sequence. When the bandwidth Δy becomes large, the topology of the Gaussian comb is no longer optimal and the comb sprawls (node 19). For the details of the simulations see the text.

For the case of 3 RBFs with non-linear fitting parameters, the algorithm has given poor performance. As is discussed in the previous section, the optimisation problem (6) in this case is sensitive to small deviations in an initial guess or change of the bandwidth Δy. Thus, it has not been possible to finish the traversal of the graph successfully (a Trust Region Algorithm [3] for the constrained optimisation was used), and the results for this approach are not presented.

The obtained results allow us to conclude that the problem of VLSI circuit delay indeed can be solved using (i) exact solution for a single gate's delay PDF and (ii) decomposition of non-Gaussian functions into Gaussian mixtures. The detailed discussion and conclusions are as follows.

(i) The exact formula for an output logic gate delay, the convolution of $\max(X_1, X_2)$ and X_0 for Gaussian RVs X_i ($i = 0, 1, 2$), allows one to build a closed-loop algorithm for forward traversal of a delay through a timing graph G. The requirement for this is that non-Gaussian PDFs of delays are presented via Gaussian mixtures, sums of RBFs of Gaussian form. The decomposition is equivalent to solving the minimisation problem (6). We have considered two different strategies for this problem.

(ii) Within the first strategy, for m RBFs it is required to determine $3m$ parameters, $3m - 1$ of which are coefficients in the *arguments* of the RBFs. The advantage is that only a few of RBFs is enough to obtain a fit with a desired accuracy, but the drawback is sensitivity to small changes in the parameters such as initial guess, choice of bandwidth Δy, etc.

(iii) The second strategy relies on a pre-set grid of equally separated Gaussian kernels $\varphi(x)$ with the same shape parameter, a *Gaussian comb*. In such a case, only coefficients that are linear multipliers to RBFs should be determined. The obvious advantage of this approach is that the optimisation problem allows the exact solution, however the required number m of RBFs increases dramatically, which increases (linearly to m) computation costs.

(iv) The comparison with MC simulations shown in Fig. 8 proofs the concept: the timing graph can be forward traversed with the relative error less than 0.01% by decomposing real PDFs into corresponding Gaussian mixtures at each node. However, when the topology of the Gaussian comb is fixed, it leads to sprawling of the latter as the bandwidth Δy becomes large.

(v) In principle, the optimisation problem of finding the weights w_i for the Gaussian comb can be solved together with the problem of finding optimal number m_* and shape parameter c_* of kernels in the comb. This should not only prevent the comb sprawling but also speed up the graph traversal procedure noticeably. At the same time, the optimisation problem for the 3-RBF case should be analysed rigorously to avoid slipping in local minima. This can be an alternative to the Gaussian comb decomposition. These issues will be addressed in a separate study.

Acknowledgement. This work has emanated from research supported in part by Synopsys, Ireland, and a research grant from Science Foundation Ireland (SFI) and is co-funded under the European Regional Development Fund under Grant Number 13/RC/2077.

References

1. Bhasker, J., Chadha, R.: Static Timing Analysis for Nanometer Designs. A Practical Approach. Springer, Heidelberg (2009). https://doi.org/10.1007/978-0-387-93820-2
2. Blaauw, D., Chopra, K., Srivastava, A., Scheffer, L.: Statistical timing anlaysis: from basic principles to state of the art. IEEE Trans. Comput.-Aided Des. Integr. Circ. Syst. **4**(8), 589–607 (2008)
3. Byrd, R.H., Schnabel, R.B., Shultz, G.A.: A trust region algorithm for nonlinearly constrained optimization. SIAM J. Numer. Anal. **24**(5), 1152–1170 (1987). https://doi.org/10.1137/0724076
4. Chang, H., Zolotov, V., Narayan, S., Visweswariah, C.: Parameterized block-based statistical timing analysis with non-Gaussian parameters, nonlinear delay functions. In: Proceedings of DAC, pp. 71–76, June 2005. https://doi.org/10.1145/1065579.1065604
5. Forzan, C., Pandini, D.: Statistical static timing analysis: a survey. Integr. VLSI J. **42**(3), 409–435 (2009). https://doi.org/10.1016/j.vlsi.2008.10.002, special Section on DCIS2006
6. Freeley, J., Mishagli, D., Brazil, T., Blokhina, E.: Statistical simulations of delay propagation in large scale circuits using graph traversal and kernel function decomposition. In: Proceedings of SMACD, July 2018
7. Gerez, S.H. (ed.): Algorithms for VLSI Design Automation. Wiley, Hoboken (1998)

8. Lavagno, L., Markov, I.L., Martin, G., Scheffer, L.K. (eds.): Electronic Design Automation for IC Implementation, Circuit Design, and Process Technology. CRC Press, Boca Raton (2016)
9. McLachlan, G., Peel, D.: Finite Mixture Models. Wiley Series in Probability and Mathematical Statistics. Wiley, Hoboken (2000)
10. Mishagli, D.: RBF approximation for non-Gaussian delay propagation in VLSI: Code (2020). https://doi.org/10.5281/zenodo.3749750
11. Orshansky, M., Nassif, S., Boning, D.: Design for Manufacturability and Statistical Design: A Constructive Approach. Series on Integrated Circuits and Systems. Springer, Heidelberg (2008). https://doi.org/10.1007/978-0-387-69011-7
12. Ramprasath, S., Vijaykumar, M., Vasudevan, V.: A skew-normal canonical model for statistical static timing analysis. IEEE Trans. Very Large Scale Integr. (VLSI) Syst. **24**(6), 2359–2368 (2016). https://doi.org/10.1109/TVLSI.2015.2501370
13. Sapatnekar, S.: Timing. Springer, Heidelberg (2004). https://doi.org/10.1007/b117318
14. Titterington, D., Smith, A., Makov, U.: Statistical Analysis of Finite Mixture Distributions. Wiley Series in Probability and Mathematical Statistics: Applied Probability and Statistics. Wiley, Hoboken (1985)
15. Visweswariah, C., Ravindran, K., Kalafala, K., Walker, S.G., Narayan, S.: First-order incremental block-based statistical timing analysis, pp. 331–336 (2004). https://doi.org/10.1145/996566.996663
16. Visweswariah, C., et al.: First-order incremental block-based statistical timing analysis. IEEE Trans. Comput.-Aided Des. Integr. Circuits Syst. 25(10), 2170–2180 (2006). https://doi.org/10.1109/TCAD.2005.862751

Ant Colony Optimization Implementation for Reversible Synthesis in Walsh-Hadamard Domain

Krzysztof Podlaski[(✉)] [iD]

Faculty of Physics and Applied Informatics, University of Lodz, Lodz, Poland
podlaski@uni.lodz.pl

Abstract. Reversible circuits are one of the technologies that can provide future low energy circuits. The synthesis of an optimal reversible circuit for a given function is an np-hard problem. The meta-heuristic approaches are one of the most promising methods for these types of optimization problems. In this paper, a new approach for ACO reversible synthesis is presented. Usually, authors build an ACO system with the use of truth table or permutation representation of the reversible function. In this work, a Walsh spectral representation of a Boolean function is used. This allows dividing search spaces into smaller "promising" areas with well-defined transition operations between them. As a result, we can minimize the enormous search space and generate better solutions than obtained by ACO synthesis with classical reversible function representation. The proposed approach was applied to benchmark reversible functions of 4,5 and 6 variables and compared to other meta-heuristic results and best-known solutions.

Keywords: ACO · Reversible circuits · Synthesis · Spectral methods · Walsh spectrum

1 Introduction

One of the most important requirements for the development of new electronic devices is power consumption. With rapid minimization and growing demand for computation power, the low-power design is under constant research. It is well-known that with any loss of information the energy is dissipated [15]. On that base, the reversible circuits, i.e the circuits that do not lose information during computation, are recognized as one of the promising alternatives for future low-power design [3,10]. It should be mentioned that reversible logic is strictly connected with another promising technology - quantum computing.

Reversible circuits can be synthesized with the use of basic gates like not, cnot, Toffoli. However, this synthesis is very different from the synthesis of classical circuits [23]. Many heuristic methods of reversible synthesis have been proposed in the literature, to name a few: transformation based algorithm [19,20], cycle-based algorithm [24], decision diagram based algorithms [28]. Most of the

V. V. Krzhizhanovskaya et al. (Eds.): ICCS 2020, LNCS 12141, pp. 230–243, 2020.
https://doi.org/10.1007/978-3-030-50426-7_18

known methods are very redundant or can be applied to a reversible function with a small number of inputs [14]. Some of the authors proposed also the use of meta-heuristic methodologies like genetic algorithms [11,27], particle swarm optimization [4,18], and ant colony optimization [17]. All presented meta-heuristic approaches use truth-table or permutation representation of a reversible function, in this paper a new ant colony optimization approach that uses the spectral representation of a reversible function is used.

The paper is organized as follows. In Sect. 2 basic concepts connected to reversible logic are introduced. Section 3 contains a description of Walsh-Hadamard spectral methods. Section 4 is devoted to the general ACO system while in Sect. 5 a detailed description of the algorithm is presented. Section 6 contains a discussion of the results obtained. The concluding remarks are included in the last Sect. 7.

2   Reversible Logic

In this Section, the basic definitions and ideas connected to reversible logic are presented for the convenience of the reader.

Definition 1 (Balanced function). *A Boolean function* $f : \{0,1\}^n \to \{0,1\}$ *is called balanced if it takes the value 1 the same number of times as 0.*

Definition 2 (Reversible function). *A mapping* $f : \{0,1\}^n \to \{0,1\}^n$ *is called a reversible function if it is bijective.*

Definition 3 (Component function). *A reversible function* $f(x), x \in \{0,1\}^n$ *can be considered as a vector of Boolean functions* $f = (f_1, f_2, ..., f_n)$, *each of these functions* f_i *will be called component functions.*

Table 1. Truth table of an exemplary $3 * 3$ reversible function f, the columns in/out uses decimal signal encoding, while two middle ones represent signals encoded as a binary string.

In	x_1	x_2	x_3	f_1	f_2	f_3	Out
0	0	0	0	0	0	1	1
1	0	0	1	0	1	0	2
2	0	1	0	0	0	0	0
3	0	1	1	1	1	1	7
4	1	0	0	1	0	1	5
5	1	0	1	0	1	1	3
6	1	1	0	1	1	0	6
7	1	1	1	1	0	0	4

Any reversible function can be represented in many ways for example as a truth table, binary decision diagram, additionally, every reversible function is bijective onto, and as such, it can be written as a permutation of input signals. From the truth table representation of a reversible function (Table 1), we can see that the function F is a bijection, every input signal appears once as an output. Every output $f(1) = f(001) = 010 = 2$ is an element of a set of all possible inputs, that is why the function can be represented as a permutation $[1, 2, 0, 7, 5, 3, 6, 4]$. All of these representations are equivalent. The component functions f_1, f_2, f_3 are connected to appropriate columns in the truth table.

A reversible circuit is a circuit that realizes a reversible function, i.e. it performs a bijective mapping of an n-bit input signal onto an n-bit output signal, the mapping is defined by a given reversible function. The circuit can be reversible if all internal operations are reversible, which means all building blocks of a reversible circuit have to be reversible themselves. The classical digital circuits are based on gates like AND and OR, these gates are not reversible, moreover, their functions are not balanced, this implies that these gates cannot be used in a reversible circuit. Additionally, in reversible circuit fan-outs are forbidden, this implies that a reversible circuit is a cascade of reversible gates [22]. The most often used library of basic reversible gates is known as multiple control Toffoli gates (MCT) and contains three types of gates: not, cnot, Toffoli.

1. $T1(s)$ - not gate, negates the signal on line s,
2. $T2(k; s)$ - controlled not gate, negates the signal on line s if the signal on control line k is equal to 1,
3. $T3(k, l; s)$ - Toffoli gate, negates the signal on line s if the signals on controlled lines k and l are equal to 1,
4. $Tm(k_1, ..., k_{m-1}; s)$ - generalized Toffoli gate, negates the signal on line s if the signals on all $m - 1$ controlled lines $k_1, ..., k_{m-1}$ are equal to 1,

Each of the MCT gates is self-inverse. It is known that any $n * n$ reversible function can be implemented with the use of the gates from the MCT library. Moreover, the number of different gates one can use to synthesize $n * n$ reversible circuit increase with the number n of input binary variables, for example for $4 * 4$ reversible domain, there are 32 available MCT gates: $4 * T1$, $12 * T2$ and $T3$, $4 * T4$.

2.1   Gate Cost

The simplest approach to evaluate the quality of a circuit is a gate count (GC), this measure, however, treats all gates in the same way. It is rather obvious that different gates will have different implementation cost, this implementation cost can differ for technology used. The MCT gate library contains many different gates, from the simplest ones T1 (not gate) to a very complex like T5 (generalized Toffoli gate with 5 input lines and 4 lines are control ones). In the literature, there are a few approaches to describe the sophistication of reversible gates. The most recognized measures of gate costs are the so-called quantum cost (QC) [2] and T-count [1,21]. All these measures are connected to the representation of reversible gates in quantum gates. QC and T-count measures grow rapidly with the growth of the number of control lines. For gates used in the paper: T1, T2 have QC = 1, T-count = 0, T3: QC = 5, T-count = 7, T4: QC = 13, T-count = 16. In this paper, quantum cost measure is used as it has similar behavior comparing to T-count and has a nonzero cost to two basic gates not and cnot. The following synthesis procedure is designed to optimize the QC of obtained circuits.

The process of reversible synthesis of a given function f is a task of generating the optimal sequence of reversible gates, that transforms input signals into outputs with the agreement with function f. In this paper, the optimal sequence means the sequence with the lowest total quantum cost. However the meta-heuristic methodology is used and by default part of the solutions can be suboptimal, the goal of the presented approach is to find solutions as near to optimal as possible.

3   Walsh-Hadamard Transform

In the previous section two different representation of reversible function was mentioned (truth table and permutation). In this paper, an additional - spectral representation is used. In the domain of Boolean functions a few generalized Fourier type transforms are well-known, i.e. Reed-Muller, Arithmetic, and Walsh [26]. Each of these transforms can be used to define spectral representation with different properties and was used for some time in the theory of Boolean functions. In this paper the Walsh transform in Hadamard order is used, called also Walsh-Hadamard transform.

Definition 4 (Walsh transform). *In n variable Boolean domain the Walsh-Hadamard transform is defined by the Kronecker product $\otimes$ of basic Walsh matrix*

$$W(n) = \bigotimes_n W(1), \qquad where \ W(1) = \begin{pmatrix} 1 & 1 \\ 1 & -1 \end{pmatrix}. \tag{1}$$

To apply Walsh-Hadamard transform to a Boolean function, we have to apply integer encoding of Boolean function.

Definition 5 (Integer encoding). *Integer encoding of a Boolean value x is defined as follows:*

$$x \rightarrow \begin{cases} 1 & when \ x = 0 \\ -1 & when \ x = 1. \end{cases} \tag{2}$$

Definition 6 (Walsh spectrum). *The Walsh-Hadamard spectrum of n variable Boolean function $f(x_1, x_2, ..., x_n)$ is represented as a vector of integer values S_f defined as:*

$$S_f = W(n)f^c. \tag{3}$$

where: f^c is a column truth-vector of function f in integer encoding.

Example 1 Let $f = (0,0,1,1,0,1,0,1)^T$ be a truth-vector, then a vector of the form $f^c = (1,1,-1,-1,1,-1,1,-1)^T$ represents integer encoded version of f. In 3 variable domain Walsh-Hadamard transform $W(3)$ have the form:

$$W(3) = \begin{pmatrix} 1 & 1 & 1 & 1 & 1 & 1 & 1 & 1 \\ 1 & -1 & 1 & -1 & 1 & -1 & 1 & -1 \\ 1 & 1 & -1 & -1 & 1 & 1 & -1 & -1 \\ 1 & -1 & -1 & 1 & 1 & -1 & -1 & 1 \\ 1 & 1 & 1 & 1 & -1 & -1 & -1 & -1 \\ 1 & -1 & 1 & -1 & -1 & 1 & -1 & 1 \\ 1 & 1 & -1 & -1 & -1 & -1 & 1 & 1 \\ 1 & -1 & -1 & 1 & -1 & 1 & 1 & -1 \end{pmatrix}. \tag{4}$$

Walsh-Hadamard spectrum S_f of function f is $S_f = (0,4,4,0,0,-4,4,0)^T$.

The elements of spectral vector S_f are often called Walsh coefficients of function f in Hadamard order. In the Walsh domain of three variables, these coefficients are often designated as $S_f = (S_0, S_3, S_2, S_{23}, S_1, S_{13}, S_{12}, S_{123})^T$. These spectral coefficients represent the correlation of function values with input variables. The zero-order coefficient S_0 represents the difference of the number of occurrence of 0 and 1 values in a truth table column, for balanced functions S_0 is always 0. The first order coefficients $\{S_1, S_2, S_3\}$ represent the correlation of function values with the values of input variables x_1, x_2, x_3 respectively. The second-order coefficients $\{S_{12}, S_{13}, S_{23}\}$ are connected with the correlation of values of the function f and xor products: $x_1 \oplus x_2$, $x_1 \oplus x_3$, $x_2 \oplus x_3$. The third-order coefficient S_{123} represents the correlation of function f and $x_1 \oplus x_2 \oplus x_3$.

In the general case we can write properties of Walsh coefficients:

- all coefficients s have an integer value, $-2^n \leq s \leq 2^n$,
- a sum of absolute values of any two coefficients cannot exceed 2^n,
- S_0 - is connected to a constant part of the function, this coefficient is equal to 0 for balanced functions,
- S_i represents the correlation of the function in consideration with the value of variable x_i,
- S_{ij} represents the correlation of the function and xor product $x_i \oplus x_j$, where $i \neq j$ $1 \leq i,j \leq n$,
- $S_{ij...m}$ represents the correlation between the function and xor product $x_i \oplus x_j \oplus ... \oplus x_m$.

Definition 7 (Reverse Walsh transformation). *From Walsh-Hadamard spectrum S_f an original integer encoded Boolean function f^c can be obtained by reverse Walsh-Hadamard transform in the form:*

$$f^c = W^{-1}(n)S_f, \qquad where: W^{-1}(n) = 2^{-n}W(n). \tag{5}$$

As reversible function can be treated as a vector of component functions, one can always derive Walsh transformation of a reversible function by application of Walsh transform to each of component functions independently.

Definition 8 (Walsh spectrum of reversible function). *The Walsh spectrum of* $n*n$ *reversible function* F *can be obtained by application of Walsh matrix to integer encoded function* F^c

$$S_F = W(n)F^c. \tag{6}$$

It should be noted that the spectral form $S_F = S_{f_1}, S_{f_2}, ...S_{f_n}$ of a reversible function $F = (f_1, f_2, ..., f_n)$ is a vector of spectral columns of component functions, i. e. $S_{f_i} = W(n)f_i^c$.

3.1 Spectral Invariant Operations

Definition 9 (Spectral invariant operations). *An operation on Boolean function* f *that preserve absolute values of Walsh spectral coefficients of the function is called a spectral invariant*

The set of spectral invariant operations have been used in Boolean logic for many years [9,12,13,16]. The set of invariant operations is built from the following function transformations:

1. negation of the function - changes sign of all spectral coefficients,
2. negation of an input variable - changes the sign of spectral coefficients connected with this variable (for example when $x_2 \rightarrow \overline{x_2}$ then first-order coefficient change sign $S_2 \rightarrow -S_2$, and similarly appropriate second-order coefficients S_{12}, S_{13} and so on,
3. permutation of input variables - exchanges the coefficients connected with appropriate variables,
4. replacement of a variable x_i with $x_i \oplus x_j$ for $i \neq j$,
5. replacement of the function truth vector f with $f \oplus x_i$.

All invariant operations can be implemented by application of appropriate reversible gates, T1 gates implement operations 1 and 2 while using T2 gates one can build operations 4 and 5. The operations 3 can be implemented by a so-called swap gate, swap gate can be build from three T2 gates. The presented relations divide the set of all reversible gates into a set of spectral invariant operations (these gates have a quantum cost equal to 1) and the rest of reversible gates that can modify the spectrum of the function in consideration.

3.2 Walsh-Hadamard Spectrum and Reversible Gates Operation

As was shown above, the simplest reversible gates T1 and T2 are connected to spectral invariant operations. That means the rest of the reversible gates have to modify the spectrum of a Boolean function. Every reversible function can be represented as a permutation matrix, in particular, any reversible gate is connected to a permutation matrix. Suppose we have a reversible circuit of the form presented below.

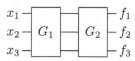

The function $F = (f_1, f_2, f_3)$ is represented by such a reversible gates G_1 and G_2, and can be written as:

$$F = G_2 G_1 I, \tag{7}$$

where I represent $3 * 3$ identity function and G_1, G_2 are permutation matrices representing reversible gates G_1, G_2. Applying Walsh-Hadamard transform to the resulting function F we have

$$WF = WG_2G_1I = WG_2W^{-1}WG_1W^{-1}WI, \tag{8}$$

$$S_F = WG_2G_1I = WG_2W^{-1}WG_1W^{-1}S_I = \widetilde{G_2}\widetilde{G_1}S_I, \tag{9}$$

where S_I denote Walsh spectral representative of identity function and $\widetilde{G_2}$ and $\widetilde{G_1}$ are Walsh-Hadamard representatives of reversible gates G_2, G_1 respectively. For any n the S_I is the simplest spectrum matrix $n * 2^n$, in each column, there is only one nonzero value. For $n = 3$ S_I have the form:

$$S_I = \begin{pmatrix} 0 & 0 & 0 \\ 0 & 0 & 8 \\ 0 & 8 & 0 \\ 0 & 0 & 0 \\ 8 & 0 & 0 \\ 0 & 0 & 0 \\ 0 & 0 & 0 \\ 0 & 0 & 0 \end{pmatrix}. \tag{10}$$

Taking into account permutation matrices of reversible gates T1, T2, T3 we can easily using matrices W and W^{-1} derive their representatives in the spectral domain, below representatives of selected gates in Walsh domain are presented:

$$T1(1) \rightarrow \begin{pmatrix} 1 & 0 & 0 & 0 & 0 & 0 & 0 & 0 \\ 0 & 1 & 0 & 0 & 0 & 0 & 0 & 0 \\ 0 & 0 & 1 & 0 & 0 & 0 & 0 & 0 \\ 0 & 0 & 0 & 1 & 0 & 0 & 0 & 0 \\ 0 & 0 & 0 & 0 & -1 & 0 & 0 & 0 \\ 0 & 0 & 0 & 0 & 0 & -1 & 0 & 0 \\ 0 & 0 & 0 & 0 & 0 & 0 & -1 & 0 \\ 0 & 0 & 0 & 0 & 0 & 0 & 0 & -1 \end{pmatrix}, \quad T2(1;2) \rightarrow \begin{pmatrix} 1 & 0 & 0 & 0 & 0 & 0 & 0 & 0 \\ 0 & 1 & 0 & 0 & 0 & 0 & 0 & 0 \\ 0 & 0 & 0 & 0 & 0 & 0 & 1 & 0 \\ 0 & 0 & 0 & 0 & 0 & 0 & 0 & 1 \\ 0 & 0 & 0 & 0 & 1 & 0 & 0 & 0 \\ 0 & 0 & 0 & 0 & 0 & 1 & 0 & 0 \\ 0 & 0 & 1 & 0 & 0 & 0 & 0 & 0 \\ 0 & 0 & 0 & 1 & 0 & 0 & 0 & 0 \end{pmatrix}, \tag{11}$$

$$T3(1,2;3) \rightarrow \begin{pmatrix} 1 & 0 & 0 & 0 & 0 & 0 & 0 & 0 \\ 0 & \frac{1}{2} & 0 & \frac{1}{2} & 0 & \frac{1}{2} & 0 & -\frac{1}{2} \\ 0 & 0 & 1 & 0 & 0 & 0 & 0 & 0 \\ 0 & \frac{1}{2} & 0 & \frac{1}{2} & 0 & -\frac{1}{2} & 0 & \frac{1}{2} \\ 0 & 0 & 0 & 0 & 1 & 0 & 0 & 0 \\ 0 & \frac{1}{2} & 0 & -\frac{1}{2} & 0 & \frac{1}{2} & 0 & \frac{1}{2} \\ 0 & 0 & 0 & 0 & 0 & 0 & 1 & 0 \\ 0 & -\frac{1}{2} & 0 & \frac{1}{2} & 0 & \frac{1}{2} & 0 & \frac{1}{2} \end{pmatrix}. \tag{12}$$

As all T1 are diagonal and T2 a permutation matrices, their application to S_I will not modify absolute values in each column, these operations can only change their positions. All other gates T3, T4, ..., Tn will change the values in the spectrum and very often change number or zeros in the spectral columns, the more zero values are in the spectrum the more linear the function is.

In Walsh domain, the functions are treated as a spectral matrix, all the possible input signals are taken into account, singular input is connected to a matrix row. The columns of the spectral matrix are connected to the spectrum of component functions. The action of the reversible gate on the actual state is represented as matrix state multiplication, which means all columns in the state matrix are treated independently, i.e. all component functions can be treated the same way.

4   Ant Colony Optimization

Ant colony optimization (ACO) is a biologically inspired meta-heuristic algorithm introduced by Dorigo in [5,7]. The main idea of the algorithm is based on the social behavior of ants during a food search. The communication in the colony is based on pheromone residue that is left by each colony member when traveling between nest and food source. At first, ACO was introduced to solve traveling salesman problem later the method was applied to many other combinatorial problems [6,8].

Usually, the ACO algorithm is defined on a graph that represents states in a search space, each edge in the graph represents actions (transitions between states) that can be taken during the walk of an artificial ant. Every action (edge) has an assigned value that represents the cost of the choice (very often the distance between the states).

The optimization procedure used in this paper is based on a colony of m artificial ants, in an iteration, every ant walks independently on a problem graph, at each node the ant has to choose an edge that leads to the nest state. When an ant reaches the end, either by reaching the final state node or the limit of steps in one iteration, pheromones residues on visited nodes are updated. The pheromone update rule and decision reasoning depend on ACO implementation.

At the start of the procedure, each node has been given an initial value τ_0. During the update procedure we allow the pheromones to evaporate in time.

$$\tau_{ij} \leftarrow \rho\tau_{ij}, \tag{13}$$

where $\rho \in [0, 1)$ is an evaporation parameter. The pheromone update procedure is done after evaporation, at the end of each iteration, every ant from the colony update pheromone by an additional deposit $\Delta\tau_{ij}^k$:

$$\Delta\tau_{ij}^k(t) = R/C^k(t), \tag{14}$$

where R is a constant and $C^k(t)$ represents the cost of the solution obtained by k-th ant in an iteration t, usually, C^k represents the sum of costs assigned to edges used in the solution. The final pheromone update procedure, that takes into account evaporation and new deposits have the form:

$$\tau_{ij}(t+1) = \rho\tau_{ij}(t) + \sum_k \Delta\tau_{ij}^k(t). \tag{15}$$

When ant moves over the graph the choice of an outgoing edge can be taken either randomly or based on actual pheromones residue and partial heuristic reasoning. In the procedure, a parameter ζ, the probability of pure random choice at each graph node, is used. In the case of pheromone-based choice the decision is based on weighted probability:

$$p_{ij}(t) = \frac{\tau_{ij}(t)^\alpha \eta_{ij}^\beta}{\sum_{i,j} \tau_{ij}(t)^\alpha \eta_{ij}^\beta}, \tag{16}$$

where $p_{ij}(t)$ is the probability of choosing j edge when an ant is in node i in iteration t, $\tau_{ij}(t)$ represents the actual pheromone deposit in node i attached to edge j at the moment t. The parameter η_{ij} represents heuristic reasoning, this part always depends on the problem to be solved. Two additional parameters α, β describe the relative importance of the pheromone and heuristic factors.

5   Algorithm Implementation

In this paper ACO is used to optimize the process of reversible function synthesis, moreover, all the reversible functions are represented with the use of the Walsh-Hadamard spectrum. The procedure starts with an initial function to be synthesized. Each ant will try to convert the function spectrum into a spectrum S_I by application of reversible gates. The solution is a sequence of gates that represents a function in consideration.

In the proposed ant colony optimization procedure state (node of the graph) is connected to the spectrum of a reversible function, while the actions (edges of the graph) are the gates that can be added to the solution. Each edge has an attached value that represents the quantum cost of the appropriate gate.

5.1   Heuristic Reasoning for Designed ACO

As was mentioned before in the process of decision making the heuristic knowledge about the word of reversible functions is taken into account (16). The target state S_I has only one nonzero value, that means during synthesis procedure we should maximize the number of zeros of the actual state.

Suppose we have two states F_1, F_2 represented by spectral matrices S_{F_a} and S_{F_b}. Additionally, there is the reversible gate G_x that transforms one state into another one, i. e. $S_{F_a} = \widetilde{G_x} S_{F_b}$. The heuristic factor for edge G_x connected to the state F_a will have the form:

$$\eta_{ax} = 1 + \frac{\text{count\_zeros}(S_{F_b}) - \text{count\_zeros}(S_{F_a})}{\min\left(\text{count\_zeros}(S_{F_b}), \text{count\_zeros}(S_{F_a})\right)} \tag{17}$$

$$\pm \frac{\sigma}{\text{qc}(G_x)}, \tag{18}$$

where σ is a scaling parameter, $\text{count\_zeros}(.)$ represent the function that returns the number of zeros in the argument, $\min(.,.)$ is the function that returns the minimum value from given arguments, $\text{qc}(.)$ represents the quantum cost of a reversible gate. The sign $\pm$ corresponds to the sign of $\text{count\_zeros}(S_{F_b})$ − $\text{count\_zeros}(S_{F_a})$.

The form of the heuristic factor (17) represents the main parameters that have to be taken into account during the synthesis procedure. The main part $\frac{\text{count\_zeros}(S_{F_b}) - \text{count\_zeros}(S_{F_a})}{\min(\text{count\_zeros}(S_{F_b}), \text{count\_zeros}(S_{F_a}))}$ depends on a change of the number of zeros in two connected states S_{F_a} and S_{F_b}, if the application of the gate increase the number of zeros the factor has positive value when number of zero decreases the factor is negative and so decrease the probability of choosing the edge. The second element of the heuristic factor $\pm \frac{\sigma}{\text{qc}(G_x)}$ distinguishes the gates that generate state with the same number of zeros in the spectrum but differs in quantum cost.

5.2   Pheromone Update Procedure

In the pheromone update procedure, the deposit from k-th ant given by Eq. (14), the sum of quantum costs of all the gates in the solution found by an ant is used for assessment of the cost of the solution C^k.

5.3   Additional Parameters of ACO Used in the Implementation

In the implementation of the ACO algorithm, the values of parameters that were used are presented in Table 2.

5.4   Creation of ACO System

The ACO is based on the graph that represents the search space. It has to be mentioned that the graph in consideration is an enormous one, for n variables, there are $2^n!$ reversible functions. It is impossible to build and keep in memory all possible functions as nodes and transition gates edges. For that reason, our search space will be built during the optimization process, whenever an ant visits a new node, important node information will be created, that means assign all edges, generate all heuristic factors, initialize pheromone deposit, etc. This lazy node initialization allows us to save memory and speed up the initialization phase. In the starting initialization phase, only the target state (node connected to S_I) and a state connected to a function f in consideration are created. The node connected to f takes the role of the nest, it means at the beginning of a new iteration every ant starts from this node.

Table 2. Parameters used in implemented ACO algorithm

Name	Short description	Value
α	The parameter connected to the influence of pheromone in decision making (16)	1
β	The parameter connected to the influence of heuristic in decision making (16)	1
ζ	The probability of random choice at every node	0.5
m	The size of the colony	30
R	The scale factor used for deposit (14)	1
ρ	The pheromone evaporation factor (13)	0.8
σ	The scale factor in heuristic rule (17)	1

During the first tests of the algorithm, it was noticed that the ACO behaves better (gives better results) when more than one final state was given. For that reason, in the initialization phase, more than one target state was created. For every gate a state that can be obtained from state S_I by application of a single gate was initialized, moreover, this states had a very simple assignment of the pheromones for edges: 1 for the edge that leads to S_I and 0 to others. This creates an additional set of states that directly leads to the final state. This procedure could be extended to next states with the distance of two or three gates from S_I, it has to be noted that while for $n = 4$ there are only 32 nearest neighbors of S_I the number of neighbors in a distance of 3 gates can be estimated to 5000–10000 and it not seem reasonable to initialize them all.

6   Numerical Results

The proposed algorithm has been implemented in python, and run to synthesize selected benchmark functions. At first ACO algorithm was used to synthesize one of functions used in publication [17]: $f = [2, 9, 7, 13, 10, 4, 2, 14, 3, 0, 12, 6, 8, 15, 11, \ 15]$. On that function, the impact of algorithm parameter values was analyzed and their final values selected (see Table 2).

In the literature, authors use many different functions as well as different cost measures to test developed synthesis algorithms. Therefore, the algorithm was tested against two sets of functions: one the set of known benchmarks and results obtained via heuristic algorithms implemented in revkit tool [25] (Table 3), the other set of functions taken from [17] in order to compare two ACO based approaches (Table 4). The results from Table 3 shows that the results obtained are not always optimal, however, most of them are near-optimal.

In the article that uses the ACO approach [17], we can find results of synthesis that takes into account the number of gates. In order to be able to compare

Table 3. Results of the presented ACO algorithm for selected benchmark functions compared with circuits obtained with revkit tool [25].

Benchmark function	ACO Walsh		Known solutions	
	GC	QC	GC	QC
4_49	35	88	36	84 [29]
4b_15g_1	27	67	34	90 [29]
4b_15g_2	32	91	35	83 [29]
4b_15g_3	25	81	32	96 [29]
4b_15g_4	26	80	35	91 [29]
4b_15g_5	35	102	34	80 [29]
aj-e11	25	53	34	90 [29]
App2.2	22	83	26	86 [29]
decode42	25	53	34	90 [29]
dmasl	24	68	30	78 [29]
gyang	30	96	16	76 [29]
hwb4	20	40	21	37 [29]
mini-alu	13	45	12	68 [29]
mod10_171	22	72	13	65 [29]
msaee	26	64	38	98 [29]
nth_prime4	18	60	26	70 [29]
oc5	31	127	26	82 [29]
oc6	30	66	36	88 [29]
oc7	20	54	32	92 [29]
oc8	24	54	38	98 [29]
hwb5	34	81	33	71 [27]
hwb6	48	105	47	107 [24]
nth_prime6_inc	59	281	57	485 [27]

Table 4. Results of the presented ACO algorithm compared to the results from [17] when minimal gate count is taken as the goal of the synthesis.

Function	Permutation based ACO [17]	Walsh based ACO
$[1, 0, 3, 2, 5, 7, 4, 6]$	3	3
$[7, 0, 1, 2, 3, 4, 5, 6]$	3	3
$[0, 1, 2, 3, 4, 6, 5, 7]$	3	3
$[0, 1, 2, 4, 3, 5, 6, 7]$	4	3
$[0, 1, 2, 3, 4, 5, 6, 8, 7, 9, 10, 11, 12, 13, 14, 15]$	7	8
$[1, 2, 3, 4, 5, 6, 7, 0]$	3	3
$[1, 2, 3, 4, 5, 6, 7, 8, 9, 10, 11, 12, 13, 14, 15, 0]$	3	3
$[0, 7, 6, 9, 4, 11, 10, 13, 8, 15, 14, 1, 12, 3, 2, 5]$	4	4
$[3, 6, 2, 5, 7, 1, 0, 4]$	6	6
$[1, 2, 7, 5, 6, 3, 0, 4]$	6	6
$[4, 3, 0, 2, 7, 5, 6, 1]$	5	5
$[7, 5, 2, 4, 6, 1, 0, 3]$	5	5
$[6, 2, 14, 13, 3, 11, 10, 7, 0, 5, 8, 1, 15, 12, 4, 9]$	11	10
$[2, 9, 7, 13, 10, 4, 2, 14, 3, 0, 12, 6, 8, 15, 11, 15]$	11	11
$[6, 4, 11, 0, 9, 8, 12, 2, 15, 5, 3, 7, 10, 13, 14, 1]$	13	12
$[13, 1, 14, 0, 9, 2, 15, 6, 12, 8, 11, 3, 4, 5, 7, 10]$	10	9

both approaches, the cost measure was changed from quantum cost (QC) to the number of gates in sequence (GC). For the cases presented in Table 4 the cost of all gates was set to 1 instead of quantum cost.

The results are shown in Table 4 show that the results obtained are comparable, in more difficult tasks often better to those obtained in [17]. This could be the result of function representation used, as the Walsh spectrum contains more global information on the function and heuristic factor of ant decision policy is probably better suited to the synthesis task.

7   Conclusions

In the paper, a new approach for meta-heuristic reversible synthesis is presented. The most important change is connected with different function representation used. The Walsh-Hadamard spectrum of a function contains some global properties of the function, i.e. the correlation of the function values with input variables. This property allows the algorithm to create a choosing policy based on the linearity of the function, the number of zeros in spectral representation of the function in consideration. Additionally knowing that some of the simplest reversible gates preserve absolute values of the spectrum the artificial ants in the ACO algorithm more often use high-cost gate only when the gate simplifies the function. The results of the presented algorithm were compared to best-known solutions, this comparison shows that even though the results obtained didn't always reach global optimal solutions, they were near the optimal ones. It has to be noted that the global optimal reversible circuits are known only for functions with the input variable numbers up to 4, for the functions with a higher number of inputs the global optimal solutions are not known. Therefore all methods that give a near-optimal solution for a high number of variables are important.

7.1 Future Areas of Research

The results obtained are promising, the extensions of the method presented are considered. As was mentioned all columns of component functions are treated independently, it is possible to store in the graph only component function states while all the decisions would be taken with the use of global knowledge, the final decision would be the sum of factors for each component function. This could lead to an exchange of information between similar functions that share the same component function spectral column. Additionally, it is possible to use two-directional synthesis, i.e. one nest can be placed in target node and search for a given function, while the other nest works the same way as presented, the pheromones could be exchanged between these two colonies. This could lead to better exploration and exploitation of search space in consideration.

References

1. Amy, M., Maslov, D., Mosca, M., Roetteler, M.: A meet-in-the-middle algorithm for fast synthesis of depth-optimal quantum circuits. IEEE Trans. Comput.-Aided Des. Integr. Circuits Syst. **32**(6), 818–830 (2013)
2. Barenco, A., et al.: Elementary gates for quantum computation. Phys. Rev. A **52**, 3457–3467 (1995)
3. Conte, T.M., DeBenedictis, E.P., Gargini, P.A., Track, E.: Rebooting computing: the road ahead. Computer **50**(1), 20–29 (2017)
4. Datta, K., Sengupta, I., Rahaman, H.: Particle swarm optimization based reversible circuit synthesis using mixed control Toffoli gates. J. Low Power Electron. **9**(3), 363–372 (2013)
5. Dorigo, M.: Optimization, learning and natural algorithms. Ph.D. thesis, Politecnico di Milano, Italy (1992)
6. Dorigo, M., Blum, C.: Ant colony optimization theory: a survey. Theor. Comput. Sci. **344**(2), 243–278 (2005)
7. Dorigo, M., Maniezzo, V., Colorni, A.: Ant system: optimization by a colony of cooperating agents. IEEE Trans. Syst. Man Cybern. Part B (Cybern.) **26**(1), 29–41 (1996)
8. Dorigo, M., Stützle, T.: Ant Colony Optimization. A Bradford Book. BRADFORD BOOK (2004)
9. Edwards, C.R.: The application of the Rademacher-Walsh transform to boolean function classification and threshold logic synthesis. IEEE Trans. Comput. **100**(1), 48–62 (1975)

10. Frank, M.P.: Throwing computing into reverse. IEEE Spect. **54**(9), 32–37 (2017)
11. Hadjam, F.Z., Moraga, C.: RIMEP2: evolutionary design of reversible digital circuits. ACM J. Emerg. Technol. Comput. Syst. (JETC) **11**(3), 27 (2014)
12. Hurst, S.L.: The Logical Processing of Digital Signals. Crane Russak & Company Inc./Edward Arnold, New York/London (1978)
13. Karpovsky, M.G., Stanković, R.S., Astola, J.T.: Spectral Logic and Its Applications for the Design of Digital Devices. Wiley, Hoboken (2008)
14. Kerntopf, P., Perkowski, M., Podlaski, K.: Synthesis of reversible circuits: a view on the state-of-the-art. In: Proceedings of the 12th IEEE Conference on Nanotechnology (IEEE-NANO), pp. 1–6. IEEE (2012)
15. Landauer, R.: Irreversibility and heat generation in the computing process. IBM J. Res. Dev. **5**(3), 183–191 (1961)
16. Lechner, R.J.: A transform approach to logic design. IEEE Trans. Comput. **100**(7), 627–640 (1970)
17. Li, M., Zheng, Y., Hsiao, M.S., Huang, C.: Reversible logic synthesis through ant colony optimization. In: Proceedings of the Design, Automation & Test in Europe Conference & Exhibition (DATE), pp. 307–310. IEEE (2010)
18. Manna, P., Kole, D.K., Rahaman, H., Das, D.K., Bhattacharya, B.B.: Reversible logic circuit synthesis using genetic algorithm and particle swarm optimization. In: Proceedings of the International Symposium on Electronic System Design (ISED), pp. 246–250. IEEE (2012)
19. Maslov, D., Dueck, G.W., Miller, D.M.: Techniques for the synthesis of reversible Toffoli networks. ACM Trans. Des. Autom. Electron. Syst. **12**(4), 42–es (2007)
20. Miller, D.M., Maslov, D., Dueck, G.W.: A transformation based algorithm for reversible logic synthesis. In: Proceedings of the 40th Annual Design Automation Conference, pp. 318–323. ACM (2003)
21. Miller, D.M., Soeken, M., Drechsler, R.: Mapping NCV circuits to optimized Clifford$+T$ circuits. In: Yamashita, S., Minato, S. (eds.) RC 2014. LNCS, vol. 8507, pp. 163–175. Springer, Cham (2014). https://doi.org/10.1007/978-3-319-08494-7_13
22. Nielsen, M.A., Chuang, I.L.: Quantum Computation and Quantum Information. Cambridge University Press (2010)
23. Saeedi, M., Markov, I.: Synthesis and optimization of reversible circuits – a survey. ACM Comput. Surv. **45**(2), 21:1–21:34 (2013)
24. Saeedi, M., Zamani, M.S., Sedighi, M., Sasanian, Z.: Reversible circuit synthesis using a cycle-based approach. J. Emerg. Technol. Comput. Syst. **6**(4), 1–26 (2010)
25. Soeken, M., Frehse, S., Wille, R., Drechsler, R.: RevKit: an open source toolkit for the design of reversible circuits. In: De Vos, A., Wille, R. (eds.) RC 2011. LNCS, vol. 7165, pp. 64–76. Springer, Heidelberg (2012). https://doi.org/10.1007/978-3-642-29517-1_6
26. Stanković, R.S., Astola, J.T.: Spectral Interpretation of Decision Diagrams. Springer, Heidelberg (2003). https://doi.org/10.1007/b97562
27. Wang, X., Jiao, L., Li, Y., Qi, Y., Wu, J.: A variable-length chromosome evolutionary algorithm for reversible circuit synthesis. J. Multiple-Valued Log. Soft Comput. **25**(6), 643–671 (2015)
28. Wille, R., Drechsler, R.: Bdd-based synthesis of reversible logic. Int. J. Appl. Metaheuristic Comput. **1**(4), 25–41 (2010)
29. Wille, R., Große, D., Teuber, L., Dueck, G.W., Drechsler, R.: RevLib: an online resource for reversible functions and reversible circuits. In: International Symposium on Multi-Valued Logic, pp. 220–225 (2008), RevLib. http://www.revlib.org

COEBA: A Coevolutionary Bat Algorithm for Discrete Evolutionary Multitasking

Eneko Osaba[1(✉)], Javier Del Ser[1,2], Xin-She Yang[3], Andres Iglesias[4,5], and Akemi Galvez[4,5]

[1] TECNALIA, Basque Research and Technology Alliance (BRTA),
48160 Derio, Spain
eneko.osaba@tecnalia.com
[2] University of the Basque Country (UPV/EHU), 48013 Bilbao, Spain
[3] Middlesex University London, The Burroughs, London NW4 4BT, UK
[4] Universidad de Cantabria, 39005 Santander, Spain
[5] Toho University, Funabashi, Japan

Abstract. Multitasking optimization is an emerging research field which has attracted lot of attention in the scientific community. The main purpose of this paradigm is how to solve multiple optimization problems or tasks simultaneously by conducting a single search process. The main catalyst for reaching this objective is to exploit possible synergies and complementarities among the tasks to be optimized, helping each other by virtue of the transfer of knowledge among them (thereby being referred to as Transfer Optimization). In this context, Evolutionary Multitasking addresses Transfer Optimization problems by resorting to concepts from Evolutionary Computation for simultaneous solving the tasks at hand. This work contributes to this trend by proposing a novel algorithmic scheme for dealing with multitasking environments. The proposed approach, coined as Coevolutionary Bat Algorithm, finds its inspiration in concepts from both co-evolutionary strategies and the metaheuristic Bat Algorithm. We compare the performance of our proposed method with that of its Multifactorial Evolutionary Algorithm counterpart over 15 different multitasking setups, composed by eight reference instances of the discrete Traveling Salesman Problem. The experimentation and results stemming therefrom support the main hypothesis of this study: the proposed Coevolutionary Bat Algorithm is a promising meta-heuristic for solving Evolutionary Multitasking scenarios.

Keywords: Transfer Optimization · Evolutionary Multitasking · Bat Algorithm · Multifactorial Optimization · Traveling Salesman Problem

1 Introduction

By using as its inspiration concepts from Transfer Learning [34] and Multitask Learning [4], Transfer Optimization is an incipient knowledge field, which has

© Springer Nature Switzerland AG 2020
V. V. Krzhizhanovskaya et al. (Eds.): ICCS 2020, LNCS 12141, pp. 244–256, 2020.
https://doi.org/10.1007/978-3-030-50426-7_19

congregated an active scientific community in recent years [28]. The principal idea behind this field is to exploit what has been learned through the optimization of one specific problem or task, when tackling of another related or unrelated optimization task. Due to its relative youth, Transfer Optimization has not been studied as deeply as other research areas. It has not been until these last years when the transferability of knowledge among tasks has become a priority among researchers from the Evolutionary Computation arena.

Within the Transfer Optimization paradigm, three separated categories can be identified: *sequential transfer, multitasking* and *multiform optimization*. The first of these classes refers to those situations in which tasks are faced sequentially, assuming that for solving a new problem/instance, the knowledge collected when solving previous tasks is used as external information [11]. The second of these categories (*Multitasking*) deals with different optimization tasks simultaneously by dynamically scrutinizing existing complementarities and synergies among them [16,39]. Finally, *multiform optimization* aims at solving a single problem by resorting to different alternative problem formulations, which are optimized simultaneously. In all these categories, there is a clear consensus in the community on the capital importance of the correlation among the tasks to be solved for positively capitalizing on the transfer of knowledge over the search [17].

Among the three divisions pointed out above, *multitasking* is the one that has arguably grasped most attention by the community. The study presented in this manuscript is focused on this specific category. Specifically, we focus on multitasking optimization through the perspective of Evolutionary Multitasking (EM, [27]). In short, EM tackles the simultaneous optimization of several optimization tasks by relying on concepts and methods from Evolutionary Computation [1,8]. In the last years, a particular flavor of EM grounded on the so-called Multifactorial Optimization strategy (MFO, [17]) has shown a superior efficiency when dealing with different environments involving several continuous, discrete, single-optimization and multi-objective optimization problems and tasks [14,18,38,43]. The majority of the literature related to this area is focused on a solver belonging to this flavor: the Multifactorial Evolutionary Algorithm (MFEA, [17]). Unfortunately, alternatives to MFEA still remain scarce to date.

Bearing this in mind, the research work presented in what follows revolves on a novel EM meta-heuristic algorithm that adopts the Bat Algorithm (BA, [41]) at its core. Specifically, we present a Coevolutionary Bat Algorithm (COEBA) for discrete evolutionary multitasking. Through this proposal, we take a step further over the state of the art by elaborating on a new research direction in two different directions. On the one hand, we contribute to the EM area by introducing a new efficient meta-heuristic scheme. It is important to point out here that, unlike most articles published so far around EM, COEBA does not find its inspirational source in the MFO paradigm. On the other hand, COEBA is the first attempt at using BA for Transfer Optimization.

It is also relevant to underscore here that the experimentation carried out in this paper considers a less studied discrete environment comprising different instances of the Traveling Salesman Problem (TSP, [21]). Concretely, we assess the performance of the proposed COEBA by comparing its performance to that obtained by MFEA. Our main purpose with this performance comparison is to elucidate that COEBA embodies a promising alternative to deal with EM scenarios. To this end, we have chosen 8 different TSP instances, giving rise to 15 multi-tasking environments with varying degrees of phenotypical relationship.

The rest of the paper is organized as follows. Section 2 introduces the background related to both Evolutionary Multitasking and the Bat Algorithm. Next, Sect. 3 exposes in detail the main features of the proposed COEBA. The experimentation setup, analysis and discussion of the results are given in Sect. 4. The study ends in Sect. 5 with conclusions and future research directions.

2   Background

This section is dedicated to providing a brief background on Evolutionary Multitasking (Sect. 2.1) and the Bat Algorithm (Sect. 2.2).

2.1   Evolutionary Multitasking

In recent years, EM has arisen as a promising paradigm for facing simultaneous optimization tasks. There are two main features that motivated the first formulation of EM. The first one is the parallelism inherent to the population of individuals, which eases the management of diverse concurrent optimization tasks faced simultaneously. Thanks to this feature, latent synergies between tasks can be automatically harnessed during the solving process [28]. The second feature is the continuous transfer of genetic material between the individuals, which allows all tasks to benefit from each other, even for those that are not strongly correlated with the rest of the pool [17].

It is widely accepted that the concept of EM was only materialized through the vision of the MFO until late 2017 [6]. Today, this nascent research stream is receiving interesting contributions in terms of new algorithmic schemes, such as the Coevolutionary Multitasking scheme proposed in [5], or the multitasking multi-swarm optimization described in [37]. Additional alternatives to MFEA have also been proposed, such as the multifactorial brain storm optimization algorithm presented in [45], the Multifactorial Differential Evolution in [10] or the hybrid particle swarm optimization-firefly algorithm introduced in [40]. Despite these recently proposed methods, MFO and its related MFEA have monopolized the research activity around this field since its inception. In fact, the authors of MFEA have recently introduced an adaptive variant of MFEA, coined as MFEA-II, thereby eliciting the momentum played by this algorithm in the field [2].

Going into mathematical details, we can formulate EM as an environment in which K tasks or problems should be optimized in a simultaneous fashion. This environment is characterized by the existence of as many search spaces as tasks. Thus, for the k-th task, its objective function T_k is characterized as $f_k : \Omega_k \to \mathbb{R}$, where Ω_k represents the search space of T_k. Let us assume that all tasks are minimization problems, so that the main objective of EM is to find a set of solutions $\{\mathbf{x}_1, \ldots, \mathbf{x}_K\}$ such that $\mathbf{x}_k = \arg\min_{\mathbf{x}\in\Omega_k} f_k(\mathbf{x})$. A crucial aspect to properly understand EM is that all individuals $\mathbf{x}_p$ in the population P to be evolved belong to a unified search space Ω^U that relates to Ω_1 to Ω_K by means of a encoding/decoding mapping functions $\xi_k : \Omega_k \mapsto \Omega^U$. Therefore, each individual $\mathbf{x}_p \in \Omega^U$ in P can be decoded $(\xi_k^{-1}(\mathbf{x}_p))$ to represent a task-specific solution $\mathbf{x}_{p,k}$ for each of the K tasks. Shifting our attention on MFO and MFEA, four different definitions are associated with each individual $\mathbf{x}_p$ of the population P: Factorial Cost, Factorial Rank, Scalar Fitness and Skill Factor. With the intention accommodating this work to the extension requisites, we refer interested readers to [2,17] for additional deeper details on how these definitions are exploited during the search over the unified space Ω^U.

Several significant works have been recently published around EM and MFO. In [44], authors present an influential application of the MFEA to different discrete problems. This paper also introduces the discrete unified encoding, used as a reference in subsequent works. A related study is [47], where MFEA was applied to the Vehicle Routing Problem. Gong et al. presented in [14] and improved version of the MFEA, endowing the algorithm with a dynamic resource allocating strategy. An interesting discrete MFEA has been also developed in [38] for the composition of semantic web services. Gupta et al. presented in [18] a multi-objective variant of MFEA, giving evidence of its efficiency on a real-world manufacturing process design problem. Finally, the work in [43] follows a similar strategy by enhancing MFEA with the incorporation of opposition-based learning. Further theoretical studies on EM and MFEO can be found in [22,46].

2.2   Bat Algorithm

BA is a nature-inspired metaheuristic based on the echolocation system of bats. In the nature, bats emit ultrasonic pulses to the surrounding environment with navigation and hunting purposes. After the emission of these pulses, bats listen to the echoes, and based on them they can locate themselves and also identify and locate preys and obstacles. Besides that, each bat is able to find the most "nutritious" areas performing an individual search, or moving towards a "nutritious" location previously found by any other component of the swarm. It is important to mention that some rules have to be previously established with the aim of making an appropriate adaptation [41]:

1. All bats use echolocation to detect the distance, and they are assumed to be able to distinguish between an obstacle and a prey.
2. All bats fly randomly at speed v_i and position $\mathbf{x}_i$, emitting pulses with a fixed frequency f_{min}, varying wavelength λ and loudness A_i to search for a prey. In this idealized rule, it is assumed that every bat can adjust in an automatic way the frequency (or wavelength) of the pulses, emitted at a rate $r \in [0,1]$. This automatic adjustment depends on the proximity of the targeted prey.
3. In the real world, the bats' emissions loudness can vary in many different ways. Nevertheless, we assume that this loudness can vary from a large positive A_0 to a minimum constant value A_{min}.

Since its proposal, BA has emerged as one of the most successful meta-heuristic solvers. It has been applied to a manifold of problems such as logistic [32], industry [24], or medicine [19]. The literature behind BA is huge and diverse, as manifested by comprehensive surveys on practical applications of BA [12,42].

3   Coevolutionary Bat Algorithm for Multitasking

Following concepts previously embraced by other alternatives in the literature [5], one of the main characteristics of the designed COEBA is its multi-population nature. By this we mean that COEBA is a method composed by a defined number of populations, or *demes* [25], comprised by the same number of individuals. More specifically, the number of groups is equal to K, i.e. the number of tasks to be optimized. Additionally, each of the K subpopulations concentrates on solving a specific task T_k. This means that bats corresponding to the k-th deme are only evaluated on task T_k.

As in MFEA, a unified representation Ω^U is used for encoding individuals. However, the most innovative aspect of COEBA is that each subpopulation has its own search space. This involves a slight size readjustment when different demes exchange individuals among them. We will hereafter use the TSP to show this size readjustment problem. Hence, we denote the size of each problem T_k (i.e. the number of *cities*) as D_k. Let us assume that individual $\mathbf{x}_i$ is encoded as a permutation of the integer set $\{1, 2, \ldots, D_k\}$. In this way, when $\mathbf{x}_p^k \in \Omega_k$ is migrated to a subpopulation in which the size of task T_k' to be solved is $D_k' < D_k$, only integers lower than D_k are considered, thus reducing the phenotype of the individual. These integers maintain the same order as in $\mathbf{x}_p^k$. The reverse procedure applies if $D_k' > D_k$. In that case, and considering that each time an individual $\mathbf{x}_p^k$ is transferred to a deme it replaces an alternative bat $\mathbf{x}_p^{k'}$, all integers between D_k and D_k' are inserted in $\mathbf{x}_p^k$ in the same positions as in $\mathbf{x}_p^{k'}$. This multiple search space strategy enhances the exploitation of the search over the demes, making the movement operators more effective.

With all these considerations in mind, Algorithm 1 shows the pseudocode of the designed COEBA. As can be seen, in the initialization process a number X of individuals are randomly generated. After initialization, each individual is evaluated over all K tasks. Then, within an iterative process, each subpopulation is built by choosing the top X/K individuals for the corresponding task (the same bat can be selected by different tasks). Once demes are composed, each one is evolved independently by following the concepts of the discrete version of the BA [32]. To be more concise, the distance between two different bats is measured by means of the Hamming Distance, namely, the number of non-corresponding elements in the sequence. Furthermore, the *inclination* mechanism is also used [31]. Thanks to this feature (lines 10–14 in Algorithm 1), the method intelligently selects the movement function suited for each bat at every iteration, depending on its specific situation regarding the leading bat of the swarm. As is shown in Algorithm 1, *2-opt* and *insertions* are used as movement functions.

Algorithm 1: Pseudocode of the proposed COEBA

---

1   Randomly generate an initial population of X bats
2   **for** *each bat x_i in the population* **do**
3 | Initialize the pulse rate r_i, velocity v_i and loudness A_i

4   Evaluate each of the individual for all the K optimization tasks
5   Build the K number of subpopulations
6   **while** *termination criterion not reached* **do**
7 | **for** *each population k* **do**
8 | **for** *each bat x_i in the subpopulation* **do**
9 | Generate new solution
10 | **if** $v_i^t < n/2$ **then**
11 | $x_i \leftarrow 2 - opt(x_i^{t-1}, v_i^t)$
12 | **else**
13 | $x_i \leftarrow insertion(x_i^{t-1}, v_i^t)$
14 | **if** *rand* $> r_i$ **then**
15 | Select one solution among the best ones
16 | Generate a new bat by selecting the best neighbor of the chosen bat
17 | **if** *rand* $< A_i$ and $f(x_i) < f(x_*)$ **then**
18 | Accept the new solution
19 | Increase r_i and reduce A_i

20 | **if** *iteration is multiple of migr* **then**
21 | Two random individuals are migrated from k to another randomly selected subpopulation

22   Return the best individual in P for each task T_k

Moreover, every *migr* iterations, each group transfers two individuals to a randomly selected population. These two bats are selected by following this criterion: the first one is selected uniformly at random among the best 10 individuals of the population, while the second one is drawn from the complete subpopulation. These two individuals substitute two randomly chosen bats, not considering the 10 best ones of the deme where the replacement is done. Finally, COEBA finishes its execution after I iterations, returning as its solution the best bat of each subpopulation. Any other stopping criterion can be adopted with no further consequences to the design of the algorithm.

4   Experimentation and Results

To assess the performance of the designed COEBA solver, an extensive experimentation has been carried out, which is described in depth in this section. As such, Subsect. 4.1 elaborates on the group of TSP instances used in the experiments, whereas Subsect. 4.2 details the experimentation setup. Finally, the obtained results are analyzed and critically examined in Subsect. 4.3.

4.1 Benchmark Problems

As introduced in Sect. 1, the experiments performed in this work consider the TSP as their benchmark problem to be optimized simultaneously. Readers interested on the formulation and theoretical aspects of this classical problem are referred to [3] or [26]. Arguably, TSP has become one of the most often used problems for performance analysis of discrete optimization algorithms. A plethora of meta-heuristic solvers have been applied to the TSP, or to any of its variants, from traditional techniques such as the Genetic Algorithm [15], Ant Colony Optimization [9] or Tabu Search [13], to modern discrete solvers such as Firefly Algorithm [20], Cuckoo Search [33], or the Water Cycle Algorithm [30]. Before proceeding further, it is important to bear in mind that the goal of the experiments is not to reach the optimal solution of the TSP instances under consideration, but to statistically compare the performance of both MFEA and COEBA when using the same instances and evaluation conditions.

This being said, the performance of COEBA and its counterpart MFEA has been measured over 8 TSP instances, which are combined to yield 15 different test scenarios. All instances have been retrieved from the TSPLIB repository [36]. Specifically, the first 8 instances of the Padberg/Rinaldi benchmark have been employed: pr76, pr107, pr124, pr136, pr144, pr152, pr226, and pr264.

4.2 Experimental Setup

For the sake of fairness in the comparisons, similar parameters and operators have been used for both MFEA and COEBA. This way, we can objectively conclude which solver reaches better outcomes using similar evaluation conditions. To ensure the replicability of this study, parameters employed for the implemented algorithms are depicted in Table 1. For this parameter setting, not only

Table 1. Parameter values set for MFEA and COEBA.

MFEA		COEBA	
Parameter	Value	Parameter	Value
Population size	200	Population size	200
Crossover Function	Order Crossover [7]	Short movement function	2-opt [23]
Mutation Function	2-opt [23]	Short movement function	Insertion [31]
Crossover Prob.	0.9	Initial A_i^0	Random number in [0.8,1.0]
Mutation Prob.	0.1	Initial r_i^0	Random number in [0.0,0.4]
migr	100	migr	100
		α & γ	0.98

works focused on MFEA and BA have been considered [17,42,44], but also good practices reported in the community for tackling routing problems [29]. In addition, all bats are initialized uniformly at random. As the termination criterion, each algorithm is stopped after $I = 500 \cdot 10^3$ objective function evaluations.

As mentioned before, 15 different test scenarios have been built for the experimentation. Each of these multitasking configurations implies that both COEBA and MFEA should face the resolution of all the tasks assigned to that scenario simultaneously. Among these test cases, 10 of them are composed by 4 TSP instances, 4 scenarios are comprised by 6 TSP instances, and the last one includes all the 8 instances. Table 2 summarizes all the considered configurations. The main rationale for building these tests scenarios is twofold: i) to reach conclusions over a diverse and heterogeneous set of multitasking scenarios, involving each TSP instance in exactly the same number of cases, and ii) to exploit the possible genetic complementarities of the instances.

Table 2. Summary of the 15 tests cases built for the experimentation.

Test case	pr76	pr107	pr124	pr136	pr144	pr152	pr226	pr264
Test_Case_4_1	×	×	×	×				
Test_Case_4_2					×	×	×	×
Test_Case_4_3	×	×					×	×
Test_Case_4_4			×	×	×	×		
Test_Case_4_5	×		×	×		×		
Test_Case_4_6		×			×	×		×
Test_Case_4_7	×	×		×		×		
Test_Case_4_8			×		×		×	×
Test_Case_4_9	×			×	×		×	
Test_Case_4_10		×	×			×		×
Test_Case_6_1	×	×	×	×	×	×		
Test_Case_6_2			×	×	×	×	×	×
Test_Case_6_3	×	×			×	×	×	×
Test_Case_6_4	×	×		×	×		×	×
Test_Case_8	×	×	×	×	×	×	×	×

Finally, all tests have been carried out on an Intel Xeon E5 – 2650 v3 computer, with 2.30 GHz and a RAM of 32 GB. Moreover, each test case has been run 20 times to account for the statistical significance of performance gaps encountered during the experimentation.

4.3   Results and Discussion

Table 3 depicts the comparisons in the results reached by COEBA and MFEA. Due to lack of space, we omit all the average outcomes for each test case. Instead, we show graphically the comparison using two colored circles. A green circle (○) implies that COEBA has performed better than MFEA in terms of fitness average. On the contrary, a red circle (●) means that MFEA has achieved better results on average. Using *Test_Case_4_3* as an example, and considering Table 2, we observe that COEBA performs better in pr76, pr226, and pr264, while MFEA is better in pr107. Thus, analyzing the content of the Table 3, we conclude that COEBA clearly outperforms MFEA over these EM scenarios, being superior to MFEA in all but 4 TSP instances evolved jointly. It is also crucial to highlight that COEBA obtains better outcomes in all the eight instances evolved jointly in *Test_Case_8*.

Table 3. Comparison of the results for the 15 tests cases built for the experimentation. (○) means COEBA outperforms MFEA. (●) means MFEA performs better.

Test Case	COEBA vs. MFEA comparison
Test_Case_4_1	○-○-○-○
Test_Case_4_2	○-○-○-○
Test_Case_4_3	○-●-○-○
Test_Case_4_4	○-○-○-○
Test_Case_4_5	○-○-○-○
Test_Case_4_6	○-○-●-○
Test_Case_4_7	○-○-○-○
Test_Case_4_8	○-○-○-●
Test_Case_4_9	○-○-○-○
Test_Case_4_10	○-○-○-○
Test_Case_6_1	○-○-●-○-○-○
Test_Case_6_2	○-○-○-○-○-○
Test_Case_6_3	○-○-○-○-○-○
Test_Case_6_4	○-○-○-○-○-○
Test_Case_8	○-○-○-○-○-○-○-○

For extending the coverage and insights provided by this experimentation, we depict in Table 4 the outcomes obtained by COEBA and MFEA for the 8 TSP instances that compose *Test_Case_8*. We show the average, best and standard deviation of results for each instance. Furthermore, we also provide the best known optima reported for each TSP instance in the literature. These results confirm that the proposed COEBA is a promising meta-heuristic for solving EM environments, outperforming MFEA in terms of both average and best outcomes in this context. Even though it is not the goal of this work, it is also relevant to note that the difference between the optimal outcomes and the average results obtained by COEBA ranges between 0.4% and 5.6%, thereby showing that our proposal not only performs competitively for multitasking environments, but also gets close to optimality of the tasks under consideration.

Table 4. Results (best/average/standard deviation of the fitness over 20 runs) obtained by COEBA and MFEA for the 8 instances in the *Test_Case_8*, and results of the Wilcoxon Rank-Sum test. Best results between COEBA and MFEA achieved over each TSP instances are highlighted in bold.

Method	pr76	pr107	pr124	pr136	pr144	pr152	pr226	pr264
	108602.4	**44927.3**	**59380.8**	**99741.1**	**59045.5**	**74819.1**	**81425.7**	**51924.3**
COEBA	108234.0	44610.0	59087.0	99741.1	58771.0	74000.0	81048.0	51079.0
	402.54	242.27	226.89	534.30	244.37	420.50	248.55	458.87
	113116.5	47110.5	62104.2	106729.3	62179.2	76117.3	84586.3	54031.7
MFEA	111073.0	46052.0	61419.0	104998.0	60534.0	74294.0	82320.0	52728.0
	2355.08	858.99	601.06	1461.53	1770.25	1756.13	6065.24	3489.31
Optima	108159.0	44303.0	59030.0	96772.0	58537.0	73682.0	80369.0	49135.0
Wilcoxon test	◯	◯	◯	◯	◯	◯	◯	◯

In order to buttress our conclusions with the statistical significance of these identified gaps, the Wilcoxon Rank-Sum test has been applied, rendering the results depicted in the last row of Table 4. The confidence interval has been set to 95%. We have compared the outcomes obtained for all the 8 TSP instances separately. Accordingly, the last row of Table 4 represent the outcomes of these statistical tests. Specifically, a green circle (◯) means that COEBA outperforms MFEA with statistical significance. On the contrary, the red circle (●) would have indicated the non-existence of evidences for ensuring the statistical significance of a gap between MFEA and COEBA. As can be seen in this table, Wilcoxon Rank-Sum test confirms that COEBA significantly outperforms MFEA in all the 8 instances embedded in this test scenario. The obtained average z-value is -2.68, with an average p-value equal to 0.00888. Considering that the critical z_c value is -1.64, and because $-2.68 < -1.64$ and $0.00888 < 0.05$, these outcomes support the significance of the performance differences at 95% confidence level. Thus, the difference is significant at this confidence level, thereby concluding that the COEBA is statistically better than MFEA for this test scenario.

5   Conclusions and Future Work

This manuscript has elaborated on the design, implementation and validation of a novel approach for solving evolutionary multitasking environments, wherein tasks are optimization problems. For reaching this goal, we have introduced the Coevolutionary Bat Algorithm (COEBA), which finds its source of inspiration from the concepts of evolutionary co-evolution and the discrete adaptation of the Bat Algorithm. A subpopulation is devoted for the optimization of each problem, with a migration policy that allow exchanging genotype information and exploiting synergies among problems. For showcasing the application of the proposed multitasking approach, an experimental setup has been devised embracing instances of the Traveling Salesman Problem as benchmark problems to be jointly solved. We have compared the outcomes attained by COEBA with the ones furnished by Multifactorial Evolutionary Algorithm (MFEA) over 15 different multitasking test cases. The results validate our hypothesis: COEBA is a promising meta-heuristic for addressing multitasking scenarios.

Several research lines have been planned to gain insight beyond the findings reported in this study. In the short term, we will gauge the scalability of COEBA by analyzing its performance and computational efficiency when simultaneously solving test cases comprising TSP instances of larger dimensionality. We also plan to design additional search mechanisms (such as alternative migration strategies), all targeted at reinforcing the transfer of knowledge among related tasks (*positive transfer*), and lowering the genotype exchange among those tasks that are not related to each other (correspondingly, *negative transfer*). In the longer term, we will explore the application of COEBA to problems stemming from other research fields [35] with discrete optimization problems at their core, such as community detection in social networks.

Acknowledgments. Eneko Osaba and Javier Del Ser would like to thank the Basque Government for its support through the EMAITEK and ELKARTEK programs. Javier Del Ser receives support from the Consolidated Research Group MATHMODE (IT1294-9) granted by the Department of Education of the Basque Government. Andres Iglesias and Akemi Galvez thank the Computer Science National Program of the Spanish Research Agency and European Funds, Project #TIN2017-89275-R (AEI/FEDER, UE), and the PDE-GIR project of the European Union's Horizon 2020 programme, Marie Sklodowska-Curie Actions grant agreement #778035.

References

1. Bäck, T., Fogel, D.B., Michalewicz, Z.: Handbook of Evolutionary Computation. CRC Press (1997)
2. Bali, K.K., Ong, Y.S., Gupta, A., Tan, P.S.: Multifactorial evolutionary algorithm with online transfer parameter estimation: MFEA-II. IEEE Trans. Evol. Comput. **24**(1), 69–83 (2020)
3. Bellmore, M., Nemhauser, G.L.: The traveling salesman problem: a survey. Oper. Res. **16**(3), 538–558 (1968)
4. Caruana, R.: Multitask learning. Mach. Learn. **28**(1), 41–75 (1997)
5. Cheng, M.Y., Gupta, A., Ong, Y.S., Ni, Z.W.: Coevolutionary multitasking for concurrent global optimization: with case studies in complex engineering design. Eng. Appl. Artif. Intell. **64**, 13–24 (2017)
6. Da, B., et al.: Evolutionary multitasking for single-objective continuous optimization: benchmark problems, performance metric, and baseline results. arXiv preprint arXiv:1706.03470 (2017)
7. Davis, L.: Job shop scheduling with genetic algorithms. In: International Conference on Genetic Algorithms and their Applications, vol. 140 (1985)
8. Ser, J., et al.: Bio-inspired computation: where we stand and what's next. Swarm Evol. Comput. **48**, 220–250 (2019)
9. Dorigo, M., Gambardella, L.M.: Ant colony system: a cooperative learning approach to the traveling salesman problem. IEEE Trans. Evol. Comput. **1**(1), 53–66 (1997)
10. Feng, L., et al.: An empirical study of multifactorial PSO and multifactorial DE. In: IEEE Congress on Evolutionary Computation, pp. 921–928 (2017)

11. Feng, L., Ong, Y.S., Tan, A.H., Tsang, I.W.: Memes as building blocks: a case study on evolutionary optimization+ transfer learning for routing problems. Memetic Comput. **7**(3), 159–180 (2015)
12. Fister Jr., I., Yang, X.S., Fister, I., Brest, J., Fister, D.: A brief review of nature-inspired algorithms for optimization/kratki pregled algoritmov po vzoru iz narave za optimizacijo. Elektrotehniski Vestnik **80**(3), 116 (2013)
13. Gendreau, M., Laporte, G., Semet, F.: A tabu search heuristic for the undirected selective travelling salesman problem. Eur. J. Oper. Res. **106**(2), 539–545 (1998)
14. Gong, M., Tang, Z., Li, H., Zhang, J.: Evolutionary multitasking with dynamic resource allocating strategy. IEEE Trans. Evol. Comput. **23**(5), 858–869 (2019)
15. Grefenstette, J., Gopal, R., Rosmaita, B., Van Gucht, D.: Genetic algorithms for the traveling salesman problem. In: Proceedings of the first International Conference on Genetic Algorithms and their Applications, pp. 160–168. Lawrence Erlbaum, New Jersey (1985)
16. Gupta, A., Ong, Y.S.: Genetic transfer or population diversification? Deciphering the secret ingredients of evolutionary multitask optimization. In: IEEE Symposium Series on Computational Intelligence, pp. 1–7 (2016)
17. Gupta, A., Ong, Y.S., Feng, L.: Multifactorial evolution: toward evolutionary multitasking. IEEE Trans. Evol. Comput. **20**(3), 343–357 (2015)
18. Gupta, A., Ong, Y.S., Feng, L., Tan, K.C.: Multiobjective multifactorial optimization in evolutionary multitasking. IEEE Trans. Cybern. **47**(7), 1652–1665 (2016)
19. Ibrahim, S., Thangamani, M.: Enhanced singular value decomposition for prediction of drugs and diseases with hepatocellular carcinoma based on multi-source bat algorithm based random walk. Measurement **141**, 176–183 (2019)
20. Kumbharana, S.N., Pandey, G.M.: Solving travelling salesman problem using firefly algorithm. Int. J. Res. Sci. Adv. Technol. **2**(2), 53–57 (2013)
21. Lawler, E.L., Lenstra, J.K., Kan, A.R., Shmoys, D.B.: The Traveling Salesman Problem: A Guided Tour of Combinatorial Optimization. Wiley, New York (1985)
22. Li, G., Zhang, Q., Gao, W.: Multipopulation evolution framework for multifactorial optimization. In: Genetic and Evolutionary Computation Conference Companion, pp. 215–216 (2018)
23. Lin, S.: Computer solutions of the traveling salesman problem. Bell Syst. Tech. J. **44**(10), 2245–2269 (1965)
24. Lu, Y., Jiang, T.: Bi-population based discrete bat algorithm for the low-carbon job shop scheduling problem. IEEE Access **7**, 14513–14522 (2019)
25. Luque, G., Alba, E.: Parallel Genetic Algorithms: Theory and Real World Applications, vol. 367. Springer, Heidelberg (2011)
26. Miller, C.E., Tucker, A.W., Zemlin, R.A.: Integer programming formulation of traveling salesman problems. J. ACM **7**(4), 326–329 (1960)
27. Ong, Y.-S.: Towards evolutionary multitasking: a new paradigm in evolutionary computation. In: Senthilkumar, M., Ramasamy, V., Sheen, S., Veeramani, C., Bonato, A., Batten, L. (eds.) Computational Intelligence, Cyber Security and Computational Models. AISC, vol. 412, pp. 25–26. Springer, Singapore (2016). https://doi.org/10.1007/978-981-10-0251-9_3
28. Ong, Y.S., Gupta, A.: Evolutionary multitasking: a computer science view of cognitive multitasking. Cogn. Comput. **8**(2), 125–142 (2016)
29. Osaba, E., Carballedo, R., Diaz, F., Onieva, E., Masegosa, A., Perallos, A.: Good practice proposal for the implementation, presentation, and comparison of meta-heuristics for solving routing problems. Neurocomputing **271**, 2–8 (2018)

30. Osaba, E., Del Ser, J., Sadollah, A., Bilbao, M.N., Camacho, D.: A discrete water cycle algorithm for solving the symmetric and asymmetric traveling salesman problem. Appl. Soft Comput. **71**, 277–290 (2018)
31. Osaba, E., Yang, X.S., Diaz, F., Lopez-Garcia, P., Carballedo, R.: An improved discrete bat algorithm for symmetric and asymmetric traveling salesman problems. Eng. Appl. Artif. Intell. **48**, 59–71 (2016)
32. Osaba, E., Yang, X.S., Fister Jr., I., Del Ser, J., Lopez-Garcia, P., Vazquez-Pardavila, A.J.: A discrete and improved bat algorithm for solving a medical goods distribution problem with pharmacological waste collection. Swarm Evol. Comput. **44**, 273–286 (2019)
33. Ouaarab, A., Ahiod, B., Yang, X.: Discrete cuckoo search algorithm for the travelling salesman problem. Neural Comput. Appl. **24**, 1659–1669 (2013). https://doi.org/10.1007/s00521-013-1402-2
34. Pan, S.J., Yang, Q.: A survey on transfer learning. IEEE Trans. Knowl. Data Eng. **22**(10), 1345–1359 (2009)
35. Precup, R.E., David, R.C.: Nature-Inspired Optimization Algorithms for Fuzzy Controlled Servo Systems. Butterworth-Heinemann (2019)
36. Reinelt, G.: TSPLIB: a traveling salesman problem library. ORSA J. Comput. **3**(4), 376–384 (1991)
37. Song, H., Qin, A., Tsai, P.W., Liang, J.: Multitasking multi-swarm optimization. In: IEEE Congress on Evolutionary Computation, pp. 1937–1944 (2019)
38. Wang, C., Ma, H., Chen, G., Hartmann, S.: Evolutionary multitasking for semantic web service composition. arXiv preprint arXiv:1902.06370 (2019)
39. Wen, Y.W., Ting, C.K.: Parting ways and reallocating resources in evolutionary multitasking. In: IEEE Congress on Evolutionary Computation, pp. 2404–2411 (2017)
40. Xiao, H., Yokoya, G., Hatanaka, T.: Multifactorial PSO-FA hybrid algorithm for multiple car design benchmark. In: IEEE International Conference on Systems, Man and Cybernetics, pp. 1926–1931 (2019)
41. Yang, X.S.: A new metaheuristic bat-inspired algorithm. In: González, J.R., Pelta, D.A., Cruz, C., Terrazas, G., Krasnogor, N. (eds.) Nature Inspired Cooperative Strategies for Optimization (NICSO 2010). Studies in Computational Intelligence, vol 284. Springer, Heidelberg (2010). https://doi.org/10.1007/978-3-642-12538-6_6
42. Yang, X.S., He, X.: Bat algorithm: literature review and applications. Int. J. Bio-Inspired Comput. **5**(3), 141–149 (2013)
43. Yu, Y., Zhu, A., Zhu, Z., Lin, Q., Yin, J., Ma, X.: Multifactorial differential evolution with opposition-based learning for multi-tasking optimization. In: IEEE Congress on Evolutionary Computation, pp. 1898–1905 (2019)
44. Yuan, Y., Ong, Y.S., Gupta, A., Tan, P.S., Xu, H.: Evolutionary multitasking in permutation-based combinatorial optimization problems: realization with TSP, QAP, LOP, and JSP. In: IEEE Region 10 Conference, pp. 3157–3164 (2016)
45. Zheng, X., Lei, Y., Gong, M., Tang, Z.: Multifactorial brain storm optimization algorithm. In: Gong, M., Pan, L., Song, T., Zhang, G. (eds.) BIC-TA 2016. CCIS, vol. 682, pp. 47–53. Springer, Singapore (2016). https://doi.org/10.1007/978-981-10-3614-9_6
46. Zhou, L., et al.: Towards effective mutation for knowledge transfer in multifactorial differential evolution. In: IEEE Congress on Evolutionary Computation, pp. 1541–1547 (2019)
47. Zhou, L., Feng, L., Zhong, J., Ong, Y.S., Zhu, Z., Sha, E.: Evolutionary multitasking in combinatorial search spaces: a case study in capacitated vehicle routing problem. In: IEEE Symposium on Computational Intelligence, pp. 1–8 (2016)

Convex Polygon Packing Based Meshing Algorithm for Modeling of Rock and Porous Media

Joaquín Torres[1]([⊠]), Nancy Hitschfeld[1]([⊠]), Rafael O. Ruiz[2],
and Alejandro Ortiz-Bernardin[3]

[1] Departamento de Ciencias de la Computación,
Universidad de Chile, Santiago, Chile
{jtorres,nancy}@dcc.uchile.cl
[2] Departamento de Ingenieria Civil, Universidad de Chile, Santiago, Chile
rafaelruiz@uchile.cl
[3] Departamento de Ingenieria Mecánica, Universidad de Chile, Santiago, Chile
aortizb@uchile.cl

Abstract. In this work, we propose new packing algorithm designed for the generation of polygon meshes to be used for modeling of rock and porous media based on the virtual element method. The packing problem to be solved corresponds to a two-dimensional packing of convex-shape polygons and is based on the locus operation used for the advancing front approach. Additionally, for the sake of simplicity, we decided to restrain the polygon rotation in the packing process. Three heuristics are presented to simplify the packing problem: density heuristic, gravity heuristic and the multi-layer packing. The decision made by those three heuristic are prioritizing on minimizing the area, inserting polygons on the minimum Y coordinate and pack polygons in multiple layers dividing the input in multiple lists, respectively. Finally, we illustrate the potential of the generated meshes by solving a diffusion problem, where the discretized domain consisted in polygons and spaces with different conductivities. Due to the arbitrary shape of polygons and spaces that are generated by the packing algorithm, the virtual element method was used to solve the diffusion problem numerically.

Keywords: Polygonal meshes · Geometric packing · Virtual element method · Computational geometry

1 Introduction

In the last decade, the use of finite element methods (FEM) have been the common engineering practice to design and evaluate the performance of different

Supported by the Chilean National Fund for Scientific and Technological Development (FONDECYT) through grants CONICYT/FONDECYT No. 1181506, No. 11180812, and No. 1181192.

V. V. Krzhizhanovskaya et al. (Eds.): ICCS 2020, LNCS 12141, pp. 257–269, 2020.
https://doi.org/10.1007/978-3-030-50426-7_20

systems. However, in many applications, the FEM has shown some limitations related to the complexity involved in the mesh generation, specially for problems in which the domain is defined by an arrangement of irregular sub-domains. In particular, these situations are found in problems related to the flux of fluid and heat in porous media, fracture mechanics of conglomerate rocks, stability of tailing dams, concrete modeling, amount others. Here, one of the issues is to deal with the random nature of the sub-domains, requiring the statistical study of the problem with multiple simulations. As consequence, the computational burden increases since it is required to draw the geometry and create the mesh several times.

With the advent of the virtual element method (VEM) [15] very general polytopal meshes (polygons can even be non-convex) can now be used to simulate problems based on Galerkin methods in a manner similar to FEM. In porous media, microstructure, rock accumulation, among others, the bidimensional domain is composed naturally of arbitrary polytopal shapes, which in a simulation can be represented by virtual elements. In this regard, the mentioned issues could be solved using the packing perspective together with numerical methods that employ polytopal meshes (i.e., Virtual Element Method). In a general perspective, packing is an optimization problem on how to organize the content of a container as densely as possible. A particular example of packing is the geometric packing; this packing comprises fitting geometric figures as much as possible inside a container. For example, packing polygons inside a rectangle container, or packing tetrahedra inside a cube container. Then, these polygons/polyhedrons could be used to define the sub-domains at the same time that could be used as a mesh under VEMs schemes.

Different packing strategies have been developed in the past employing different packing geometries. For example, adopting a circle and sphere packing [3,8,10] and exploring applications of circle packing through simulations of discrete earthquakes [16] or employing packing geometries based on square-like or rectangular-like shapes [6,7]. However, the use of convex polygon shapes had been received a limited attention, being the advancing front approach [4] one of the seminal works in this matter.

In this work, we propose a new packing algorithm designed for the generation of polygonal meshes to be used for modeling of rock and porous media based on VEM. Here, the packing problem to be solved corresponds to a two-dimensional packing of convex-shape polygons, where the designed algorithm is based on the locus operation used for the advancing front approach [4]. Additionally, for the sake of simplicity, we decided to restrain the polygon rotation in the packing process. In the following sections, the locus operation principle is presented first. Then, it is explained how the new algorithm works, and subsequently, to demonstrate the potential and the feasibility of the proposed algorithm, a diffusion problem is solved using the virtual element method on a domain discretized with a packing-based mesh.

2   Advancing Front Approach

From the computational complexity theory, the packing problem is considered as a NP problem [2], which indicates the imperative use of an heuristic to establish a proper solution. The heuristic employed here is based on the so called Advancing Front Technique [4] to allocate a new polygon on the boundary of the current polygon cluster. The algorithm determines the position of the new polygon by applying an operation to all active polygons in the cluster. The concept of active polygons is used to identify the polygons that belong to the top layer of polygons in the packing container. In particular, the algorithm builds a locus over each of these active polygons, which is defined as the resulting polygon after sliding the new polygon around the active polygons. This operation is similar to the Minkowski addition [1]. However, the Minkowski addition extends the polygon in a particular direction, contrary to the locus that extents the polygon in all directions. The locus generated from the active polygons of the layer (loci) help to identify the possible positions of the new polygon on the cluster. In this regard, the intersection of two loci results in the possible positions to allocate the center of the new polygon. Note that this approach enforce the intersection of the polygon with the active layer. Figure 1 shows a scheme to exemplify the algorithm steps to build the locus polygons. In the figure, it is possible to observe the locus of each polygon and the possible positions of the new polygon defined by the intersection between the locus. The computational cost to obtain the locus is of order $O(n + m)$ using a cross product to determine intersection on each step with n and m the number of vertices of P and Q, respectively.

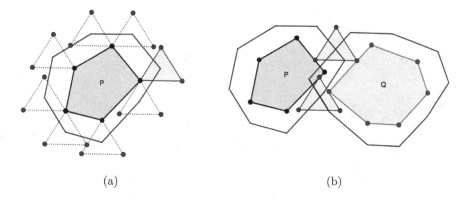

(a)   (b)

Fig. 1. Example of the locus algorithm with two polygons P and Q. a) Positions where the polygon moves around and the resulting locus. b) Intersection points of the two loci and the possible positions to allocate the center of the new polygon.

Note that the advancing front approach has been used before for packing circles, specifically using the locus of polygons to fill the domain [5]. Similarly, the algorithm has also been used to perform a sequential sphere packing [9].

3   Convex Polygon Packing Algorithm

As it was stated before, the packing algorithm is considered a NP problem which requires an heuristic approach. For this purpose, we propose three different approaches on how to generate the mesh. These approaches are:

1. Density heuristic: Aims to place the polygons on a position that minimize the area generated between the packed polygons.
2. Gravity heuristic: Prioritizes the Y position overall. It tries to place the polygon as low as it can.
3. Multi-layer packing: This packing works differently, groups the input on different batches of polygons, so it simulates multiple layers piled up. In each of these layers, the multi-layer packing uses either the density heuristic or the gravity packing heuristic.

Figure 2 shows a scheme to exemplify a comparison between the density heuristic and the gravity heuristic. We show the decision the algorithm takes when inserting a polygon on the layout. On both images, the polygons with striped lines are the positions where the next polygon could be placed. The green polygon the position selected to place the polygon.

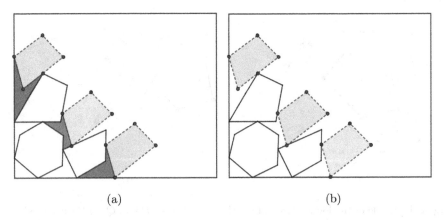

(a)   (b)

Fig. 2. Gravity heuristic and density heuristic decision example. (a) Decision made by the area heuristic choosing the middle position because the empty space left by the polygon has the smallest area. (b) Decision made by the gravity heuristic choosing the bottom position.

3.1   Input Construction

The input construction refers to the definition of any new polygon that will be introduced in the container. Here, two restrictions are imposed: only convex-shape polygons are generated, and the rotation is restrained. Additionally, two variables are introduced to control the roundness and size. Then, the polygons are generated adopting a strategy based on circumscribed circles [13] following three steps: (1) create a circle, (2) placing points randomly on the circle perimeter, and (3) connecting the points in a counterclockwise order. As a summary, the input of this algorithm is a list of radius and number of vertices, while the output corresponds to a list of convex polygons.

3.2   Inserting a New Polygon

Once the new polygon is defined, it is necessary to define the algorithm to allocate it. The first step corresponds to insert the first polygon, which comprises of inserting the polygon at the bottom left corner of the layout (container). For this purpose, the algorithm searches for the minimum x and y coordinates of the polygon. Then, the polygon is translated to the corner applying a translation. In the second step, the process is repeated until there is no polygon left to pack or cannot pack the remaining polygons. Inside the loop, the next polygon on the list is take and located in the best position it could be placed depending on the heuristic decision. If the polygon cannot be inserted, the algorithm returns null, and it is discarded.

Additionally, it is important to highlight that at the moment to insert a new polygon, the algorithm iterates over the active polygons to identify if the new element suits. Thus, the algorithm does the following steps to determine the best position:

1. It looks for the neighbors of the polygons and checks if the distance between them is less than the longest diagonal of the inserted polygon.
2. Then, it obtains the locus polygon of both packed polygons and intersects those loci to get the intersection points.
3. Finally, it tests the obtained position with the used heuristic. If the result of the evaluation is better than the current best one, it saves the position; otherwise, the position is discarded.

After the algorithm allocates the polygon, it checks if the neighbor polygons of the inserted polygon are not suitable for following iterations. For example, if none of the next polygons can be place near them, then that polygon is not suitable. This check helps the algorithm to discard polygons and also improve the computational cost.

3.3   Neighborhood Data Structure

The allocation of new polygons could have a significant computational cost. In order to alleviate this computational burden, we propose the construction of a neighborhood data structure that is consulted in each iteration.

In particular, the data structure consists of a graph with nodes defined by the centroid of each polygon and the links between the polygon and its neighbors. There are concave or convex polygons between polygons, those polygons are named "spaces" and also those spaces are inserted to the graph. The graph is initialized inserting the container. The container is accounted employing two nodes, one that represents the container itself and another that represents the empty space. After inserting the container, the graph is updated in the same fashion after each polygon insertion: (1) identify the space that contains the inserted polygon, (2) generate two new spaces from allocating the polygon, (3) add the new spaces and the polygon as new nodes, and (4) update the neighborhood links. Figure 3 shows (a) the mesh of an intermediate iteration of the algorithm and (b) the graph build from the polygon insertions.

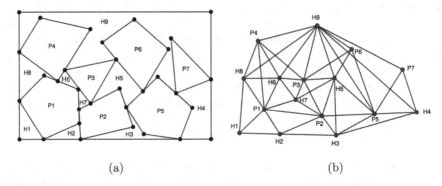

(a)   (b)

Fig. 3. Packing mesh and its graph data structure. Each of the polygons is named with a P, and the spaces with H. (a) Intermediate stage of the algorithm. (b) Graph representation of the intermediate stage (a)

In each step of the packing algorithm, the next inserted polygon or the target polygon is processed in the same way. After we insert the target polygon, a space inside is divided into two spaces. So we need to look for that space to update the graph. Note that this space can be a concave or convex polygon. So to find this space we use the Shimrat's algorithm [14], which finds if a point lies inside a polygon. The Shimrat's algorithm uses ∞ to cast a ray, for this purpose, we use a point far away from the layout and then casting a ray from there to the centroid of the target polygon. Then we use this algorithm on each space across the graph, if this ray intersects even times, the centroid of the target polygon is outside that space, and if the ray intersects odd times, the centroid of the target polygon is inside.

After getting the space to divide or the overlapping space, the algorithms divide this space into two new ones. Thus for this division, we need to find the intersection points between the target polygon and the overlapping space. For this intersection, we use a simple algorithm of order $O(n^2)$, but it can be improved with an algorithm that uses a sweep line approach [17] to order

$O(nlog(n))$. Then the algorithm uses intersection points to build the new spaces, iterating across the overlapping hole in clockwise order from one intersection point to the other intersection point. Then calculates the same route but starting with the second intersection point. The same algorithm is applied to the target polygon generating 4 routes. Although we have these four routes, the algorithm does not know how to connect these routes correctly. For example, it can merge two routes and have a wrong representation of the layout generating a space containing the target polygon. This error can cause an incorrect construction of the mesh. We solve this problem efficiently with the Shimrat's algorithm. In this way, the algorithm connects two random routes, one from the target polygon and from the overlapping space. If the space built contains the centroid of the target polygon, this means we are connecting the wrong routes, so we change one of these and get the correct polygon. We get the other space from the two remaining routes.

Finally, with the new two spaces, we update the graph structure linking the spaces to the polygons linked with the overlapping space. We get the neighbor polygons with the intersection points between the space and the neighbor polygons. Also, we add the target polygon into the graph, link it with the neighbor polygons, and the two new spaces. This concludes an insertion step of the algorithm.

4   Complexity Analysis

In this section, we analyze the computational order of the algorithm by computing the computational complexity of each step and, finally, the order of the whole algorithm. Being p the number of packed polygons and n the number of vertices of the resulting mesh, the order of each step is:

1. First, we look for pairs of polygons and test first if they are close. This test costs $c * p$ being c the number of neighboring polygons, but the number of neighbors is negligible compared to the number of polygons of the mesh. So looking for pairs of polygons cost $O(p)$.
2. After we have a pair of polygons, we get the locus of the two chosen polygons. Getting the locus costs $O(n)$ [4].
3. We got the loci for the possible positions for the target polygon. Then, we intersect the loci and get the points where we can allocate the centroid of the target polygon. The intersection of the loci costs $O(n)$ because the loci are convex.
4. Now that we got a possible position, we test the position on each heuristic:
   - Density Heuristic: For this heuristic, we need to first find the overlapping space with the target polygon. This search cost $O(n^2)$. Then, we build two spaces generated when inserting the target polygon. This step costs $O(r)$ with r the number of segments of the route. In the worst case, the route generated includes all the links of the graph, so it costs $O(n)$. In consequence, this step cost $O(n^2)$.

- Gravity Heuristic: This heuristic uses the Y coordinate to test. When we create a polygon, we previously stored the centroid on the polygon representation. In consequence, getting the Y coordinate is constant, so this step is order $O(1)$.
- Multi-layer packing: This packing uses one of the previous heuristics to test the position. So the order of this heuristic is the same as the one used to pack multi-layers.

5. After we got the best position, we insert the target polygon on the mesh and we update the graph. To update the graph, we need to find the overlapping space, build the two new spaces when placing the target polygon, and update the links. Because the spaces of the layout can be concave, we can not use the intersection algorithm between convex polygons. We use the simplest algorithm of order $O(n^2)$, and then updating the graph is just updating the neighbors of the divided space. This update can cost at most the number of segments of the graph that is order $O(n)$. In conclusion, this step cost $O(n^2)$.
6. Finally, checking if the target polygon closed the surrounding of their neighbors costs $O(c)$, and as the number of neighbors is negligible compared to the number of polygons of the mesh, then the step is order $O(1)$.

In summary, being p the number of polygons inserted by the algorithm, the algorithm costs $O(p^2 n)$ for the gravity heuristic, and costs $O(p^2 n^2)$ for the density heuristic. The number of polygons p is directly related to the number of vertices n so we can replace it. The final order for the gravity heuristic packing is $O(n^3)$, and for the density heuristic packing is $O(n^4)$.

5   Performance Experiments

To test the performance of the algorithm, we designed a variety of tests. The first batch of tests comprised a comparison between the two heuristics. The experiments considered an increment on the number of vertices of the packed polygons, changing the size of the packed polygons, packing polygons close to a regular polygon vs convex polygons generated randomly. These experiments test time and efficiency of the heuristics, with efficiency the percentage of space of the container covered by the inserted polygons. We ran all these experiments on an Intel Core i5-8400 CPU.

5.1   Heuristic Comparison

The results of the experiments were that the algorithm with the density heuristic reaches a maximum efficiency of 75% and with the gravity heuristic reaches a maximum efficiency of 80%. Also, the density algorithm takes 4000 s when packing polygons of 20 vertices, but the gravity heuristic takes 40 s to pack the same list. That difference in time happens in all the experiments, concluding that the gravity heuristic reaches its goal in less time. Figure 4 shows two examples of resulting meshes of each heuristic. The packing on the left shows the result of using the density heuristic and on the right the mesh resulting of applying the gravity heuristic. The clear differences are that the generated meshes using the density heuristic contains spaces that have more area but zones of the mesh where it is more dense. Instead of using the gravity heuristic produces meshes with more uniformly distributed polygons. These differences reflect the results of the experiments showing why the density heuristic reaches less efficiency.

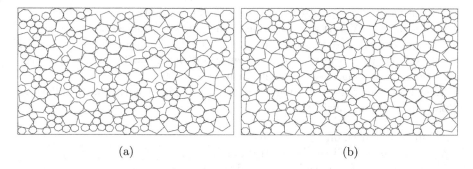

Fig. 4. Generated meshes when packing three different polygons of different size and number of vertices. (a) Mesh generated using the density heuristic and (b) mesh generated using the gravity heuristic.

5.2   Multi-Layer Packing

The last approach designed is the Multi-layer packing. This algorithm inserts layer by layer the polygons on the container. We inserted the polygon in ascending order by size of the polygon, i.e., the first layer comprised the polygons with the biggest radius and on each following layer the following biggest polygon. We tested the same experiments on this approach, reaching a maximum of 78% using the density heuristic for each layer. Figure 5 shows an example of a mesh got by the multi-layer packing algorithm. On that example we packed the polygons on three layers each with different size and a different number of vertices. The size of the polygon decrease on each passing layer. In a qualitative analysis these new meshes seems to be worse than the one layered meshes in space efficiency.

  We compare a one layer container versus a multi-layer container, both experiments using the density heuristic. We made five different experiments: (a) packing with polygons of 5 vertices all same size, (b) packing with polygons with 5 or 7 vertices all same size, (c) packing polygons with 5 vertices and two different sizes, (d) packing polygons with 20 vertices all same size, and (e) packing polygons with 20 vertices and two different sizes. The multi-layer packing used different polygons on each of the layers, we used the polygons on ascending order by size, decreasing the size on each layer, i.e., on experiment (c) we used on the first layer the bigger polygons and then on the second layer used the smaller ones. Figure 6 show the result of the experiments described and it show a similar behaviour of both approaches. The difference is that the multi-layer packing needs a bigger container to be more efficient in covered space. The multi-layer packing algorithm generates different meshes but with a 2% loss in efficiency.

6   Preliminary Simulation Results

In a recent work about polygonal meshes [12], a C++ library for the virtual element method [15] (VEM) was developed. Therein polygonal meshes were generated from a constrained Voronoi diagram [11] of the domain. The present work aims to extend the variety of meshes generated in [12] to be able to perform simulations of packing-based problems using the VEM.

To test a mesh generated using the proposed packing algorithm, the following diffusion problem is considered:

1. A unit square domain ($1\,\mathrm{m} \times 1\,\mathrm{m}$).
2. A $0\,^{\circ}\mathrm{C}$ applied to all the boundary nodes.
3. A conductivity of $1\,\mathrm{W}/(\mathrm{mK})$ assigned to all the polygons.
4. A conductivity of $0.1\,\mathrm{W}/(\mathrm{mK})$ assigned to all the spaces.
5. A heat source that varies according to the following equation:

$$b(x, y) = 32y(1 - y) + 32x(1 - x)$$

The foregoing diffusion problem is solved numerically using the VEM. This method can handle arbitrary polytopal meshes including non-convex polygons, and thus it is very appealing for simulations that use packing-based meshes.

Figure 7(a) shows the domain and mesh used in the diffusion problem. Figure 7(b) depicts the VEM temperature field. Finally, the heat flux is shown in Fig. 8.

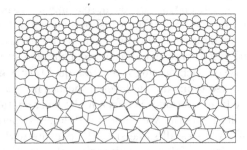

Fig. 5. Generated mesh when packing on a three layered container. The algorithm inserts the polygons depending in how we configure them. Here each layer has polygons with different size and a different number of vertices.

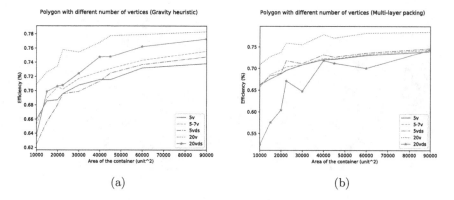

Fig. 6. Efficiency results from testing the packing with (a) gravity heuristic on one container versus (b) multi-layer packing. We ran the five experiments on each algorithm. The experiments are divided into to categories, the experiments that change number of vertices: (1) "5vs" polygons of 5 vertices, (2) "5–7v" polygons with 5 or 7, and (3) "20v" polygons with 20 vertices; and the experiments that pack two different sizes of polygons: (4) "5vds" polygons with 5 vertices and two different sizes, and (5) "20vds" polygons with 20 vertices and two different sizes.

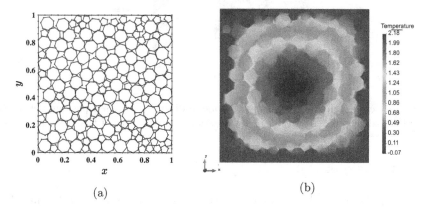

Fig. 7. Heat conduction in a unit square domain discretized with a packing mesh: (a) Packing mesh and (b) temperature field.

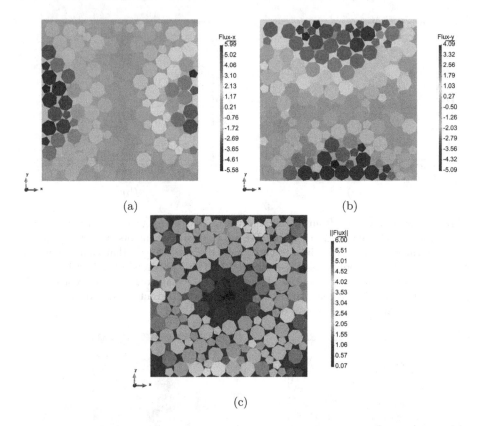

Fig. 8. Heat conduction in a unit square domain discretized with a packing mesh: (a) flux in the x-direction, (b) flux in the y-direction and (c) norm of the flux.

7   Conclusions

The experimental results showed that the gravity heuristic is better in CPU time and covers more area than the density heuristic. The gravity heuristic packing algorithm takes less CPU time than the density, behaving according the theoretical order shown in Sect. 4. It also improves the efficiency by 5% with respect to the density heuristic: 80% vs 75%, respectively.

The experiments testing the multi-layer packing algorithm showed an interesting new type of resulting meshes. However, there is a decrease of 2% in efficiency compared to the gravity heuristic packing, but this decrease is not significant.

Thus, the meshes resulting from the packing algorithms can be useful for modeling of rock and porous media. As an example, we illustrated its potential by solving a diffusion problem, where the discretized domain consisted in polygons and spaces with different conductivities. Due to the arbitrary shape of polygons and spaces that are generated by the packing algorithm, the virtual element method was used to solve the diffusion problem numerically. This shows that, for the virtual element method, a new variety of problems can be solved using packing-based meshes.

References

1. Agarwal, P.K., Flato, E., Halperin, D.: Polygon decomposition for efficient construction of Minkowski sums. Comput. Geom. **21**(1–2), 39–61 (2002)
2. Berkey, J.O., Wang, P.Y.: Two-dimensional finite bin-packing algorithms. J. Oper. Res. Soc. **38**(5), 423–429 (1987)
3. Collins, C.R., Stephenson, K.: A circle packing algorithm. Comput. Geom. **25**(3), 233–256 (2003)
4. Feng, Y., Han, K., Owen, D.: An advancing front packing of polygons, ellipses and spheres. In: Discrete Element Methods: Numerical Modeling of Discontinua, pp. 93–98. ASCE Library (2002)
5. Feng, Y., Han, K., Owen, D.: Filling domains with disks: an advancing front approach. Int. J. Numer. Meth. Eng. **56**(5), 699–713 (2003)
6. Hopper, E., Turton, B.C.: An empirical investigation of meta-heuristic and heuristic algorithms for a 2D packing problem. Eur. J. Oper. Res. **128**(1), 34–57 (2001)
7. Huang, W., Chen, D., Xu, R.: A new heuristic algorithm for rectangle packing. Comput. Oper. Res. **34**(11), 3270–3280 (2007)
8. Li, Y., Ji, S.: A geometric algorithm based on the advancing front approach for sequential sphere packing. Granular Matter **20**(4), 1–12 (2018). https://doi.org/10.1007/s10035-018-0829-7
9. Löhner, R., Oñate, E.: An advancing front technique for filling space with arbitrary separated objects. Int. J. Numer. Meth. Eng. **78**(13), 1618–1630 (2009)
10. Mohar, B.: A polynomial time circle packing algorithm. Discr. Math. **117**(1–3), 257–263 (1993)
11. Okabe, A., Boots, B., Sugihara, K., Chiu, S.N.: Spatial Tessellations: Concepts and Applications of Voronoi Diagrams. Wiley Series in Probability and Statistics, 2nd edn. Wiley (2009)
12. Ortiz-Bernardin, A., Alvarez, C., Hitschfeld-Kahler, N., Russo, A., Silva-Valenzuela, R., Olate-Sanzana, E.: Veamy: an extensible object-oriented C++ library for the virtual element method. Numer. Algorithms **82**, 1–32 (2018)
13. Pinelis, I.: Cyclic polygons with given edge lengths: existence and uniqueness. J. Geom. **82**(1–2), 156–171 (2005). https://doi.org/10.1007/s11075-018-00651-0
14. Shimrat, M.: Algorithm 112: position of point relative to polygon. Commun. ACM **5**(8), 434–435 (1962)
15. da Veiga, B.L., Brezzi, F., Cangiani, A., Manzini, G., Marini, L.D., Russo, A.: Basic principles of virtual element methods. Math. Models Methods Appl. Sci. **23**(01), 199–214 (2013)
16. Williams, G.B.: Earthquakes and circle packings. J. D'Analyse Mathématique **85**(1), 371–396 (2001)
17. Žalik, B.: Two efficient algorithms for determining intersection points between simple polygons. Comput. Geosci. **26**(2), 137–151 (2000)

Computational Science in IoT and Smart Systems

Modelling Contextual Data for Smart Environments. Case Study of a System to Support Mountain Rescuers

Radosław Klimek$^{(\boxtimes)}$ ⓘ

AGH University of Science and Technology,
al. Mickiewicza 30, 30-059 Krakow, Poland
rklimek@agh.edu.pl

Abstract. Context-aware pervasive systems are complex, due to the need to gather detailed environmental information and to perform a variety of context reasoning processes in order to adapt behaviours accordingly. These operations are merged seamlessly. We show the feasibility and vitality of a fully designed system for mountain rescue operations, with various aspects of the contextual processing in middleware, as well as analyse its context life cycle. The system is verified through intensive experiments with a rich set of categorised context data. The contextual processing is shown in different weather scenarios. The service is geared towards software development, converging IoT (Internet of Things) and cloud computing with specific reference to smart application scenarios.

Keywords: Streaming sensor data · Modelling contextual information · Middleware · IoT · Rescuing activity

1 Introduction

Context-aware systems are analysing complex information which is relevant to a monitored entity and falls into a wide range of data categories [6,15]. However, the context understanding presented in the well-known paper by Dey and Abowd [6], seems too general by today's standards, and to be used practically requires categorisation, see paper by Zimmerman et al. [15], introducing a form of interpretation, which allows us to govern the context complexity. In this article, we have categorised the used domain context, which describe the mountain environment. Smart decisions based on various situations and operational scenarios are taken autonomously and pro-actively. System operations are transparent to the sensed entities. Decisions are taken by middleware, which seamlessly binds together all elements.

The first contribution is the categorisation of contextual data for the requirements of mountain environments, especially focusing on supporting mountain rescuers. It is an enabler when reducing complexity, and prioritising activities. Another contribution is a simulation experiment on contextual data processing

© Springer Nature Switzerland AG 2020
V. V. Krzhizhanovskaya et al. (Eds.): ICCS 2020, LNCS 12141, pp. 273–287, 2020.
https://doi.org/10.1007/978-3-030-50426-7_21

to better understand the nature of defined data, established categories, decision processes, and threats as regular languages. A separate contribution is to holistically verify, thanks to an environment simulator, via a series of experiments, the designed system, which supports mountain rescue operations. Previously, only separate system components were tested, see [9], basing on randomly generated datasets. Presently, the mountain environment simulator produces different datasets. The simulator reflects all the most important aspects of real mountain environments. Five different weather scenarios were proposed and performed. All objectives authenticate system feasibility, credibility and vitality. The system is both an enabler and a provider in increasing understanding of the significance of context-aware decisions, which are based on redundancy, spatial proximity, context transition, context sharing, and other context features. It was designed to help, among other IT engineers, better understand the specifics of context-aware systems.

2   Related Works

The basic definition of context was provided by a paper by Dey and Abowd [6]. A paper by Zimmermann et al. [15] identified many context taxonomies, user and role, process and task, location and time, amongst others. The first definition seems too general today, since it does not help to govern the complexity of modern software systems. Thus, we introduced the categorisation, which seems the most appropriate for mountain environments. When choosing data for a context, we act in accordance with Crowley's suggestions, see paper [5], that is, only focusing on relevant elements and relationships. A survey by Augusto et al. [3] investigates the notion of context from a historical perspective, as well as showing the relationship between Artificial Intelligence (AI) and Intelligent Environments (IE). A paper by Alegre et al. [1] provides a comprehensive and detailed survey concerning engineering aspects for context-aware applications. It discusses developing methodologies, as well as engineering and conceptualisation for context-aware systems, constituting a solid base for designing their own systems. Hong et al., in a paper [7], state that only a small number of research papers provide development guidelines for context-aware systems, while reducing system complexity can only be achieved by using an appropriate system infrastructure, and context modelling techniques The lack of design techniques is also stated in paper [1].

A paper by Marconi et al. [10] describes a project co-financed by the European Commission, to provide a ground and aerial robotic platform, which supports search and rescue activities in mountain environments. The project does not discuss the fundamental aspects of constant activity monitoring. We are going to show that such an analysis is possible, and can be effective. Our approach is an extension to the aforementioned project, or the beginning of a new one.

This paper is a continuation of [9], where a context-aware and pro-active system to support mountain rescuers was proposed. The current work goes one step

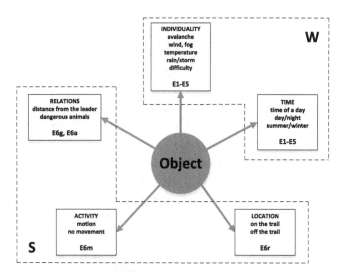

Fig. 1. Assumed context model (with indicated threat levels, see Table 2), see also [9], and division into weather context W and non-weather context S

further, because the whole system was built, redesigned as an independent and entire component. In previous works, only message streaming brokers and SAT solvers were tested using randomly generated data. Now, the system was subjected to holistic and comprehensive verification. For this purpose, the mountain environment simulator was created [12]. The obtained contextual data is both important and distinctive for a context-aware system, for example: redundancy, spatial proximity, context transition, context sharing, among others. It is also worth noting that we did not meet many simulators of this type, an exception is a paper by Aronica et al. [2], describing a simulator for rescue operations in marine environments.

3 Preliminaries

Information from this section is based on [9, Section III], however assumptions are revised, are clearer, and some understatements and ambiguities are removed. We establish *context information categories*, see Fig. 1, which influence the monitored object. This context situation is a subject of system predictions.

A division into *weather context W* (see Table 1, threats: E2–E5) and *non-weather context S* (a dangerous animal on paths, distance from the leader, lack of movement, and on/off the trail, threats: E6a, E6g, E6m, E6r) was implemented. The context data results directly from sensor data and is available for reasoning purposes after the filtration process. Table 2 shows detected threat levels for two context categories, both weather and non-weather categories.

Table 1. Weather context information, see also [9]

Context		Labels	Information
Avalanche		**A1–A5**	Increasingly difficult conditions
Weather conditions	Wind	**W1–W3**	Increasingly difficult conditions
	Fog	**F1–F3**	
	Temperature	**T1–T3**	
	Rain, storm	**R1–R3**	
Difficulty levels		**D1–D4**	Increasingly difficult levels for trails

Table 2. Increasing threat levels (top row: weather threat symbols, icon colours and danger names; bottom row: non-weather threat symbols and surrounding shapes)

Weather	**E1**	**E2**	**E3**	**E4**	**E5**
	Green	Yellow	Orange	Red	Black
	Low	Medium	Increased	High	Very high
Non-weather		**E6g**	**E6r**	**E6m**	**E6a**
		Pentagon	Circle	Square	Triangle

4   Contextual Data Processing

Context creates its *context life cycle*, that is, the sequence of stages (gathering, modelling, repositoring, reasoning, distribution, and visualisation), which structures processes of contextual pieces of data metamorphosis. Context data goes through particular stages. Starting with data gathering, its pre-processing, or modelling data is located in the repository. After the logical reasoning process, data is distributed in various system locations and visualised. The context data is updated periodically. Figure 2 shows a workflow [11] for the operations of the designed system, however, it is focused on tasks and data flows involving contextual data processing.

Raw data from weather sensors, tourist locations from BTS stations or GPS data, as well as animal geolocation, are placed in *Sink*. This data is then filtered and modelled, and then tabularised and placed in *Repository*. After tabularisation, one is able to determine precise weather conditions on specific routes (or their fragments), or assign each tourist to a specific route. Levels of alerts and avalanches are defined manually by mountain rescuers. *Repository* contains all the tabulated data prepared to make decisions based on logical inference. Recommendations are being prepared for each monitored tourist, according to threat

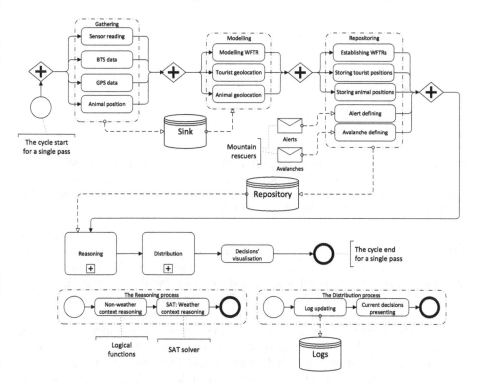

Fig. 2. Workflow for context processing, a single pass: gathering, modelling, repositoring, reasoning, distribution, and visualisation. (WFTR means wind, fog, temperature and rain, as a basic set of weather factors.)

levels, see Table 2. Visualisation occurs on available devices, such as monitors or smartphones, along with the possibility to send text messages.

From a single tourist perspective, threat signals generated by the workflow are described by the regular expression $L_i \equiv (E,)^+$, where the comma technically separates the single workflow iterations, and $E \equiv N|S|W|S\cdot W$, where N means no threat, $S \equiv E6a|E6g|E6m|E6r$, and $W \equiv E2|E3|E4|E5$. An example for L_i is a finite sentence N, $E6a$, N, $E6mE2$, $E3$, ... which ends when the object leaves the monitored area. However, the workflow generates threat warnings for every tourist at the same time, thus $L \equiv L_1 \cup L_2 \cup \ldots L_n$, where n is the total number of tourists once observed or currently being observed in the monitored area. Thus, every L_i and L are regular languages, and are generated by type-3 grammars [8].

We conducted a simple yet interesting simulation regarding the contextual data under consideration, and the results are shown in Fig. 3. The assumptions are as follows: duration time 24 h, sampling every 0.5 h, which gives 48 iterations. Normal distribution for the tourist population, population peak is 200 people at 2:00 pm, and standard deviation 4 h. We examine three time periods: morning

Fig. 3. Simulation of contextual data processing, and two categories (W, S), average values: left – weather context information, middle – non-weather context information, and right – cumulatively all pieces of contextual data. (Blue and violet show similar values and fuse together for the left and middle figures.) (Color figure online)

5:00 am–11:00 am, noon-afternoon 11:00 am–5:00 pm, and evening-night 5 pm–5 am. Probability for E2–E5 (considered together) is 20% for the 1st time period, increases by 30% for the 2nd period, and by 100% for the 3rd period. Probability for E6a is 5%, E6g is 5% (while 30% tourists are in groups), E6m is 5%, and E6r is 10%. Threats for both categories are calculated independently. If we have any threat which belongs to S, subsequent ones are not calculated, see [9, Algorithm 3], which is a result of conditional checking. The use of a particular context data is shown in Fig. 3. Individuality and Time are equally used to calculate the type-W threat. On the other hand, Relation, Activity and Location are used to calculate the type-S threat, however, Relation is used more often (for E6a, E6g). The largest number of threats W occurs with a large number of tourists, see the 11:00 am–5:00 pm period at the right figure. At night, the 5 pm–5 am period, which is the longest one, there are very few tourists, but the threat probability is doubled, which gives a relatively large number of threats. The number of Individuality and Time readings will always be the largest when compared to others, because they concern every tourist. At night, the 5 pm–5 am period, there are very few tourists and the number of S threats must be small.

5   Mountain Environment Simulator

5.1   Basic Assumptions

The developed simulator [12] enables generating extensive data, which mirror real mountain conditions within the monitored area and applies to its numerous aspects, for example: weather conditions, tourists' location (trail), walking speed, probability of changing or continuing walking along the same route on trail intersections or the probability of getting lost. Other important aspects of the simulator's work are related to animal migrations. Figure 4 presents screenshots from the simulator system, together with the monitored area. Figure 4 also presents exemplary screenshots of different administrative panels of the simulator. They enable us to influence mountain conditions, and in effect, the datasets generated by the system, by setting particular parameters.

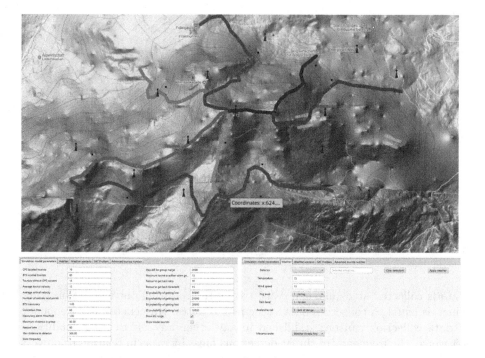

Fig. 4. Simulator screenshots: top – map of the monitoring system before tourist enters (the mountain routes are prepared using QGIS [13] and the base map is from *Google Maps*), bottom, left – admin panel, the simulation tab [12], bottom, right – admin panel, the weather tab [12]

Let us analyse five different weather conditions, with a separate simulation process being prepared for each case:

1. summer, rather bad weather conditions, but they improve, possible periodical fluctuations;
2. summer, very good weather conditions, but at some point, they significantly worsen (until the end of the simulation process);
3. summer, very good weather conditions, but they worsen for short periods of time;
4. winter, difficult weather conditions, with periodical fluctuations;
5. winter, very difficult weather conditions, but at some point, they significantly improve (until the end of the simulation process).

Each simulation process takes around one hour. and processes 25–30 times faster than real life. In other words: a one-hour simulation is equal to processes which, in a real life mountain scenario, would take approximately thirty hours. Figure 5 presents considered weather scenarios. The total number of weather threats for all tourists within a monitored area was shown. (As a threat we consider all levels from E2 to E5 excluding E1 which describes a normal situation.)

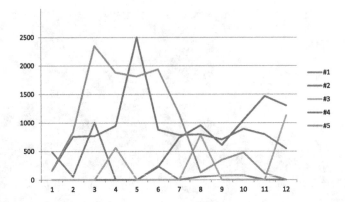

Fig. 5. Total number of detected threats for five different scenarios as a result of weather condition fluctuations. (The entire simulation period is divided into twelve intervals.)

Data is collected at regular intervals, twelve times during one simulation hour, which is equal to collecting data every five minutes. (In general, the frequency of data collection can be established in any other way.) If weather conditions get worse, the frequency of threat detections must increase in relation to objects within the monitored area. We also assume regular division of tourists on routes, as well as the fact that weather changes appear at the same time within the whole monitored area.

Scenario #1 shows weather fluctuations which stabilise themselves with time. At the beginning, both #2 and #3 have excellent weather conditions. In the first example they rapidly worsen, while the second example only experiences some local deviations. Both #4 and #5 are related to winter conditions. The first example describes normal weather fluctuations, while the second example is related to very bad weather conditions which gradually improve. Each diagram shows the overall presence of threats. In the case of winter scenarios, there are rather higher threat levels, which range E2 to E5, while during summer there are lower levels from the same range. It is not necessary to show the internal structure of those threats.

5.2   Simulation Results

Table 3 shows a general overview of simulation processes, and what happened on the routes within a monitored area. A lot of emphasis has been put on making this image both realistic and reliable. All numbers refer to the total number of events, i.e. events which happened during the hour-long simulation. Only a group of four rows, starting from the third row, concerns the current number of tourists.

"Low BTS location accuracy situations" means situations where the accuracy of the position, determined by the data from BTS stations, is too low. Broadly

Table 3. Simulation processes' general summary

	#1	#2	#3	#4	#5
Total number of tourists	3308	3344	3080	3575	3298
Tourists who left the area	3088	3123	2870	3351	3060
Current/last number of tourists	220	221	210	224	238
BTS located tourists	136	140	129	132	144
GPS located tourists	84	81	81	92	94
Tourists who denied GPS data	29	33	16	30	26
Animal threats	628	629	519	417	343
Weather threats	2003	6349	2478	10527	11195
Avalanche risk alarms	20	6	27	18	17
"No movement" situations	84	364	118	836	620
"Out of route" situations	85	372	122	848	623
Low BTS location accuracy situations	1152	967	637	222	640
"One weather detector" situations	3891	4027	3831	4029	4176
Tourists who lost their group leader	149	225	130	470	360

speaking, the algorithm works in such a way, see [4], that, by knowing the distance between two stations, it determines two intersection points of the circles which have their centres in the exact location of stations, and radiuses equal to the distance between those stations. Having those two points, we can calculate their distance to the third station. Moreover, knowing the distance between a tourist and the third station, determined from the strength of signal, we can decide which one of the two predetermined points is closer to our result. Discrepancy between the distance determined from the algorithm, and the distance determined from the strength of a signal is treated as an inaccuracy. In the case of the discrepancy being too high, a report is sent to the system, and there is the possibility to send a BTS drone.

The general image presented above is supplemented by the presentation of weather threats which appeared in every simulation process (level E1 describes a normal situation) and the total number of threats on each route, see Table 4. The data shown proves that the mountain environment was simulated in a realistic way. Numerous simulation aspects concerning weather and non-weather threats were considered.

Redundancy as repetition of information, or inclusion of additional information to improve the quality of processing, occurs in the system when locating objects in the monitored area. The basic way to localise a tourist is to analyse the data from a BTS station. GPS data is obviously much more precise in relation to geolocation, but this data can only be obtained from users who agreed, after entering the monitored area, to such a means of sending data regarding

Table 4. Total number of recorded weather threats

	#1	#2	#3	#4	#5
By routes					
Route1	742	1906	682	2437	2087
Route2	0	1	0	1	4
Route3	55	572	215	1151	1180
Route4	683	1135	487	2487	5053
Route5	64	283	159	765	1368
Route6	32	206	40	523	270
Route7	49	229	57	457	282
Route8	378	2017	838	2706	951

	#1	#2	#3	#4	#5
By emergency level					
E2	1774	4438	1259	3352	3301
E3	229	1797	567	4669	3114
E4	0	114	298	1314	1696
E5	0	0	354	1192	3084

their location. The process of positioning for each object is possible thanks to the comparison of data from BTS and GPS, if the latter exists. The use of redundancy to resolve the location issues has been shown as a subset of data on the current/last number of tourists in Table 3.

The benefits of redundancy also apply when considering tourist groups. People can visit a monitored area individually, but can also be organised into registered groups. There are no rules on how large a group can be, but the typical size of a group is 3, 4 or 5 people. (In a small number of cases, there are also two-person groups.) Then, even if one member agreed to send GPS data, it may be helpful to localise other BTS-oriented tourist positions. Table 5 shows the localisation data gathered for a one-hour simulation. It has been proved that the redundancy of information may be successfully used, mainly in relation to tourist locations, which makes the system more effective and precise.

On the other hand, Fig. 6 supplements the above image and presents the volatility of the different kinds of data regarding observed tourists within registered groups. The figure shows data for one simulation scenario, however, the images for the remaining scenarios are very similar. All analysed scenarios prove the credibility of the simulation processes. Fluctuations connected with groups

Table 5. Tourists registered in groups, or redundancy for the group localisation case

	#1	#2	#3	#4	#5
Total number of tourists in groups	103	82	102	31	102
Number of groups	24	22	28	9	24
Number of BTS located tourists	90	77	92	27	92
Number of GPS located tourists	13	5	10	4	10
Locations improved	49	18	33	9	40

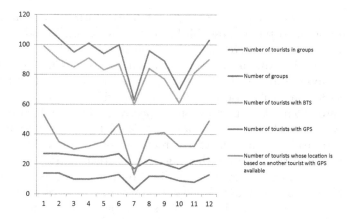

Fig. 6. General view of tourists in groups in the case of simulation #1. (The entire simulation period is divided into twelve intervals.)

and localisations which follow are a subject of natural volatility. If a tourist remains within a group, the information about the location of other group members, if followed by GPS, is used to prove the BTS location. Thus, it is another example and proof of the reasonable use of data redundancy in the system.

Spatial proximity means nearness or closeness in space. It is an important and compulsory aspect of the analysis of an intelligent system. In our system, it is implemented in relation to weather conditions, precisely speaking, when downloading data which is generated by meteorological stations located on the routes or in their nearest surrounding. If in the close neighbourhood, there are a few stations, but we only require the data from the nearest one, and if they are within a similar distance, we only require the data from the one the object is approaching. During the simulation there were numerous situations where we had to choose between 1–3 stations. Table 6 presents the results concerning meteorological station readings. It needs to be emphasised that although all tourists,

Table 6. Spatial proximity for weather stations

	#1	#2	#3	#4	#5
Number of analysed tourists	334	365	324	401	382
Total number of events	1620	1723	1614	1737	1772
Including stations: 1	427	486	462	474	453
Including stations: 2	909	979	885	983	1033
Including stations: 3	284	258	267	280	286
Average number of events per tourist	4.9	4.7	5.0	4.3	4.6

without exception, were subjected to the same rules of situational evaluation for weather conditions, for the purposes of this particular experiment, that is spatial proximity, we randomly selected a certain representative subset of tourists. In order to illustrate the experiment, every fifth tourist was chosen. Apart from the number of analysed tourists, Table 6 also includes the number of all events for weather data for particular tourists and differentiation of events when 1, 2 or 3 stations were taken into consideration, respectively. The obtained results are representative and credible, and this statement is related not only to an average number of data readings when the tourist remained on the route, but also to the fact that we are mostly dealing with taking two weather stations into consideration. The station closest to the walking direction was chosen as the one which possessed the most useful data to evaluate the tourists' most recent situation.

Context transition means dynamically switching environments which surround and influence object state and behaviour via the pervasive smart system operations that follow. Basically, tourists' context may change regularly, because it is influenced by weather changes, and, indirectly, also by the degree of a particular rule difficulty, time of day, etc. Contextual changes were observed individually, in relation to every single tourist, along its way which may include a set of routes. Table 7 presents the results of research over contextual changes in relation to tourists who finished their hiking and left the monitored area. The data obtained proves the stability of the simulation process, because of the similar values of particular variables. Of course, in real winter condition situations, routes are visited by a fewer number of tourists, however, we did not decide to decrease these values for winter scenarios, since our goal was to verify the simulation processes, alongside each comparable input value for each scenario.

Figure 7 shows the context transitions for each simulation. Every five minutes of the simulation, all important variables are saved, which gives an image of the environment within the system. For that reason, we are able to calculate an average number of context changes per tourist. Also, for this experiment, the obtained results prove the stability and reliability of the simulation processes. In the initial period of each simulation, the average number of changes is lower, which can be explained by having fewer tourists on routes. When the simulation starts, there are no tourists on the routes. That changes as the simulation continues, until it reaches a natural value. (There are numerous well-known national parks which are closed during some designated periods, or as a result of

Table 7. Context transition for tourists who finished their excursions

	#1	#2	#3	#4	#5
Number of analysed tourists	3088	3123	2870	3351	3060
Average number of transitions per object	23.26	22.34	24.50	20.36	21.58
Minimum/maximum value		3/250	4/166	3/148	3/146	3/127
Standard deviation		20.82	19.41	20.61	17.11	17.67

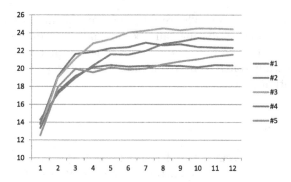

Fig. 7. Context transitions during the simulation, or average number of transitions per object. (The entire simulation period is divided into twelve intervals.)

catastrophes. Opening them and giving access to tourists meets the initial conditions of our simulation scenarios.) All results concerning the context transition prove the experiments success and credibility. In the future, the data may be also subjected to deeper analysis, typical for context-aware and pro-active systems.

Context sharing means overlapping and participation of the information, or knowledge, by different objects. Thus, groups of objects which share context, also share knowledge of how things are perceived within these groups, see also [15]. Figure 8 shows how context sharing is perceived in our system. The results are illustrated in the following way: we have twelve intervals of the simulation process, variables which describe objects are stored as a subject of simulation. Each object is described by using variables to define their context (a touristic route, data from a weather station, difficulty level, BTS/GPS availability). Context sharing was expressed as a percentage in relation to contextual variables. Thus, the most common case is context sharing 25%, with the rest being related to a rather lower number of tourists. 100% coverage is quite rare. The results obtained are fully natural and prove the credibility of the whole simulation process. The particular percentage groups are pairwise disjointed. The figure is limited to

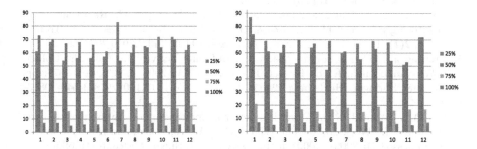

Fig. 8. Context sharing, or number of tourists grouped into four percentages: top – summer (#1), bottom – winter (#5). (The entire simulation period is divided into twelve intervals.)

winter and summer scenarios. The results were very similar for the remaining seasons. Moreover, context sharing can also be analysed more precisely in relation to particular mountain routes. Again, we conclude that all obtained results are not fundamentally different from our images of naturalness and the credibility of the simulation process.

6   Conclusions

We have shown that a sensor-based context-aware system has the ability to sense mountain environments, supporting rescue operations effectively. We have identified contextual elements in accordance with context and system requirements [14]. The number of tourists assumed in all simulation scenarios seems to be quite vast, considering the size of the monitored area. The effectiveness of the proposed solutions were validated by numerous experiments. Some concerns may arise due to the fact that all experiments were carried out on a local host, i.e. both physical phenomena collected by sensors as well as location data together with the monitoring system itself were located in one place. The designed system is also a source of rich analysis focused on contextual data processing. We intend to developed this system to enable defining arbitrary weather scenarios and providing contextual data analysis on demand.

References

1. Alegre, U., Augusto, J.C., Clark, T.: Engineering context-aware systems and applications. J. Syst. Softw. **117**(C), 55–83 (2016). https://doi.org/10.1016/j.jss.2016.02.010
2. Aronica, S., et al.: An agent-based system for maritime search and rescue operations. In: Omicini, A., Viroli, M. (eds.) Proceedings of the 11th WOA 2010 Workshop, Dagli Oggetti Agli Agenti, CEUR Workshop Proceedings, Rimini, Italy, 5–7 September 2010, vol. 621. CEUR-WS.org (2010)
3. Augusto, J., Aztiria, A., Kramer, D., Alegre, U.: A survey on the evolution of the notion of context-awareness. Appl. Artif. Intell. **31**(7–8), 613–642 (2017). https://doi.org/10.1080/08839514.2018.1428490
4. Calabrese, F., Colonna, M., Lovisolo, P., Parata, D., Ratti, C.: Real-time urban monitoring using cell phones: a case study in Rome. IEEE Trans. Intell. Transp. Syst. **12**(1), 141–151 (2011). https://doi.org/10.1109/TITS.2010.2074196
5. Crowley, J.L.: Context driven observation of human activity. In: Aarts, E., Collier, R.W., van Loenen, E., de Ruyter, B. (eds.) EUSAI 2003. LNCS, vol. 2875, pp. 101–118. Springer, Heidelberg (2003). https://doi.org/10.1007/978-3-540-39863-9_9
6. Dey, A.K., Abowd, G.D.: Towards a better understanding of context and context-awareness. In: Workshop on The What, Who, Where, When, and How of Context-Awareness (CHI 2000) (2000). http://www.cc.gatech.edu/fce/contexttoolkit/
7. Hong, J.Y., Suh, E.H., Kim, S.J.: Context-aware systems: a literature review and classification. Expert Syst. Appl. **36**(4), 8509–8522 (2009). https://doi.org/10.1016/j.eswa.2008.10.071

8. Hopcroft, J.E., Motwani, R., Ullman, J.D.: Introduction to Automata Theory, Languages, and Computation. Addison-Wesley (2006)
9. Klimek, R.: Exploration of human activities using message streaming brokers and automated logical reasoning for ambient-assisted services. IEEE Access **6**, 27127–27155 (2018). https://doi.org/10.1109/ACCESS.2018.2834532
10. Marconi, L., et al.: The SHERPA project: smart collaboration between humans and ground-aerial robots for improving rescuing activities in alpine environments. In: 10th IEEE International Symposium on Safety, Security, and Rescue Robotics (SSRR), College Station, Texas, USA, 5–8 November 2012, pp. 1–4. IEEE (2012)
11. Object Management Group: Business Process Model and Notation (BPMN) version 2.0. Technical report, Object Management Group (2011)
12. Olesek, A.: Simulation of a mountain environment for the analysis of contextual models and data, Engineering diploma thesis, supervisor: Radosław Klimek, AGH University of Science and Technology (2019)
13. QGIS Development Team: Website: QGIS - free and open geographic information system (2018). https://qgis.org/. Accessed 6 Dec 2018
14. Vieira, V., Tedesco, P., Salgado, A.C.: Designing context-sensitive systems: an integrated approach. Expert Syst. Appl. **38**(2), 1119–1138 (2011). https://doi.org/10.1016/j.eswa.2010.05.006. Intelligent Collaboration and Design
15. Zimmermann, A., Lorenz, A., Oppermann, R.: An operational definition of context. In: Kokinov, B., Richardson, D.C., Roth-Berghofer, T.R., Vieu, L. (eds.) CONTEXT 2007. LNCS (LNAI), vol. 4635, pp. 558–571. Springer, Heidelberg (2007). https://doi.org/10.1007/978-3-540-74255-5_42

Fuzzy Intelligence in Monitoring Older Adults with Wearables

Dariusz Mrozek[1]([✉])(iD), Mateusz Milik[1], Bożena Małysiak-Mrozek[2]([✉])(iD), Krzysztof Tokarz[2], Adam Duszenko[1], and Stanisław Kozielski[1]

[1] Department of Applied Informatics, Silesian University of Technology, Akademicka 16, 44-100 Gliwice, Poland
dariusz.mrozek@polsl.pl
[2] Department of Graphics, Computer Vision and Digital Systems, Silesian University of Technology, Akademicka 16, 44-100 Gliwice, Poland
bozena.malysiak@polsl.pl, krzysztof.tokarz@polsl.pl

Abstract. Monitoring older adults with wearable sensors and IoT devices requires collecting data from various sources and proliferates the number of data that should be collected in the monitoring center. Due to the large storage space and scalability, Clouds became an attractive place where the data can be stored, processed, and analyzed in order to perform the monitoring on large scale and possibly detect dangerous situations. The use of fuzzy sets in the monitoring and detection processes allows incorporating expert knowledge and medical standards while describing the meaning of various sensor readings. Calculations related to fuzzy processing and data analysis can be performed on the Edge devices which frees the Cloud platform from performing costly operations, especially for many connected IoT devices and monitored people. In this paper, we show a solution that relies on fuzzy rules while classifying health states of monitored adults and we investigate the computational cost of rules evaluation in the Cloud and on the Edge devices.

Keywords: Internet of Things · Cloud computing · Edge computing · Wearable sensors · Fuzzy sets · Fuzzy rules · Monitoring · Older adults

1 Introduction

Older adults are more likely to suffer from various accidents that can happen in their daily lives due to often poorer motor coordination, reduced gait and

This work was supported by Microsoft Research within Microsoft Azure for Research Award grant, pro-quality grant for highly scored publications or issued patents (grant No 02/020/RGJ19/0166), the professorship grant (02/020/RGPL9/0184) of the Rector of the Silesian University of Technology, Gliwice, Poland, and partially, by Statutory Research funds of Department of Applied Informatics, Silesian University of Technology, Gliwice, Poland (grant No BK/SUBB/RAu7/2020).

V. V. Krzhizhanovskaya et al. (Eds.): ICCS 2020, LNCS 12141, pp. 288–301, 2020.
https://doi.org/10.1007/978-3-030-50426-7_22

balance function, and weakened reflex. Falls, for example, are the leading cause of death from injury in the elderly [1]. Older adults often require more attention from family members or caregivers, and should more frequently monitor their health, especially after some disease-related incidents, like a heart attack or stroke. However, even after these health-related incidents, older adults want to come back to the normal, active life and stay as much self-reliant as possible. This gives them a sense of being part of society and the sense of being needed. Still, disease-related incidents or age-related problems cause many fears among older people, like *What happens to me when the incident happens, again?* or *Will I be able to call emergency in case of danger?*, and cast doubts on whether they are able to handle their daily duties. Such incidents also raise many questions among family members - *Are my parents safe at the moment?* or *How do they feel today when it is so hot or humid?*. Daily phone calls made by family members to their older relatives are nice, but they can also cause older adults to feel being controlled. These questions and doubts motivate the efforts to build noninvasive, subtle and unobtrusive systems that would allow to keep an eye on older people without disturbing them in their lives and react only in case of dangers.

Nowadays the Internet of Things (IoT) is the main technology used for helping people to deal with many everyday activities. We can observe growing attention for using IoT also in the area of personalized healthcare, including elderly people care [22]. Wearable devices connected to the Internet can act as the main technological layer to gather information about major life parameters as heartbeat, body temperature, the blood pressure and saturation, and even the ECG. A mobile phone can be used as the universal platform utilizing internal and external sensors and a communication module that allows sending the information about the health status as well as the current location of the user, body position, falls, strokes [8]. The connection between the smartphone and the wearable devices can be established using one of many wireless protocols like ZigBee, WiFi or Bluetooth. The most interesting is the Bluetooth Low Energy [23] (BLE) protocol thanks to its energy efficiency and simple implementation.

In telemedical systems, gathered data are sent to the data center for big data-enabled analysis with tailored tools and algorithms [6,13,17]. The big number of continuously monitored parameters of many patients causes a large amount of data that must be sent to the medical monitoring center. The situation when data from many patients are sent simultaneously or many users are retrieving data at the same time can cause network congestion and database overload [5,10]. The answer to this problem can be the Edge computing technology. Processing data on the Edge of the network rather than in the Cloud data center keeps analyzing close to the patients and helps to eliminate unnecessary latency [20].

The analysis of life and activity parameters of people, especially older ones, requires careful examination of incoming signals but also needs some flexibility [12]. Obtained values of the parameters (e.g., heart rate) may fall into certain ranges, which are defined by the existing medical standards and decide whether the obtained value is normal or abnormal, but still, it is important to know how much abnormal the situation is and whether it is getting better or worse. For this reason, we decided to use fuzzy sets and fuzzy logic while creating flexible rules for monitoring older adults and alerting caregivers in case of danger. The application of fuzzy rules gives us not only the information on whether a given rule is satisfied or not but also how much it is satisfied (the degree of truth). Moreover, further monitoring of the degree of truth for the rule allows to observe its changes (increase or decrease), which can be valuable information in the decision-making process (e.g., regarding possible reactions for the health state of the monitored person).

In this paper, we show how the fuzzy rules implemented on the Edge IoT devices can be applied in the monitoring of older people. We show the architecture of the system, where the detection of dangerous health states occurs on the distributed Edge devices, freeing the central Cloud-based system from additional data processing.

2 Related Works

Currently, many studies are being carried out on the application of fuzzy logic in the field of IoT. Taking into account the conducted research, we can distinguish two or three basic research approaches to using fuzzy logic in IoT. Firstly, fuzzy logic is used to reduce the amount of data flowing from the sensors (incoming frequency), which improves the performance of data analysis. Secondly, several systems also use the fuzzy logic to convert sensor data using specific fuzzy rules to the linguistic values of the relevant linguistic variables, and thirdly, fuzzy rules are also used to control home or industrial automation systems, monitoring or notification.

Emerging systems are most often created and developed in research institutes, although commercial solutions also appear, such as Vitruvius [3,4]. Vitruvius is a platform that allows users to generate applications in real time, using sensors installed in vehicles. Fuzzy logic has been used here (contributing to the first group of approaches) to reduce the frequency of data transfers. The amount of data flowing from sensors (sent from mobile clients to the server) was reduced and the data quality and the analysis performance were improved (the reduction of the input data ranged between 51.54% and 53.6%). We obtained similar results when performing fuzzy fusion of data while monitoring effectiveness of sport training [21] and when joining multiple data streams with the use of hopping umbrella [16].

In the works [7,11], both, the first and the second approach, were combined together. In [7] Dilli et al. used the fuzzy logic to classify and select the most appropriate IoT resource to the customer's request on the basis of QoS attributes. Authors also proposed adding fuzzy logic to the initial classification of resources in order to reduce the computational costs generated by MCDA algorithms they use. In the work [11], the authors applied fuzzy logic in a fire detection system. The use of fuzzy logic allows to optimize and limit the number of rules that should be checked in order to make the right decisions. The reduction of rules decreases the activity of fire sensors without significantly affecting performance. It also leads to the extended life of batteries used in sensors.

In the works [2, 18] dominates the second approach. Distribution of data in IoT usually depends on the application and requires routing protocols that take into account the context that must include auto-configuration functions. In the work [2], Chen presented a smart agent-based tracking system based on the IoT architecture using fuzzy cognitive maps (FCM) and fuzzy rules for describing the product life cycle. In [18], Araujo et al. proposed an approach to choosing the IoT route using fuzzy logic to meet the requirements of specific applications. In this case, the fuzzy logic is used to mathematically describe imprecise information expressed by a set of linguistic rules.

References [9, 14, 19] combine the second and the third approach. In the work [14], the authors used the fuzzy logic in two stages: in the first one, the fuzzy logic is used to transform the outside temperature, internal temperature and air humidity taken from sensors to linguistic values (according to pre-defined fuzzy rules). In the next step, the fuzzy rules were used to determine the best moment in which heating or climatic systems should be switched on or switched off. Similarly, in the work [19] Santamaría et al. implemented fuzzy logic in two stages of data processing. In the first stage, data read from sensors are transformed according to fuzzy rules into appropriate linguistic values corresponding to the states of human activity (resting, walking, running). In the second step, based on these states and current readings from wearable sensors, anomalies are detected (e.g., increased heart rate) and an appropriate message is generated to the user. In the work [9], the fuzzy logic is used to infer about the health state of the production machine that is being monitored. Raw data from sensors (e.g., the amount of smoke, temperature) are sent to the Cloud. Then, fuzzy rules are used to convert these raw data to the appropriate linguistic terms (low, high, typical, etc.), and in the next step, a corresponding message is generated to the user reporting the health state of the machine. Our solution also falls in the second and the third approach, but in contrast to presented solutions, the fuzzy data analysis and classification is performed on the Edge, which will be implemented in the architecture presented in the next section.

3 Cloud-Based IoT System for Monitoring Older People

Monitoring people at large scale requires not only collecting information on the selected physiological parameters but also having a logical layer that is able to decide about possible dangers. Due to a wide range of users (monitored people, caregivers, and other alerted people) and wide scaling capabilities we decided to develop our monitoring system with the Cloud computing components. The system for monitoring older people that we have developed is composed of three parts (see Fig. 1):

1. a smart band with sensors that read parameters of the monitored person,
2. a smartphone with an application that is able to receive sensor readings from the smart band,
3. a Cloud-based system that manages connected devices, stores data, and sends notifications.

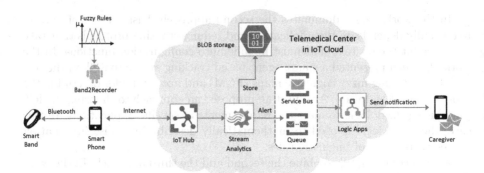

Fig. 1. Architecture of the Cloud-based system for monitoring older adults with fuzzy rules implemented in the smartphone on the Edge.

We relied our solution on the Xiaomi Mi Band 2 smart bands. The smart band consists of several sensors. For the purpose of the project, we monitored a heart rate (hr) and the number of steps taken in a unit of time (steps per minute, spm) by a person wearing the smart band (activity). Sensor readings are sent from the smart band to the smartphone working under control of the Android operating system through the Bluetooth Low Energy (BLE) protocol. The Bluetooth protocol has appropriate communication profiles defined for the purpose of communication between devices. During the communication between the smart band and the smartphone, we used the Health Device Profile (HDP) that defines the way of communication with devices such as scales, glucometers, thermometers, and other medical devices. The smartphone has a dedicated application installed, called *Band2Recorder*, that can communicate with the smart band and send data to the Cloud.

The role of the *Band2Recorder* application is to initiate readings and receive data from the Mi Band 2 smart band. In order to read measurements made by the smart band, the mobile application registers special *Listener* objects that are waiting to receive data. While reading the heart rate, it is necessary to initiate the measurement in the smart band from the level of the mobile application by sending a specific command to it. Reading the number of steps does not require additional measurement initiation, the smart band sends messages containing the number of steps performed when the monitored person moves and performs some activity.

Before the *Band2Recorder* application sends the sensor data to the Cloud, they are labeled according to the fuzzy rules implemented in the Fuzzy Rules module of the application. – If the number of steps per minute and the heart rate meet the premises of a fuzzy rule, appropriate label is added to a sent message indicating that the sent data deviate from the norm and somebody should be notified. The detection of danger may be also immediately reported to the monitored person through the application installed on the smartphone without exchanging additional messages with the Cloud center, which reduces latency and informs the person on the possible danger. The application can be easily extended to automatically call medical services to send an ambulance to the patient in particular circumstances, but this possibility has not been implemented and tested yet. Fuzzy rules (described in more details in Sect. 4) are defined in the IoT Device module in the Cloud, exactly in the IoT Hub service, and are distributed to the IoT devices while initiating the Device-to-Cloud communication. The *Band2Recorder* application communicates with the Cloud using the Internet, which is necessary for the operation of the designed system. The application also supplements the sensor data with the information on the location of the monitored person (latitude and longitude obtained from the smartphone), which helps to localize the person in emergency situations. Communication between the mobile application and the Cloud is carried out using the MQTT (Message Queue Telemetry Transport) data protocol. The protocol does not guarantee high speed and bandwidth, which is not needed when sending small data packets but ensures higher transmission reliability.

The IoT Hub, which is the Cloud gateway for IoT devices, receives data from the mobile application located on the Edge, distributes the received data to subsequent services and hosts definitions of used fuzzy rules and linguistic values. Additionally, from the IoT Hub, the *Band2Recorder* application gets the definitions of fuzzy rules. The IoT Hub contains the IoT Device module that stores the definitions of the fuzzy rules that specify in which situations sensor readings are labeled as unusual. The advantage of the solution applied is the fact that the Edge devices will receive a notification about each change in the definition of fuzzy rules and they will always work on the newest version of the rules.

The Stream Analytics Unit processes the data stream passed by the IoT Hub. As an alternative to the Edge-based rules evaluation, the Stream Analytics unit can evaluate fuzzy rules within its job. However, it increases the utilization of its resources and the network traffic (as it will be shown in Sect. 5). In both variants (the Edge-based and the Cloud-based rule evaluation), messages labeled as alerting (indicating increased risk levels) are transferred to a queue in the Service Bus service. They are consumed and analyzed by the Logic Apps module, which creates and sends appropriate notifications to a defined caregiver or a family member. Notifications are adjusted to the risk level describing the importance of the detected problem and to the class of danger. All messages collected on the IoT Hub are also placed in the BLOB storage space for further analysis. BLOB is a binary repository that stores data exactly as they were sent to the Cloud. By storing and analyzing all of the transferred data, the caregiver can get a broader view of the situation after he was notified on the occurrence of the unusual state. Further analysis of the collected data with other (e.g., ML-based) data exploration methods may also allow detection of early symptoms of incoming problems. The caregiver has the possibility to check the values of monitored variables read by the smart band just before the occurrence of abnormal heart rate values and continue monitoring after he was notified.

4 Fuzzy Sets in the Monitoring of Older People

The architecture presented in Fig. 1 implements fuzzy rules to classify possibly dangerous situations and send alert notifications to the caregiver. Each of the fuzzy rules has the following general form:

IF x_1 is Φ_1 AND x_2 is Φ_2 AND ... AND x_p is Φ_p, THEN y is Ψ

where $x_i \in X_i$ are monitored parameters, $y \in Y$ is a health risk, and $\Phi_1, \ldots, \Phi_p$ and Ψ are fuzzy sets.

There are mainly two physiological parameters monitored with the use of the smart band, i.e., heart rate (HR) and performed activity. The activity is classified on the basis of the number of steps per minute (SPM). Values of both monitored parameters are transformed to linguistic values of appropriate linguistic variables defined by fuzzy sets. Figure 2 shows the definitions of the fuzzy sets used in the classification of the state of the monitored person. With the presented linguistic values we have created the fuzzy rules shown in Listing 1.1.

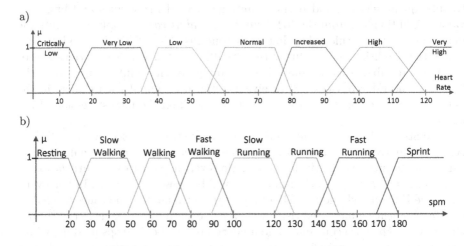

Fig. 2. Defined linguistic variables with linguistic values for Heart rate (a) and performed activity (b) on the basis of the number of steps taken per minute.

```
1 {Rule #1 - faint or death}
2 IF HeartRate IS criticallyLow
3 THEN risk IS veryHigh
4
5 {Rule #2 - probable ventricular tachycardia, VT}
6 IF (HeartRate IS High OR HeartRate is veryHigh) AND Activity
       IS Resting
7 THEN risk IS high
8
9 {Rule #3 - overload, too intensive activity}
```

```
10 IF HeartRate IS veryHigh AND Activity IS fastWalking
11 THEN risk is medium
12
13 {Rule #4 - probability of faint}
14 IF (HeartRate IS Low) AND (Activity IS fastWalking OR Activity
       IS slowRunning OR Activity IS fastRunning OR Activity IS
       sprint)
15 THEN risk IS veryHigh
16
17 {Rule #5 - probable bradycardia}
18 IF (HeartRate IS Low OR HeartRate IS veryLow) AND (Activity IS
       resting)
19 THEN risk IS medium
20
21 {Rule #6 - resting}
22 IF (HeartRate IS Normal) AND (Activity IS resting)
23 THEN risk IS low
24 ...
```

Listing 1.1. Partial fuzzy rules used while monitoring older people.

T-norm is used for the AND operator and t-conorm is used for the OR operator used in premises in the IF condition of each fuzzy rule. In our solution, we used minimum t-norm defined as follows:

$$T_{min}(\mu_{\Phi_k}(x_i), \mu_{\Phi_l}(x_j)) = \min(\mu_{\Phi_k}(x_i), \mu_{\Phi_l}(x_j)), \tag{1}$$

and maximum t-conorm:

$$T_{max}(\mu_{\Phi_k}(x_i), \mu_{\Phi_l}(x_j)) = \max(\mu_{\Phi_k}(x_i), \mu_{\Phi_l}(x_j)), \tag{2}$$

where $T : [0,1] \times [0,1] \to [0,1]$, $x_i \in X_i, x_j \in X_j$ are values of any two monitored parameters, μ_{Φ_k} and μ_{Φ_l} and membership functions that define fuzzy sets Φ_k and Φ_l.

On the basis of the rules, each of the sensor readings $e \in E$ (where E is a data stream) gets an appropriate label of health risk (severity level). This is achieved according to Algorithm 1.

What is important, fuzzy rules and linguistic variables they rely on are defined in the Cloud, in the IoT Hub Cloud gateway. The IoT Hub holds the register of many IoT devices that can be connected to the Cloud, each per monitored person. The fuzzy rules and linguistic variables are defined in the IoT Device module of the IoT Hub, which is created per a physical IoT Device connected to the Cloud. This means that the rules and definitions of linguistic variables can be adjusted to a monitored person. For example, some rules or parts of the premises can be skipped for the older adults that are in good health condition, e.g., jogging regularly.

5   Experimental Results

Labeled sensor readings containing the information on the heart rate, activity, the location of the monitored person (latitude and longitude), and additional

Algorithm 1. Risk evaluation with fuzzy rules

```
 1: for each e ∈ E do
 2:     id ← e.deviceId
 3:     L_HR ← getLingValues(HR, id)          //HR and SPM are the internal
 4:     L_SPM ← getLingValues(SPM, id)        //names of linguistic variables
 5:     max ← 0
 6:     label_HR ← null
 7:     for each linguistic value l ∈ L_HR do
 8:         calculate membership degree μ_l(e.hr)
 9:         if μ_l(e.hr) > max then
10:             max ← μ_l(e.hr)
11:             label_HR ← l
12:         end if
13:         μ_HR = max
14:     end for
15:     max ← 0
16:     label_SPM ← null
17:     for each linguistic value l ∈ L_SPM do
18:         calculate membership degree μ_l(e.spm)
19:         if μ_l(e.spm) > max then
20:             max ← μ_l(e.spm)
21:             label_SPM ← l
22:         end if
23:         μ_SPM = max
24:     end for
25:     risk ← evaluateRules_{1-6...}()
26:     μ_risk ← T_{min/max}(μ_HR, μ_SPM)
27:     return risk, μ_risk
28: end for
```

data (required to connect to the Cloud, authenticate in the system, identify the monitored person) can be transmitted to the Cloud constantly or in case of danger. In both situations the data is stored in the Cloud storage space. The data is used twofold. Current sensor readings labeled as critical or important (severity level) trigger sending the notifications to the defined caregivers. All data are also stored in the BLOB Storage. Historical data with sensor readings are stored for future re-use, training and data analysis with the use of machine learning models.

Validation of fuzzy rules for the incoming sensor readings, even if simple, takes some time and may pose a pressure on the stream processing units in the Cloud. Especially, that rules can be adjusted for particular persons. We tested how evaluation of fuzzy rules evaluated by the stream processing units in the Cloud influences their operational capabilities. To this purpose, during our experiments with older people, we measured the size of each message (event) sent to the Cloud by the IoT device with *Band2Recorder* application. Each message always took 862B. With certain periods between sending successive messages from the device and assuming a uniform data transfer between the device and the Cloud, we can easily estimate the network traffic and the consumption of the Cloud storage space.

Figure 3 shows how much data will be transferred through the network and then stored in the Cloud within 24-h monitoring time for various periods between sending successive messages containing sensor readings. As can be observed, sending data every minute generates 1.2 MB of data that must be transferred within 24 h. This is not much, taking into account that the caregiver or the family member can be notified quite quickly in the situation of danger (with a 1-min latency). Longer periods between taking data from the smart band and between data transfers (less frequent readings) allow to further decrease the network traffic and the storage space consumed (e.g., 0.06 MB for 20-min periods), but increase the latency and should be used in situations when we monitor a general daily health state of the person without the necessity to react immediately.

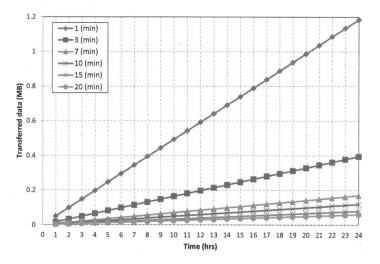

Fig. 3. Consumption of the storage space during every 24 h of monitoring for various time periods between successive sensor readings.

Figure 4 shows the total level of the network traffic and the consumption of the storage space within the 30-days monitoring time for various time periods between successive sensor readings. Consequently, Fig. 5 shows the number of messages sent to the Cloud within the 30-days monitoring time for various time periods between successive sensor readings. Both figures are useful when planning hardware infrastructure and a level of services upon which the monitoring system working in the Cloud is built. For example, the S1 tier of the IoT Hub, which allows to connect any number of IoT devices and to receive 400,000 messages daily, costs 25 USD per month (as for January 22, 2020, on the basis of [15]).

Then, having the characteristics of data transmission, we created a generator application, which is a digital twin simulating a given number of IoT devices.

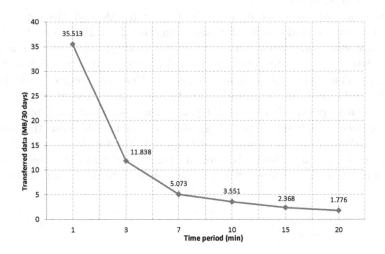

Fig. 4. The total level of the network traffic and the consumption of the storage space within the 30-days monitoring time for various time periods between successive sensor readings.

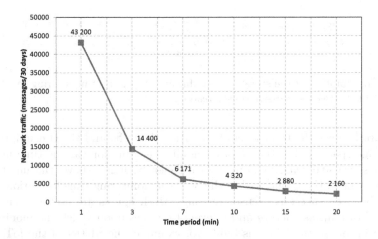

Fig. 5. The number of messages sent to the Cloud within the 30-days monitoring time for various time periods between successive sensor readings.

We connected the digital twin to our architecture in the place of the smartphone with the *Band2Recorder* application. With the use of the digital twin, we were able to test the system for a variable number of connected devices. We monitored the maximum utilization of streaming units (%) in two cases: (1) when fuzzy rules were implemented in the stream processing job and evaluated directly on stream processing units, and (2) when fuzzy rules were implemented at the Edge (were implemented within the digital twin). Results of these tests are presented in Fig. 6. In such an experimental environment, we simulated the load generated by several thousands of IoT devices (up to 10,000). Results show that, while processing events already labeled at the Edge, the utilization of resources of the Stream Processing Unit is quite constant, At the same time, for the increasing number of IoT devices the additional overhead related to evaluation of fuzzy rules in the Cloud (implemented directly in the Stream Processing Unit) causes slow but constant increase in the maximum utilization of streaming units (Max % SU utilization). This increase leads to the necessity of scaling the system on processing units faster, than in the case when fuzzy rules are implemented on the Edge device. Although we could expect such a behavior, we can observe the dynamics of the process for both implementations.

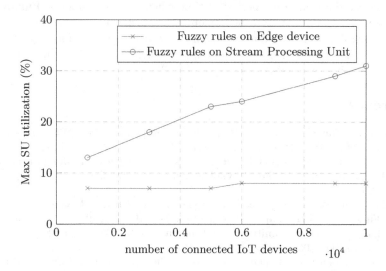

Fig. 6. Maximum utilization of streaming units (%) for fuzzy rules implemented on the Stream Processing Unit and on the Edge device.

6   Discussion and Conclusions

Evaluation of fuzzy rules on the smartphone, which is the Edge device, puts the data pre-processing at the proximity of data sources and moves the burden of the pre-processing from the Cloud to the Edge. If we decide to transmit only the messages for important notification purposes without collecting the whole history in the Cloud storage space, we could additionally reduce (dozens to hundreds of times) the network traffic and the storage space consumed. Our results correspond to those reported in related works [3,4,16]. The reduction rate would depend on the number of dangerous situations detected and the number of notifications sent to the caregiver.

Implementation of fuzzy rules on the Edge devices allows using fuzzy techniques for data processing close to the wearable units, where the data are collected as sensor measurements. As a consequence, the data can be labeled very quickly on personal devices and in such a form transmitted to the Cloud data center. This frees the Cloud data center from the analysis and classification of data coming from many such devices. Such an implementation also postpones the need for scaling the system resources. Linguistic values represented by fuzzy sets allow assigning sensor readings to meaningful terms and, by applying fuzzy rules, use them in reasoning about the current health state of the monitored person. Our experiments confirmed that even with a relatively short periods between data transmissions the volume of data sent to the Cloud within 24 h is relatively low, which in contrast to several use cases mentioned in [20] gives the comfort of collecting all data at least for some groups of monitored people or collecting the data for some period of time when the risk appears. Finally, our conclusions on the advantages of implementing fuzzy intelligence on the Edge can be generalized. This group of techniques can be applied not only in monitoring older adults but also in other domains, where IoT devices can improve or optimize ongoing processes.

References

1. Burns, E., Kakara, R.: Deaths from falls among persons aged ≥65 years - United States, 2007–2016. MMWR Morb. Mortal Wkly. Rep. **67**, 509–514 (2018)
2. Chen, R.Y.: Intelligent IOT-based tracing system for backward design using FCM and fuzzy rule. In: 2013 Fourth Global Congress on Intelligent Systems, pp. 229–233 (2013)
3. Cueva-Fernandez, G., Espada, J.P., García-Díaz, V., Crespo, R., Garcia-Fernandez, N.: Fuzzy system to adapt web voice interfaces dynamically in a vehicle sensor tracking application definition. Soft Comput. **20**(8), 3321–3334 (2015). https://doi.org/10.1007/s00500-015-1709-2
4. Cueva-Fernandez, G., Espada, J.P., García-Díaz, V., Gonzalez-Crespo, R.:Fuzzy decision method to improve the information exchange in a vehicle sensor tracking system. Appl. Soft Comput. **35**, 708 – 716 (2015)
5. Cupek, R., Ziebinski, A., Drewniak, M., Fojcik, M.: Knowledge integration via the fusion of the data models used in automotive production systems. Enterp. Inf. Syst. **13**(7–8), 1094–1119 (2019)
6. Cupek, R., Ziebinski, A., Zonenberg, D., Drewniak, M.: Determination of the machine energy consumption profiles in the mass-customised manufacturing. Int. J. Comput. Integr. Manuf. **31**(6), 537–561 (2018)
7. Dilli, R., Argou, A., Pilla, M.L., Pernas, A.M., Reiser, R.H.S., Yamin, A.C.: Fuzzy logic and MCDA in IoT resources classification. In: SAC, New York, USA, pp. 761–766 (2018)

8. Divyanshu, D., Amod, A., Anshudha, S., Nupur, A., Pradeep, D., Ambarish, P.: Smart phone applications as a source of information on stroke. J. Stroke **16**(2), 86–90 (2014)
9. Gerald, P., Novilla, A., Balute, A.A., Gonzales, D.: The use of fuzzy logic for online monitoring of manufacturing machine: an intelligent system. Circ. Comput. Sci. **2**, 31–39 (2017). https://doi.org/10.22632/ccs-2017-252-61
10. Lin, B.S., Hsiao, P.C., Cheng, P.H., Lee, I.J., Jan, G.E.: Design and implementation of a set-top box-based homecare system using hybrid cloud. Telemed. e-Health **21**(11), 916–922 (2015). pMID: 26075333
11. Maksimovic, M., Vujovic, V., Perisic, B., Milosevic, V.: Developing a fuzzy logic based system for monitoring and early detection of residential fire based on thermistor sensors. Comput. Sci. Inf. Syst. **12**, 63–89 (2015)
12. Małysiak-Mrozek, B., Lipińska, A., Mrozek, D.: Fuzzy join for flexible combining big data lakes in cyber-physical systems. IEEE Access **6**, 69545–69558 (2018). https://doi.org/10.1109/ACCESS.2018.2879829
13. Małysiak-Mrozek, B., Stabla, M., Mrozek, D.: Soft and declarative fishing of information in big data lake. IEEE Trans. Fuzzy Syst. **26**(5), 2732–2747 (2018). https://doi.org/10.1109/TFUZZ.2018.2812157
14. Meana-Llorián, D., García, C.G., G-Bustelo, B.C.P., Lovelle, J.M.C.,Garcia-Fernandez, N.: IoFClime: the fuzzy logic and the Internet of Things to control indoor temperature regarding the outdoor ambient conditions. Future Gener. Comput. Syst. **76**, 275–284(2017)
15. Microsoft: Azure IoT Hub - Pricing. https://azure.microsoft.com/pl-pl/pricing/details/iot-hub/. Accessed 22 Jan 2019
16. Mrozek, D., Tokarz, K., Pankowski, D., Malysiak-Mrozek, B.: A hopping umbrella for fuzzy joining data streams from IoT devices in the Cloud and on the Edge. IEEE Trans. Fuzzy Syst. 1 (2019). https://doi.org/10.1109/TFUZZ.2019.2955056
17. Mrozek, D.: Scalable Big Data Analytics for Protein Bioinformatics. CB, vol. 28. Springer, Cham (2018). https://doi.org/10.1007/978-3-319-98839-9
18. Araujo, H.D.S., et al.: A proposal for IoT dynamic routes selection based on contextual information. Sensors **12**, 353 (2018)
19. Santamaría, A.F., Raimondo, P., Rango, F.D., Serianni, A.: A two stages fuzzy logic approach for Internet of Things (IoT) wearable devices. In: 2016 IEEE27th Annual International Symposium on Personal, Indoor, and Mobile RadioCommunications (PIMRC), pp. 1–6, September 2016. https://doi.org/10.1109/PIMRC.2016.7794563
20. Shi, W., Cao, J., Zhang, Q., Li, Y., Xu, L.: Edge computing: vision and challenges. IEEE Internet Things J. **3**(5), 637–646 (2016)
21. Wachowicz, A., Małysiak-Mrozek, B., Mrozek, D.: Combining data from fitness trackers with meteorological sensor measurements for enhanced monitoring of sports performance. In: Rodrigues, J.M.F., et al. (eds.) ICCS 2019. LNCS, vol. 11538, pp. 692–705. Springer, Cham (2019). https://doi.org/10.1007/978-3-030-22744-9_54
22. Wan, J., et al.: Wearable IoT enabled real-time health monitoring system. EURASIP J. Wirel. Commun. Netw. **2018**(1), 1–10 (2018). https://doi.org/10.1186/s13638-018-1308-x
23. Zhang, T., Lu, J., Hu, F., Hao, Q.: Bluetooth low energy for wearable sensor-based healthcare systems. In: 2014 IEEE Healthcare Innovation Conference (HIC), pp. 251–254, October 2014. https://doi.org/10.1109/HIC.2014.7038922

Deep Analytics for Management and Cybersecurity of the National Energy Grid

Ying Zhao$^{(\boxtimes)}$ (iD)

Naval Postgraduate School, Monterey, CA 93943, USA
yzhao@nps.edu

Abstract. The United States's energy grid could fall into victim to numerous cyber attacks resulting in unprecedented damage to national security. The smart concept devices including electric automobiles, smart homes and cities, and the Internet of Things (IoT) promise further integration but as the hardware, software, and network infrastructure becomes more integrated they also become more susceptible to cyber attacks or exploitation. The Defense Information Systems Agency (DISA)'s Big Data Platform (BDP), deep analytics, and unsupervised machine learning (ML) have the potential to address resource management, cybersecurity, and energy network situation awareness. In this paper, we demonstrate their potential using the Pecan Street data. We also show an unsupervised ML such as lexical link analysis (LLA) as a causal learning tool to discover the causes for anomalous behavior related to energy use and cybersecurity.

Keywords: Big data platform · Deep analytics · Cybersercurity · Usage patterns · Anomaly detection · Lexical link analysis · Causal learning

1 Introduction

The United States' energy grid is evolving towards smart grid of future, which incorporates the digital technology to improve reliability, security and efficiency of the electric system through bi-directional information exchange, distributed generation, and storage resources for a fully automated power delivery network. The smart and integrated grids as shown in Fig. 1 seek efficiency through common communication standards and integrated networks, meeting the demand for the rapid growth in a cost-effective manner [16]. To further the concept smart devices including electric automobiles, smart homes and cities, and the Internet of Things (IoT) promise further integration. Better ways to manage the energy grid through better tools can also lead to better manage our energy resources and reduce greenhouse gasses.

V. V. Krzhizhanovskaya et al. (Eds.): ICCS 2020, LNCS 12141, pp. 302–315, 2020.
https://doi.org/10.1007/978-3-030-50426-7_23

Fig. 1. The concept of smart or integrated grids seeks efficiency through common communication standards and an integrated grid (Electrical Power Research Institute) [20]

This energy and smart grid not only need tools for better management and automation, but also face risks and vulnerability that bring unprecedented challenges and damage to national security:

1. Threats typical to smart grids and IoT devices are that they are not only vulnerable to physical faults and attacks but to cyber attacks as in the Internet, since the rise of the Internet and the integration via the Open Systems Interconnection model (OSI) allows an integration of standards throughout the different types of networks and devices Threats and vulnerabilities therefore are similar. The energy assets would naturally be susceptible to cyber attacks such as a Distributed Denial of Service (DDOS), worms, viruses and similar cyber exploits which might be the reasons for the Ukrainian power grid problems [16] and the Massachusetts gas explosions [18]. An attacker could take control of the company's "Supervisory Control and Data Acquisition" (SCADA) distribution management system [16]. The U.S. Energy Information Administration (EIA) notes that the rise of electric vehicles and smart devices has as one obstacle which is the cybersecurity [17]. Correlations between different categories of sensor big data could potentially act as "red flags" or early-warning signs of the cyber breach and vulnerability.
2. Distribution automation (DA) is a concept of smart grid which focuses on the operation and system reliability at the distribution level. The conventional centralized control management strategy is less effective for the smart grid due to the unidirectional power flow and requirement of control of a distributed grid. It is imperative to predict hotspots and areas of greatest concern which will lead to a reduction in waste and inefficiency or expose security vulnerabilities. This also calls for big data and deep analytics to be operated in a distributed fashion.
3. Risks include unanticipated operational conditions, for example, for a grid-connected microgrid, severe weather conditions or grid blackouts may trigger an unintentional islanding accident [2], which threats the safety operation and causes technical challenges.

New and emerging technologies offer opportunities to better analyze energy sensor data to improve our understanding of where energy is wasted and how to identify and best respond to risks. The IoT significantly increases the volume and velocity of big data through the concept such as "smart cities". Big data analytics can reduce risks and show scalable solutions for detecting patterns and anomalies from collective intelligence and from the distributed data sources [1].

In the past, data mining techniques [11], energy and entropy theories, wavelet transform [6], machine learning algorithms such as predictive maintenance, electric device health monitoring, and power quality monitoring have been used for energy and smart grids [1]. The support vector machine (SVM) is used for an islanding detection [2]. A deep learning is used for IoT device as deep sparse coding [3]. Localized fault characterization uses a hybridization of evolutionary learning and clustering techniques [4,5]. SVM, AdaBoost, and extreme learning machine (ELM) are used for online detection of risky events in power system [7,8] Innovation of big data also comes to energy grid management and cybersecurity for example, real-time social sensor data using Twitter, Facebook could provide new insight using the location data [9,10].

Unique methods illustrated in this paper can be used for both energy management and cybersecurity because we show data collection and analysis from sensors as the key components for constantly monitoring and detecting the threats from collective and distributed data sources. Specifically, this paper addresses how a "Big Data Platform" (BDP), lexical link analysis (LLA) to discover anomalies, potential threats, and vulnerabilities using the Pecan Street data set as a use case. The key contribution of this paper is causal learning since human is necessary in the process for the validation of decision making.

2   Pecan Street Big Data

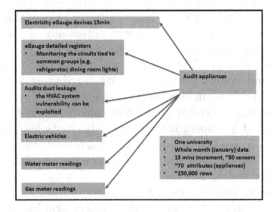

Fig. 2. Pecan Street data

The data source used in this research was obtained by the Pecan Street organization [12]. Pecan Street collects energy usage for a smart city which means there is a conscious and curated effort to record the right data for energy consumption in a methodical manner. The organization host and maintain one of the largest databases of consumer electricity and water use in the world. 750 million records are collected daily as circuit-level use data from multiple sources available through their Dataport website. The data track appliance level consumer behavior.

We extracted sample research data from Pecan Street Dataport. Figure 2 shows the depiction of a sample Pecan Street data. The sample data consists of participants' electricity usage data (per kWh) with 69 data fields including one key field (user id or data id and timestamp) and the remaining 67 fields listing various equipment used on site (e.g., furnace, kitchen, lights, dish washer, dryer, etc.). We selected one month of data consisting of 250,000 records in 15 min data blocks for 100 participants (users or data ids) as follows:

- air1: air conditioner 1
- air2: air conditioner 2
- air3: air conditioner 3
- aquarium1: aquarium 1
- bathroom1: bathroom 1
- bathroom2: bathroom 2
- bedroom1: bedroom 1
- ...

The "air1" field records electricity usage for air conditioner 1 for 31 days in for a January. Baselines could be set and monitored to find anomalies, for example, an insider threat could be the unauthorized running of energy grid servers in January not August which would increase the air conditioner usage. Figure 3 shows an example of Pecan Street data.

3 Big Data Platform (BDP)

The Big Data Platform (BDP), which has been developed by the Defense Information Systems Agency (DISA) [19], runs on Amazon Web Services (AWS) including a mix of big data standard and customized tools for data ingestion, management, security, exploration, and analysis. These functions are supported by open source tools including Apache Spark [25], Apache Storm [26], Hadoop Map/Reduce, Kibana [27], NodeJS [28], and R-Shiny [29].

BDP is designed for real-time processing of Big Data beginning at ingestion and ultimately presenting useful data visualizations that may alert decision makers of energy leaks and security vulnerabilities. The BDP has the strict compliance with the DISA security standards which provide a secure system security that can be an advantage to store big data such as Pecan Street and National Energy Grid. There are also analytics in BDP which can perform more complicated calculations on a larger data set. For the Pecan Street sample data set, we first applied BDP to provide initial useful information. We later applied unsupervised machine learning algorithms k-means and lexical link analysis (LLA) to discover patterns and anomalies in the data set.

4 Application of BDP

We first ingested and parsed the Pecan Street data into the BDP system. After ingesting, the data were available to the analytics tools inside BDP such as Unity. Kibana is used to create a display of metrics, heatmaps, graphs, and charts. The BDP system is designed to display results quickly for real-time big data so that trends and outliers can be discovered quickly. BDP uses a catalogue or taxonomy shared by multiple users in the same domain. For example, in a typical cybersecurity environment, people often use similar network monitoring tools to collect data and monitor activities, therefore, the data fields are similar and can be shared across multiple locations. The characteristic applies to the energy grid as well. This is an unique advantage of using BDP. Figure 4 shows the average electricity usage each hour over one month for the Pecan Street data set. It is interesting to note that the spikes in use are not regular. Figure 5 shows a dashboard of graphs and metrics representing electricity usage over 24 h for different areas of the data set. These could be updated in near real-time for monitoring activities should energy grid data hosted in such a secure data center. For example, why do the outside lighting plugs have higher average electricity usage around the noon time?

A	B	C	D	E	F	G	H	I	J	K	L	M	N	O	P	Q
dataid	local_15min	use	air1	air2	air3	airwindowunit1	aquarium1	bathroom1	bathroom2	bedroom1	bedroom2	bedroom3	bedroom4	bedroom5	car1	clotheswasher1
93	1/1/2013 0:00	0.856266667	0.0414													0
93	1/1/2013 0:15	0.850666667	0.041													0
93	1/1/2013 0:30	0.857533333	0.040933333													0
93	1/1/2013 0:45	0.846266667	0.036933333													0
93	1/1/2013 1:00	0.5238	0.0366													0
93	1/1/2013 1:15	0.445533333	0.038													0
93	1/1/2013 1:30	0.447533333	0.038													0
93	1/1/2013 1:45	0.5276	0.038133333													0
93	1/1/2013 2:00	0.5628	0.038333333													0
93	1/1/2013 2:15	0.544333333	0.038866667													0
93	1/1/2013 2:30	0.547066667	0.038666667													0
93	1/1/2013 2:45	0.550266667	0.038133333													0
93	1/1/2013 3:00	0.5446	0.0382													0
93	1/1/2013 3:15	0.513866667	0.038066667													0
93	1/1/2013 3:30	0.442533333	0.038													0
93	1/1/2013 3:45	0.448666667	0.038													0
93	1/1/2013 4:00	0.4476	0.038													0
93	1/1/2013 4:15	0.436	0.038													0
93	1/1/2013 4:30	0.4458	0.038													0
93	1/1/2013 4:45	0.447466667	0.038													0
93	1/1/2013 5:00	0.441933333	0.038													0
93	1/1/2013 5:15	0.501933333	0.038													0
93	1/1/2013 5:30	1.143733333	0.043733333													0.003733333
93	1/1/2013 5:45	1.102666667	0.043266667													0.0034
93	1/1/2013 6:00	0.5606	0.038066667													0

Fig. 3. Pecan Street data example

4.1 Unsupervised Machine Learning

We first applied the K-means clustering algorithm from MATLAB and clustered the 250 K records into 10 clusters as shown in Fig. 6. K-means requires a chosen k and k = 10 in our case for simplicity. Figure 6 is a radar graph showing cluster

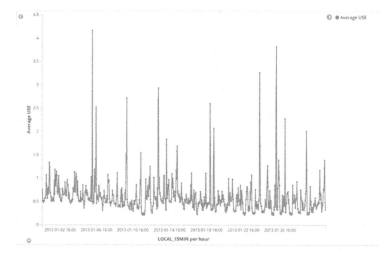

Fig. 4. Average electricity usage each hour over one month

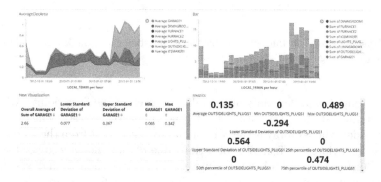

Fig. 5. A dashboard shows graphs and metrics representing electricity usage over 24 h for different areas of the data set. BDP allows to update such graphs in real-time for monitoring activities.

center values, i.e., average usages within clusters for the 67 areas labeled in the circle. These clusters represent the discovered patterns. The characteristics of the clusters show behavior patterns of the users and time periods in which characteristics of usage patterns can be summarized in the following examples:

- Cluster 7 (series7): Average high usages within the cluster attribute to the areas of "use", "grid", "drye1", "furnace1", "poollight1", and "waterheater1".
- Cluster 6 (series6): Average high usages attribute to the areas of "use", "car1", "gen", and "grid".
- Cluster 5 (series5): Average high usages attribute to the areas of "gen" and "grid" (negative – giving back to the grid).

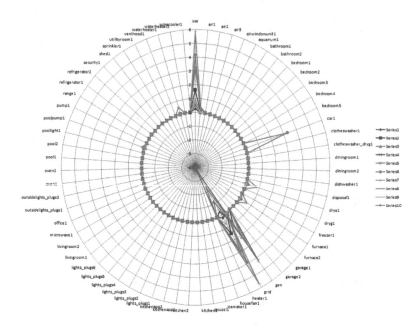

Fig. 6. Unsupervised learning and cluster characteristics using the k-means algorithm and displayed using a radar graph and k-means.

We then computed an anomaly index value for each of the data points, which is the minimum distance of a data point to the 10 cluster centers. The higher an anomaly index, the far away is the corresponding data point from the 10 patterns ("normal behaviors"). Figure 7 shows the value of the anomaly index for each cluster. Cluster 7, 6, and 5 have the highest 3 values of anomaly indexes, which are the potential candidates for further investigation for the areas of energy management and cybersecurity.

The anomaly detection system detected 3 anomaly profiles which gives three different reasons for these users to be different from the population as a whole. The information shows human analysts behavioral patterns for attributes and source of the anomaly, for example "gen" means there is a generator at home and negative "grid" means the generator gives energy back to the grid. The combination might indicate a different usage pattern for some users. If some anomaly patterns are trending collectively from various locations, they are the opportunities for human decision makers and consequences of the anomalies detected can be alerted to the human analysts via the BDP server.

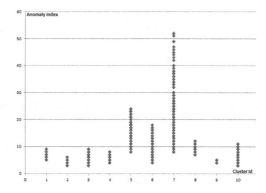

Fig. 7. The values of the anomaly index for the clusters

5 Lexical Link Analysis (LLA)

LLA is an unsupervised ML method [13,14] which describes the characteristics of a complex system using a list of attributes or features, or specific vocabularies or lexical terms. Because the potentially vast number of lexical terms from big data, LLA can be viewed as a deep model for big data. LLA can describe a system using feature pairs as bi-gram lexical terms extracted from data. LLA automatically discovers word pairs, and displays them as networks.

Bi-grams allow LLA to be extended to numerical or categorical data. For example, using structured data, such as attributes from the Pecan Street data set, we discretize numeric attributes and categorize their values to paired features. The feature pair model can further be extended to a context-concept-cluster model [21]. A context can represent a location, a time point, or an object shared across data sources. For example, for the Pecan Street data, the data id and time point can be contexts.

5.1 LLA Outputs for the Pecan Street Data Set

In order to use LLA, we first generate word feature networks for the data set. The value for an attribute in Fig. 3, such as "grid" is discretized into three bins when applying LLA as a word feature: 1) less than (lt) the mean ($\bar{m}$) of the feature minus one standard deviation ($\bar{m} - \sigma$), 2) between (bt) the mean minus one standard deviation ($\bar{m} - \sigma$) and the mean plus one standard deviation ($\bar{m} + \sigma$), and 3) more than (mt) the mean plus one standard deviation ($\bar{m} + \sigma$). A node in LLA represents a discretized feature. For example, $grid\_mt\_1.8$ means if the "grid" (i.e., grid usage of electricity in a 15 min interval for a data id) is more than 1.8.

Probability and lift are the two measures in LLA defined in Eq. (1) and Eq. (3) to measure the strength of an association between two word features.

$$prob_{ij} = \frac{word\ features\ i, j\ together}{word\ feature\ j} \tag{1}$$

$$prob_i = \frac{word\ feature\ i}{all\ word\ features} \qquad (2)$$

$$lift_{ij} = \frac{prob_{ij}}{prob_i} \qquad (3)$$

Figure 8 shows the output of LLA for a word feature $grid\_mt\_1.8$'s associations with other features using the "lift" as the association strength measure listed as follows (filtered using "lift" > 4):

- $dry e1\_mt\_1.0$: "dryer1" (dryer 1)'s usage of electricity is more than 1.0 in a 15 min interval
- $car11\_mt\_2.0$: "car1" (car 1)'s usage of electricity is more than 2.0 in a 15 min interval
- $air1\_mt\_0.6$: "air1" (air conditioner 1)'s usage of electricity is more than 0.6 in a 15 min interval
- $air2\_mt\_0.6$: "air2" (air conditioner 2)'s usage of electricity is more than 0.6 in a 15 min interval
- $waterheater1\_mt\_2.0$: "waterheater1" (water heater 1)'s usage of electricity is more than 2.0 in a 15 min interval
- $poolpump1\_mt\_1.5$: "poolpump1" (pool pump 1)'s usage of electricity is more than 1.5 in a 15 min interval
- $poolpump1\_bt\_0.6\_1.5$: "poolpump1" (pool pump 1)'s usage of electricity is between 0.6 and 1.5 in a 15 min interval
- $oven1\_mt\_0.3$: "oven1" (oven 1)'s usage of electricity is more than 0.3 in a 15 min interval
- $dataid\_5357$: data id (user) 5357

Figure 11 shows $gen\_mt\_1.7$ (i.e., a generator, such as solar, alternative, and renewable energy with inverter interfaced distributed generators (IIDGs), generates electricity more than 1.7 in a 15 min interval) is associated with $grid\_lt\_-0.9$ (i.e., grid usage of electricity is less than -0.9, negative, giving back to the grid in a 15 min interval for a data id). These results are similar to the k-means result in Fig. 6.

LLA allows a drill down search as shown in Fig. 9. When clicking both nodes $grid\_mt\_1.8$ and $waterheater1\_mt\_2.0$: 373 data records in the Pecan Street data set have both characteristics $grid\_mt\_1.8$ and $waterheater1\_mt\_2.0$ and they are listed in the LLA search result. 373 data records have the characteristics $waterheater1\_mt\_2.0$, 100% of them also have the characteristics $grid\_mt\_1.8$. 16,434 data records have the characteristics $grid\_mt\_1.8$ out of the total 120,847 data records. So the lift is 7.4.

LLA also discovers interesting associations, for example, $grid\_mt\_1.8$ is associated with a specific user (data id) of "5357" in Fig. 8. As another example, Fig. 11 shows the time points of a day are associated with $gen\_bt\_0.6\_1.7$ (i.e., generater) generates electricity between 0.6 and 1.7 in a 15 min interval).

5.2   Discussion: Discovering Causal Associations Using LLA

A unique requirement of anomaly detection for energy management and cyber-security is causality analysis because human analysts need to understand causes behind any observable anomaly effects. This calls a systematic approach of deep analytics that is also causality analysis, i.e., linking an anomaly effect, e.g., grid usage of electricity in a 15 min interval is more than 1.8 (*grid_mt_1.8*), to the

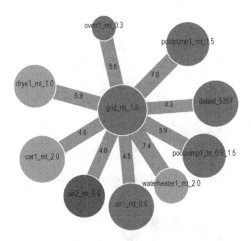

Fig. 8. Causal level 1

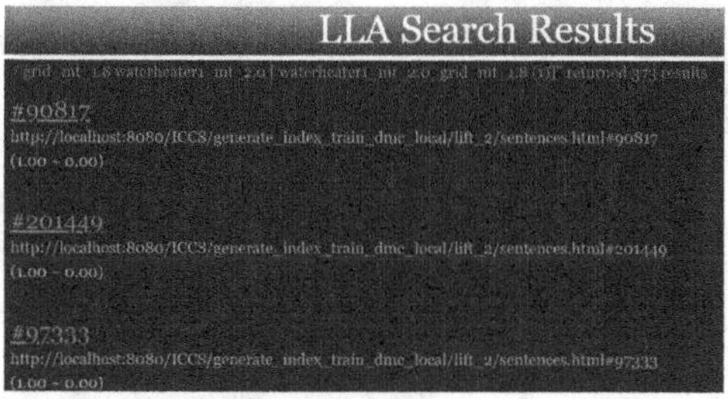

Fig. 9. Drill down

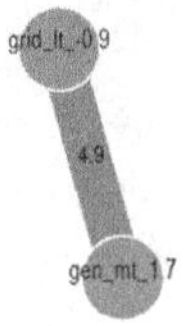

Fig. 10. Grid usage of electricity is less than −0.9: negative, giving back to the grid

causes, e.g., specific users or time points. The key factors for causal learning includes the three layers of a causal hierarchy [23,24] - association, intervention and counterfactuals (Fig. 10).

The common consensus is that data-driven analysis or data mining can discover initial statistical correlations and associations from big data. Human analysts need to validate and understand if the associations make sense and what are the real causes and effects.

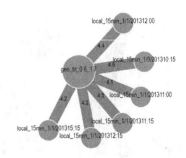

Fig. 11. Local time causal relations

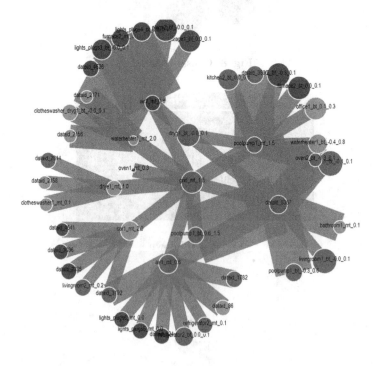

Fig. 12. Causal associations level 2

In a real-life application, one often wants to predict causes based on the data of effects, i.e., computing and validating the probability (P) of a potential cause (C) given an effect (E), i.e., $P(C|E)$. Effects are often observable data, e.g., $grid\_mt\_1.8$ in the Pecan Street data set. Causes, e.g., specific users and time points need to be discovered from the observable data. $P(C|E)$ is difficult to discover because causes are often hidden, anomalous, and capricious. In machine learning practices, the associations, correlations or probabilistic rules are typically cross-validated using separate or new data sets. Causal learning requires the intervention and counterfactual reasoning. An intervention reasoning tries to answer the question: What will happen if one takes an action? For example, instead of examining $P(C|E)$, if E is actionable or $P(C|do(E))$ [23] can be examined. The intervention more than just mining the existing data.

Counterfactual reasoning tries to answer the question: What if I had acted differently? If $P(C|E)$ is high-probability rule discovered from data, $P(C|Not\ E)$, $P(Not\ C|E)$, and $P(Not\ C|Not\ E)$ are the counterfactuals needed in the reasoning. Traditionally, the counterfactual is defined as the effect of an action for an entity and for the same entity without the action.

LLA calculates the lift measure that is one of the counterfactual reasoning in causal learning [22].

In the Pecan Street data set, although the linked features as shown in Fig. 8 make sense to human analysts, the specific user or data id or time points might be more detailed causes for energy management and cybersecurity. In Fig. 12, each cause feature nodes can be expanded to another level to reveal more causes such as more data ids (users) linked to the first level causes as shown in Fig. 8.

In LLA, when $lift_{E,C_i} > 1$, C_i is a potential cause for E. However, if another cause C_j is a confounder of C_i and E, then $lift_{C_j,C_i} > 1$. So if $lift_{C_j,C_i} > 1$ for some C_j and $lift_{C_j,E} > 1$, then C_j not C_i is the cause of E. In Fig. 12, only $dataid\_5357$ directly links to $grid\_mt\_1.8$ and the first level features $poolpump1\_mt\_1.5$, $poolpump1\_bt\_0.6\_1.5$, $air1\_mt\_0.6$, and $air12\_mt\_0.6$. $dataid\_5357$ is a real cause and we can eliminate $poolpump1\_mt\_1.5$, $poolpump1\_bt\_0.6\_1.5$, $air1\_mt\_0.6$, and $air12\_mt\_0.6$. Other causes $waterheater1\_mt\_2.0$, $drye1\_mt\_1.0$, $car1\_mt\_2.0$, and $oven1\_mt\_0.3$ are independent causes with no confounders.

6   Conclusion

We demonstrated that BDP and deep analytics using the Pecan Street data for anomaly detection and causality analysis of resource management, cybersecurity, and energy network situation awareness. We also demonstrated unsupervised learning algorithms to discover the usage patterns and anomalies. We also defined an anomaly index and showed its values for the clusters and time points. We showed LLA as an innovative approach to discover causal associations. The information can help business users to see the patterns and detect abnormal activities for the management and cybersecurity of an energy grid.

Acknowledgement. Thanks to the support of the Office of Naval Research for supporting the research, Quantum Intelligence, Inc. for support and collaboration, and the Pecan Street project for providing the data set. The views and conclusions are those of the authors and should not be interpreted, expressed or implied of the U.S. Government.

References

1. Zhang, Y., Huang, T., Bompard, E.F.: Big data analytics in smart grids: a review. Energy Inform. **1**(1), 1–24 (2018). https://doi.org/10.1186/s42162-018-0007-5
2. Rezaul, A.M., Muttaqi, K.M., Bouzerdoum, A.: Evaluating the effectiveness of a machine learning approach based on response time and reliability for islanding detection of distributed generation. IET Renew. Power Gener. **11**(11), 1392–1400 (2017)
3. Zico, K.J., Batra, S., Ng, A.Y.: Energy disaggregation via discriminative sparse coding. In: Proceedings of the 23rd International Conference on Neural Information Processing Systems, vol. 1, pp. 1153–1161. Vancouver (2010)
4. De Santis, E., Rizzi, A., Sadeghian, A.: A learning intelligent system for classification and characterization of localized faults in smart grids. In: 2017 IEEE Congress on Evolutionary Computation (CEC), San Sebastian, 5–8 June 2017
5. Wang, X., McArthur, S., Strachan, S., Kirkwood, J., Paisley, B.: A data analytic approach fault diagnosis and prognosis for distribution automation. IEEE Trans. Smart Grid **9**(6), 6265–6573 (2017)
6. Mishra, D.P., Samantaray, S.R., Joos, G.: A combined wavelet and data-mining based intelligent protection scheme for microgrid. IEEE Trans. Smart Grid **7**(5), 2295–2304 (2016)
7. Wang, J., Xiaofu, X., Zhou, N., Li, Z., Wang, W.: Early warning method for transmission line galloping based on SVM and AdaBoost bi-level classifiers. IET Gener. Transm. Distrib. **10**(14), 3499–3507 (2016)
8. Zhang, Y., Yan, X., Dong, Z.Y., Zhao, X., Wong, K.P.: Intelligent early warning of power system dynamic insecurity risk toward optimal accuracy-earliness tradeoff. IEEE Trans. Ind. Inform. **13**(5), 2544–2554 (2017)
9. Bauman, K., Tuzhilin, A., Zaczynski, R.: Using social sensors for detecting emergency events: a case of power outages in the electrical utility industry. ACM Trans. Manage. Inf. Syst. **8**, 2–3 (2017)
10. Sun, H., Wang, Z., Wang, J., Huang, Z., Carrington, N.L., Liao, J.: Data-driven power outage detection by social sensors. IEEE Trans. Smart Grid **7**(5), 2516–2524 (2017)
11. Singh, S., Yassine, A.: Mining energy consumption behavior patterns for households in smart grid. IEEE Trans. Emerg. Top. Comput. **7**(3), 404–419 (2017)
12. Pecan: https://www.pecanstreet.org/. Accessed 13 Apr 2020
13. Zhao, Y., Gallup, S.P., MacKinnon, D.J.: System self-awareness and related methods for improving the use and understanding of data within DoD. Softw. Qual. Prof. **13**(4), 19–31 (2011). http://asq.org/pub/sqp/
14. Zhao, Y., Mackinnon, D.J., Gallup, S.P.: Big data and deep learning for understanding DoD data. J. Defense Softw. Eng. (2015). Special Issue: Data Mining and Metrics, July/August 2015, pp. 4–10. Lumin Publishing, ISSN 2160–1577. http://www.crosstalkonline.org/storage/flipbooks/2015/201507/index.html
15. Zhang, Z.: Smart Grid in America and Europe: Similar Desires. Different Approaches. Public Utilities Fortnightly, vol. 149, p. 1 (2011)

16. Vianna, G.: Vulnerabilities in the North American Power Grid: Global Security Studies, Fall 2016, vol. 7, no. 4 (2016)
17. Chase, N.: Autonomous Vehicles: Uncertainties and Energy implications: 2018 U.S. Energy Information Administration Independent Statistics & Analysis (2018). https://www.eia.gov/conference/2018/pdf/presentations/nicholas_chase.pdf
18. Associated Press: The Latest: Pressure Sensors Focus of Gas Explosions Probe, 16 September 2018. https://www.nytimes.com/aponline/2018/09/16/us/ap-us-gas-explosions-the-latest.html
19. BDP: https://www.disa.mil/newsandevents/2016/Big-Data-Platform. Accessed 13 Apr 2020
20. IoT: http://smartgrid.epri.com. Accessed 13 Apr 2020
21. US patent 8,903,756: System and method for knowledge pattern search from networked agents (2014). https://www.google.com/patents/US8903756
22. Zhao Y., MacKinnon, D., Jones, J.: Causal learning using pair-wise associations to discover supply chain vulnerability. In: the Proceedings of the 11th International Conference on Knowledge Discovery and Information Retrieval (KDIR 2019), 17–19 September 2019, Vienna, Austria (2019). https://www.insticc.org/Primoris/Resources/PaperPdf.ashx?idPaper=80705
23. Mackenzie, D., Pearl, J.: The Book of Why: The New Science of Cause and Effect. Penguin, New York (2018)
24. Pearl, J.: The Seven Pillars of Causal Reasoning with Reflections on Machine Learning (2018). http://ftp.cs.ucla.edu/pub/stat_ser/r481.pdf
25. Apache Spark: https://spark.apache.org/. Accessed 13 Apr 2020
26. Apache Storm: http://storm.apache.org/. Accessed 13 Apr 2020
27. Kibana: https://www.elastic.co/products/kibana. Accessed 13 Apr 2020
28. NodJS: https://nodejs.org/en/. Accessed 13 Apr 2020
29. R-shiny: https://shiny.rstudio.com/. Accessed 13 Apr 2020

Regression Methods for Detecting Anomalies in Flue Gas Desulphurization Installations in Coal-Fired Power Plants Based on Sensor Data

Marek Moleda[1,2]([✉]) [iD], Alina Momot[2] [iD], and Dariusz Mrozek[2] [iD]

[1] TAURON Wytwarzanie S.A., Jaworzno, Poland
marek.moleda@gmail.com
[2] Silesian University of Technology, ul. Akademicka 16, 44-100 Gliwice, Poland
dariusz.mrozek@polsl.pl

Abstract. In the industrial world, the Internet of Things produces an enormous amount of data that we can use as a source for machine learning algorithms to optimize the production process. One area of application of this kind of advanced analytics is Predictive Maintenance, which involves early detection of faults based on existing metering. In this paper, we present the concept of a portable solution for a real-time condition monitoring system allowing for early detection of failures based on sensor data retrieved from SCADA systems. Although the data processed in systems, such as SCADA, are not initially intended for purposes other than controlling the production process, new technologies on the edge of big data and IoT remove these limitations and provide new possibilities of using advanced analytics. This paper shows how regression-based techniques can be adapted to fault detection based on actual process data from the oxygenating compressors in the flue gas desulphurization installation in a coal-fired power plant.

Keywords: Predictive maintenance · Power plant · SCADA · Anomaly detection

1 Introduction

Operational Technology (OT) systems refer to hardware and software solutions that are capable of changing various industrial processes through the direct monitoring and controlling of physical devices, procedures, routes, and events in a

This work was supported by the Polish Ministry of Science and Higher Education as part of the Implementation Doctorate program at the Silesian University of Technology, Gliwice, Poland (contract No 0053/DW/2018), and partially, by the professorship grant (02/020/RGPL9/0184) of the Rector of the Silesian University of Technology, Gliwice, Poland, and partially, by Statutory Research funds of Department of Applied Informatics, Silesian University of Technology, Gliwice, Poland (grant No BK/SUBB/RAu7/2020 and grant No BKM/SUBB-MN/RAu7/2020).

© Springer Nature Switzerland AG 2020
V. V. Krzhizhanovskaya et al. (Eds.): ICCS 2020, LNCS 12141, pp. 316–329, 2020.
https://doi.org/10.1007/978-3-030-50426-7_24

factory. Internet of Things (IoT) and integration of IT and OT systems provide new opportunities for intelligent use of advanced analytical solutions to increase the reliability of various production processes, including those performed in power plants and the entire energy sector [8,14]. A significant risk factor in the energy sector is unplanned downtime, which causes production breaks and, thus, loss of revenue. Increasing the reliability of equipment operation is a critical factor in reducing the losses mentioned above, and one of the methods to achieve this goal is the use of predictive maintenance. Predictive maintenance reduces planned and unplanned downtime by early detection of faults and better planning of repairing actions. In this work, we present how the regression-based techniques can be used for real-time condition monitoring on the basis of sensor data retrieved from SCADA (supervisory control and data acquisition) systems [5] and failure logs. The research objective of this work is to adjust the data mining process to achieve satisfactory results with a minimum effort from analysts and system engineers. In order to prove the portability of the presented solution, we use the algorithm previously applied for the feed pumps [13] for the analysis of data coming from oxygenating compressors. We optimize the steps of the algorithm to achieve satisfactory prediction results. Experiments are carried out on real data from sensors monitoring the oxygenating compressor, which is a component of the flue gas desulphurization installation in a coal-fired power plant. Oxygenating compressors are indispensable in the process of producing gypsum from sulfur dioxide as a side effect of coal combustion. We present a model-based anomaly detection approach with the potential to be easily applied to other devices. We also show that general-purpose regression-based heuristic algorithms can enhance the benefits of the processing of data from OT repository, various data streams, and related big data sets, including various logs.

2 Background and Related Works

Predictive maintenance is applied in various branches of industry, especially in those of them, in which failures may cause considerable economic consequences.

2.1 Predictive Maintenance

Predictive maintenance is a concept that involves planning maintenance work based on the health of the equipment. This approach is possible when we have complete data on the operation of the device and expert knowledge on how to use it for analysis [1,4,22]. The task consists of the analysis of archival events, diagnostic results, and device statistics by the production engineer or production data analyst. OT systems, including SCADA repositories, can be used as sources of data for predictive maintenance. Then, we can codify the expert knowledge in the form of rules and classifiers to get essential indicators, monitor devices, and discover anomalies, as it is presented in [7,16–18]. In this way, we can create a

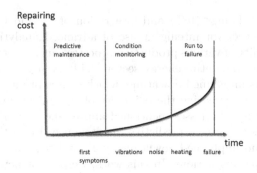

Fig. 1. Repairing costs depending on the time of detection of the fault.

predictive maintenance system based on real-time condition monitoring. Predictive maintenance is a reasonable compromise between *preventive maintenance* [3,12] and *run to failure* maintenance. The primary purpose of the methods used in predictive maintenance is the early detection of faults [15,23,26] based on the first symptoms to minimize the repairing costs (as shown in Fig. 1). Other potential benefits are:

- Changing the operational models to *Fix as required*, in which the service is done before failure is expected.
- Minimizing the probability of unplanned downtime by real-time health monitoring.
- Reducing unnecessary repairs of equipment being in good condition.

2.2   Model-Based Prediction

The concept of model-based prediction is to create a black-box model [23] that represents a specific process, as it is shown in Fig. 2.

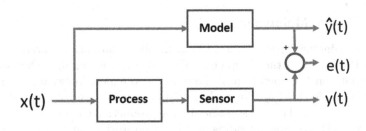

Fig. 2. Model-based prediction system.

The model created for the component (in this case it is a measurement) is designed to predict the indication of sensor $y(t)$ based on known input $x(t)$. The difference between the actual value $y(t)$ and the predicted $\hat{y}(t)$ is a measure of the process deviation from the model value. The deviation from the expected value is assumed to be a symptom of a changing health condition of the device, i.e., an imminent failure or an improvement in performance after repair. The essence of the proposed method is to create a model based on data from the failure-free period and to monitor the $e(t)$ deviation value.

2.3 Modeling Based on Regression

Machine learning (ML) is one of the techniques frequently applied in predictive maintenance performed in the industrial environment. However, the spectrum of used ML models varies and is specific for the application area. Machine learning-based algorithms generally can be divided into two main classes:

- supervised - where information on the occurrence of failures is present in the training data set;
- unsupervised - where process information is available, but no maintenance related data exists.

Supervised approaches require the availability of a data set $S = \{x_i, y_i\}_{i=1}^{n}$, where a couple $\{x_i, y_i\}$ contains the information related to the i-th process iteration. Vector $x_i \in \mathbb{R}^{1 \times p}$ contains information related to the p variables associated with available process information [21]. Depending on the type of y we distinguish: classification models (if categorical labels are predicted) and regression models (if the results are continuous values).

Supervised learning is successfully used in the area of predictive maintenance to classify faults by building fault detectors. In the literature, these detectors rely on various Artificial Intelligence (AI) techniques, such as Artificial Neural Networks [2,6], k-Nearest Neighbours [24], Support Vector Machines [10,20], or Bayesian networks [19,25], frequently using some methods for reducing dimensionality of data, such as Principle Component Analysis [11,27].

2.4 The Coding of Electrical Components

To obtain a uniform system of marking devices and installations in the power plant, the KKS coding system (Kraftwerk Kennzeichensystem) is used. The KKS marking system had been used since the early 1980s by power plant constructors and power plant operators to name and identify all components of a power plant. The code structure is shown in Fig. 3 with the example of a temperature meter.

3 Regression-Based Anomaly Detection

Our failure prediction model applies regression algorithms in predictive maintenance tasks. For the set of all input data, where $X = [x_1, x_2, x_3, \ldots, x_m]^T$ is

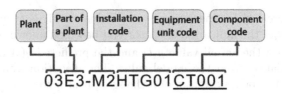

Fig. 3. Coding of temperature meter using KKS notation

the vector of individual measurements from m sensors, we estimate our response variable as a polynomial function of other variables, i.e.:

$$\hat{y}_i = \sum_{d=1}^{d_{max}} \sum_{\substack{j=1 \\ j \neq i}}^{m} a_{jd}^{(i)} x_j^d + a_0^{(i)} + \varepsilon^{(i)} \qquad \forall i \in \{1, 2, \ldots, k\}, k \leq m, \qquad (1)$$

where $\varepsilon^{(i)}$ is the i-th independent identically distributed normal error and coefficients $a_j^{(i)}$ are calculated using the method of least squares [9]. In such a way, we build a bag of predictive models for the investigated (monitored) variables (signals) and predict the values of the variables based on other signals that are correlated with the investigated one.

The difference between the observed and the predicted values is not an absolute measure that can be used when comparing the values with other signals. To normalize the results, we introduced the NRE (normalized relative error) coefficient [13] (which is a multiple of the mean standard deviation) to measure the degree of deviation for the i-th variable in a data set:

$$NRE_i = \frac{|y_i - \hat{y}_i| - MAE_i}{RMSE_i}, \qquad (2)$$

where y_i represents the observed values, $\hat{y}_i$ represents the predicted values, $RMSE_i$ is the root mean square error, and MAE_i is the mean absolute error.

By selecting a variable with the maximum value of the NRE_{max} (called *maximum normalized relative error*) we can identify the signal which is probably the cause of the upcoming fault:

$$NRE_{max} = \max(NRE_1, NRE_2, .., NRE_k). \qquad (3)$$

This makes it possible to diagnose a source of the anomaly quickly.

Withe the use of regression algorithms and the NRE metrics, we create a bag of models for detecting anomalies. In our case, we focus on creating a reproducible process that is easy to use to monitor various equipment in the power plant. In order to achieve this goal, we automate data processing tasks and reduce the necessary analytical efforts.

3.1   Description of Input Data and Correlation Between Signals

The data used in the experiment comes from monitoring two oxygenating compressors from flue gas desulphurization installations. Each of the devices is described by ten parameters, such as power, bearing temperature, vibrations, oil pressure. These parameters are measurements collected by sensors and are stored in the SCADA system. Available process data covers the period from January 2017 to May 2019, in which the compressors were in operation for about one-third of that time. Data was obtained off-line for two compressors labeled as 1HTG01 and 2HTG01. Analyzing Pearson's correlation coefficient for individual signals, we can distinguish groups of strongly correlated signals. The existence of strong correlations (i.e., values close to 1 or -1) means significant relationships between different variables and justifies the use of the regression model for anomaly prediction. A graphical matrix of correlations for a period of six months is shown in Fig. 4.

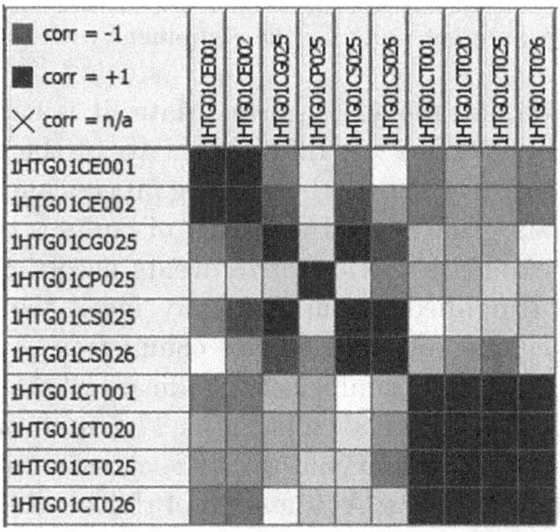

Fig. 4. Correlation matrix for the training set.

3.2 Data Preprocesing and Integration

Data collection. Data preprocessing covers the selection of a group of measurements that will be part of the input set. In the case of a single prediction task, we can easily determine the group of measurement units related to the device (e.g., using technical documentation). However, when we want to use the prediction model to cover the whole unit containing tens of thousands of measurement points, the work is no longer trivial. The coding method described in Sect. 2.4 is helpful in this case. The notation method allows for easy separation of all subcomponents associated with the master equipment, as it is shown in Fig. 5. In this way, we can significantly reduce the set of potential input signals for further analysis (e.g., correlation analysis, constant value filtering, etc.).

Automatic Data Labelling. To learn the model of how the proper signal looks like, we have to provide the data describing the state when the investigated device is healthy. To filter out the periods in which failures occurred from the input data

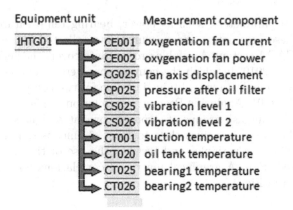

Fig. 5. Grouping measurement units for the equipment with the use of KKS code.

set and train the prediction model on correct data, it is necessary to label the records with data from a service log. Information about the occurrence of faults is needed to extract a training set for the training process and to further evaluate and visualize the results. Analysis of the impact of faults is time-consuming and requires engineering expertise to categorize events accordingly. Therefore, the labeling process was simplified to automatically mark the data set based on fault registration dates and repairing service completion. A column containing true/false data is added to the training set to indicate whether the device was in a fault condition at that time, as shown in Fig. 6. The input data is filtered based on the priority to isolate less significant events such as service works, including oil change or periodic inspections. Automation of the labeling process requires a standardized way of recording service works and good quality of data. However, possible errors can be compensated for by the size of the training set, and their impact on the quality of the algorithm can be marginalized.

3.3   Model Optimization

Optimisation step is employed in our process to adjust the parameters of the model to give the best results, and at the same time to avoid overfitting situation. The parameters we optimize are the polynomial degree and a set of input features. The optimization was verified in the k-fold cross-validation process, where we divided the input set into five equal parts. By minimizing the mean squared error value while optimizing the polynomial degree on a 3-month test set, we obtained the results shown in Table 3.3.

Polynomial Fitting. For most variables, the best prediction results were obtained for a low polynomial degree. The higher the polynomial degree, the more sensitive the changes in the input set are. On the other hand, for the higher polynomial degrees, the prediction model exposes more abnormal situations, but it operates very unstable and is susceptible to noise (Table 1).

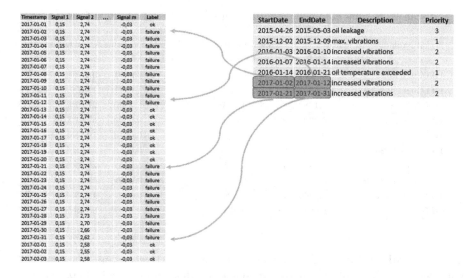

Fig. 6. Method for automatic labeling.

Table 1. Mean square error (MSE) depending on the degree of polynomial.

Signal name	polynomial degree				
	1	2	3	4	5
1HTG01CP025	**0,0001**	0,0002	0,0002	0,0002	0,0011
1HTG01CS025	2,7962	**1,2322**	1,8621	29,408	264,76
1HTG01CS026	2,6363	**2,1789**	2,9565	16,536	34,689
1HTG01CG025	**0,00001**	0,00001	0,00002	0,0001	0,0002
1HTG01CT025	**0,0087**	0,0111	0,0196	0,0286	0,1813
1HTG01CT026	**0,0067**	0,008	0,0173	0,0106	0,0281
1HTG01CT001	**12,433**	12,643	14,953	19,346	30,349
1HTG01CT020	0,0818	**0,0745**	0,0791	0,1373	1,1435
1HTG01CE001	**0,1006**	0,1084	0,1286	0,1659	3,1217
1HTG01CE002	8,0852	8,4233	**8,0673**	30,904	154,71

Feature Selection. In the next optimization step, we select the input features for the polynomial degree calculated earlier for all k signal models in a bag. By choosing the right parameters it is possible to increase the accuracy of the model by about 20% (i.e., reduce the mean squared error by 20% on average) as it is shown in Table 3.3. For this purpose, we tested two algorithms - Backward elimination and Genetic algorithm. The use of both methods gave the same results in terms of the number of features, the mean square error, and improvement in prediction accuracy. The time of the Backward elimination (29,785 combinations) compared to the Genetic algorithm (27,535 combinations) is also similar (Table 2).

Table 2. Results obtained by optimizing the number of analyzed features.

Signal name	All features		Backward elimination		Genetic algorithm		Improvement
	No. feat	MSE	No. feat	MSE	No. feat	MSE	%
1HTG01CP025	9	0,0001	2	0,0001	2	0,0001	9,76%
1HTG01CS025	9	1,2322	7	1,0971	7	1,0971	12,3%
1HTG01CS026	9	2,1789	5	1,7033	5	1,7033	27,92%
1HTG01CG025	9	0,00001	3	0,00001	3	0,00001	57,77%
1HTG01CT025	9	0,0087	7	0,0073	7	0,0073	19,15%
1HTG01CT026	9	0,0067	6	0,0053	6	0,0053	26,18%
1HTG01CT001	9	12,433	4	10,352	4	10,352	20,09%
1HTG01CT020	9	0,0745	3	0,0499	3	0,0499	49%
1HTG01CE001	9	0,1006	7	0,099	7	0,099	1,58%
1HTG01CE002	9	8,067	3	4,8747	3	4,8747	65,49%

Time Window Length. When designing a predictive system, an essential parameter of the built predictive model is a time window, which provides a training set for the model. The time window may also influence the predictive capabilities of the model. Therefore, we also investigated this parameter in our research. Visual differences in the results depending on the length of the time window are shown in Fig. 7. Longer time windows covering wide data range (3 months in our experiments) adapt more slowly to sudden changes such as renovation (in January 2019 in Fig. 7), but clearly shows a disparity between failure periods and proper operation. In the case of short time windows (1 month), the model adapts quickly to the variability of the environment, i.e., it can correct itself soon after a renovation. On the other hand, in the case of a long-term malfunction, it can treat the fault as the correct state.

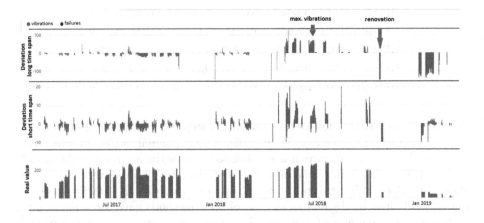

Fig. 7. Deviations from normal state obtained for long- and short time window.

4   Testing Prediction Capabilities

4.1   Evaluation Function and Error Distribution

While detecting anomalies, one of the quality factors assessing the performance of the built predictive model is to obtain a high error ratio for the failure period comparing to the normal operation period. This is achieved in our model. For example, the error distribution (deviation from the actual value) for the 2GHG01 compressor is shown in Fig. 8. The red color is used to indicate normal operation state, and the blue color is used to indicate emergency conditions (the stated period was extended by an earlier two weeks to assess the ability to detect pre-emergency conditions). While the red histogram is close to a normal distribution with an average of around 0, the blue histogram is characterized by a significant diagonality and a shift in the average value.

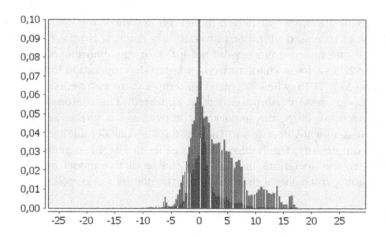

Fig. 8. Distribution of deviations for the fault and normal states. (Color figure online)

4.2   Detection Capability of the Model

Short Term Prediction. One of the failures, which was recorded on 23 February for 1GTG01 compressor, was bearing damage. The damage occurred at a time when increased vibration levels were recorded, resulting in a significant overhaul. However, before the failure itself, we can notice a significant increase in the deviation of the estimated values. About an hour before the device was switched off on 20 February, a considerable increase in the deviation is visible (more than 100% of the average value in normal operation). The physical effect of the failure observed had a reflection on the actual pressure drop behind the oil filter. The fault occurred three days later on the compressor startup, as it is visible in Fig. 9.

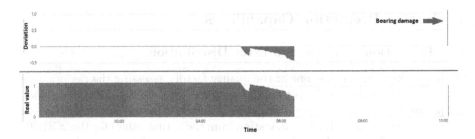

Fig. 9. Deviation and actual oil pressure behind the filter before bearing failure.

Long Term Prediction. During the period considered, several significant events were recorded in the fault logs. For the compressor 1HTG01, the bearing was damaged in February 2017, followed by reports of loud operation and bearing vibrations, as a result of which the device was overhauled in April 2017. After the renovation, the prediction model did not show any deviations despite the reports of increased vibrations around July-August. However, the analysis of actual vibration measurements did not indicate any abnormal operation. The compressor 2HTG01 was characterized by failure-free operation for an extended period until May 2018, when increased oil temperature was recorded, and in the following weeks, several vibration defects appeared. The compressor was overhauled in October 2018, the impact of the overhaul is visible in the chart in Fig. 10 as a significant decrease in the error (to a considerable negative value). For both compressors, the failure points visible in Fig. 10 as registered events coincide with the deviations indicated by the predictive model for the investigated variable, which shows that the predictive model works well.

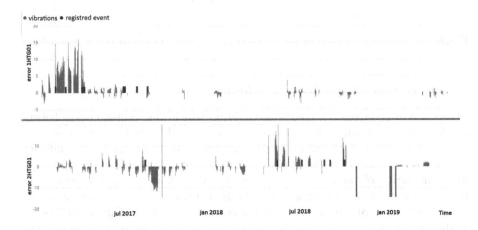

Fig. 10. Vibration deviation with the indication of faults and service actions.

4.3   Comparison of Model Results

The results obtained (the values of NRE) were calculated by dividing the deviation value by the local root mean square error (equivalent to standard deviation) as in Eq. 2, and then the values for which the standard deviation exceeded the threshold value 6 (determined experimentally) for less than 15 min were filtered out. That means that the alarm for the predicted incoming fault is raised only when the $NRE > 6$ for at least 15 min. The same threshold was used for the condition monitoring of boiler feed pumps described in [13], bringing good quality prediction results. In Table 3, we show the effectiveness of the predictive models built for compressors 1HTG01 and 2HTG02, and for boiler feed pumps investigated in our previous work [13]. The results confirm that the presented approach to anomaly detection through regression-based predictive analytics achieved a good level of specificity, which translates into a small number of false positives (type I errors). Sensitivity can be improved by developing an evaluation method or improving optimization steps (Fig. 11).

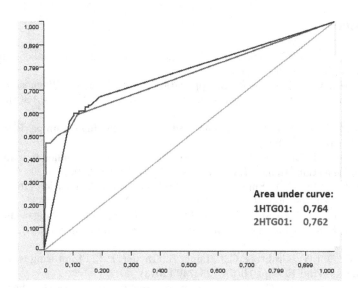

Fig. 11. ROC curves for anomaly detection predictive models built for compressors 1HTG01 and 2HTG02.

Table 3. Performance of the predictive models built for various devices in power plant.

	Accuracy	AUC	Sensitivity	Specificity
Boiler feed pump	0,86	0,89	0,67	0,95
Compressor 1HTG01	0,79	0,764	0,58	0,91
Compressor 2HTG01	0,93	0,762	0,47	0,98

5   Discussion and Conclusions

The obtained results confirm good portability of the presented approach, which after some adjustments, can be used for detecting anomalies and early fault prediction for various types of equipment located in power plants. The regression-based algorithm, previously used to predict faults in the boiler feed pumps, was successfully applied to the oxygenating compressors in the flue gas desulphurization installation. The results obtained for different devices are comparable. The fact that they were achieved using simple data mining methods and little effort encourage further works on improving model quality. Comparing to our previous work, we proposed methods of automation of feature selection, data collection, and grouping based on specific coding of devices and installations in the power industry. The unquestionable advantage of the algorithm is the absence of data labeling compared to most classification-based methods mentioned in the related literature. Data labeling in the current work is not a necessary element and was done only to assess the quality of the model.

References

1. Afefy, I.H.: Reliability-centered maintenance methodology and application: a case study. Engineering **2**(11), 863 (2010)
2. Awadallah, M.A., Morcos, M.M.: Application of ai tools in fault diagnosis of electrical machines and drives-an overview. IEEE Trans. Energy Convers. **18**(2), 245–251 (2003)
3. Bachus, L., Custodio, A.: Know and Understand Centrifugal Pumps. Elsevier Science, Amsterdam (2003)
4. Bevilacqua, M., Braglia, M.: The analytic hierarchy process applied to maintenance strategy selection. Reliab. Eng. Syst. Saf. **70**(1), 71–83 (2000)
5. Boyer, S.A.: SCADA: Supervisory Control and Data Acquisition. International Society of Automation, Research Triangle Park (2009)
6. Cococcioni, M., Lazzerini, B., Volpi, S.L.: Robust diagnosis of rolling element bearings based on classification techniques. IEEE Trans. Ind. Inform. **9**(4), 2256–2263 (2012)
7. Feng, Y., Qiu, Y., Crabtree, C.J., Long, H., Tavner, P.J.: Monitoring wind turbine gearboxes. Wind Energy **16**(5), 728–740 (2013)
8. Fortino, G., Savaglio, C., Zhou, M.: Toward opportunistic services for the industrial Internet of Things. In: 2017 13th IEEE Conference on Automation Science and Engineering (CASE), pp. 825–830. IEEE (2017)
9. Freedman, D.A.: Statistical Models: Theory and Practice, Revised edn. Cambridge University Press, Cambridge (2009)
10. Konar, P., Chattopadhyay, P.: Bearing fault detection of induction motor using wavelet and support vector machines (SVMs). Appl. Soft Comput. **11**(6), 4203–4211 (2011)
11. Malhi, A., Gao, R.X.: PCA-based feature selection scheme for machine defect classification. IEEE Trans. Instrum. Measur. **53**(6), 1517–1525 (2004)
12. Marscher, W.D., et al.: Avoiding failures in centrifugal pumps. In: Proceedings of the 19th International Pump Users Symposium. Texas A&M University. Turbomachinery Laboratories (2002)

13. Moleda, M., Momot, A., Mrozek, D.: Predictive maintenance of boiler feed water pumps using SCADA data. Sensors **20**(2), 571 (2020)
14. Moleda, M., Mrozek, D.: Big data in power generation. In: Kozielski, S., Mrozek, D., Kasprowski, P., Małysiak-Mrozek, B., Kostrzewa, D. (eds.) BDAS 2019. CCIS, vol. 1018, pp. 15–29. Springer, Cham (2019). https://doi.org/10.1007/978-3-030-19093-4_2
15. Nouri, M., Fussell, B.K., Ziniti, B.L., Linder, E.: Real-time tool wear monitoring in milling using a cutting condition independent method. Int. J. Mach. Tools Manuf. **89**, 1–13 (2015)
16. Qiu, Y., Chen, L., Feng, Y., Xu, Y.: An approach of quantifying gear fatigue life for wind turbine gearboxes using supervisory control and data acquisition data. Energies **10**(8), 1084 (2017)
17. Qiu, Y., Feng, Y., Sun, J., Zhang, W., Infield, D.: Applying thermophysics for wind turbine drivetrain fault diagnosis using SCADA data. IET Renew. Power Gener. **10**(5), 661–668 (2016)
18. Qiu, Y., Feng, Y., Tavner, P., Richardson, P., Erdos, G., Chen, B.: Wind turbine SCADA alarm analysis for improving reliability. Wind Energy **15**(8), 951–966 (2012)
19. Qu, J., Zhang, Z., Gong, T.: A novel intelligent method for mechanical fault diagnosis based on dual-tree complex wavelet packet transform and multiple classifier fusion. Neurocomputing **171**, 837–853 (2016)
20. Rojas, A., Nandi, A.K.: Detection and classification of rolling-element bearing faults using support vector machines. In: 2005 IEEE Workshop on Machine Learning for Signal Processing, pp. 153–158. IEEE (2005)
21. Susto, G.A., Schirru, A., Pampuri, S., McLoone, S., Beghi, A.: Machine learning for predictive maintenance: a multiple classifier approach. IEEE Trans. Ind. Inform. **11**(3), 812–820 (2014)
22. Swanson, L.: Linking maintenance strategies to performance. Int. J. Prod. Econ. **70**(3), 237–244 (2001)
23. Tautz-Weinert, J., Watson, S.J.: Using SCADA data for wind turbine condition monitoring-a review. IET Renew. Power Gener. **11**(4), 382–394 (2016)
24. Tian, J., Morillo, C., Azarian, M.H., Pecht, M.: Motor bearing fault detection using spectral kurtosis-based feature extraction coupled with k-nearest neighbor distance analysis. IEEE Trans. Ind. Electron. **63**(3), 1793–1803 (2015)
25. Wu, J., Wu, C., Cao, S., Or, S.W., Deng, C., Shao, X.: Degradation data-driven time-to-failure prognostics approach for rolling element bearings in electrical machines. IEEE Trans. Ind. Electron. **66**(1), 529–539 (2018)
26. Yang, C., Liu, J., Zeng, Y., Xie, G.: Real-time condition monitoring and fault detection of components based on machine-learning reconstruction model. Renew. Energy **133**, 433–441 (2019)
27. Zhang, X., Xu, R., Kwan, C., Liang, S.Y., Xie, Q., Haynes, L.: An integrated approach to bearing fault diagnostics and prognostics. In: Proceedings of the 2005, American Control Conference, 2005, pp. 2750–2755. IEEE (2005)

Autonomous Guided Vehicles for Smart Industries – The State-of-the-Art and Research Challenges

Rafal Cupek[1(✉)], Marek Drewniak[2], Marcin Fojcik[3], Erik Kyrkjebø[3],
Jerry Chun-Wei Lin[3], Dariusz Mrozek[1], Knut Øvsthus[3],
and Adam Ziebinski[1]

[1] Silesian University of Technology, Gliwice, Poland
{rcupek,dariusz.mrozek,aziebinski}@polsl.pl
[2] AIUT Sp. z o.o. (Ltd.), Gliwice, Poland
mdrewniak@aiut.com.pl
[3] Western Norway University of Applied Sciences, Bergen, Norway
{Marcin.Fojcik,erik.kyrkjebo,knut.ovsthus}@hvl.no,
jerrylin@ieee.org

Abstract. Autonomous Guided Vehicles (AGVs) are considered to be one of the critical enabling technologies for smart manufacturing. This paper focus on the application of AGVs in new generations of manufacturing systems including: (i) the fusion between AGVs and collaborative robots; (ii) the application of machine to machine communication for integrating AGVs with the production environment and (iii) AI-driven analytics that is focused on the data that is produced and consumed by AGV. This work aims to evoke discussion and elucidate the current research opportunities, highlight the relationship between different subareas and suggest possible courses of action.

Keywords: Autonomous Guided Vehicles (AGV) · Collaborative robotics · Machine to Machine (M2M) communication · AI-driven analytics

1 Introduction

The growing popularity of Autonomous Guided Vehicles (AGV), which are used in manufacturing, has not only been the result of their technical features but also from their ability to cooperate. Cooperative-based internal logistics permits increased production flexibility. Because AGV are in the executive part of the internal logistics, their cooperation with other information systems and manufacturing equipment is particularly important. AGV have become a critical enabling technology for agile production systems. Modern production systems are characterised by a demand for a high degree of flexibility in order to cope with the frequent changes that result from orders that are changed by customers, low material buffers, the agile production technologies that are performed by robotised production stations and the many variants of production technology that can be used [1]. All of the factors mentioned above require the production process to be supported online by highly advanced information services, which are

V. V. Krzhizhanovskaya et al. (Eds.): ICCS 2020, LNCS 12141, pp. 330–343, 2020.
https://doi.org/10.1007/978-3-030-50426-7_25

performed during the successive steps in the production chain. This means that the production activities cannot be centrally planned but have to be optimised locally in order to handle the ongoing production tasks, the available materials, the production equipment and the technologies. The new generation of manufacturing execution systems has to support the autonomy and distribution of the decision-making processes [2].

The increasing use of AGV in manufacturing has far-reaching consequences for the industrial communication systems. Unlike for other machines, the use of wired networks is not possible in the case of AGV, which have to move through large internal and external areas. On the other hand, production support algorithms based on machine learning require huge volume of data provided by AGV. In this way, a communication with AGV, becomes convergent rather with IoT than with the classical industrial communication. In this context, artificial intelligence that is based on supervised machine learning can be used to optimise the internal logistical tasks that are performed by AGV. This requires a multi-criteria optimisation that takes into account the multiple goals that are connected with the requirements to move materials and semi-products. This process must be performed in a distributed, dynamic and autonomous way [3]. Moreover, AGV also have to communicate with the production stands, Manufacturing Execution Systems (MES) and other AGV. For this reason, the machine learning approach has to be combined with Machine-to-Machine (M2M) communication. This requires, on the one hand, adjusting the information that is collected from the production systems into a form that is suitable for automatic processing, which can be performed by using artificial intelligence algorithms. The results that are produced by these algorithms have to be converted into a form that can be exchanged with and used by the production systems [4]. In turn, the precise information that is owned by AGV includes important parameters about the materials, semi-finished products and finished products that are being transported. Such data can be beneficial for the Manufacturing Execution System (MES) and for the Business Intelligence (BI) solutions that support the long-term optimisation of the production processes.

The goal of this paper is to summarise the existing research results and to show the most significant challenges related to the effective use of AGV, which are considered to be a part of a new generation of the manufacturing ecosystem. The paper is organised as follows: the second section presents the research challenges related to the fusion between AGV and collaborative robots. The research issues associated with the application of machine to machine communication for integrating AGV with the production environment are presented in section three. The fourth section summarises the state of the art in AI-driven analytics that are focused on the data that is produced and consumed by AGV. The conclusions are presented in Sect. 5.

2 Research Challenges That Are Related to the Fusion Between AGV and Collaborative Robots

In small series manufacturing, the production tools have to be adjusted to specific products and the process organisation must follow these changes in order to avoid or reduce any losses that would result from non-productive time gaps [5]. The logistics tasks must be performed in a distributed, dynamic and autonomous way. AGV are key components that are necessary for developing flexible and efficient internal transport systems [6].

Fig. 1. Autonomous Guided Vehicle produced by AIUT Ltd.

Traditional industrial robot systems usually perform dedicated tasks on specific assembly stations and are often isolated from other stations and human operators using physical barriers such as fences. Introducing safe, collaborative robots [7] into production systems (that also work together with human operators) opens up the possibility of using robots for a wider variety of tasks and permits the physical barriers to the rest of the production processes to be removed. Such an option enables robot automation to be introduced more widely into factories – but these robots are still usually dedicated to specific assembly stations. Combining collaborative robots [8, 9] with AGV (Fig. 1) merges the dynamic logistic benefits of AGV with the flexibility of collaborative robots to enable a more extensive use between many assembly stations. This could also increase the efficiency of the production staff. However, the autonomous operation of AGV requires that several issues have to be resolved. The integration of an AGV, a collaborative robot and various sensors requires that data fusion methods [10] be used to achieve a docking functionality and to recalibrate the AGV and robot to that particular assembly station.

Virtual sensing [11], which is based on computational models and is widely used to optimise an operation or product quality in industry, is a non-invasive method that is used to measure the parameters in dynamic systems. Monitoring an industrial process using data fusion and virtual sensing techniques [12] supports developing methods that permit the changes in the area of production lines to be detected. The use of fusion techniques allows more information about the state of a system to be obtained from several sensors. Multi-sensor data fusion [7] permits the working status of a process and machinery to be acquired by integrating sensors into manufacturing systems. The data fusion from AGV, robots, assembly stations and production monitoring system using virtual sensing can enable new methods and algorithms to be created in order to optimise the short series production process in industry.

The first main challenge is to obtain the fusion of the data between AGV and the collaborative robots, which could be achieved by developing a distributed computer system architecture for integrating AGV [13], collaborative robots [9] and the required sensors [14]. A navigation system [15] supports the movements of AGV between assembly stations and helps to position an AGV for docking to an assembly station.

Using distance sensors enables the docking precision to be increased. To determine the accuracy of the measurements [16, 17], the selected sensors must be tested (camera, lidar, gyroscope, accelerometer, optical encoders). A vision-based recognition system further confirms that an AGV has arrived at the correct production station. Using methods that are dedicated to sensor fusion allows algorithms to be developed in order to increase the accuracy of the distance measurements [18]. The proposed methods also enable algorithms to be developed that allow precise docking to the assembly station.

Distance sensors (e.g. inclinometers, cameras, including Time-of-Flight (ToF) cameras, optical encoders) [14, 19, 20] can also be used to increase the re-calibration of a collaborative robot to the mounting site on the assembly station. Using recognition methods based on a vision system increases the precision of the measurements. This approach allows the information covered by distance sensors to be combined with the vision system [18] to develop an algorithm that increases the accuracy of the distance measurements. The algorithms that are obtained allow the robot's displacement regarding the reference coordinate system at the mounting place at the assembly station to be calculated. In addition, robot-based calibration can be used to set the reference points for each axis. The displacement of the robot arm to the reference points at the assembly station and the use of precise force measurement on selected axes enables a more accurate determination of the current position of a collaborative robot.

The next main challenge is to develop methods to support the cooperation between a collaborative robot and the production staff. One of the major challenges with physical Human-Robot Interaction (pHRI) is how to handle the possible or perceived risk of human-robot collisions [8]. Human motion and intention estimates that based on sensors [9] can ensure that the human states are continuously monitored and the human motion data together with the force/torque (FT) measurements from a collaborative robot can serve as inputs to a collision-handling (CH) system. A CH system can continuously monitor and assess the risk of unwanted interactions and collisions and implement the necessary reaction mechanisms, e.g. lower the speed of the robot, switch to a gravity compensation and compliance mode or to adhere to an admittance reflex strategy to drive away from the unwanted contact force [8]. A CH system should be implemented at the lowest control level in order to ensure that handling collisions takes priority over task execution. A collaborative robot control scheme for pHRI for robots that are mounted on AGV allows an overall system architecture to be built that takes into account the changes in roles between a human and robot in the interaction (leader-follower, collaboration) [21, 22] and how to dynamically share the tasks in any human-robot cooperative load transport [23]. The overall system architecture should also allow the seamless, shared cooperative pHRI control for a wide range of possible applications.

Although the approaches to cooperative pHRI control can be independent of any predefined trajectories or paths for the robot [24], the available information on the task goal should be exploited in the overall pHRI control scheme. This can be accomplished by dividing tasks into sub-activities that can serve as feedforward reference inputs to the control system and that can try to make the most of the flexibility, knowledge and sensory skills of a human and the efficiency, strength, endurance and accuracy of a robot. Information on the task goal can also serve as an essential parameter for when a human or a robot should take the lead in a cooperative pHRI operation [22] and will be dependent on the particular pHRI application.

The overall physical interaction between humans and robots is a multi-dimensional dynamic control problem that is hard to define explicitly using rigid state-space models that have fixed parameters. Therefore, learning strategies for pHRI are now emerging as an essential research challenge and many learning and adaptation approaches are currently being proposed. They are primarily focused on two aspects of pHRI – adaptive impedance control, and learning the desired trajectories [8]. Adaptive impedance controllers are focused on learning or adapting to the impedance gains and feedforward torques of the controller to improve the robot capabilities during an interaction. Some recent results have used neural networks or biomimetic approaches to learning [25]. Learning the desired trajectories for a pHRI has, in recent years, been primarily focused on using hidden Markov models (HMM) to find the appropriate motion patterns for human-robot cooperation. A novel approach for a mobile bi-manual platform is taken in [26], where semantic task structures are learned during a joint task execution and where the semantic labels on the task segments are given by the human partner using speech recognition.

3   Machine to Machine Communication for Integrating AGV with the Production Environment

Machine to Machine (M2M) communication in the new generation of manufacturing systems can be considered on three levels: (i) the low-level communication protocols that support a reliable and time determined exchange of information between devices and between a device and production system [27, 28] – in the case of AGV, low-level wireless communication protocols must take into account the movement of an AGV, which might result in a high degree of electromagnetic interference and also the need to maintain communication despite its moving between different network segments; (ii) the communication middleware, which based on the lower-level networks that support independent logical communication channels between an AGV and any cooperating devices or systems including syntax based data format conversion and information exchange management such as MQTT, CoAP [29], OMADM, LwM2M, XMPP [30] and OPC UA [31] – in the case of AGV, the communication middleware must support a high degree of flexibility, automatic reconfiguration and different types of information modes including cyclic real-time communication with control systems, event-based communication between the control subsystems and also batch communication with the management systems that are focused on aggregated data exchange and (iii) the ontology that describes the information models [4, 32] – in the case of AGV, the ontology should facilitate and ensure unambiguous information exchange with other AGV, different production stations as well as with the production support systems such as an MES (Manufacturing Execution System). Below, we illustrate the scope of the M2M issues starting with the ontology domain and ending with wireless communication.

The ontology that is used for AGV has to correspond with the information models that are used in the new generation of manufacturing systems. The ongoing changes that are called the Fourth Industrial Revolution are the primary interest of manufacturing companies, production technologies suppliers and the advisory groups that are supported by government agencies. The latter is attempting to structure the changes to harmonise the introduction of modern ICT technologies in manufacturing on a global scale. Examples of such global organisations are the German Platform Industrie 4.0; the NIST (National Institute of Standards and Technology), which is similar to the US Department of Commerce or the SMLC (Smart Manufacturing Leadership Coalition), which operates in South Korea [33]. All these solutions are characterized by increased autonomy and a departure from the centralized hierarchical system to the horizontal cooperative model. Since, detailed analysis is beyond the scope of this work, the authors discuss only one selected features of one, but commonly recognized approach the Reference Architecture Model for Industry 4.0 RAMI4.0 [34].

The vertical axis of RAMI4.0 (Fig. 2) organises the system structure and the functions of the individual layers, which are from the bottom up (i) the Asset layer, which represents reality, i.e. the asset that exists in the physical world; (ii) the Integration layer, which represents the transition from the physical world to the information world; (iii) the Communication layer, which describes the Industry 4.0-compliant access to the information and functions of a connected asset by other assets; (iv) the Information layer, which describes the data that is used, generated or modified by the technical functionality of an asset; (v) the Functional layer, which describes the (logical) functions of an asset (technical functionality) connected with/associated with its role in the Industry 4.0 system and (vi) the Business layer, which describes the commercial view including the general organisational boundaries.

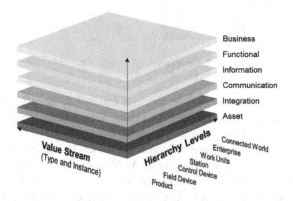

Fig. 2. A Reference Architecture Model for Industry 4.0 (RAMI 4.0) [34]

RAMI4.0 defines the Administration Shell (Fig. 3), which is the glue that enables the different physical entities that are used for production, including the equipment, materials, production technology and related software for controlling production and the human staff that is involved in the production to be joined. A Communication Layer that is based on a service-oriented architecture (SOA) is responsible for providing the seamless connection between the entities while the Information Layer is responsible for supporting the meta-information that enables the information to be interpreted correctly by taking the required presentation context into account. The Administration Shell can allow access for both the fundamental elements of the system and for the highly complex aggregates that consist of many components. In the case of AGV, we can indicate two levels for the application of the Administration Shell – one for a single vehicle that is responsible for executing the transportation tasks and the second for a fleet of AGV that offers holistic transportation services. Moreover, for each level, the information model must be presented according to its context of use, which means that multiple Administration Shells have to be implemented on each aggregation level. The Administration Shell should be implemented by communication middleware such as OPC UA. The OPC UA is an object-based and service-oriented communication middleware that not only supports the exchange of information but also organises the information models [35]. It is based on the client-server communication principle. Secure communication uses a Secure Channel. The Service Set is established by an OPC client, which is then managed by the services that are collected in the Session Service Set. The address space of the OPC UA server exposes both the data and the relevant information models that can be browsed by any connected client.

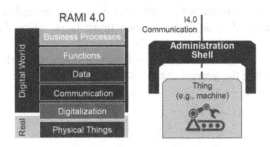

Fig. 3. RAMI4.0 administration shell [34]

OPC UA is also useful as a modelling tool that can be used in accordance with other standards that have been prepared by organisations such as W3C, ISA, OMAC, PLCopen, VDMA and others. Such an approach has resulted in several domain-specific information models that have been created under OPC UA, which are dedicated to a given type of application. In the context of AGV, two such models seem to be particularly valuable – (1) the OPC Foundation and AutomationML Companion Specification "AutomationML for OPC UA" [36], which can be used for a technical description of an AGV and its components and (2) "OPC UA for ISA-95 Common Object Model" [37], which is more relevant for communication with an MES. The address space of the OPC UA server can be browsed by any OPC UA clients that are connected. A reference mechanism is used to express the dependencies between the parts of the model. References link OPC objects with their type definition, other objects and the variables that are responsible for the physical data representation. The data types define the interpretation of the values of variables, which are also connected by references. The types of all of the references are known to clients and therefore the kind of relationship between the connected nodes can be easily determined. The number of references between two particular nodes is not limited, so different dependencies can be expressed in one model. Clients can use the browsing services to discover all of the data items and model-based types of information about the items that are managed by the OPC UA server.

A Data Access mechanism supports effective communication between OPC UA servers and clients. Each OPC UA client defines a set of subscriptions for the selected OPC UA servers, which are then performed as parts of individual sessions. The number of Monitored Items for each subscription is agreed upon between the client and the server. These mechanisms enable clients to select the required data. In this model, a client subscribes to variables with a given Sampling Interval [ms] according to which the OPC UA server only sends new information to the client when there is a change in the source of information, which is checked cyclically with the frequency that is defined by the sampling interval. The information is bound with its quality status and two timestamps – one from the source and the second from the server. Data transfers are optimised by grouping them within the sessions for efficiency [38].

The OPC Historical Access (HA) provides an interface with a historical archive that is connected with/associated with a given Object node and can be used by various applications such as the HMI, Report Generation, Analysis etc. The data that is provided by the historical interface consists of historical records or calculated values such as the min, max, average etc. A Client can read or annotate the historical data. Because OPC UA is only a communication interface, the historical data should be stored in an external database. Both Data Access and Historical Access communication modes use a common memory model. Each client can freely select the information in the server's address space and request access to the data under the requirements that they have defined. A common memory model does not support the integrity of the information between clients. In cases in which horizontal data integrity is essential, an Event-based communication Alarms & Events mechanism should be used. In Event-based communication, all subscribed clients receive precisely the same messages about any state changes of the Object that was selected for the subscription.

The physical connection with AGV has to be provided by Wireless Sensors and Actuators Networks (WSAN). WSAN consist of interconnected sensors and actuators that collaborate in order to monitor and control a system. The sensors measure the physical quantities and report this to the actuators, which then act on them. Currently, WSAN design generally uses a sink to collect and process the collected data and to send commands to the actuators. A WSAN handles network functions such as routing and scheduling medium access. The increased application of WSAN wireless technologies in industry has given rise to a plethora of protocol designs [39]. However, due to the added delay as the data transmits through the sink, future network design will include peer-to-peer communication between the sensor nodes and actuator nodes. This will be discussed further below.

Transmitting through sinks creates challenges. Current industrial sensor networks (that also include WSAN) face a challenge due to the limited interoperability between different systems. One option is to relay the sensor data (using a sink solution) between the stovepipe solutions. Three challenges emerge. First, since the two solutions are disjoint and both offer a similar functionality, there is a duplication of functionality. In addition to the added cost, this solution also adds a delay as the data must be relayed through interconnection points. The relay point is located on the edge of the network, thus creating an excessive delay compared to a peer-to-peer connection. Second, the main challenge facing the network designer is ensuring the predictable maximum delay end-to-end. Interconnecting two disjoint technologies in which each separate network segment operates independently from each other obstructs such guarantees. Furthermore, since each stovepipe solution resides in disjoint domains and operates in isolation, it is not possible to establish an optimal path. Third, each separate network segment (stovepipe) has its own management system, which means that no coherent management can be developed. The only viable solution is to transform the current disjoint technologies into a common infrastructure similar to the transformation that occurred in the telecom industry in the 1990s.

Due to the co-location of the sensors and the controllers on an autonomous unit, peer-to-peer communication between the units is preferable. Transmitting data through a sink delays the data too much. In addition, this solution makes the autonomous unit dependent on the infrastructure. Communication between an autonomous unit and fixed infrastructure requires special attention because the units detach from an access point (or base station) and re-attach to a new location (or station). Layer 3 technologies (IP) offer solutions for moving units, but further investigations/research is required to understand its implications on the delay as well as its robustness. Moreover, security will be an issue as both the de-detachment and re-attachment of the unit must follow strict security procedures. These procedures are likely to add a delay.

Based on reconfigurable I/Os and communication channels a real-time simulator can usually be easily connected to the practical relays from different vendors. Real-time simulators are used to validate the hardware-in-the-loop (HIL), develop algorithms for adaptive protection, design system schemes for integrity protection and perform remedial action schemes [40].

4 AI-Driven Analytics Focused on Data Produced and Consumed by AGV

AI-driven analytics is one of the domains that plays a significant role in maintaining a fleet of AGV and a production cycle. These analytics cover the development and use of ML algorithms to analyse the behaviour of AGV and to detect any anomalies, possible problems or failures. Machine learning provides many algorithms that fall into two main classes: (i) supervised – where information on the occurrence of faults is present in the training data set and (ii) unsupervised – where the process information is available, but no maintenance-related data exists. The supervised approaches [41] are divided into (i) classification models – if the categorical labels are predicted and (ii) regression models – if the results are continuous values. Classification and regression may need to be preceded by a relevance analysis, which attempts to identify the attributes that are significantly relevant to the classification and regression processes [42]. Supervised learning is successfully used in the area of predictive maintenance to classify faults by building fault detectors. In the literature, these detectors rely on various AI techniques such as Artificial Neural Networks [43], k-Nearest Neighbours,

Support Vector Machines [44], Bayesian networks [45] or Principle Component Analysis [46].

Unsupervised learning techniques primarily work based on algorithms that detect outliers. Outliers can be detected using statistical tests that assume a distribution or probability model for the data or by using distance measures in which objects that are remote from any other cluster are considered to be outliers. Building models that do not require labelled data is possible because of techniques such as auto-encoders, Deep Belief Networks or statistical analysis [47].

Industrial IoT systems usually generate large volumes of data in various formats and states (e.g. historical, data streams), which raises the challenges of Big Data. AI-driven analytics has to deal with these challenges (Fig. 4). These challenges led to the creation and popularisation of data lake systems. Data lakes are repositories that keep the data in its original format. In recent years, the sudden proliferation of data in industrial computer systems has increased the pressure to introduce data lake-based analysis methods. More and more data are considered to be useful sources of information for making critical decisions. The vast volume of data that must be processed and the variety of formats that the data is stored in is a significant research challenge [48]. The uncertainty of data complicates data analysis and the inclusion of expert knowledge in data processing offers many advantages [49].

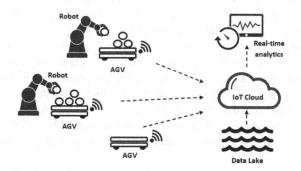

Fig. 4. Remote monitoring, managing, and detecting failures in AGVs connected to the control center through real-time AI-based analytics.

Standard condition monitoring techniques rely on inspecting and observing the physical properties of AGV. The methods that are used include visual monitoring (contaminant, leaks, thermograph), audible monitoring and physical monitoring (temperature, vibration). Using real-time analysis of production data and advanced data exploration, we can implement remote condition monitoring and predictive maintenance tools to detect the first signs of failure long before the appearance of the early alarms that precede AGV failures in a short period [50]. As discrete production lines become more and more complicated, predictive maintenance has become a vital task for the engineers that are responsible for production support. Many potential technological and technical problems can be detected based on early signs that are first noticeable in changes in energy consumption. The current data can be compared with the information related to the energy consumption profiles in an appropriate production context. This comparison allows maintenance tasks to be planned and, as a result, reduces losses related to production breakdowns.

Thus, several challenges should be considered for the implementation of AGV in real-world domains and applications. Based on the AI-driven methodologies, pure data can produce more accurate performance. Most of the data that is collected from AVG may be incomplete and inconsistent and especially those data are collected by mobile or senor devices. Currently, the pre-processing step of data (i.e. labelling the data) in AI-driven analysis must be performed manually. It is necessary to build an intelligent model to refine the collected data in order to obtain excellent performance. To have better support for decision making, it is also a challenge to implement an automatic decision-support system that is based on the results produced by the AI-driven methodologies. Moreover, the results of AVG that are produced should consider the multi-objective criteria in order to achieve global optimisation for different tasks and domains. It is a considerable challenge to develop an optimised AI-driven system for AVG for intelligent decision-making.

5   Conclusions

The use of Autonomous Guided Vehicles (AGV) in production systems has many advantages as it allows production lines to be automated and accelerates logistics. However, it also raises many challenges that provide a space for future research. In this paper, authors tried to evoke a discussion on selected issues related to application of AGVs in flexible manufacturing systems including: (i) the fusion between AGV and collaborative robot with focus on flexibility and interoperability; (ii) the new models for machine to machine communication for AGVs which allow them to cooperate with production environment and at the same time use solutions developed for IoT; (iii) AI-driven analytics focused on the data produced and consumed by AGV that have to be adapted to the pipe-line processing of data collected from many distributed sources.

Acknowledgements. This work was supported by the Polish National Centre of Research and Development from the project "Hybrid system of automated internal logistics supporting adaptive manufacturing" (grant agreement no POIR.01.01.01-00-0460/19-01). The project is realized as Operation 1.1.1.: "Industrial research and development work implemented by enterprises" of the Smart Growth Operational Programme from 2014-2020 and is co-financed by the European Regional Development Fund and partially by Statutory Research funds of Department of Applied Informatics and Department of Distributed Systems and Informatic Devices, Silesian University of Technology, Gliwice, Poland (grant No BK/2020).

References

1. Maskell, B.: The age of agile manufacturing. Supply Chain Manag. Int J. **6**(1), 5–11 (2001)
2. Cupek, R., Ziebinski, A., Huczala, L., Erdogan, H.: Agent-based manufacturing execution systems for short-series production scheduling. Comput. Ind. **82**, 245–258 (2016)
3. Wan, J., Yi, M., Li, D., Zhang, C., Wang, S., Zhou, K.: Mobile services for customization manufacturing systems: an example of industry 4.0. IEEE Access **4**, 8977–8986 (2016)

4. Cupek, R., Ziebinski, A., Drewniak, M., Fojcik, M.: Knowledge integration via the fusion of the data models used in automotive production systems. Enterp. Inf. Syst. **13**(7–8), 1094–1119 (2019)
5. European Commission: A Manufacturing Industry Vision 2025. European Commission (Joint Research Centre) Foresight Study, Brussels (2013)
6. Andreasson, H., et al.: Autonomous transport vehicles: where we are and what is missing. IEEE Robot. Autom. Mag. **22**(1), 64–75 (2015)
7. Realyvásquez-Vargas, A., Arredondo-Soto, K.C., García-Alcaraz, J.L., Márquez-Lobato, B.Y., Cruz-García, J.: Introduction and configuration of a collaborative robot in an assembly task as a means to decrease occupational risks and increase efficiency in a manufacturing company. Robot. Comput. Integr. Manuf. **57**, 315–328 (2019)
8. Haddadin, S., Croft, E.: Physical human–robot interaction. In: Siciliano, B., Khatib, O. (eds.) Springer Handbook of Robotics, pp. 1835–1874. Springer, Cham (2016). https://doi.org/10.1007/978-3-319-32552-1_81
9. Kyrkjebø, E.: Inertial human motion estimation for physical human-robot interaction using an interaction velocity update to reduce drift. In: Companion of the 2018 ACM/IEEE International Conference on Human-Robot Interaction, pp. 163–164. ACM (2018)
10. Mrozek, D., Tokarz, K., Pankowski, D., Malysiak-Mrozek, B.: A hopping umbrella for fuzzy joining data streams from IoT devices in the cloud and on the edge. IEEE Trans. Fuzzy Syst. **28**(5), 916–928 (2020). https://doi.org/10.1109/tfuzz.2019.2955056
11. Liu, L., Kuo, S.M., Zhou, M.: Virtual sensing techniques and their applications. In: 2009 International Conference on Networking, Sensing and Control, Okayama, pp. 31–36 (2009). https://doi.org/10.1109/icnsc.2009.4919241
12. Lee, M.C., Park, M.G.: Artificial potential field based path planning for mobile robots using a virtual obstacle concept. In: Proceedings 2003 IEEE/ASME International Conference on Advanced Intelligent Mechatronics (AIM 2003), Kobe, Japan, vol. 2, pp. 735–740 (2003). https://doi.org/10.1109/aim.2003.1225434
13. Ziebinski, A., Cupek, R., Piech, A.: Distributed control architecture for the autonomous mobile platform. In: AIP Conference Proceedings, vol. 2040, no. 1, p. 080012. AIP Publishing (2018)
14. Grzechca, D., Paszek, K.: Short-term positioning accuracy based on mems sensors for smart city solutions. Metrol. Meas. Syst. **26**(1), 95–107 (2019)
15. Roth, H., Schilling, K.: Navigation and docking manoeuvres of mobile robots in industrial environments. In: IECON 1998, Proceedings of the 24th Annual Conference of the IEEE Industrial Electronics Society (Cat. No. 98CH36200), Aachen, Germany, vol. 4, pp. 2458–2462 (1998)
16. Ziebinski, A., Cupek, R., Nalepa, M.: Obstacle avoidance by a mobile platform using an ultrasound sensor. In: Nguyen, N.T., Papadopoulos, G.A., Jędrzejowicz, P., Trawiński, B., Vossen, G. (eds.) ICCCI 2017. LNCS (LNAI), vol. 10449, pp. 238–248. Springer, Cham (2017). https://doi.org/10.1007/978-3-319-67077-5_23
17. Ziebinski, A., Bregulla, M., Fojcik, M., Kłak, S.: Monitoring and controlling speed for an autonomous mobile platform based on the hall sensor. In: Nguyen, N.T., Papadopoulos, G. A., Jędrzejowicz, P., Trawiński, B., Vossen, G. (eds.) ICCCI 2017. LNCS (LNAI), vol. 10449, pp. 249–259. Springer, Cham (2017). https://doi.org/10.1007/978-3-319-67077-5_24
18. Ziebinski, A., Cupek, R., Erdogan, H., Waechter, S.: A survey of ADAS technologies for the future perspective of sensor fusion. In: Nguyen, N.-T., Manolopoulos, Y., Iliadis, L., Trawiński, B. (eds.) ICCCI 2016. LNCS (LNAI), vol. 9876, pp. 135–146. Springer, Cham (2016). https://doi.org/10.1007/978-3-319-45246-3_13

19. Grzechca, D., Wróbel, T., Bielecki, P.: Indoor localization of objects based on RSSI and MEMS sensors. In: 2014 14th International Symposium on Communications and Information Technologies (ISCIT), pp. 143–146. IEEE (2014)

20. Grzechca, D., Hanzel, K.: The positioning accuracy based on the UWB technology for an object on circular trajectory. Int. J. Electron. Telecommun. **64**(4), 487–494 (2018)

21. Thobbi, A., Gu, Y., Sheng, W.: Using human motion estimation for human-robot cooperative manipulation. In: 2011 IEEE/RSJ International Conference on Intelligent Robots and Systems, San Francisco, CA, pp. 2873–2878 (2011)

22. Jarrassé, N., Charalambous, T., Burdet, E.: A framework to describe, analyze and generate interactive motor behaviors. PLoS ONE **7**(11), e49945 (2012). https://doi.org/10.1371/journal.pone.0049945

23. Mörtl, A., Lawitzky, M., Kucukyilmaz, A., Sezgin, M., Basdogan, C., Hirche, S.: The role of roles: physical cooperation between humans and robots. Int. J. Robot. Res. **31**(13), 1656–1674 (2012)

24. Kyrkjebø, E.: Motion coordination of mechanical systems: leader-follower synchronization of Euler-Lagrange systems using output feedback control (2007)

25. Yang, C., Ganesh, G., Haddadin, S., Parusel, S., et al.: Human-like adaptation of force and impedance in stable and unstable interactions. IEEE Trans. Robot. **27**(5), 918–930 (2011)

26. Medina, J.R., Lawitzky, M., Mörtl, A., Lee, D., et al.: An experience-driven robotic assistant acquiring human knowledge to improve haptic cooperation. In: 2011 IEEE/RSJ International Conference on Intelligent Robots and Systems, pp. 2416–2422 (2011)

27. Fadlullah, Z.M., Fouda, M.M., Kato, N., Takeuchi, A., Iwasaki, N., Nozaki, Y.: Toward intelligent machine-to-machine communications in smart grid. IEEE Commun. Mag. **49**(4), 60–65 (2011)

28. Varghese, A., Tandur, D.: Wireless requirements and challenges in Industry 4.0. In: 2014 International Conference on Contemporary Computing and Informatics (IC3I), pp. 634–638. IEEE (2014)

29. Thota, P., Kim, Y.: Implementation and comparison of M2M protocols for internet of things. In: 2016 4th International Conference on Applied Computing and Information Technology/3rd International Conference on Computational Science/Intelligence and Applied Informatics/1st International Conference on Big Data, Cloud Computing, Data Science & Engineering (ACIT-CSII-BCD), pp. 43–48. IEEE (2016)

30. Elgazzar, M.H.: Perspectives on M2M protocols. In: 2015 IEEE Seventh International Conference on Intelligent Computing and Information Systems (ICICIS), pp. 501–505. IEEE (2016)

31. Durkop, L., Czybik, B., Jasperneite, J.: Performance evaluation of M2M protocols over cellular networks in a lab environment. In: 2015 18th International Conference on Intelligence in Next Generation Networks, pp. 70–75. IEEE (2015)

32. Kang, H.S., et al.: Smart manufacturing: past research, present findings, and future directions. Int. J. Precis. Eng. Manuf. Green Technol. **3**(1), 111–128 (2016)

33. Lin, S.W., et al.: Whitepaper zu "Architecture Alignment and Interoperability" von Plattform Industrie 4.0 und Industrial Internet Consortium, pp. 1–19, December 2017. https://www.plattformi40.de/I40/Redaktion/DE/Downloads/Publikation/whitepaperiicpi40.html

34. Schweichhart, K.: Reference Architectural Model Industrie 4.0 (RAMI 4.0), AG1 Standardization & Reference Architecture, Plattform Industrie 4.0. https://ec.europa.eu/futurium/en/system/files/ged/a2schweichhartreference_architectural_model_industrie_4.0_rami_4.0.pdf

35. Cupek, R., Folkert, K., Fojcik, M., Klopot, T., Polaków, G.: Performance evaluation of redundant OPC UA architecture for process control. Trans. Inst. Meas. Control **39**(3), 334–343 (2017)

36. AutomationML e.V. and OPC Foundation: OPC UA Information Model for AutomationML - Release 1.00.00
37. OPC Unified Architecture for ISA-95 Common Object Model Companion Specification Release 1.00
38. Cupek, R., Ziebinski, A., Franek, M.: FPGA based OPC UA embedded industrial data server implementation. J. Circuits Syst. Comput. **22**(08), 1350070 (2013)
39. Somappa, A.A.K., Øvsthus, K., Kristensen, L.M.: Implementation and deployment evaluation of the DMAMAC protocol for wireless sensor actuator networks. Procedia Comput. Sci. **83**, 329–336 (2016)
40. Adrah, C.M., Kure, Ø., Liu, Z.: Communication network modeling for real-time HIL power system protection test bench. In: 2017 IEEE PES PowerAfrica, pp. 295–300. IEEE (2017)
41. Susto, G.A., Schirru, A., Pampuri, S., McLoone, S., Beghi, A.: Machine learning for predictive maintenance: a multiple classifier approach. IEEE Trans. Ind. Inf. **11**, 812–820 (2014)
42. Han, J., Pei, J., Kamber, M.: Data Mining: Concepts and Techniques. Elsevier, San Francisco (2011)
43. Awadallah, M.A., Morcos, M.M.: Application of AI tools in fault diagnosis of electrical machines and drives-an overview. IEEE Trans. Energy Convers. **18**, 245–251 (2003)
44. Konar, P., Chattopadhyay, P.: Bearing fault detection of induction motor using wavelet and Support Vector Machines (SVMs). Appl. Soft Comput. **11**, 4203–4211 (2011)
45. Qu, J., Zhang, Z., Gong, T.: A novel intelligent method for mechanical fault diagnosis based on dual-tree complex wavelet packet transform and multiple classifier fusion. Neurocomputing **171**, 837–853 (2016)
46. Malhi, A., Gao, R.X.: PCA-based feature selection scheme for machine defect classification. IEEE Trans. Instrum. Meas. **53**, 1517–1525 (2004)
47. Moleda, M., Momot, A., Mrozek, D.: Predictive maintenance of boiler feed water pumps using SCADA data. Sensors **20**(2), 571 (2020)
48. Malysiak-Mrozek, B., Lipinska, A., Mrozek, D.: Fuzzy join for flexible combining big data lakes in cyber-physical systems. IEEE Access **6**, 69545–69558 (2018)
49. Małysiak-Mrozek, B., Stabla, M., Mrozek, D.: Soft and declarative fishing of information in big data lake. IEEE Trans. Fuzzy Syst. **26**(5), 2732–2747 (2018)
50. Cupek, R., Ziebinski, A., Zonenberg, D., Drewniak, M.: Determination of the machine energy consumption profiles in the mass-customised manufacturing. Int. J. Comput. Integr. Manuf. **31**(6), 537–561 (2018)

IoT-Based Cow Health Monitoring System

Olgierd Unold[1]([⊠])[ID], Maciej Nikodem[1][ID], Marek Piasecki[1][ID], Kamil Szyc[1][ID],
Henryk Maciejewski[1][ID], Marek Bawiec[1][ID], Paweł Dobrowolski[2][ID],
and Michał Zdunek[3]

[1] Department of Computer Engineering, Wrocław University of Science
and Technology, Wrocław, Poland
{olgierd.unold,maciej.nikodem,marek.piasecki,kamil.szyc,
henryk.maciejewski,marek.bawiec}@pwr.edu.pl
[2] Department of Control Systems and Mechatronics, Wrocław University
of Science and Technology, Wrocław, Poland
pawel.dobrowolski@pwr.edu.pl
[3] CORP For Farm Animals, ul. Polna 62, 51-361 Wilczyce, Poland
michalzdunek@gmail.com

Abstract. Good health and wellbeing of animals are essential to dairy
cow farms and sustainable production of milk. Unfortunately, day-to-
day monitoring of animals condition is difficult, especially in large farms
where employees do not have enough time to observe animals and detect
first symptoms of diseases. This paper presents an automated, IoT-based
monitoring system designed to monitor the health of dairy cows. The
system is composed of hardware devices, a cloud system, an end-user
application, and innovative techniques of data measurements and analysis algorithms. The system was tested in a real-life scenario and has
proved it can effectively monitor animal welfare and the estrus cycle.

Keywords: Agriculture 4.0 · Smart farming · Precision livestock
farming · Cow health monitoring · Wireless sensor networks

1 Introduction

Nowadays, the agriculture industry is facing several challenges, including demo-
graphics, food waste, and scarcity of natural resources. According to [5], roughly
800 million people worldwide suffer from hunger, and 650 million will still be
undernourished by 2030. The publication reports that by 2050 humanity will
have to produce 70% more food. To solve the scarcity problem, many complementary actions must be taken. Among them, traditional farm and agricultural
methods must be replaced and supported by new advancements in technology,
including sensors, devices, robots, and information technology.

Internet of things (IoT), already mature and effective technology, seems to
be one of the remedies for low efficiency and productivity in agriculture and

© Springer Nature Switzerland AG 2020
V. V. Krzhizhanovskaya et al. (Eds.): ICCS 2020, LNCS 12141, pp. 344–356, 2020.
https://doi.org/10.1007/978-3-030-50426-7_26

livestock. In [8], the following areas of IoT application in agriculture are listed: agricultural field monitoring, greenhouse monitoring, agricultural drones, livestock monitoring, smart irrigation control, agriculture warehouse monitoring, and soil monitoring. The objective of this paper is to present a new IoT-based livestock monitoring system dedicated to the automated measurement of dairy cow health state in a conventional loose-housing cowshed.

2   Related Works

The most common form of cow behavior assessment is a visual observation, but intensive and sustainable detection is limited due to time constraints and lack of human resources [10]. Introducing precision dairy-monitoring (PDM) technology for monitoring dairy cattle allows to avoid disturbing natural behavioral expression [9], to reduce human resources needs, and overall costs [1].

The IoT technology has already been shown to feature potential benefits in dairy cattle farming. In [3] commercially marketed IoT technologies were compared with traditional, visual observation. The tested devices were attached to the animal's legs and ears. Some of the tested PDM sensors had a high correlation between direct observations of feeding, rumination, and lying behaviours In turn, Wang et al. [19] examined the accuracy of a 3-axial accelerometer attached to the legs of a cow in classifying animal behavior.

Making use of the IoT enables to monitor of udder health, estrus events, feet, and leg health, and metabolic health [7,16,17]. The solutions proposed in the literature use PDM devices such as collars, ear tags, and leg bands for behavior monitoring [3,4].

Detection of cows ready for insemination is still considered to be a significant problem in dairy farming [15]. There are many changes in cow behavior indicating estrus, like standing to be mounted by fellows, restlessness, sniffing the vulva of another cow, flehmen, licking, or resting with the chin on the back of another cow [13,18]. Estrus is accompanied by alterations in feeding and rumination [12]. Early estrus detection can be supported by an IoT-based monitoring system [14].

The main objective of the proposed cow health monitoring system is to classify dairy cow behavior with particular emphasis on estrus. The monitoring device for in situ data collection is located in a neck collar with a counterweight.

3   Architecture of Cow Health Monitoring System

The dairy cow monitoring system (called the CowMonitor system) is composed of hardware components, the cloud system, and the end-user application (Fig. 1). Hardware components include the monitoring device (called the CowDevice) and the infrastructure devices: the Hub, WiFi access points, and routers. The cloud system is composed of the Server hosting database and application server.

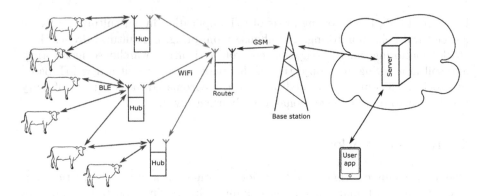

Fig. 1. Architecture of IoT-based cow health monitoring system. Different arrow colors denote different wireless communication technologies: BLE (blue), WiFi/Ethernet (red), GSM/Internet (black). (Color figure online)

3.1 CowDevice

Every animal wears a CowDevice attached to the collar. The device is a dedicated, battery-powered microcontroller device equipped with low-power inertial sensors (accelerometer and magnetometer) and wireless communication. The device is based on nRF 52832 system on chip—an integrated ARM Cortex 4F processor and a Bluetooth Low Energy (BLE) v.5 compatible radio transceiver. The device performs regular accelerometer and magnetometer measurements at the frequency of 20 Hz. The set of measured parameters was chosen experimentally to ensure it will be possible to infer individual animal activity based on the sensor readings. Initially, raw measurements were stored on an SD card. They were downloaded every week, analysed offline, and used to design preliminary algorithms. This enabled to design a proprietary data aggregation algorithms that process raw measurements from sensors in non-overlapping time windows of 10 s. Aggregation enables us to extract only the vital information and lowers the amount of data that needs to be transmitted from the CowDevice down to below 20 bytes every 10 s. This amount is small enough to use BLE communication while the aggregated values still allow classifying animal activity (e.g., rumination, feeding, walking, etc.) and decide estrus.

CowDevices use opportunistic data transmission [2,6] to ensure efficient, energy preserving, and robust transmission of measurements. In this method, the values are transmitted inside a payload of BLE advertisement messages. Advertisements are short messages broadcasted periodically in connectionless communication mode thus ensuring an average power consumption below 300 μA.' However, because advertisements are broadcasted without acknowledgments from the Hubs, they do not ensure reliable data transmission. It is quite likely that a single advertisement is not correctly received. Therefore, the CowDevice transmits advertisements periodically until new aggregate values are available. Every 10 s, the payload is updated with the new aggregate values, and advertisement transmission is restarted. This method improves the overall reliability of the

Fig. 2. (On left) CowDevices (gray) attached to the collar of the animal and (on right) its construction

transmission at the costs of increased energy consumption (compared to single transmission), increased congestion in the communication channel (due to a larger amount of communication), and consequently, a higher probability of collision. Despite the disadvantages, the opportunistic data transmission has proved to be an efficient method for information transmission and was shown to apply to dense networks composed of a large number of BLE devices [11] (Fig. 2).

Aggregated measurements are broadcasted periodically every 250 ms inside the payload of the advertisement messages. Except for the measurements, the advertisements also include additional information regarding device status and packet sequence numbers. Every set of the aggregated measurements is transmitted approximately 40 times (every 250 ms for 10 s), ensuring reliable transmission of all the measurements. The CowDevice runs on the battery that enables its operation for up to 5 years without replacement.

3.2   Hub

The Hub is a dedicated gateway device that receives the BLE packets and forwards them to the server located in the cloud. The Hub (Fig. 3) is build using Raspberry Pi Zero W single-board computer that is equipped with built-in BLE and WiFi connectivity. Every Hub picks up advertisements broadcasted by the CowDevices, eliminates duplicate measurements (based on the sequence number included in the payload) and forwards the unique set of aggregated measurements to the server for future processing. The hubs also measure radio signal parameters (e.g., Received Signal Strength Indicator - RSSI), record the MAC address of the transmitters, and calculate the number of advertisements received from each MAC. The MACs are checked against a whitelist so that Hubs only process the BLE advertisements coming from known CowDevices. This allows to reduce the workload of the Hub and improves its performance. The MAC and the number of advertisements is aggregated and forwarded to the Server providing more insight on cow activity, and enabling rough estimation of cow localization on the farm. Additionally, the MAC is used to assign measurements to the particular animal.

Fig. 3. The Hub in IP57 enclosure

The Hubs ensure reliable transmission of the measurements to the Server using the application-level protocol with retransmissions. This feature enables effective data transmission even if serial communication is used (e.g., GSM) or the central Server is down.

3.3 Server

The server performs an essential role in our system. It is located in the cloud and communicates with the Hubs over the Internet (either the GSM connection or a cable). It keeps the information about the databases, manages the upcoming files, runs algorithms to calculate characteristics, raises alarms, and enables REST (REpresentational State Transfer) communication with hubs and end-user applications.

We needed to store information about the whole system. We used databases to keep data about farmers (for example id, name, credentials), cowsheds (for example size, GPS coordinates), cows (for example id, group), CowDevices (for example id, mac, name), hubs (for example id, mac, location in cowshed), data from sensors and finally characteristics and alarms calculated by our algorithms.

The server receives data from CowDevice sensors. Hubs constantly send them via REST communication by using compressed files. The files are stored, unpacked, analyzed, and after all necessary operations, they are archived again. The server analyzes incoming messages; first, it removes duplicates (it can happen when different hubs received information from the same CowDevice), next it checks if message contains correct pieces of information (CowDevice can send frame with device status - for instance, to inform about not working module) and finally it backups old data to an external server.

The server also uses the scheduler, which is responsible for running algorithms a few times in an hour. Each algorithm uses historical data from sensors (usually a few hours back) and previous results of algorithms (usually a few days back). We called the results of the algorithms as characteristics. These characteristics are used to generate graphs and raise alarms. The server controls which data should be analyzed and how to interpret results. Our system will also be allowed to localized animals in cowsheds based on RSSI signal (it is one of an algorithm we work on).

The last significant role of the server is controlling communications with hubs and end-user Android application. Communication is carried out by REST. Endpoints allow reading information (for example, characteristics, alarms, graphs) about farms, a herd of cows, groups, and particular cow and checking status of the system (by administration). They also allow adding new animals to the system or connecting cow devices with a particular cow. They also control communications with hubs by allowing them to send files.

3.4 Android Application

The end-user Android application displays information about each individual cow and each group of cows to the farmer (Fig. 4). It is an example of how it can look like in a commercial solution, so its functionality is limited. It allows monitoring the status of a herd of cows by farmers. Thanks to the information presented in graphs and alarms, farmers can better monitor the health of their cows and faster response to any disorders. This is important, especially when there are many animals in the herd, and the farmer cannot monitor each animal every day.

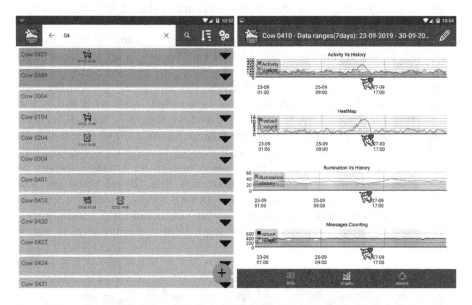

Fig. 4. Screenshots from the application. (On left) List of cows with (marked red) and without (marked green) active alarms. (On right) Plots with characteristics for selected cow. (Color figure online)

The application allows checking the actual state of the herd using a list of all the cows with additional information. It was important that the information is presented in such a way that it is straightforward and easy to understand. This includes color-coding and alarm icons (Fig. 4) that encode different types of alarms including health alarm (usually correlated with problems with feeding), estrus related alarm (usually suggest that a cow is ready for insemination), or the device-related alarm (which informs the farmer about CowDevice malfunction).

The farmers can also check detailed information about the particular cow. This is needed, especially when they would like to get new knowledge of why the system decided to raise the alarm. The application uses graphs to present information about cow status. For example, activity, rumination, heat, or the number of incoming messages (for the system diagnosis purpose).

The application also allows us to add new animals to the system and link it with the CowDevice. It uses the Representational State Transfer (REST) communication with the server and BLE to setup parameters of the CowDevices. The farmer needs to set information about the cow, including its ear tag number and breeding group. The farmer can also edit or remove the cows from the system.

4   Use Case

During the development of the CowMonitor system, over three years (2017–2019), several measurements were carried out at three different dairy farms in Poland. The largest test installation was carried out at the "AgroTak Zagrodno" dairy farm in Lower Silesia in Poland. Most of the illustrations in this section are based on the measurements taken there during the last 6 months of 2019. During this period, depending on the month, from 81 to 119, cows were monitored using CowDevice sensors.

The three-axis accelerometer of the CowDevice is used as the primary data source. For research purposes, all three axes were tested, but later it turned out that the two axes (horizontal X: diametrically along the cow's body, from head to tail, and vertical Y: top-down) provide enough information to distinguish all basic behaviors of the cow. The Bluetooth channel bandwidth is very limited, so only selected aggregating characteristics can be sent for further analysis. To enable optimal selection of transmitted parameters, first, the reference data was collected for 2 months, in the form of recordings with a video monitoring system. Then all measurement data from accelerometer were hand-tagged using a specially developed application. An example graph from this application is shown in Fig. 5.

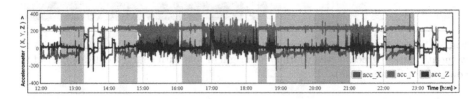

Fig. 5. Results of hand-tagging 12 h from the life of cow number 393. The graph presents the measurements from the X, Y, Z axis of the accelerometer. Periods of rumination are marked with yellow background (Color figure online)

The selection of relevant features was done with the use of classification tree analysis. Then, the standard decision tree learning algorithm C4.5 was used to enable the detection of selected basic behaviors of a dairy cow (Fig. 6).

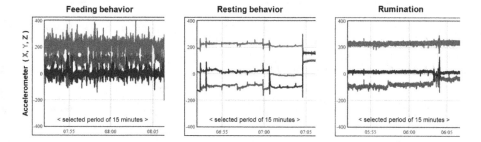

Fig. 6. Characteristic graph shapes related to behaviors detected by classification trees used in the CowMonitor system

Finally, the algorithms processing the measurement data, have been implemented in the form of a three-level architecture, where:

- Level "0" - calculates statistical characteristics to identify typical short term micro-behaviors (feeding, ruminating, resting, mounting other cows, etc.). Measurement data from 10 s are frequency filtered and aggregated into a selected set of characteristics. This level is processed directly on board of the CowDevice.
- Level "1" - evaluates the current state of the cow (aggregates data for each clock hour, against the background of the history of the previous few hours). Parametrized decision trees are used on this level, to allow extraction and aggregation of useful cow wellbeing criteria.
- Level "2" - elaborates a reference model taking into account simple stereotypical characteristics of a particular cow (estimated from the last few days and used by level 1). Level "1" and "2" are processed on the cloud server.
- Work is also underway to develop a more accurate reference model, calculated on the basis of the data from the previous 2–6 months (individual modelling of each cow, in place of one universal stereotype model).

The main task to achieve, at this stage of the research, was the most reliable and precise diagnosis of current cow state and behavior. Particular emphasis was placed on the estrus time detection and supporting the farmer in taking insemination decisions. This goal has been achieved successfully. Cow state is easily observable through a series of complementary charts which are illustrated in several following figures (Fig. 7, 8, and 9).

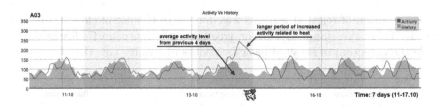

Fig. 7. Chart presenting characteristic A03, which generally estimates the level of overall cow's activity, with a visible increase at the time of estrus. "Heat Alarm" is marked below the chart at October 14.

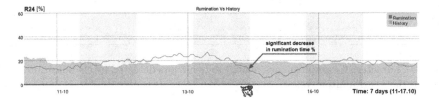

Fig. 8. Chart presenting characteristic R24 (percentage of rumination time in the last 24 h). Rumination time is a complementary information to confirm the heat. Significant reduction at October 15 is a side effect of increased estrus activity.

These graphs show one of the cows (numbered 140) in October 2019, when:

- the first signs of estrus were detected by CowMonitor characteristics around 5pm on October 14,
- later, the system generated "Heat Alarm", the same evening at 9 pm,
- finally, the estrus was confirmed by a vet, the next day morning.

The innovative achievement is the development of a new way to detect the estrus-specific behavior of a cow in the form of characteristic HL03 (Fig. 9). The central concept of this indicator is somehow analogous to multidimensional scanning. It aggregates both short and log-termed features of cow activity to a compound measurement estimating estrus occurrence. It allows distinguishing, with very high efficiency, the rising trend of estrus activity, from the increases in the activity caused by other factors. The current value of this indicator (red

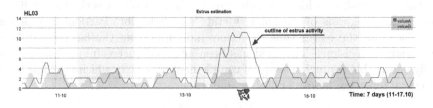

Fig. 9. Characteristic HL03 estimating the level of estrus-specific activities over last 3 h (value proportional to the probability of estrus) (Color figure online)

color) is displayed against the background of historical values from the previous few 1–4 days (green). This type of visualization allows taking into account the physical characteristics of individual cows. Figure 10 merges the sequence of estrus occurrences successfully detected over the six months.

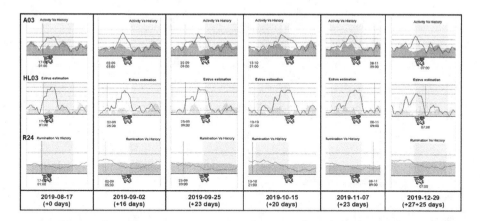

Fig. 10. Charts illustrating successful estrus detection in subsequent 6 months (August–December 2019). Penultimate estrus (on December 4), was not measured due to a power outage in the farm buildings.

The second important task of the CowMonitor system (in addition to the recognition of estrus) is the early detection and diagnosis of critical diseases. After the measurements were carried out, the breeding and veterinary documentation was reviewed to check whether there is a correlation between the alarms generated by the system and confirmed cases of diseases. Mastitis disease was selected as the main subject of the study. During the 6 months of measurements, 10 subclinical or clinical mastitis cases were recorded in the observed herd. In almost all cases, this disease manifests itself in a very sharp decrease in rumination. Figure 11 presents typical shapes of rumination charts for such a situation. In eight cases, the alarm was generated before the mastitis was diagnosed. In the ninth case, the symptoms are visible in the CowMonitor but were too delicate to cause an alarm. Only in one case, no changes in the rumination were observed at all. These first results, with an average disease detection efficiency of around

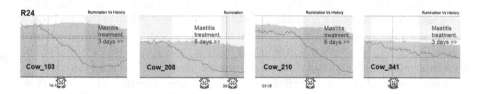

Fig. 11. Large decrease in the rumination characteristic to the mastitis disease

90%, are very promising. However, the data about a larger number of cases need to be collected to perform reliable statistical tests.

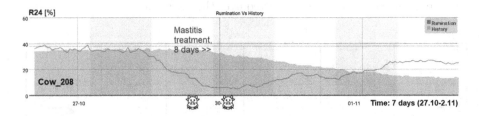

Fig. 12. First detection of mastitis for cow number 208, on October 29th

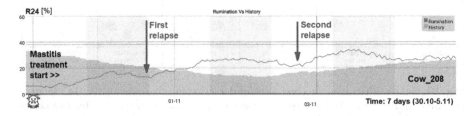

Fig. 13. Monitoring mastitis treatment: two sequential relapses on October 31st and November 3rd

The high utility of the CowMonitor system appeared in the difficult case of mastitis in cow number 208. After accurate diagnosis on October 29, appropriate treatment was started (Fig. 12). Unfortunately, the first type of treatment did not end with a full recovery. There were two sequential relapses during the following days (Fig. 13). CowMonitor rumination charts, allow accurate monitoring of treatment progress. The vet or dairy farmer can immediately change the type of treatment, when progress is too slow or when the symptoms of relapse occur. Without the CowMonitor system, precise monitoring of treatment progress is much more difficult.

5   Conclusions

The developed system efficiently and accurately monitors behavior of the dairy cows and allows to detect a particular physiological status like estrus and some health problem (e.g., mastitis). This task is supported by IoT infrastructure consisting of hardware devices, the cloud system, and the end-user application. New innovative techniques of data measurement, including the aggregate behavior indicator, enable us to precisely discriminate cow activities. The prototype devices presented were evaluated in real setup and have been so efficient that are now being transformed to a commercial system.

In the future, the system can be extended with new algorithms that could detect symptoms of other diseases, for example, lameness. In particular, work is underway on the use of a magnetometer for low-energy estimation of cow mobility, and their correlation with the diagnosis of diseases manifested by walking changes. It is worth noting that the location of each CowDevice can be roughly estimated by the measurements collected by every Hub. This can be used to detect cows changing the breeding group, simplify the location of a cow when needed (e.g., for veterinary treatment) or provide more information about the animal activity and movement that may support health diagnostic. The use of BLE v.5.1, which supports Angle-of-Arrival localization, may further improve the location accuracy and provide more benefits to the system. Additionally, one universal behavior stereotype model is planned to be replaced by individual modelling of each cow.

Funding. This research was jointly funded by the Wroclaw University of Science and Technology and Polish Agency for Enterprise Development (POIR.02.03.02-02-0009/17).

References

1. Adamczyk, K., Cywicka, D., Herbut, P., Trześniowska, E.: The application of cluster analysis methods in assessment of daily physical activity of dairy cows milked in the voluntary milking system. Comput. Electron. Agric. **141**, 65–72 (2017)
2. Aguilar, S., Vidal Ferré, R., Gomez, C.: Opportunistic sensor data collection with bluetooth low energy. Sensors **17**, 159 (2017). https://doi.org/10.3390/s17010159
3. Borchers, M., Chang, Y., Tsai, I., Wadsworth, B., Bewley, J.: A validation of technologies monitoring dairy cow feeding, ruminating, and lying behaviors. J. Dairy Sci. **99**(9), 7458–7466 (2016)
4. Caja, G., Castro-Costa, A., Knight, C.H.: Engineering to support wellbeing of dairy animals. J. Dairy Res. **83**(2), 136–147 (2016)
5. De Clercq, M., Vats, A., Biel, A.: Agriculture 4.0: the future of farming technology. In: Proceedings of the World Government Summit, Dubai, UAE, pp. 11–13 (2018)
6. Del Campo, A., Cintioni, L., Spinsante, S., Gambi, E.: Analysis and tools for improved management of connectionless and connection-oriented BLE devices coexistence. Sensors (Switz.) **17**(4), 792 (2017). https://doi.org/10.3390/s17040792
7. Dolecheck, K., et al.: Behavioral and physiological changes around estrus events identified using multiple automated monitoring technologies. J. Dairy Sci. **98**(12), 8723–8731 (2015)
8. Mirani, A.A., Memon, M.S., Rahu, M.A., Bhatti, M.N., Shaikh, U.R.: A review of agro-industry in IoT: applications and challenges. Quest Res. J. **17**(01), 28–33 (2019)
9. Müller, R., Schrader, L.: A new method to measure behavioural activity levels in dairy cows. Appl. Anim. Behav. Sci. **83**(4), 247–258 (2003)
10. Nielsen, P.P., Fontana, I., Sloth, K.H., Guarino, M., Blokhuis, H.: Validation and comparison of 2 commercially available activity loggers. J. Dairy Sci. **101**(6), 5449–5453 (2018)
11. Nikodem, M., Bawiec, M.: Experimental evaluation of advertisement-based bluetooth low energy communication. Sensors **20**(1), 107 (2020). https://doi.org/10.3390/s20010107, https://www.mdpi.com/1424-8220/20/1/107

12. Pahl, C., Hartung, E., Mahlkow-Nerge, K., Haeussermann, A.: Feeding characteristics and rumination time of dairy cows around estrus. J. Dairy Sci. **98**(1), 148–154 (2015)
13. Palmer, M.A., Olmos, G., Boyle, L.A., Mee, J.F.: Estrus detection and estrus characteristics in housed and pastured holstein-friesian cows. Theriogenology **74**(2), 255–264 (2010)
14. Roelofs, J., Van Erp-van der Kooij, E.: Estrus detection tools and their applicability in cattle: recent and perspectival situation. Anim. Reprod. (AR) **12**(3), 498–504 (2015)
15. Roelofs, J., López-Gatius, F., Hunter, R., Van Eerdenburg, F., Hanzen, C.: When is a cow in estrus? Clinical and practical aspects. Theriogenology **74**(3), 327–344 (2010)
16. Rutten, C.J., Velthuis, A., Steeneveld, W., Hogeveen, H.: Invited review: sensors to support health management on dairy farms. J. Dairy Sci. **96**(4), 1928–1952 (2013)
17. Shahriar, M.S., et al.: Detecting heat events in dairy cows using accelerometers and unsupervised learning. Comput. Electron. Agric. **128**, 20–26 (2016)
18. Sveberg, G., et al.: Behavior of lactating holstein-friesian cows during spontaneous cycles of estrus. J. Dairy Sci. **94**(3), 1289–1301 (2011)
19. Wang, J., He, Z., Ji, J., Zhao, K., Zhang, H.: IoT-based measurement system for classifying cow behavior from tri-axial accelerometer. Ciência Rural **49**(6) (2019)

Visual Self-healing Modelling for Reliable Internet-of-Things Systems

João Pedro Dias[1,2]([envelope]) [ORCID], Bruno Lima[1,2] [ORCID], João Pascoal Faria[1,2] [ORCID],
André Restivo[1,3] [ORCID], and Hugo Sereno Ferreira[1,2] [ORCID]

[1] DEI, Faculty of Engineering, University of Porto, Porto, Portugal
{jpmdias,bruno.lima,jpf,arestivo,hugosf}@fe.up.pt
[2] INESC TEC, Porto, Portugal
[3] LIACC, Porto, Portugal

Abstract. Internet-of-Things systems are comprised of highly heterogeneous architectures, where different protocols, application stacks, integration services, and orchestration engines co-exist. As they permeate our everyday lives, more of them become safety-critical, increasing the need for making them testable and fault-tolerant, with minimal human intervention. In this paper, we present a set of self-healing extensions for Node-RED, a popular visual programming solution for IoT systems. These extensions add runtime verification mechanisms and self-healing capabilities via new reusable nodes, some of them leveraging *meta-programming* techniques. With them, we were able to implement self-modification of *flows*, empowering the system with self-monitoring and self-testing capabilities, that search for malfunctions, and take subsequent actions towards the maintenance of health and recovery. We tested these mechanisms on a set of scenarios using a live physical setup that we called *SmartLab*. Our results indicate that this approach can improve a system's reliability and dependability, both by being able to detect failing conditions, as well as reacting to them by self-modifying *flows*, or triggering countermeasures.

Keywords: Internet-of-Things · Runtime verification · Self-healing ·
Software engineering · Visual programming

1 Introduction

The Internet-of-Things (IoT) is a network of programmable uniquely identifiable devices, known as *things*, that can sense (*i.e.*, sensors) and change (*i.e.*, actuators) their environment [22]. Within the nature of IoT systems, there are several particularities that, although not new or unique, congregate at an unprecedented scale in terms of interconnected devices, people, systems, and information resources, leading to an ever-increasing complexity that developers must address. These systems—typically built with heterogeneous parts, mostly resulting from the integration of different, and, sometimes, already existent, systems (*i.e.*, systems of systems [8])—are not only logically distributed but also geographically, and commonly have to deal with power constraints and real-time needs.

© Springer Nature Switzerland AG 2020
V. V. Krzhizhanovskaya et al. (Eds.): ICCS 2020, LNCS 12141, pp. 357–370, 2020.
https://doi.org/10.1007/978-3-030-50426-7_27

The wide range of IoT application scenarios urges the need for tools that allow users with reduced technical knowledge to configure and adapt their systems to their needs. These requirements lead to the birth of several different *low-code* and visual programming solutions that try to reduce the inherent complexity of programming and configuring these systems. Ray et al. [29] identify several visual programming solutions tailored to this domain. To get a grasp on the popularity of these solutions, we surveyed open-source tools (hosted on GitHub), using the number of stars as the primary metric. We can observe that the most popular visual programming solution for this domain is, by far, Node-RED (9600 stars), followed by XOD (583), Ardublock (376, Arduino-only development), Snap4Arduino (99, Arduino-only), Wyliodrin (84), Intel IoT Services Orchestration Layer (80), miniBloq (72), and NETLabToolkit (17, Arduino-only).

As systems' complexity increases, it inevitably results in people becoming *"overwhelmed by the effort to properly control the assembled collection,"* [28] increasing the probability of human-induced errors and failures; developing becomes hard, labor-intensive, and expensive, no matter how *low-code* the infrastructure is [16]. IoT is acknowledged to be a particular example of these complex systems [23], where recovering from faults becomes challenging [14]. As a result of this inherent complexity, researchers have argued that there is an imminent need for *autonomic components* [3,4,31]. From single devices (*e.g.*, smart locks) to whole systems (*e.g.*, smart homes), components should be capable of self-management, reducing the need for frequent human interactions [18]. This becomes essential in mission-critical systems, or when devices are deployed in remote locations (*e.g.*, wildfire control) or hard to access areas (*e.g.*, inside walls).

Ganek and Corbi [13] identify four desired *self* properties, namely: *self-configuring*, *self-healing*, *self-optimization*, and *self-protection*. All these characteristics require a certain degree of *runtime introspection* [19] from the system. Monitoring has been the most common approach for understanding a running system [1,12,15]; this technique allows one to retrieve operational data about running systems by using several distinct methods, but it is usually done by external tools and without a *feedback loop*. Some authors do propose the usage of runtime verification as a way to detect malfunctions and failures of system elements and their interactions [1], which act as a *lightweight* verification mechanism, complementing techniques such as model checking and testing. The main difference lies in providing the missing *feedback loop*, allowing taking actions as soon as some incorrect behavior is detected [21]. This verification mechanism can be used as a foundation for self-healing IoT systems.

Our work focuses on the principle that systems should be able to reconfigure themselves to recover from failures introduced by faulty parts. To achieve this, the running system must be able to model itself so that it can identify the faulty components during its operation (*i.e.*, runtime), without the need for human inspection. Our main contribution is the ability to visually model diagnosis and recovery/maintenance of health mechanisms to improve IoT systems' reliability, thus enabling them to be *self-healing*. These mechanisms have been developed, applied, and tested, as extensions that we named Self-Healing Extensions for

Node-RED (SHEN). We validated our approach by executing a set of scenarios on top of a live, physical setup, called *SmartLab*. The in-place based system was first upgraded with the designed extensions; then, a set of common scenarios was executed, and the resulting system behavior observed. Our experiments show the feasibility of the approach, pointing to improvements in terms of system reliability and dependability, despite several limitations and challenges that this particular VPL language poses, and which limit the full potential of our approach.

The structure of the remaining paper is as follows: Sect. 2 provides an overview of the main concepts, Sect. 3 explores related work, Sect. 4 describes our approach for runtime verification and self-healing as Node-RED extensions, Sect. 5 describes the experimental phase, Sect. 6 discusses the limitations, challenges, and benefits of our approach and Node-RED itself, and lastly, Sect. 7 provides some final remarks.

2 Preliminaries

The following paragraphs introduce some fundamental building blocks and key concepts of this work, focusing on IoT. The Node-RED tool is presented (Sect. 2.1) along with additional details about its functioning and known limitations. The current practices, in terms of validation and verification, are briefly presented and discussed (Sect. 2.2). Lastly, the concepts of autonomic computing, more specifically, self-healing, are presented (Sect. 2.3).

2.1 Node-RED

Node-RED is an open-source mashup-based[1] approach for developing IoT systems. Its "programs" are a set of *flows*, which consist of *nodes* connected by *wires*. Several *node templates* are usually provided that can be used (*e.g.*, drag-and-dropped) into a *flow canvas*. Once the developer creates or updates a flow, it must be *deployed*; a process that persists the new flow version and (re-)starts the whole system [7]. More recently, *flows* acquired the ability to be version controlled and exported. The portfolio of available *nodes* can be extended via *plugins* that implement new ones, either in (1) JavaScript, or (2) by the composition of existent *nodes* in the form of *sub-flows*. Input *nodes* typically subscribe to external services, listen for data on a specific port, or start processing HTTP requests. Once the data is processed by a given *node*, either from an external service or from an upstream *node*, a method is called with the resulting data on downstream *nodes* that can either generate additional events or push the results to outside services or systems [7]. Mashup tools are known to lower the barrier of application development significantly [24].

Despite its features and popularity, this tool still presents several limitations to our objective. There are no proper mechanisms for debugging and testing

---

[1] Mashup-based developed systems are the result of composing or mashing up existing services, components, and devices [26].

flows, beyond adding special nodes having logging capabilities. The message passing mechanism is not typed, which means simple connection errors are not detected before they are deployed. Meta-facilities, such as *reflection* and *reification*, are not available for usage in the flows, which might be due to it not leveraging the usage of a formally defined meta-model as a way of representing its abstractions [27]. The tool is also designed as a centralized *orchestrator*, in the sense that every flow—particularly every message passing activity—must be executed by it, even if several nodes gather or publish their information to external systems. One contributing factor to this limitation is the non-usage of model-based techniques, which leads to a platform-dependent specification, hindering the ability to generate target-specific code. Its design favors the modeling of the system's overall behavior as a *dataflow*, but the behavior of each particular component is mostly opaque and must be implemented manually. As such, it is harder to inspect, simulate, analyze, and change flows as a whole when compared to model-based systems—including during *runtime*.

The result is that, although Node-RED presents an easy platform to prototype simple systems, it quickly falls behind once the complexity starts increasing. Ray's survey findings [29] concur with our analysis, arguing that although several domains of applications already take great advantage of the use of recent advances in Visual Programming Languages, the emerging field of IoT is still lingers far behind other sectors.

2.2 Validation and Verification of IoT Systems

The complexity of our target systems affects not only their design and development processes but also implies a greater complexity of their verification and validation procedures. Traditional approaches for testing software-only systems are mostly limited and insufficient by overlooking fundamental factors about interaction with the real world, and mostly ignoring the hardware counterpart [9]. Of the available solutions, most focus solely on a specific platform, language, or standard, hindering overall improvement or extension, and do not provide out-of-the-box functionality [9]. The lack of IoT-specific testing systems can also lead to the adoption of poor testing practices; a closer examination allows the identification of recurring behaviors in these applications and a set of corresponding testing strategies [10]. Pontes et al. [25] proposed a pattern-based approach to IoT testing by identifying five specific test patterns, namely: *Test Periodic Readings, Test Triggered Readings, Test Alerts, Test Actions,* and *Test Actuators.* They claim that once these are available as test patterns, the overall process of testing becomes easier, as they can be reused to test recurrent behaviors in different scenarios.

2.3 Self-healing Systems

Ghosh et al. [14] describe systems with self-healing capabilities to be those that can deal with disruptions in their operation by (1) detecting system failures and possibly diagnosing the root cause of the problem, (2) determining a fix (*i.e.,* maintenance of health), and (3) recovering (even if only to a less capable but safe and healthy state). Self-healing may use models (*external* or *internal*) that monitor the system's behavior (*probes*), allowing it to adapt to environmental or operational circumstances. These approaches can be *intrusive*, if implemented internally within the system itself, or *non-intrusive*, if they consider the guarded system as a complete unit; they are *closed-loop* when they try to avoid all *a priori* known failure sources (*i.e.*, all possible states are known before recovery), or *open-loop* otherwise [28]. The typical recovery mechanisms employed include reconfiguration and replication of components (hardware and software) and degradation of the quality-of-service (QoS) [1].

3   Related Work

Athreya et al. [5] suggest devices should be able to manage themselves both in terms of configuration (self-configuration) and resource usage (self-optimization), proposing a measurement-based learning and adaptation framework that allows the system to adapt itself to changing system contexts and application demands. Although their work has some considerations about resilience to failures (*e.g.*, power outages, attacks), it does not address self-healing concerns.

The concept of *responsible objects*, introduced by Angarita et al. [3], states that *things* should be self-aware of their context (passage of time, the progress of execution and resource consumption), and apply *smart* self-healing decisions taking into account component transaction properties (backward and forward recovery). Their approach shows limitations, *viz.* (1) when applied to time-critical applications, as it is not clear how much time we should wait for a transaction to finish, (2) some processes, such as those triggered by emergencies, cannot be compensated, and (3) when is it acceptable to perform *checkpoints* in a continuously running system that cannot be *rolled-back*? It also disregards the typical capability of devices (*e.g.*, limited memory, power) that might challenge the implementation of transactions.

Aktas et al. [1] are amongst the first to purpose runtime verification mechanisms to identify issues by resorting to a complex event processing (CEP) technique and *"applying rule-based pattern detection on the events generated real-time"*. They do not address *self-healing* and only convey a summary of problems or possible problems to human operators. Leotta et al. [20] also present runtime verification as a testing approach by using UML state machine diagrams to specify the system's expected behavior. However, their solution depends on the definition of a formal specification of the complete system, which is unfeasible for highly-dynamic IoT environments (*e.g.*, dynamic network topology).

We could not find any work that focuses on bringing runtime verification mechanisms for visual programming environments. This is not unexpected, as Leotta et al. [20] point out that *"software testing (in IoT) has been mostly overlooked so far, both by research and industry,"* and later corroborated by Seeger et al. [30], claiming that most of the research being conducted in visual programming for IoT has been disregarding failure detection and recovery.

4   System Architecture and Behavior

We encapsulate runtime verification and self-healing mechanisms by extending Node-RED with new *nodes*. The following subsections detail our approach for those that are subsequently used in the validation scenarios (Sect. 5.1), but they are part of a more extensive palette [11].

4.1   Visual Runtime Verification

Node-RED has several limitations regarding testing and debugging of *flows*, from not providing *out-of-the-box nodes* capable of doing these tasks, to some design decisions of the programming environment itself.

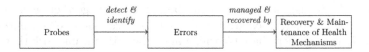

Fig. 1. Self-healing sequence. The use of runtime verification *probes* the system for errors, and the self-healing is accomplished by the activation of system recovery and/or maintenance of health mechanisms.

Regarding runtime verification capabilities, we created *nodes* that allow inspecting the system under test (SUT), *i.e.*, *probing* the system (Fig. 1), including the test patterns presented in Pontes et al. [25] and detailed in Sect. 5.1. Some devices and services (*e.g.*, message brokers, datastores, third-party services) can only be tested by implementing *black-box* reachability checking, such as the new MQTTBrokerTimeout node that asserts if the broker is still alive.

4.2   Visual Self-healing Approach

Following the self-healing loop described by Psair et al. [28], our *detection* component is composed by nodes that allow *runtime testing* and provide *diagnosis* information (Fig. 1), after which the *recovery* process is accomplished by nodes that implement *maintenance of health* and *recovery* mechanisms:

Replacement. Replace a faulty component with a duplicate spare one;
Balancing. Reduce or manage the load of a component to avoid damage;
Isolation. Isolate the failing component to keep the system in a *healthy state*;
Persistence. Assume that a failure does not cause further system degradation;
Redirection. Change the flow during a failure to a recovery routine and then back to the original;
Relocation. Move a system component (along with its dependencies) from a faulty host to a healthy one;
Diversity. Switch between different approaches or processes during runtime.

Supporting these features requires *meta-facilities* that allow changing a system's behavior during runtime. As Node-RED does not formally provide them, we found a workaround by resorting to its *external* REST API from *inside* our nodes, thus gaining the ability to create, delete and change the configuration of *flows* and other *nodes*. This is exemplified by the `SetFlowStatus` node, which allows toggling flows *on* and *off*, thus providing the necessary capabilities for *redirection*, *replacement*, and *isolation*. We were then able to change between instances of message brokers and create a *balancing* mechanism. Other self-healing mechanisms were implemented by adding *secondary* flows and *sub-flows*, that are triggered when some precondition is met.

5   Experimental Scenarios and Results

Validating new solutions for runtime verification and self-healing requires scenarios representative of the characteristics, issues, and challenges of real-world IoT environments, such as heterogeneity and real-time needs. We carried experiments on *SmartLab*, an experimental *testbed* with four actuators and three sensing devices (each having more than one sensor) deployed in a laboratory (Fig. 2), responsible for a set of user-interaction features.

5.1   Scenarios

We devised three scenarios to demonstrate both the necessity of runtime verification as well as self-healing mechanisms. Although these scenarios do not cover all possibilities, we believe them to be sufficient to show the complexity, challenges, and, in this case, Node-RED limitations and trade-offs.

Unavailability of Message Broker. MQTT is the base of most of our *SmartLab* communications; thus, it needs a message broker. Typically, the defined

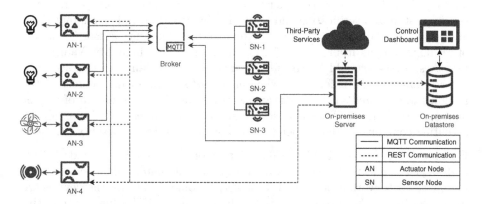

Fig. 2. System component diagram, showing the main system parts, along with the different devices (actuators and sensors) and the enabling communication protocols.

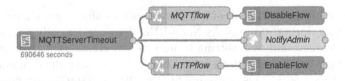

Fig. 3. `MQTTServerTimeout` example, for detecting and healing (using a replacement strategy) a potential unavailability of the broker.

flows are triggered when a new message is received (the flow subscribes to a specific topic). In this scenario (*cf.* Fig. 3), the message broker is both the *bottleneck* and a *single point of failure* (SPOF) of the system; if it fails, the functionality of the system is compromised. To verify its availability (*i.e.*, health status), a *heartbeat* pattern was followed: when the broker stops sending its periodic signal, it is assumed that some fault occurred. The same logic can be easily applied to other publish-subscribe protocols. When this kind of fault is detected, a *redirection strategy* is followed, ensuring the continuation of communications. In our scenario, we trigger a change from the *MQTT-dependant flow* to the alternative *HTTP-based flow*.

Erroneous Sensor Readings. *SmartLab* relies on the readings from different sensors so that it can act according to user-defined rules. As an example, if smoke is detected, an alarm or another notification mechanism should be triggered (and possibly trigger some contention mechanism like sprinklers). These procedures depend on the *timeliness* and *correctness* of readings. Sensor malfunctioning can display an array of different behaviors, such as outputting *out-of-bound* or *out-of-spec* values; these can lead to wrong decisions and may end up having nefarious effects to the point of impacting the well-being of humans. Several strategies can be used separately or in combination to detect sensor malfunction. Sensors that provide periodic readings can be verified by analyzing the expected periodicity (*cf.* Sect. 2.2). Other errors, such as *out-of-bounds* and *out-of-spec* readings require customized verification and tailored failure conditions. Fortunately, these are usually available; *e.g.*, the DHT11 temperature/humidity sensor is capable of readings ranging from 0 °C to 50 °C, and 20% to 80% humidity. Values outside these ranges should be considered erroneous by default. In this scenario, an *isolation strategy* is followed; when an *out-of-spec* problem is detected, the readings are ignored via the `TestAndFilter` node. In the presence of redundant sensors, other readings may still be used by the system; otherwise, all the actuating components that depend on that sensor cease their activity (Fig. 4).

Connectivity Issues. Devices that are part of our *SmartLab* provide HTTP *and* MQTT connectivity. These devices (especially actuators) depend on receiving messages to work as supposed. However, in some situations, the devices are not accessible by the protocol used by default (*e.g.*, MQTT) due to connectivity disruptions, protocol *bugs*, or other reasons, thus becoming inaccessible and eventually causing problematic side-effects (*e.g.*, sprinklers not turning on in

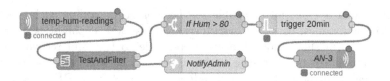

Fig. 4. `TestAndFilter` example in a flow that triggers an actuator if the humidity is above 80%, but verifies for correct sensor readings beforehand.

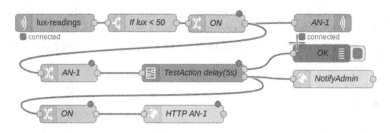

Fig. 5. `TestAction` example, where a verification is made to check if the lights turn on (request sent via MQTT) after a given interval, by checking if the luminosity lowers below 50 lux. If not, a secondary flow sends a new *on* request via HTTP.

the presence of a fire). In this case, a verification can be carried after a certain amount of time (*cf.* Sect. 2.2), asserting if the request has been processed by the device, preferentially using an alternative communication protocol. As an example, after a state change request message is sent to an actuator via MQTT, one could request its status, after a given time, to verify if the reported state corresponds to what was expected. Fixing scenarios in which the state of the system does not correspond to the expected, requires a *diversity strategy*. Having *things* that are capable of using different protocols allows us to adapt by dynamically switching to the most stable one given the systems' conditions (although usually incurring in a trade-off, such as the differences in energy consumption between MQTT and HTTP). As an example, if the light controlling device does not turn on the lights, as requested by the MQTT broker, a second request is made to the same device, this time using HTTP. This only can be implemented if both the device and the system can communicate using several different protocols. For this, we implemented a `TestAction` node that connects to the trigger and actuator *nodes* and checks if actions are triggered correctly. If not, a secondary flow is triggered, repeating the failed request using a different protocol. The resulting scenario implementation is depicted in Fig. 5.

5.2   Results

We showed improvements to *SmartLab* reliability and dependability both by detecting failures as they happen and recovering or maintaining the systems' health. Node-RED does not provide any *out-of-the-box* solution for dealing with failing components, nor to dynamically change the system's behavior during runtime, which is essential to enable *self-healing*. After adding such function-alities via new nodes, users can now leverage these new capabilities. Our first example scenario shows how it becomes possible to test and recover from a SPOF (exemplified as a message broker failure). The same method could be used to deal with other SPOFs, including failures of Node-RED itself, with a `RedundancyManager` node that activates *duplicated* and *inactive* flows on a different Node-RED instance (provided one is available). The second scenario shows how to isolate a system's component to ensure that its misbehaviors do not compromise the system as a whole. The last scenario shows how we can now manage several (redundant) communication protocols as an enabler of *self-healing* mechanisms, and the importance of continuously asserting the actuators outcome.

6 Discussion

Ensuring the dependability of software systems has been the goal of most fault-tolerance research in the past years [6]. In IoT, ensuring systems are secure, reliable, and compliant is becoming a paramount concern due to the recent increase in safety-critical applications. Fault-tolerance becomes more challenging due to several factors, including, but not limited to: (1) the high heterogeneity of devices, (2) the interaction and limitations of systems deployed in a physical world, (3) the fragmentation of the field, ranging from the unusually high number of communication protocols, to the different and competing standards, and (4) the intrinsic dependability on hardware that might simply fail [2]. Moreover, in a perfect environment, every actuator should possess a monitoring sensor capable of verifying its intended end state; however, real-world cost efficiency might limit their availability to critical components.

The pervasiveness and complexity of IoT have contributed to the rise of visual programming, in particular Node-RED, as the go-to solution (see Sect. 1). Never-theless, as it slowly permeates our lives, it becomes crucial to ensure proper func-tioning through self-verification and self-recovery features: *self-healing*. Although previous work attempted to tackle runtime verification and self-healing mecha-nisms to specific IoT systems (see Sect. 3), none was found to provide this kind of feature in a visual environment. Previous work also relies heavily on new systems (*e.g.*, rule-based monitoring services and CEP approaches), without attempting to integrate into the existent ecosystem of tools and platforms.

Although we chose to extend Node-RED due to its popularity, several chal-lenges limit its potential concerning our use-cases (or introduce unnecessary accidental complexity). We already discussed some in Sect. 2.1, but while imple-menting our test scenarios (Sect. 5.1), the following issues became disproportion-ately prominent, namely:

Support for labels and annotations: Nodes do not visually provide sufficient information about their connectors and internal status, making flows harder to construct, debug, and adapt. Most (if not all) nodes configuration cannot

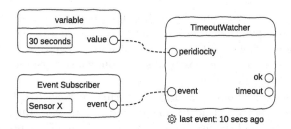

Fig. 6. Mockup of possible *node* interface with annotations, labels and multiple inputs/outputs.

be set or changed by other nodes. A solution similar to Fig. 6 seems more useful, not only in presenting this information but also in terms of flexibility regarding our goals;

Multiple inputs: Although Node-RED supports several outputs per node, they cannot have differentiated inputs (see Fig. 6). This poses both a cognitive and technical difficulty in defining and readjusting the behavior of nodes both during the design and runtime phases, including the configuration of test conditions and recovery measures;

Types and static analysis: Nodes do not have the notion of types; this allows the user to incorrectly connect two nodes, where the destination expects a different data type than the one sent by the origin one. This leads to common (and simple) errors that only make themselves noticeable after deployment of the whole flow, possibly introducing severe inconsistencies in the system;

Debugging: Besides the provided logging capabilities of Node-RED, using *debug nodes*, no other debugging technique is available. This means that breakpoints, *node* inspection, value history, and other apparatus are absent, severely hindering the ability for the developer to understand what went wrong in the internal logic of a node;

Meta-programming: Formal mechanisms of *introspection* and *reification*, essential for effective meta-programming, are non-existent. This limits the possibility of adjusting *flows* in runtime and forces us to rely on external APIs that were not designed for this particular purpose and which might easily break.

Despite these limitations, it was possible (up to a certain extent) to fulfill our goals mostly by using its visual notation, as seen in Sect. 5 and discussed in Sect. 5.2. It should be noted that all implemented strategies fall into the *forward error recovery* category, *i.e., "continue from an erroneous state by making selective corrections to the system state"* [17]. Exploration of *backward error recovery* techniques is harder due to the dependency of system state *checkpoints*, that needs to capture a mix of device internal states, concurrent communication protocols messages, and controller state.

To further improve the self-healing capabilities of systems such as the presented *SmartLab*, devices should have extra features such as diverse communication channels (*e.g.*, Wi-Fi and ZigBee), remote management capabilities (*e.g.*, independent watchdogs that allow to gracefully restore a device), and capability announcement, which would empower dynamic usage of redundant devices. We observe these features are mostly absent from consumer-grade devices, most probably due to cost efficiency.

Several challenges remain unaddressed by this work, such as (1) dealing with concurrent inputs that can lead to unexpected states (*e.g.*, the system decides to turn on the lights and the user manually turns them off), which may result in false assertions by the runtime verification mechanisms, (2) auto-discovery and configuration of new devices in the system (*e.g.*, a new mobile device can be used as a redundant sensing node while it remains in the system network), and (3) what are the reasonable operational states that the system should converge to in the case of failure (*e.g.* if the system has to decide between shutting down the smoke alarm or the surveillance system, which one should take prevalence?). Supporting and articulating with other *self-** aspects is also an open challenge towards fully autonomic systems; this includes *self-protection*, *self-optimization*, and *self-configuration* [13].

7   Conclusion

IoT systems are perhaps one of the most significant examples of heterogeneous architectures in existence. Different protocols, different application stacks, different integration services, and different orchestration engines, all must come together in a technological solution that allows both an organic growth from end-users, as well as dealing with security and privacy concerns at unprecedented levels. The consequence is that the system is required to keep functioning at minimal levels, even when parts of it become non-compliant, faulty, or even under attack. Requiring the end-user to address these challenges is unrealistic, as most of them are not developers. Even most system integrators cannot keep up with the pace of release devices, which very seldom adhere to open standards.

In this paper, we argue that an IoT system that attempts to tackle the presented challenges must be capable of *self-healing*. This is not a small feat, as most of the research being conducted in integration tools for IoT recurrently disregard failure detection and recovery. We fulfill these desiderata with SHEN, Self-Healing Extensions for Node-RED. As this very popular tool lacks built-in testing and self-healing capabilities, we use it as a case-study for common failure and recovery scenarios, and (1) show how to leverage *meta-programming* techniques to allow self-modification of *flows* via a custom *plugin*, (2) explore common self-healing patterns and how they can be solved by such techniques, (3) provide them as reusable *nodes* for others to incorporate in their systems, and (4) discuss which challenges remain open and which might need rethinking architectural and design decisions.

To validate our claims, we applied SHEN to the existing *SmartLab*, and proceed to show its behavior for three different scenarios, *viz.* (1) Unavailability of Message Broker, (2) Erroneous Sensor Readings and (3) Connectivity Issues.

We conclude that we can improve the system's reliability and dependability, both by being able to detect failing conditions, as well as reacting to them by self-modification of defined *flows*. Future work includes (a) the extension of the SHEN palette with more runtime verification's and self-healing mechanisms, and (b) case studies over various degrees of systems complexity, and in different contexts and scales.

Acknowledgement. This work was partially funded by the Portuguese Foundation for Science and Technology (FCT), under the research grants SFRH/BD/144612/2019 and SFRH/BD/115358/2016.

References

1. Aktas, M.S., Astekin, M.: Provenance aware run-time verification of things for self-healing Internet of Things applications. Concurr. Comput. **31**(3), 1–9 (2019)
2. Aly, M., Khomh, F., Gueheneuc, Y.G., Washizaki, H., Yacout, S.: Is fragmentation a threat to the success of the internet of things? IEEE Internet Things J. **6**(1), 472–487 (2019)
3. Angarita, R.: Responsible objects: towards self-healing Internet of Things applications. In: Proceedings - IEEE International Conference on Autonomic Computing, ICAC 2015, pp. 307–312 (2015)
4. Ashraf, Q.M., Habaebi, M.H.: Introducing autonomy in Internet of Things. In: 14th International Conference on Applied Computer and Applied Computational Science (ACACOS 2015) (2015)
5. Athreya, A.P., DeBruhl, B., Tague, P.: Designing for self-configuration and self-adaptation in the Internet of Things. In: Proceedings of the 9th IEEE International Conference on Collaborative Computing: Networking, Applications and Worksharing, COLLABORATECOM 2013, pp. 585–592 (2013)
6. Avizienis, A., Laprie, J.C., Randell, B.: Fundamental Concepts of Dependability. Technical Report Series University of Newcastle Upon Tyne Computing Science, vol. 1145, no. 010028, pp. 7–12 (2001)
7. Blackstock, M., Lea, R.: Toward a distributed data flow platform for the Web of Things (Distributed Node-RED). In: Proceedings of the 5th International Workshop on Web of Things - WoT 2014, pp. 34–39 (2014)
8. Delicato, F.C., Pires, P.F., Batista, T., Cavalcante, E., Costa, B., Barros, T.: Towards an IoT ecosystem. In: Proceedings of the First International Workshop on Software Engineering for Systems-of-Systems, SESoS 2013, pp. 25–28. ACM (2013)
9. Dias, J.P., Couto, F., Paiva, A.C.R., Ferreira, H.S.: A brief overview of existing tools for testing the Internet-of-Things. In: IEEE International Conference on Software Testing, Verification and Validation Workshops, pp. 104–109, April 2018
10. Dias, J.a.P., Ferreira, H.S., Sousa, T.B.: Testing and deployment patterns for the Internet-of-Things. In: Proceedings of the 24th European Conference on Pattern Languages of Programs. EuroPLop 2019. ACM (2019)
11. Dias, J.P.: jpdias/node-red-contrib-self-healing: Replication package for ICCS 2020, April 2020. https://doi.org/10.5281/zenodo.3746414
12. Dundar, B., Astekin, M., Aktas, M.S.: A big data processing framework for self-healing Internet of Things applications. In: 2016 12th International Conference on Semantics, Knowledge and Grids (SKG), pp. 62–68. IEEE (2016)

13. Ganek, A.G., Corbi, T.A.: The dawning of the autonomic computing era. IBM Syst. J. **42**(1), 5–18 (2003)
14. Ghosh, D., Sharman, R., Rao, H.R., Upadhyaya, S.: Self healing systems—survey and synthesis. Decis. Support Syst. **42**(4), 2164–2185 (2007). Decision Support Systems in Emerging Economies
15. İnçki, K., Arı, İ., Sözer, H.: Runtime verification of IoT systems using complex event processing. In: 2017 IEEE 14th International Conference on Networking, Sensing and Control (ICNSC), pp. 625–630. IEEE (2017)
16. Janssen, P., Erhan, H., Chen, K.W.: Visual dataflow modelling - some thoughts on complexity. In: Proceedings of the 32nd eCAADe Conference (2014)
17. Jia, W., Zhou, W.: Reliability and replication techniques. In: Distributed Network Systems: From Concepts to Implementations, pp. 213–254 (2005). https://doi.org/10.1007/0-387-23840-9_9
18. Kopetz, H.: Real-Time Systems. Real-Time Systems Series. Springer, Heidelberg (2011). https://doi.org/10.1007/978-1-4419-8237-7
19. Krupitzer, C., Roth, F.M., VanSyckel, S., Schiele, G., Becker, C.: A survey on engineering approaches for self-adaptive systems. Pervasive Mob. Comput. **17**, 184–206 (2015)
20. Leotta, M., Ancona, D., Franceschini, L., Olianas, D., Ribaudo, M., Ricca, F.: Towards a runtime verification approach for Internet of Things systems. In: Pautasso, C., Sánchez-Figueroa, F., Systä, K., Murillo Rodríguez, J.M. (eds.) ICWE 2018. LNCS, vol. 11153, pp. 83–96. Springer, Cham (2018). https://doi.org/10.1007/978-3-030-03056-8_8
21. Leucker, M., Schallhart, C.: A brief account of runtime verification. J. Logic Algebraic Program. **78**(5), 293–303 (2009)
22. Minerva, R., Biru, A., Rotondi, D.: Towards a definition of the Internet of Things (iot). IEEE Internet Initiative **1**, 1–86 (2015)
23. Morin, B., Harrand, N., Fleurey, F.: Model-based software engineering to tame the IoT jungle. IEEE Softw. **34**(1), 30–36 (2017)
24. Patel, P., Cassou, D.: Enabling high-level application development for the Internet of Things. J. Syst. Softw. **103**(C), 62–84 (2015)
25. Pontes, P.M., Lima, B., Faria, J.a.P.: Test patterns for IoT. In: Proceedings of the 9th ACM SIGSOFT International Workshop on Automating TEST Case Design, Selection, and Evaluation, A-TEST 2018, pp. 63–66. ACM (2018)
26. Prehofer, C., Chiarabini, L.: From IoT mashups to model-based IoT. In: W3C Workshop on the Web of Things (2013)
27. Prehofer, C., Chiarabini, L.: From Internet of Things mashups to model-based development. In: 2015 IEEE 39th Annual Computer Software and Applications Conference, vol. 3, pp. 499–504. IEEE (2015)
28. Psaier, H., Dustdar, S.: A survey on self-healing systems: approaches and systems. Computing (Vienna/N.Y.) **91**(1), 43–73 (2011)
29. Ray, P.P.: A survey on visual programming languages in Internet of Things. Sci. Program. **2017**, 1–6 (2017)
30. Seeger, J., Bröring, A., Carle, G.: Optimally self-healing IoT choreographies (2019)
31. Vermesan, O., et al.: Internet of Things strategic research roadmap. Internet Things-Glob. Technol. Soc. Trends **1**(2011), 9–52 (2011)

Comparative Analysis of Time Series Databases in the Context of Edge Computing for Low Power Sensor Networks

Piotr Grzesik$^{(\boxtimes)}$ and Dariusz Mrozek

Department of Applied Informatics, Silesian University of Technology,
ul. Akademicka 16, 44-100 Gliwice, Poland
pj.grzesik@gmail.com, dariusz.mrozek@polsl.pl

Abstract. Selection of an appropriate database system for edge IoT devices is one of the essential elements that determine efficient edge-based data analysis in low power wireless sensor networks. This paper presents a comparative analysis of time series databases in the context of edge computing for IoT and Smart Systems. The research focuses on the performance comparison between three time-series databases: TimescaleDB, InfluxDB, Riak TS, as well as two relational databases, PostgreSQL and SQLite. All selected solutions were tested while being deployed on a single-board computer, Raspberry Pi. For each of them, the database schema was designed, based on a data model representing sensor readings and their corresponding timestamps. For performance testing, we developed a small application that was able to simulate insertion and querying operations. The results of the experiments showed that for presented scenarios of reading data, PostgreSQL and InfluxDB emerged as the most performing solutions. For tested insertion scenarios, PostgreSQL turned out to be the fastest. Carried out experiments also proved that low-cost, single-board computers such as Raspberry Pi can be used as small-scale data aggregation nodes on edge device in low power wireless sensor networks, that often serve as a base for IoT-based smart systems.

Keywords: Time series · PostgreSQL · TimescaleDB · InfluxDB · Edge computing · Edge analytics · Raspberry Pi · Riak TS · SQLite

1 Introduction

In the recent years we have been observing IoT systems being applied for multiple use cases such as water monitoring [20], air quality monitoring [24], and health monitoring [25], generating a massive amount of data that is being sent to the cloud for storing and further processing. This is becoming a more significant challenge due to the need for sending the data over the Internet. Due to that, a new computing paradigm called edge computing started to emerge [28]. The main idea behind edge computing is to move data processing from the cloud to the

© Springer Nature Switzerland AG 2020
V. V. Krzhizhanovskaya et al. (Eds.): ICCS 2020, LNCS 12141, pp. 371–383, 2020.
https://doi.org/10.1007/978-3-030-50426-7_28

devices that are closer to the source of data in order to reduce the volume of data that needs to be send to the cloud, improve reaction time to the changing state of the system, provide resilience and prevent data loss in situations where Internet connection is not reliable or even not available most of the time. To achieve that, edge computing devices need to be able to ingest data from sensors, analyze them, aggregate metrics, and send them to the cloud for further processing if required. For example, while collecting and processing environmental data on air quality, the edge device can be responsible for aggregating data and computing Air Quality Index (AQI) [22], instead of sending raw sensor readings to the environmental monitoring center. In systems with multiple sensors generating data at a fast rate, efficient storage and analytical system running on edge device becomes a crucial part. Due to the time-series nature of sensor data, dedicated time series databases seem like a natural fit for this type of workload. This paper aims to evaluate several time series databases in the context of using them in edge computing, low-cost, constrained device in form of Raspberry Pi that is processing data from environmental sensors. The paper is organized as follows. In Sect. 2, we review the related works. In Sect. 3, we describe databases selected for comparison. Section 4 describes testing environment, used data model as well as testing methodology. Section 5 contains a description of the performance experiments that we carried out. Finally, Sect. 6 concludes the results of the paper.

2   Related Works

In the literature, there is a few research concerning the comparison of various time-series databases. In the paper [27], Tulasi Priyanka Sanaboyina compared two time-series databases, InfluxDB and OpenTSDB, based on the energy consumption of the physical servers on which the databases are running under several reading and writing scenarios. The author concludes the research with claims that InfluxDB consumes less energy than OpenTSDB in comparable situations.

Bader et al. [17] focused on open source time-series databases, examined 83 different solutions during their research, and focused on the comparison of twelve selected databases, including InfluxDB, PostgreSQL and OpenTSDB among others. All selected solutions were compared based on their scalability, supported functions, granularity, available interfaces, and extensions as well as licensing and support.

In his research [21], Goldschmidt et al. benchmarked three open-source time-series databases, OpenTSDB, KariosDB and Databus in the cloud environment with up to 36 nodes in the context of industrial workloads. The main objective of the research was to evaluate selected databases to determine their scalability and reliability features. Out of the three technologies, KairosDB emerged as the one that meets the initial hypotheses about scalability and reliability.

Wlodarczyk, in his article [29], provides an overview and comparison of four offerings, Chukwa, OpenTSDB, TempoDB, and Squwk. The analysis focused on feature differences between selected technologies, without any performance benchmarks. The author identified OpenTSDB as a most popular choice for the time series storage.

Pungilă et al. [26] compared the databases to use them in the system that stores large volumes of sensor data from smart meters. During the research, they compared three relational databases, SQLite3, MySQL, PostgreSQL, one time-series database, IBM Informix with DataBlade module, as well as three NoSQL databases, MonetDB, Hypertable and Oracle BerkeleyDB. During the experiments, it was determined that Hypertable offers the most significant number of insert operations per second, but is slower when it comes to scanning operations. The authors suggested that BerkeleyDB offers a compromise when there is a need for a workload that has a balanced number of both insert and scan operations.

Fadhel et al. presented research [20] concerning the evaluation of the databases for a low-cost water quality sensing system. Authors identified InfluxDB as the most suitable solution, listing the ease of installation and maintenance, support for multiple interface formats, and HTTP GUI as the deciding factors. In the second part of the research, they conducted performance experiments and determined that InfluxDB can handle the load from 450 sensors.

In his article [23], Kiefer provided a performance comparison between PostgreSQL and TimescaleDB for storage and analytics of large scale, time-series data. The author presented that at the scale of millions of rows, TimescaleDB offers up to 20× higher ingest rates than PostgreSQL, at the same time offering time-based queries to be even 14,000× faster. The author also mentions that for simple queries, e.g., indexed lookups, TimescaleDB will be slower than PostgreSQL due to more considerable planning time.

Boule, in his work [19], described a performance comparison for insert and read operations between InfluxDB and TimescaleDB. It is based on a simulated dataset of metrics for a fleet of trucks. According to results obtained during the experiments, TimescaleDB offers a better read performance than InfluxDB in tested scenarios.

Based on the above, it can be concluded that most of the current research focuses on the use of time-series databases for large-scale systems, running in cloud environments. One exception to that is the research [20], where authors evaluate several databases in the context of a low-cost system; however, presenting performance tests only for one of them, InfluxDB. In contrast to the mentioned works, this paper focuses on the comparison of the performance of several database systems for storing sensor data at the edge devices that have limited storage and compute capabilities.

3   Time-Series Databases

Time series database (TSDB) is a database type designed and optimized to handle timestamped or time-series data, which is characterized by a low number of relationships between data and temporal ordering of records. Most of the time series workloads consist of a high number of insert operations, often in batches.

Query patterns include some forms of aggregation over time. It is also important to note that in such workloads, data usually does not require updating after being inserted. To accommodate these requirements, time-series databases store data in the form of events, metrics, or measurements, typically numerical, together with their corresponding timestamps and additional labels or tags. Data is very often chunked, based on timestamp, which in turn allows for fast and efficient time-based queries and aggregations. Most TSDBs offer advanced data processing capabilities such as window functions, automatic aggregation functions, time bucketing, and advanced data retention policies. There are currently a few approaches to building a time-series database. Some of them, like OpenTSDB or TimescaleDB, depend on already existing databases, such as HBase or PostgreSQL, respectively, while others are standalone, independent systems such as InfluxDB. In recent years, according to DB Engine ranking, as seen in Fig. 1, the growth rate of the popularity of time series databases is the highest out of all classified database types. For the experiments, databases were selected based on their popularity, offered aggregation functionalities, support for ARM architecture, SQL or SQL-like query language support as well as on their availability without commercial license.

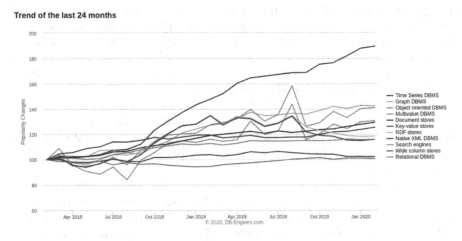

Fig. 1. Growth trend of various types of databases in the last 24 months according to DB-Engines.com [2]

3.1   TimescaleDB

TimescaleDB is an open-source, time-series database, written in C programming language and is distributed as an extension of the relational database, PostgreSQL. It is developed by Timescale Inc., which also offers enterprise support and cloud hosting in the form of Timescale Cloud offering. TimescaleDB is optimized for fast ingest and complex queries [14]. Thanks to the support for all SQL operations available in PostgreSQL, it can be used as a drop-in replacement of a traditional relational database, while also offering significant performance improvements for storing and processing time-series data. By taking advantage of automatic space-time partitioning, it enables horizontal scaling, which in turn can further improve the ingestion capabilities of the system. It stores data in structures called hypertables, which serve as an abstraction for a single, continuous table. Internally, TimescaleDB splits hypertables into chunks that correspond to a specific time interval and partition keys. Chunks are implemented by using regular PostgreSQL tables [16]. Thanks to being an extension of PostgreSQL DBMS, it supports the same client libraries that support PostgreSQL. According to the DB Engines ranking [15], it is the 8th most popular time-series database.

3.2   InfluxDB

InfluxDB is an open-source, time-series database, written in Go programming language, developed and maintained by InfluxDB Inc., which also offers enterprise support and a cloud-hosted version of the database. Internally, it uses a custom-build storage engine called Time-Structured Merge (TSM) Tree, which is optimized for time series data. It has no external dependencies, is distributed as a single binary, which in turn allows for easy deployment process on all major operating systems and platforms. InfluxDB supports InfluxQL, which is a custom, SQL-like query language with support for aggregation functions over time series data. It supports advanced data retention policies as well as continuous queries, which allow for automatic computations of aggregate data to speed up frequently used queries [5]. It uses shards to partition data and organizes them into shards groups, based on the retention policy and timestamps. InfluxDB is also a part of TICK stack [4], which is a data processing platform that consists of a time-series database in form of InfluxDB, Kapacitor, which is a real-time streaming data processing engine, Telegraf, the data collection agent and Chronograf, a graphical user interface to the platform. Client libraries in the programming languages like Go, Python, Java, Ruby, and others are available, as well as command-line client "influx". According to DB Engines ranking [3], it is the most popular time-series database management system.

3.3   Riak TS

Riak TS is an open-source, distributed NoSQL database, optimized for the time series data and built on top of Riak KV database [9], created and maintained by Basho Technologies. Riak TS is written in Erlang programming language, supports masterless, multi-node architecture to ensure resiliency to network and hardware failures. This type of architecture also allows for efficient scalability with near-linear performance increase [10]. It supports a SQL-like query language with aggregation operations over time series data. It offers both HTTP and PBC APIs as well as dedicated client libraries in Java, Python, Ruby, Erlang, and Node.js. Besides, it has a native Apache Spark [1] connector for the in-memory analytics. According to DB Engines ranking [11], it is the 15th most popular time-series database.

3.4 PostgreSQL

PostgreSQL is an open-source relational database management system written in C language and currently maintained by PostgreSQL Global Development Group. PostgreSQL runs on all major operating systems, is ACID [30] compliant and supports various extensions, namely TimescaleDB. It supports a major part of the SQL standard and offers many features, including but not limited to, triggers, views, transactions, streaming replication. It uses multi-version concurrency control, MVCC [18]. In addition to being a relational database, it also offers support for storing and querying document data thanks to JSON, JSONB, XML, and Key-value data types [6]. There are client libraries available in programming languages like Python, C, C++, Java, Go, Erlang, Rust, and others. According to DB Engines ranking [7], it is the 4th most popular database overall. It does not offer any dedicated support and optimizations for time-series data.

3.5 SQLite

SQLite is an open-source relational database, written in C language. The SQLite source code is currently available in the public domain. It is a lightweight, single file, and unlike most databases, it is implemented only as a library and does not require a separate server process. SQLite provides all functionalities directly by the function calls. Its simplicity makes it one of the most widely used databases, especially popular in embedded systems. SQLite has a full-featured SQL standard implementation with support for functionalities such as triggers, views, indexes, and many more [12]. Similar to PostgreSQL, it does not offer any specific support for time series data. Besides, it does not provide a data type for storing time, and it requires users to save it as numerical timestamps or strings. According to DB Engines ranking [13], it is the 7th most popular relational database and 10th most popular database overall.

4   Testing Environment and Data Model

The testing environment was based on a 6LoWPAN sensor network that is a part of the environment monitoring system, which consists of a group of the edge router device that additionally serves as a database and analytical engine. It is also responsible for sending aggregated metrics to the analytic system in the cloud for further processing. Another part of the network is composed of ten sensor nodes that are sending measurements such as air quality and weather condition metrics to the edge router device. Figure 2 presents the network diagram of the described system.

In this research, we focused on performance evaluation of the edge database functionality of the presented system. To simplify the testing environment and

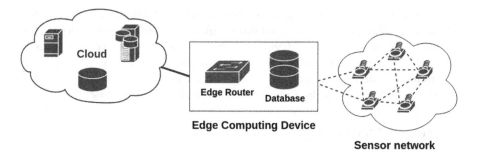

Fig. 2. Network diagram of the edge computing system.

allow for running tests multiple times in a reasonable amount of time, we developed a small Python application to serve as a generator of sensor readings instead of using data generated by the physical network. As an edge device we decided to use a Raspberry Pi single-board computer, with the following specification [8]:

- CPU - Broadcom BCM2711, Quad core Cortex-A72 (ARM v8) 64-bit SoC @ 1.5 GHz
- Memory - 4 GB LPDDR4-3200 SDRAM
- Storage - SDHC card (16 GB, class 10)
- OS - Raspbian GNU/Linux 10 (buster) with kernel version 4.19.50-v7l+

Table 1. Data model used for the performance experiments

Value	Type
Temperature	Float
Pressure	Float
Humidity	Float
PM2.5	Float
PM10	Float
NO_2	Float
Sensor ID	String
Location	String
Time	Timestamp (Integer)

4.1 Data Model

Each data point sent by the sensor consists of air quality metrics in the form of NO_2 and dust particle size metrics – PM2.5 and PM10. Besides, it also carries information about weather conditions such as ambient temperature, pressure, and humidity. Each reading is timestamped and tagged with the location of the sensor and the unique sensor identifier. Table 1 shows the structure of a single data point with corresponding data types. For the experiments, we generated data from 10 simulated sensors, where each sensor sends reading every 15 s over 24 h. It resulted in 28,800 data points used for performance testing.

4.2   Testing Methodology

For testing, a small Python application was developed separately for each of the selected databases. The application was responsible for reading simulated time-series data, inserting that data into the database and reading the data back from the database, while measuring the time it took to execute all of the described operations. Table 2 presents the list of the databases along with their corresponding client libraries. It also shows versions of the software used during the experiments.

Table 2. Database and client library versions

Database	Database version	Client library	Client library version
TimescaleDB	1.5.1	psycopg2	2.8.4
InfluxDB	1.7.9	influxdb	5.2.3
Riak TS	1.5.2	riak	2.7.0
PostgreSQL	11.5	psycopg2	2.8.4
SQLite	3.27.2	sqlite3	2.6.0

5   Performance Experiments

To evaluate the insertion and querying performance, we conducted several experiments. Firstly, we ran the test to assess the writing capabilities of all selected databases by simulating the insertion of data points in two ways: one-by-one and in batches of 10 points. The reason for that was to accommodate the fact that databases can offer better performance for batch insertions, and it is possible to buffer data before saving it to the database. In this step, for each database, we ran the simulation 50 times (except for SQLite where simulations were run 20 times due to relatively long simulation time). Secondly, we ran the experiments to evaluate the query performance of all selected solutions in three scenarios. In the first scenario, we evaluated a query for average temperature in the chosen period, grouped by location. In the second query, we tested a query for minimum and maximum values of NO_2, PM2.5, and PM10 in the selected period, once again grouped by location. In the last, third scenario, we evaluated the performance of a query that counts data points grouped by sensor ID in the selected period for which NO_2 was larger than selected value and location was equal to a specific one. Each query was executed 5000 times. The query scenarios were selected in order to test the performance of the databases for most common aggregation queries that can be used in scenarios where the analysis has to be performed directly on the edge device or when the data needs to be aggregated before sending to the cloud in order to reduce the volume of transferred data.

5.1   Insertion

In the first simulation, we evaluated the insertion performance in two different scenarios. Figure 3 presents the obtained results in the form of the average number of data points inserted per second in both scenarios. For one-by-one insertion, we observe PostgreSQL and TimescaleDB as the best performing solutions, with 260 and 230 points inserted per second, respectively. Next is Riak TS with 191 points, followed by InfluxDB with 54 points per second. On the other side of the spectrum is SQLite, with only 8 points per second inserted on average. In the second scenario, with batch insertions of 10-point batches, we observed a general trend of higher ingestion rates for all databases in comparison to single point writes with InfluxDB being 8.65 times faster, both PostgreSQL and SQLite improving 6.74 times, TimescaleDB improving 5.15 times and Riak TS noting the smallest relative increase by 2.45 times. We observed similar results as for one-by-one insertion, with PostgreSQL being the most performant database in the tested scenario with 1,756 data points ingested per second, followed by TimescaleDB with 1,187 points, InfluxDB with 474 points and Riak TS with 468 points. Once again, SQLite recorded the worst performance, with only 55 points ingested per second.

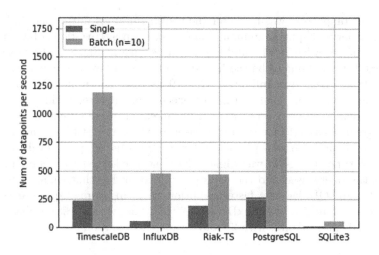

Fig. 3. Number of data points ingested per second for each tested database.

5.2 Querying

In the following experiments, we tested the reading performance for three different queries. Results are presented in the form of the average query execution time in milliseconds for each database. Due to the fact that execution for Riak TS was in all cases 20–40 times slower than for all other solutions, the results for Riak TS were removed from the further comparison to improve the readability of the presented charts. Figure 4 shows both the query used in the first scenario as well as the obtained results. In this scenario, InfluxDB emerged as the fastest solution with average query execution time of 24 ms, followed by PostgreSQL and TimescaleDB with 41 and 52 ms, respectively. SQLite was the slowest, recording average query execution time of 66 ms.

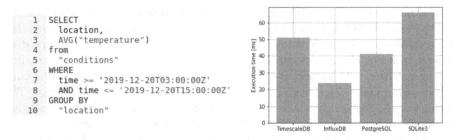

```
 1   SELECT
 2       location,
 3       AVG("temperature")
 4   from
 5       "conditions"
 6   WHERE
 7       time >= '2019-12-20T03:00:00Z'
 8       AND time <= '2019-12-20T15:00:00Z'
 9   GROUP BY
10       "location"
```

(a) Tested aggregation query   (b) Average query execution time

Fig. 4. Query and test results for the first querying scenario.

Next, a comparison was made for the results obtained during the evaluation of second query computing minimum and maximum aggregations of air quality metrics. The recorded results and queries are shown in Fig. 5. In this example, PostgreSQL turned out to be the fastest solution with average query execution time of 48 ms, next was InfluxDB with 70 ms and TimescaleDB with 72 ms. Tested query took the longest time to execute on SQLite, taking on average 81 ms. We can observe a general trend of increased query execution time with more aggregations performed in comparison to the first testing scenario.

The last experiment was performed for the third tested query, evaluating the number of times the NO_2 was higher than the predefined threshold. Figure 6 presents the query used and the results obtained during that simulation. Once again, PostgreSQL was the fastest solution with an average query execution time of 15 ms, followed by InfluxDB with 29 ms. The two slowest databases were TimescaleDB and SQLite, with 39 and 40 ms per execution on average.

5.3 Results Summary

Considering results for all presented simulations, we can observe that in almost all cases, PostgreSQL is the best performing solution for the evaluated workloads,

```
1   SELECT
2     location,
3     MIN("pm_2_5"),
4     MAX("pm_2_5"),
5     MIN("pm_10"),
6     MAX("pm_10"),
7     MIN("no2"),
8     MAX("no2")
9   from
10     "conditions"
11   WHERE
12     time >= '2019-12-20T03:00:00Z'
13     AND time <= '2019-12-20T15:00:00Z'
14   GROUP BY
15     "location"
```

(a) Tested aggregation query   (b) Average query execution time

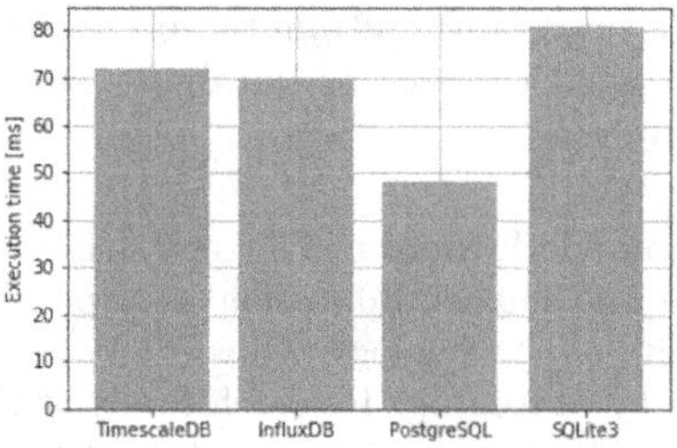

Fig. 5. Query and test results for the second querying scenario.

```
1   SELECT
2     sensor_id,
3     Count("no2")
4   from
5     "conditions"
6   WHERE
7     "no2" > 8
8     AND "location" = 'outdoor'
9     AND time >= '2019-12-20T03:00:00Z'
10     AND time <= '2019-12-20T15:00:00Z'
11   GROUP BY
12     "sensor_id"
```

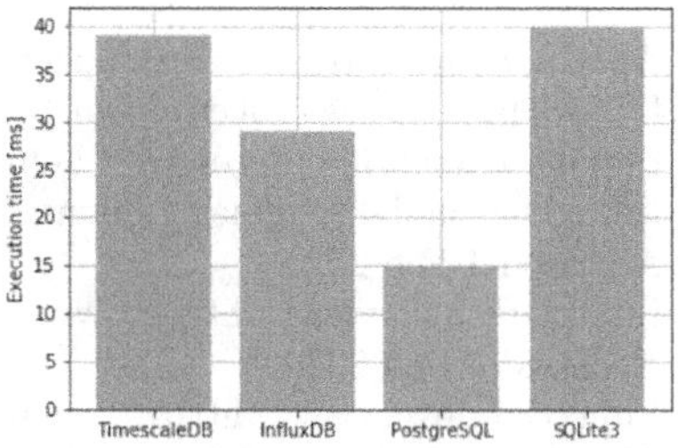

(a) Tested aggregation query   (b) Average query execution time

Fig. 6. Query and test results for the third querying scenario.

except for InfluxDB, which turned out to be faster for the first aggregation query. It was validated that batching data points for insertion causes performance gains, as high as 8.65 times more data points ingested per second for InfluxDB. With the exception of Riak TS, all databases executed tested queries on average in less than 80 ms, and the relative differences in performance for queries are not as high as in the case of insertion.

6   Concluding Remarks

The selection of a proper storage system with declarative querying capabilities is an essential element of building efficient systems with edge-based analytics. This research aimed to compare the performance of several databases in the context of edge computing in wireless sensor networks for IoT-based smart systems. We believe that experiments and analysis of the results presented in the paper complement the performance evaluation of InfluxDB presented in [20] by showcasing performance results for multiple databases and can serve as a reference when selecting an appropriate database for low-cost, edge analytics applications. As it turned out, for a smaller scale, it might make sense to choose a more traditional, relational database like PostgreSQL, which offers the best performance in all but one tested case. However, when features such as data retention policies, time bucketing, automatic aggregations are crucial for the developed solution, dedicated time-series databases such as TimescaleDB and InfluxDB become a better choice.

Acknowledgments. The research was supported by the Polish Ministry of Science and Higher Education as a part of the CyPhiS program at the Silesian University of Technology, Gliwice, Poland (Contract No. POWR.03.02.00-00-I007/17-00), by Statuatory Research funds of the Silesian University of Technology, Gliwice, Poland (Grant BKM-576/RAU2/2019 ZAD.1), and partially, by the professorship grant (02/020/RGPL9 /0184) of the Rector of the Silesian University of Technology, Gliwice, Poland.

References

1. Apache Spark. https://spark.apache.org/. Accessed 9 Jan 2020
2. DBMS popularity broken down by database model. https://db-engines.com/en/ranking_categories. Accessed 2 Feb 2020
3. InfluxDB on DB-engines ranking. https://db-engines.com/en/system/InfluxDB. Accessed 1 Feb 2020
4. InfluxDB overview. https://www.influxdata.com/products/influxdb-overview/. Accessed 2 Feb 2020
5. InfluxDB overview. https://www.influxdata.com/products/influxdb-overview/. Accessed 9 Jan 2020
6. PostgreSQL documentation. https://www.postgresql.org/about/. Accessed 9 Jan 2020
7. PostgreSQL on DB-engines ranking. https://db-engines.com/en/system/PostgreSQL. Accessed 1 Feb 2020
8. Raspberry Pi 4 datasheet. https://www.raspberrypi.org/documentation/hardware/raspberrypi/bcm2711/rpi_DATA_2711_1p0_preliminary.pdf. Accessed 4 Feb 2020
9. Riak KV documentation. https://riak.com/products/riak-kv/index.html. Accessed 9 Jan 2020
10. Riak TS datasheet. https://riak.com/content/uploads/2016/05/Riak-Riak-TS-Datasheet.pdf. Accessed 9 Jan 2020
11. Riak TS on DB-engines ranking. https://db-engines.com/en/system/Riak+TS. Accessed 1 Feb 2020
12. SQLite documentation. https://www.sqlite.org/about.html. Accessed 9 Jan 2020
13. SQLite on DB-engines ranking. https://db-engines.com/en/system/SQLite. Accessed 1 Feb 2020
14. TimescaleDB documentation. https://docs.timescale.com/latest/introduction. Accessed 9 Jan 2020
15. TimescaleDB on DB-engines ranking. https://db-engines.com/en/system/TimescaleDB. Accessed 1 Feb 2020
16. TimescaleDB: SQL made scalable for time-series data (2017). https://pdfs.semanticscholar.org/049a/af11fa98525b663da18f39d5dcc5d345eb9a.pdf
17. Bader, A., Kopp, O., Falkenthal, M.: Survey and comparison of open source time series databases. In: Mitschang, B., et al. (eds.) Datenbanksysteme für Business, Technologie und Web (BTW 2017) - Workshopband, pp. 249–268. Gesellschaft für Informatik e.V., Bonn (2017)

18. Bernstein, P.A., Goodman, N.: Concurrency control in distributed database systems. ACM Comput. Surv. (CSUR) **13**(2), 185–221 (1981)
19. Boule, B.: How to benchmark IoT time-series workloads in a production environment. https://blog.timescale.com/blog/how-to-benchmark-iot-time-series-workloads-in-a-production-environment/. Accessed 9 Jan 2020
20. Fadhel, M., Sekerinski, E., Yao, S.: A comparison of time series databases for storing water quality data. In: Auer, M.E., Tsiatsos, T. (eds.) IMCL 2018. AISC, vol. 909, pp. 302–313. Springer, Cham (2019). https://doi.org/10.1007/978-3-030-11434-3_33
21. Goldschmidt, T., Jansen, A., Koziolek, H., Doppelhamer, J., Breivold, H.P.: Scalability and robustness of time-series databases for cloud-native monitoring of industrial processes. In: 2014 IEEE 7th International Conference on Cloud Computing, pp. 602–609, June 2014
22. Kanchan, K., Gorai, A., Goyal, P.: A review on air quality indexing system. Asian J. Atmos. Environ. **9**, 101–113 (2015)
23. Kiefer, R.: TimescaleDB vs. PostgreSQL for time-series: 20x higher inserts, 2000x faster deletes, 1.2x-14,000x faster queries. https://blog.timescale.com/blog/timescaledb-vs-6a696248104e/. Accessed 9 Jan 2020
24. Liu, X., Nielsen, P.S.: Air quality monitoring system and benchmarking. In: Bellatreche, L., Chakravarthy, S. (eds.) DaWaK 2017. LNCS, vol. 10440, pp. 459–470. Springer, Cham (2017). https://doi.org/10.1007/978-3-319-64283-3_34
25. Paul, A., Pinjari, H., Hong, W.H., Seo, H., Rho, S.: Fog computing-based IoT for health monitoring system. J. Sens. **2018**, 1–7 (2018)
26. Pungila, C., Fortiş, T.F., Ovidiu, A.: Benchmarking database systems for the requirements of sensor readings. IETE Tech. Rev. **26**, 342–349 (2009)
27. Sanaboyina, T.P.: Performance evaluation of time series databases based on energy consumption. Master's thesis, Department of Communication Systems (2016)
28. Singh, S.: Optimize cloud computations using edge computing. In: 2017 International Conference on Big Data, IoT and Data Science (BID), pp. 49–53, December 2017
29. Wlodarczyk, T.W.: Overview of time series storage and processing in a cloud environment. In: 4th IEEE International Conference on Cloud Computing Technology and Science Proceedings, pp. 625–628, December 2012
30. Yu, S.: ACID properties in distributed databases. Advanced eBusiness Transactions for B2B-Collaborations (2009)

Conversational Interface for Managing Non-trivial Internet-of-Things Systems

André Sousa Lago[1], João Pedro Dias[1,2](✉), and Hugo Sereno Ferreira[1,2]

[1] DEI, Faculty of Engineering, University of Porto, Porto, Portugal
{up201303313,jpmdias,hugosf}@fe.up.pt
[2] INESC TEC, Porto, Portugal

Abstract. Internet-of-Things has reshaped the way people interact with their surroundings. In a smart home, controlling the lights is as simple as speaking to a conversational assistant since everything is now Internet-connected. But despite their pervasiveness, most of the existent IoT systems provide limited out-of-the-box customization capabilities. Several solutions try to attain this issue leveraging end-user programming features that allow users to define rules to their systems, at the cost of discarding the easiness of voice interaction. However, as the number of devices increases, along with the number of household members, the complexity of managing such systems becomes a problem, including finding out why something has happened. In this work we present Jarvis, a conversational interface to manage IoT systems that attempts to address these issues by allowing users to specify time-based rules, use contextual awareness for more natural interactions, provide event management and support causality queries. A proof-of-concept was used to carry out a quasi-experiment with non-technical participants that provides evidence that such approach is intuitive enough to be used by common end-users.

Keywords: Internet-of-Things · Conversational assistants · Software engineering · Natural Language Processing · Visual programming

1 Introduction

The Internet of Things (IoT) is usually defined as the networked connection of everyday objects with actuating and sensing capabilities, often equipped with a collective sense of intelligence [21]. The integration of such objects creates a vast array of distributed systems that can interact with both the environment and the human beings around them, in a lot of different ways [21]. This flexibility of IoT systems has enabled their use across many different product areas and markets, including smart homes, smart cities, healthcare, transportation, retail, wearables, agriculture and industry [17].

Still, one of the most visible application of IoT are *customized smart spaces*, such as *smart homes* as the current technology make it possible for consumers to create a customized IoT experience based on *off-the-shelf* products [16].

© Springer Nature Switzerland AG 2020
V. V. Krzhizhanovskaya et al. (Eds.): ICCS 2020, LNCS 12141, pp. 384–397, 2020.
https://doi.org/10.1007/978-3-030-50426-7_29

The initial popularity of devices such as single-board computers and low-cost micro-controllers, followed by widespread cloud-based solutions controlled by mobile phones, it is now commonplace to remotely interact with a myriad of devices to perform automated tasks such as turning the lights on and opening the garage door just before one arrives home [16,22]. But as the number of devices and interactions grows, so does the management needs of the system as a whole, as it becomes essential to understand and modify the way they (co)operate. In the literature this capability commonly known as *end-user programming* [6], and once we discard trained system integrators and developers, two common approaches emerge: visual programming tools and conversational assistants [22].

Visual programming solutions are usually deployed as centralized orchestrators, with access to the devices and components that comprise such systems. These platforms range from simplified *if-then* rules (*e.g.* IFTTT[1]) to exhaustive graphical interfaces (*e.g.* Node-RED[2]) through which one can visualize, configure and customize the devices and systems' behaviour [7,19,20]. Most visual approaches attempt to offer integration with third-party components (*e.g.*, google calendar), so that their services can be used as part of the system's behavioural rules.

These solutions, however, possess some disadvantages for non-technical *end-users*. Consider a Node-RED system orchestrating an user's smart home with multiple devices. Even in situations where there are only a couple of dozen rules defined, it can be challenging to understand why a specific event took place due to the overwhelming data flow that results from these. Furthermore, just one dozen rules can already lead to a system not possible to visualize in a single screen [11]. The more rules one adds, the harder it becomes to conceptually grasp what the system can do. Part of the reason is because they are built to be *imperative*, not *informative*; current solutions mostly lack in meta-facilities that enable the user or the system to *query* itself [5].

Another common, and sometimes complementary, alternative to visual programming, is the many conversational assistants in the market, such as Google Assistant, Alexa, Siri and Cortana, that are capable of answering natural language questions and which recently gained the ability to interact with IoT devices (see [18] and [15] for a comparison of these tools). Amongst the most common features they provide is allowing direct interaction with sensing and actuating devices, which enables the *end-user* to *talk* to their light bulbs, thermostats, sound systems, and even third-party services. The problem with these solutions is that they are mostly comprised of *simple* commands and queries directly to the smart devices (*e.g. is the baby monitor on?"*, *"what is the temperature in the living room?"*, or *"turn on the coffee machine"*. These limitations mean that although these assistants do provide a comfortable *interaction* with devices, a huge gap is easily observable regarding their capabilities on *managing* a system as a whole and allowing the definition of rules for how these *smart spaces* oper-

---

[1] IFTTT, https://ifttt.com.
[2] Node-RED, https://nodered.org.

ate. Even simple rules like *"close the windows everyday at 8pm"* or *"turn on the porch light whenever it rains"* are currently not possible, unless one manually defines every single one of them as a capability via a non-conversational mechanism. Furthermore, most assistant are deliberately locked to specific vendor devices, thus limiting the overall experience and integration.

One can conclude that although current smart assistants can be beneficial and comfortable to use, they do not yet have the complexity and completeness that other systems like Node-RED. Meanwhile, visual programming environments are still far too technical for the common *end user*. In this paper, we propose a system that tackles the problem of *managing* IoT systems in a conversational approach, towards shortening the existing feature gap between assistants and visual programming environments. Parts of this work are from [13] master's thesis.

The rest of this document is structured as follows: Sect. 2 provides a summary of related works which identify open research challenges; in Sect. 3 we propose our approach to support *complex* queries in conversational assistants, which implementation details are further presented in Sect. 4; we proceed in Sect. 5 to evaluate our approach using simulated scenarios and experimental studies. Finally, Sect. 6 drafts some closing remarks.

2   Related Work

There exists some work in this area that recognize the problem of controlling and managing IoT infrastructures by an *end-user* via a several approaches.

Kodali et al. [12] present an home automation system to *"increase the comfort and quality of life"*, by developing an Android app that is able to control and monitor home appliances using MQTT, Node-RED, IFTTT, Mongoose OS and Google Assistant. Their limitations lie in that the *flows* must first have been first created in Node-RED, and the conversational interface is used just to trigger them, ignoring all the *management* activities.

Austerjost et al. [3] recognized the usefulness of voice assistants in home automation and developed a system that targets laboratories. Possible applications reported in their paper include stepwise reading of standard operating procedures and recipes, recitation of chemical substance or reaction parameters to a control, and readout of laboratory devices and sensors. As with the other works presented, their voice user interface only allows controlling devices and reading out specific device data.

He et al. [9], concludes that, even with conversational assistants, most of IoT systems have usability issues when faced with complex situations. As example, the complexity of managing devices schedules rises with the number of devices and the common conflicting preferences of household members. Nonetheless, as concluded by Ammari et al. [2], controlling IoT devices is one of the most common uses of such assistants.

Agadakos et al. [1] focus on the challenge of understanding the causes and effects of an action to infer a potential sequence. Their work is based on a mapping the IoT system' devices and potential interactions, measuring expected behaviours with traffic analysis and side-channel information (e.g. power) and detecting causality by matching the mapping with the collected operational data. This approach would potentially allow the *end user* to ask *why is something happening*, at the cost of modified hardware and a convoluted side-channel analysis. They did not attempted to port their findings into a conversational approach.

Braines et al. [4] present an approach based on Controlled Natural Language (CNL) – natural language using only a restricted set of grammar rules and vocabulary – to control a smart home. Their solution supports (1) *direct question/answer exchanges*, (2) *questions that require a rationale as response* such as *"Why is the room cold?"* and (3) *explicit requests to change a particular state*. The most novel part of their solution is in trying to answer *questions that require a rational response*, however they depend on a pre-defined smart home model that maps all the possible causes to effects.

From the above analysis, the authors were not able to found any solution that would simultaneously provide: (1) a non-trivial management of an IoT system, (2) be comfortable and easy to use by a non-technical audience, and (3) allow the user to better understand how the system is functioning. By *non-trivial* we mean that it should be possible to define new rules and modify them via a conversational approach, achieving a *de facto* integration of multiple devices; not just directly interacting with its basic capabilities. The comfort would be for the user not to have to move or touch a device to get his tasks done (i.e. using voice), or edit a Node-RED visual flow. As to understanding their system's functioning, we mean the ability to grasp *how* and *why* something is happening in their smart space. This last point, combined with the other two, would ideally allow someone to simply ask why something happens.

3   Solution Overview

We propose the development of a conversational bot dedicated to the management of IoT systems that is capable of defining and managing complex system rules. Our prototype is called **Jarvis**, and is available as a reproducible package [14].

Jarvis's abilities reach across different levels of operational complexity, ranging from direct one-time actions (e.g. *turn on the light*) to repeating conditional actions (e.g. *when it is raining, close the windows*). Jarvis also lets the user easily *understand* and *modify* the rules and cooperation of multiple devices in the system, through queries like *why did the toaster turn on?* In these cases, we incorporated Jarvis with *conversational awareness* to allow for chained commands; the following dialogue exemplifies this particular feature:

> **User:** *"Why did the toaster turn on?"*
> **Jarvis:** *"You told me to turn it on at 8 AM."*
> **User:** *"Okay, change it to 7:50 AM."*
> **Jarvis:** *"Sure, toaster timer was changed."*

... the reader would note that the second user's query would not make sense on its own. We believe that such features improve the user's experience since it avoids repeating information that has already been mentioned in the conversation, and presents a more *natural* (conversational) interaction.

To ease the integration with nowadays systems and provide us an *experimental reproducible environment*, we integrated the interface with existing platforms such as Google Assistant[3] and Slack[4], amongst others. We made sure to provide the ability for Jarvis to interact both via *voice* and *text*.

4   Implementation Details

Figure 1 presents the high-level software components of Jarvis. Each component and corresponding techniques are explained in the following subsections.

User Interface
(Alexa, Slack,...)   Dialogflow   Jarvis Backend   RabbitMQ   IoT System
                                           & Databse   Message Queue   Gateway

Fig. 1. Jarvis overall architectural components.

4.1   Conversational Interface

To develop the conversational interface, we decided to opt for Dialogflow[5] as this platform provides built-in integration with multiple popular *frontends* and there exists extensive documentation for this purpose [10]. In this case, we used (1) the Slack team-communication tool, and (2) Google Assistant, so that both text and voice interfaces were covered. In the case of Google Assistant, the user may use any supported device paired with their account to communicate with Jarvis, following a known query prefix such as *"Hey Google, talk to Jarvis"*. Regardless of which type of interface is used, the result is converted to *strings* representing the exact user query and subsequently sent to Dialogflow's backend (thus overcoming potential challenges due to Speech Recognition), which are then analyzed using Natural Language Processing (NLP) techniques. Advancement of the existent NLP techniques made available by Dialogflow falls out-of-the-scope of this work.

4.2   Dialogflow Backend

Upon receiving a request, Dialogflow can either produce an automatic response or send the parsed request to a fulfillment *backend*. This component is thus responsible for parsing the incoming *strings* into a *machine understandable* format (JSON). There are a few key concepts that are leveraged in our implementation:

---

[3] Google Assistant, GoogleAssistant.

[4] Slack, https://slack.com.

[5] Dialogflow, https://dialogflow.com/.

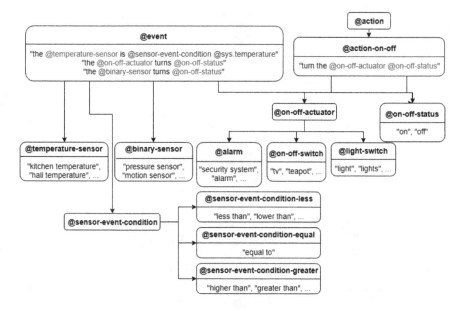

Fig. 2. Main entities defined in Jarvis' Dialogflow project.

Entity. Things that exist in a specific IoT ecosystem can be represented by different literal strings; for example, an entity identified by `toggleable-device` may be represented by *"living room light"* or *"kitchen light"*. Additionally, entities may be represented by *other* entities. Dialogflow use of the @ symbol (*i.e.* `@device`) for entities, and provides some system's defaults;

Intent. An intent represents certain type of user interaction. For instance, an intent named *Turn on/off device* may be represented by `turn the @device on` and `turn the @device off`. For a request such as *"turn the kitchen light on"*, Dialogflow understands that `@device` corresponds to *kitchen light* and provides that data to the fulfillment backend;

Context. Contexts allow intents to depend on previous requests, enabling the creation of context-aware interactions. These are what supports queries such as *"cancel that"* or *"change it to 8AM"*.

Multiple *intents, entities* and *contexts* were defined in Jarvis and the main ones are illustrated in Fig. 2. Here we provide in detail one of its *intents*:

Event Intent
Usage Creates an action that is performed upon a certain event, such as an activity of another device or a change of a device's status. **Definition** `@action:action when @event:event` **Example** *Turn the bedroom light on when the living room light turns off*

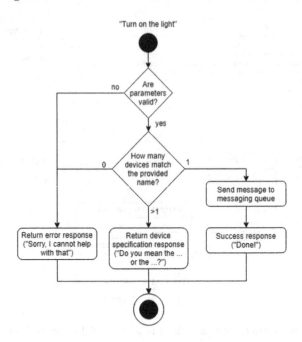

Fig. 3. Activity diagram for the parsing of the query *turn on the light.*

With the above definitions, this component takes requests and builds the corresponding objects containing all actionable information to be sent to the Jarvis backend for further processing.

4.3   Jarvis Backend

For each of the defined intents, this component has an equivalent class responsible for (a) parsing the request, (b) validating its request parameters (*e.g.* device name or desired action), and (c) generating an appropriate response. An overview is provided in Fig. 3. Should the request contain errors, an *explanatory* response is returned. When all the parameters are considered valid, but the intended device is *unclear* (*e.g.* user wants to turn on the light, but there is more than one light), the generated response specifically asks the user for further clarification in order to gain *context*. To tackle cancellation intents, we model all actions using the COMMAND design pattern, thus providing both a straightforward *execute* and *undo* mechanism, as well as an history of performed actions. For most intents, such as *direct actions* or *"why did something happen?"* queries, the effects are immediate. However, *period actions*, *events* and *causality queries* require a different design approach so that they can perform actions on the backend without the need for a request to trigger them.

Period Actions and Events. A *period action* is an intent be carried and then undone after a certain period (*e.g.* *"turn on the light from 4 pm to 5 pm"*). In these scenarios, we generate a *state machine* to differentiate between all the different action status, such as (a) nothing has executed yet (before 4PM), (b) only the first action was executed (after 4 but before 5PM), and (c) both have been executed (after 5PM). We use a combination of *schedulers* and *threads* to guarantee proper action, and abstract all these details inside the COMMAND pattern. The same strategy applies for rules such as *"turn on the light every day at 5 pm"*, with the appropriate state machine and scheduler modifications. This mechanism is (obviously) different for events that are the result of intentions such as *"turn on the kitchen light when the presence sensor is activated"*. Here, we leverage a *publish-subscribe* approach by orchestrating multiple unique and identifiable *message queues* for the devices' actions and state transitions. Upon startup of the backend, we create class listeners that subscribe the corresponding event queues of available devices, and callback the Jarvis backend when appropriate. This orchestration management is dynamic, and depends on the specific rules that are defined. In the aforementioned intent, we would add an observer to look for messages on the presence sensor's event queue with the value on.

Causality Queries. This relate to the user asking why something happened (*e.g.* *"why did the light turn on?"*). To implement them, we augment each COMMAND to determine whether it can cause a specific condition to be true. This *per se* solves some scenarios where the answer can be found by looking at the history of executed commands (*e.g.* *"because you asked me to turn it on at 3:56PM"*). However, there might exist multiple rules may have cause the condition to be true, in which case it is not enough to blame the latest logged command. Instead, there are two possible approaches: (a) return the earliest possible cause, or (b) use a heuristic to infer the most relevant one. Another non-trivial scenario is where the explanation is due to a chain of interconnected rules. Here, it seems that one can (a) reply with the complete chain of events, (b) reply with the latest possible cause, or (c) engage in a *conversation* through which the user can explore the full chain of events as they deem adequate (*e.g.* *"tell me more about things that are triggered by rain"*). In this work, we opted to use the earliest possible cause for the first scenario, and the latest for the second; more complex alternatives can be found in [1,4].

5 Experiments and Results

In order to understand how Jarvis compares to other systems, we established a baseline based on (1) a visual programming language, and (2) a conversational interface. Node-RED was picked amongst the available visual programming solution, as it is one of the most popular visual programming solutions [8]. It follows a flow-based programming paradigm, providing its users with a web-based application through which they can manage rules via *connections* between *nodes* that represent devices, events and actions [20]. Google Assistant was selected for the

Table 1. Simulated scenarios comparison.

Scenario	Jarvis	Google Assistant	Node-RED
One-time action	•	•	•
One-time action w/unclear device	•	·	·
Delayed action	•	·	•
Period action	•	·	•
Daily repeating action	•	·	•
Daily repeating period action	•	·	•
Cancel last command	•	·	·
Event rule	•	·	·
Rules defined for device	•	·	·
Causality query	•	·	·

conversational interface due to its naturality[6]. There are plenty of ways users can interact with it: (a) the standalone Google apps, (b) built-in integration with Android and Chrome OS, or (c) with standalone hardware such as the Google Home. We compare to this baseline according to two criteria: (1) the *number of different features*, and (2) their *user experience* in terms of easiness of usage and intuitiveness. For the first, we created a list of simulated scenarios to assess the ability to manage IoT systems. We then performed a (quasi-)controlled experiment with users to assess the second criteria.

5.1 Simulated Scenarios

Table 1 summarizes the comparison of our prototype to the chosen baseline. It is important to clarify that *one-time action w/ unclear device* refers to actions like *"turn on the light"* with which Jarvis asks the user to clarify which device he means based through responses such as *"do you mean the bedroom or living room light?"*. A *cancel last command* refers to the ability of undoing the last action or rule creation by specifically saying that. Finally, *rules defined for device* refers to the user performing queries that require introspection, such as *"what rules are defined for the bedroom light?"*

It is easy to observe that our prototype provides several features that are not present in either the Google Assistant or Node-RED. Both of these products do a lot more than these features, but especially with the Assistant, the advantage is clear since the only kind of IoT feature it supports is the *one-time action*. Our second conclusion is that it is possible to bring some of the features currently available in visual programming environments to a conversational interface; the converse (how to bring conversational features to Node-RED), eludes the authors.

---

[6] The work by López et al. [15] compares Alexa, Google Assistant, Siri and others, and claim that although *"Siri was the most correct device (...) Google assistant was the one with the most natural responses"*.

5.2   (Quasi-)Controlled Experiment

We performed a (quasi-)controlled experiment with 17 participants, to gain insight into how *end users* responded to a conversational approach. Our sample includes mostly participants without formal technological skills (14), with ages ranging from 18 to 51. We made sure that (a) all were familiar with basic technologies, though, such as basic usage of smartphones and the Internet, and (b) that even non-native English participants had adequate speaking and understanding skills.

Methodology. Each participant was given 5 tasks (1 control task and 4 study tasks) to be performed with the help of Jarvis, using Google Assistant as the system interface. As an example, this was one of the sets of tasks given to participants within a scenario with a *living room light, bedroom light* and a *living room motion sensor*:

- **Task 0 (control):** *Turn on the living room light;*
- **Task 1:** *Turn the living room light on in 5 min;*
- **Task 2:** *Turn the living room light on when the motion sensor triggers;*
- **Task 3:** *Check the current rules defined for the bedroom light, and then make it turn on everyday at 10pm;*
- **Task 4:** *Find out the reason why the bedroom light turned on. Ask Jarvis why it happened and decide whether the answer was explanatory.*

The only instructions given were that they should talk to the phone in a way that feels the most natural to them to complete the task at hand. Besides the tasks, participants were also given the list of IoT devices available in the simulated smart house that they would be attempting to manage through. As a way of increasing the diversity and reducing the bias of the study, we created two different sets of tasks that participants were assigned to randomly. Each set also had different devices, with different smart house topologies. The participants were assigned to one of the test sets randomly.

Variable Identification. For each of the tasks, we collected (1) if the participant was able to complete it, (2) the time taken to complete, and (3) the number of unsuccessful queries. This count was made separately for (a) queries that were not understood by the assistant's speech recognition capabilities (*e.g.* microphone malfunction, background noise), (b) queries where the user missed the intention or made a syntactic/semantic error (*e.g.* *"turn up the lighting"*), and (c) valid queries that an human could interpret, but that Jarvis was unable to.

Subjective Perception. After completing the tasks, we introduced a non-conversational alternative (Node-RED), explaining how all tasks could have been performed using that tool. We inquired the participants whether they perceived any advantages of Jarvis over such a tool, and whether they would prefer Jarvis over non-conversational tools. Finally the participants were asked if they had any suggestions to improve Jarvis and the way it handles system management.

Table 2. Experimental results (task completion rate, task time and incorrect queries), including average and median values.

Task	Done	Time (s)		IQ (Ast)		IQ (User)		IQ (Jvs)		IQ	
		Avg	Med	Avg	Med	Avg	Med	Avg	Med	Avg	Med
0 (1)	94%	6.40	6.0	0.13	0.0	0.25	0.0	0.13	0.0	0.50	0.0
1 (1)	94%	7.10	7.0	0.38	0.0	0.50	0.5	0.00	0.0	0.50	0.5
2 (1)	88%	10.00	10.0	0.75	0.5	0.63	0.5	0.25	0.0	1.00	1.0
3 (1)	100%	20.00	19.5	0.13	0.0	0.13	0.0	0.75	1.0	1.00	1.0
4 (1)	94%	9.00	8.0	0.25	0.0	0.38	0.0	0.00	0.0	0.63	0.0
0 (2)	100%	6.40	6.0	0.33	0.0	0.00	0.0	0.33	0.0	0.67	0.0
1 (2)	94%	7.60	7.0	0.11	0.0	0.00	0.0	0.44	0.0	0.56	0.0
2 (2)	100%	9.90	10.0	0.00	0.0	0.11	0.0	0.78	1.0	0.89	1.0
3 (2)	88%	19.44	19.0	0.33	0.0	0.33	0.0	0.22	0.0	0.89	1.0
4 (2)	100%	8.33	8.0	0.33	0.0	0.22	0.0	0.22	0.0	0.78	1.0

Results. Table 2 compiles the results observed during the study, each row representing a task given to the participant. Each column means:

- **Task:** number of the task (0–4) and the task set number in parenthesis (1/2);
- **Done:** percentage of participants that completed the task successfully;
- **Time:** time in seconds that participants took to complete the task;
- **IQ (Ast):** number of occurrences that queries were incorrect due to the Google Assistant not properly recognizing the user's speech;
- **IQ (User):** number of occurrences that queries were incorrect due to the user not speaking a valid query;
- **IQ (Jvs):** number of occurrences that queries were incorrect due to Jarvis not recognizing a valid query;
- **IQ:** total invalid queries, *i.e.* sum of *IQ (Ast)*, *IQ (User)* and *IQ (Jvs)*.

Discussion. The complexity of the queries increases from task 0 to task 3 since the queries require more words or interactions. This is reflected by the corresponding increase in time in both task sets. The numbers related to incorrect queries show some occurrences at the (voice) assistant level, which means the speech recognition failed to correctly translate what the participants said. Although this does not have implications on the evaluation of Jarvis, it does indicate that this sort of systems might be harder to use due if they are not truly multilingual. Directly comparing the time needed to complete a task to what would be needed to perform it in a visual programming language is meaningless; either the task is not defined, and that would require orders of magnitude longer than what we observe here, or the task is defined and the times will be obviously similar. Similarly, we also observe a few instances of incorrect queries due to grammar mistakes or semantically meaningless, *cf. IQ (User)*, and therefore did not match the sample queries defined in Dialogflow. Nevertheless, there where grammatically incorrect user queries such as *"turn on lights"* but which still carries enough information to understand what the user's intent is. We consider more serious the number of *valid* sentences that were considered incorrect queries by Jarvis, *cf. IQ (Jvs)*. These could have been caused by either a mispronunciation of a device's name, or a sentence structure that is unrecognizable by the Dialogflow configuration. This possibly represents the most serious threat to our proposal, to which we will later dedicate some thoughts on how to mitigate it. Nonetheless, the success rate of all tasks is very high (always greater than 88%), which provides evidence that the system might be intuitive enough to be used without previous instruction or formation. These points were reflected by the participants' subjective perception, were they claimed Jarvis to be easy to use, intuitive, and comfortable; ultimately, these would be the deciding factors for end-users to prefer Jarvis over a non-conversational interface. An additional observation pertaining Jarvis' answers, particularly those regarding causality queries, were state by some users, where they claimed that if the provided response was too long, it would become harder to understand it due to the sheer increase of conveyed information. A possible solution for this problem would be to use a hybrid interface that provides both visual and audio interactions, but there could be other approaches such as an interactive dialogue that shortens the sentences.

Threats to Validity. Empirical methods seem to be one of the most appropriate techniques for assessing our approach (as it involves the analysis of human-computer interaction), but it is not without liabilities that might limit the extent to which we can assess our goals. We identify the following threats: **(1) Natural Language Capabilities,** where queries like *"enable the lights"* might not be very common or semantically correct, but it still carries enough information so that a human would understand its intention. The same happen with device identification, such as when the user says *turn on the bedroom lights*, and the query fails due to the usage of the plural form. During our study, we observed many different valid queries that did not worked due to them not being covered by the Dialogflow configuration; **(2) Coverage error**, which refers to the mismatch between the *target* population and the *frame* population. In this scenario, our target population was (non-technical) end-users, while the frame population was all users that volunteered to participate; and **(3) Sampling errors** are also possible, given that our sample is a small subset of the target population. Repeating the experience would necessarily cover a different sample population, and likely attain different results. We mitigate these threats by providing a reproducible package [14] so other researchers can perform their own validation.

6   Conclusion

In this paper we presented a conversational interface prototype able to carry several different management tasks currently not supported by voice assistants, with capabilities that include: (1) Delayed, periodic and repeating actions, enabling users to perform queries such as *"turn on the light in 5 min"* and *"turn on the light every day at 8 am"*; (2) The usage of contextual awareness for more natural conversations, allowing interactions that last for multiple sentences and provide a more intuitive conversation, *e.g. "what rules do I have defined for the living room light?"*; (3) Event management, that allows orchestration of multiples devices that might not necessarily know that each other exists, *e.g. "turn on the light when the motion sensor is activated"*; and (4) Causality queries, to better understand how the current system operates, *e.g. "why did the light turn on?"*

We conducted (quasi-)controlled experiments with participants that were asked to perform certain tasks with our system. The overall high success rate shows that the system is intuitive enough to be used by people without significant technological knowledge. It also shows that most challenges lie in the natural language capabilities of the system, as it is hard to predict them any user queries that have the same intrinsic meaning. We thus conclude that incorporating recent NLP advances (that were beyond the scope of this paper) would have an high impact in terms of making it more flexible to the many different ways (correct or incorrect) that users articulate the same intentions.

Nonetheless, by doing a feature comparison, we can observe that Jarvis was able to implement many features that current conversational assistants are lacking, while simultaneously being more user-friendly than the available alternatives to IoT management (such as visual programming approaches). As future work, we believe that our approach could be improved by sometimes engaging in a longer (but fragmented) conversation with the user, particularly when providing causality explanations. This would allow the user to understand more information at his own pace, but also because it would enable them to make changes to the rules as the conversation unfolds.

Acknowledgement. This work was partially funded by the Integrated Masters in Informatics and Computing Engineering of the Faculty of Engineering, University of Porto (FEUP) and by the Portuguese Foundation for Science and Technology (FCT), under research grant SFRH/BD/144612/2019.

References

1. Agadakos, I., Ciocarlie, G., Copos, B., Lepoint, T., Lindqvist, U., Locasto, M.: Butterfly effect: causality from chaos in the IoT. In: International Workshop on Security and Privacy for the Internet-of-Things, April 2018
2. Ammari, T., Kaye, J., Tsai, J.Y., Bentley, F.: Music, search, and IoT: how people (really) use voice assistants. ACM Trans. Comput.-Hum. Interact. **26**(3) (2019). https://doi.org/10.1145/3311956
3. Austerjost, J., et al.: Introducing a virtual assistant to the lab: a voice user interface for the intuitive control of laboratory instruments. SLAS TECHNOL.: Transl. Life Sci. Innov. **23**(5), 476–482 (2018)
4. Braines, D., O'Leary, N., Thomas, A., Harborne, D., Preece, A.D., Webberley, W.M.: Conversational homes: a uniform natural language approach for collaboration among humans and devices. Int. J. Adv. Intell. Syst. **10**(3/4), 223–237 (2017)

5. Dias, J.P., Faria, J.P., Ferreira, H.S.: A reactive and model-based approach for developing internet-of-things systems. In: 2018 11th International Conference on the Quality of Information and Communications Technology (QUATIC), pp. 276–281 (2018)

6. Fischer, G., Giaccardi, E., Ye, Y., Sutcliffe, A.G., Mehandjiev, N.: Meta-design: a manifesto for end-user development. Commun. ACM **47**(9), 33–37 (2004)

7. Gennari, R., Bozen-bolzano, L.U., Melonio, A., Bozen-bolzano, L.U.: End-User Development, vol. 10303. Springer, Cham (2017). https://doi.org/10.1007/978-3-319-58735-6

8. Giang, N.K., Lea, R., Blackstock, M., Leung, V.C.: Fog at the edge: experiences building an edge computing platform. In: 2018 IEEE International Conference on Edge Computing (EDGE), pp. 9–16. IEEE (2018)

9. He, W., Martinez, J., Padhi, R., Zhang, L., Ur, B.: When smart devices are stupid: negative experiences using home smart devices. In: 2019 IEEE Security and Privacy Workshops (SPW), pp. 150–155 (2019)

10. Janarthanam, S.: Hands-on Chatbots and Conversational UI Development: Build Chatbots and Voice User Interfaces with Chatfuel, Dialogflow, Microsoft Bot Framework, Twilio, and Alexa Skills. Packt Publishing Ltd., Birmingham (2017)

11. Janssen, P., Erhan, H., Chen, K.W.: Visual dataflow modelling - some thoughts on complexity. In: Proceedings of the 32nd eCAADe Conference (2014)

12. Kishore Kodali, R., Rajanarayanan, S.C., Boppana, L., Sharma, S., Kumar, A.: Low cost smart home automation system using smart phone. In: 2019 IEEE R10 Humanitarian Technology Conference (R10-HTC)(47129), pp. 120–125 (2019)

13. Lago, A.S.: Exploring complex event management in smart-spaces through a conversation-based approach. Master's thesis, Faculty of Engineering, University of Porto (2018)

14. Lago, A.: andrelago13/jarvis: initial release, April 2020. https://doi.org/10.5281/zenodo.3741953

15. López, G., Quesada, L., Guerrero, L.A.: Alexa vs. siri vs. cortana vs. google assistant: a comparison of speech-based natural user interfaces. In: Nunes, I.L. (ed.) AHFE 2017. Advances in Intelligent Systems and Computing, vol. 592, pp. 241–250. Springer, Cham (2018). https://doi.org/10.1007/978-3-319-60366-7_23

16. Mainetti, L., Mighali, V., Patrono, L.: An IoT-based user-centric ecosystem for heterogeneous smart home environments. In: 2015 IEEE International Conference on Communications (ICC), pp. 704–709 (2015)

17. Miranda, J., et al.: From the internet of things to the internet of people. IEEE Internet Comput. **19**(2), 40–47 (2015)

18. Mitrevski, M.: Conversational interface challenges. Developing Conversational Interfaces for iOS, pp. 217–228. Apress, Berkeley (2018). https://doi.org/10.1007/978-1-4842-3396-2_8

19. Prehofer, C., Chiarabini, L.: From IoT mashups to model-based IoT. In: W3C Workshop on the Web of Things (2013)

20. Ray, P.P.: A survey on visual programming languages in internet of things. Sci. Program. **2017**, 1–6 (2017). https://doi.org/10.1155/2017/1231430

21. Xia, F., Yang, L.T., Wang, L., Vinel, A.: Internet of things. Int. J. Commun. Syst. **25**(25) (2012)

22. Zarzycki, A.: Strategies for the integration of smart technologies into buildings and construction assemblies. In: Proceedings of eCAADe 2018 Conference, pp. 631–640 (2018)

Improving Coverage Area in Sensor Deployment Using Genetic Algorithm

Frantz Tossa[1,2]($\boxtimes$), Wahabou Abdou[1], Eugène C. Ezin[2], and Pierre Gouton[1]

[1] University of Bourgogne Franche-Comté, Dijon, France
frantz.tossa@u-bourgogne.fr
[2] University of Abomey-Calavi, Abomey-Calavi, Bénin

Abstract. Wireless sensor networks (WSN) are a collection of autonomous nodes with a limited battery life. They are used in various fields such as health, industry, home automation. Due to their limited resources and constraints, WSNs face several problems. One of these problems is the optimal coverage of a observed area. Indeed, whatever the domain, ensuring optimal network coverage remains a very important issue in WSNs, especially when the number of sensors is limited. In this paper, we aim to cover a two-dimensional Euclidean area with a given number of sensors by using genetic algorithm in order to find the best placement to ensure a good network coverage. The maximum coverage problem addressed in this paper is based on the calculation of the total area covered by deployed sensor nodes. We first, define the problem of maximum coverage. For a given number of sensors, the proposed algorithm find the best position to maximize the sensor area coverage. Finally, the results show that the proposed method well maximize the sensor area coverage.

Keywords: Wireless sensor networks · Sensors deployment · Area coverage · Genetic Algorithm

1 Introduction

The rapid growth of wireless technologies and the decrease in their cost has made it possible to generalize the use of Wireless Sensor Networks (WSN). Originally used in the military field, this type of network is now present in several fields, ranging from industrial monitoring to the measurement of environmental data, home automation, fire detection, medical sector [9,11]. WSN applications have therefore critical roles in daily life. Most of these applications have the task of monitoring a target point or area, recording a reaction and transmitting it. Broadly, a WSN is a collection of sensors which communicate thanks to some wireless technologies. WSNs consist of small nodes that differ from traditional networks within their communication and sensing ranges. Sensor nodes, sense the physical phenomena located in the area and transmit the data collaboratively to sink node. A sensor node can be either sensing, transmitting node, relay

© Springer Nature Switzerland AG 2020
V. V. Krzhizhanovskaya et al. (Eds.): ICCS 2020, LNCS 12141, pp. 398–408, 2020.
https://doi.org/10.1007/978-3-030-50426-7_30

node or both of them together. Therefore, to perform their tasks, wireless sensor networks must be effective.

Deployment is one of the important aspects that have a direct impact on the WSN effectiveness. The deployment mechanisms of WSN can be classified into two main categories: deterministic and non-deterministic [1,11]. The deployment of a WSN affects almost all of its mains performance metrics, such as coverage, connectivity, and network lifetime. Among these metrics, we focus on coverage. Indeed for an effective design and employment of sensor networks in various application scenarios, the coverage relies on numerous parameters. It reflects the quality of monitoring a point or area by a sensor. In [10], coverage is defined as how well or to how much extent each point of a deployed network is under the vigilance of a sensor node. The coverage can be classified into three categories [1,2]:

- Barrier coverage: The objective is to achieve an arrangement of sensors with the task of maximizing the detection probability of a specific target penetration through the barrier.
- Point coverage: The objective is to cover a set of point (target) with known position that need to be monitored. This coverage scheme focuses on determining sensor nodes exact positions while guarantee efficient coverage application for a limited number of immobile targets.
- Area coverage: The main objective is to cover (monitor) a region and to maximize the detection rate of a specific area. In this paper, we focus on this type of coverage.

Most of the previous studies on this problem have focused on how to reduce the number of sensors to cover the area. In our case, we examine the problem of the maximum coverage of the area with a given number of sensors of the same type. In real-life situations, due to cost considerations, the number of sensors is often limited, but the requirement to cover an area as wide as possible is necessary. As indicated in [12], depending on the objectives, the coverage problem can be formulated in different ways. In this study, we aim to cover a two-dimensional detection area by using a limited number of sensors to have the best network coverage. Instead of using methods such as circle packing algorithms [5,8], we use a Genetic Algorithm (GA). In addition, due to the random deployment of the sensors, causing a probable overlap, we propose a new way as the evaluation function of the GA to know the exact surface covered by sensors. To the best of our knowledge, this method is not been used in the literature.

The rest of the paper is organized as follows: in Sect. 2 we briefly present the related work on sensor deployment and maximum coverage. In Sect. 3, the problem definition is provided, a mathematical modeling is proposed. In Sect. 4 the proposed approach is detailed and in Sect. 5, we present and discuss numerical results. Section 6 concludes this paper and gives an overview of some future works.

2   Related Works

Optimal sensor placement for good and accurate measurement is essential for data transmission in sensor networks. That's why getting the best network coverage is important. This problem is known as maximum coverage sensor deployment problem (MCSDP) and is known to be NP-hard. Several works have been carried out in various situations to solve sensor placement for best area monitoring and they have presented their solutions with various algorithms.

Ozan Zorlu and Ozgur Koray Sahingoz proposed a genetic algorithm to increase the coverage of an homogeneous wireless sensor networks [14]. The problem of maximum coverage that the authors address in this paper is based on the calculation of the total area covered by the sensors. In the paper, the covered area is tried to calculated by geometric formulas. They first describe the problem, then they formulate it for their genetic algorithm. In their proposal, they seek to deploy a minimum number of sensor nodes which should not overlap and not go beyond the observed area for maximum coverage. Their objective is to maximize the area coverage while minimizing the intersection between sensor nodes. Based on this formulation of the problem, they calculate the area covered by the deployed sensors. They reach the maximum coverage by varying the parameters of the genetic algorithm and finally show the algorithm perform well and is stable.

Based on previous work that applies genetic algorithms to sensor deployment problems, Yourim Yoon and Yong-Hyuk Kim analyze the problem, its representation and its properties and propose a methodology which they consider new and more adapted to the properties of genetic algorithms [12]. An effective evaluation method and a new standardization technique are also proposed. The authors in this paper, adopt the Boolean disk coverage model and deal with the zone coverage problem, which also addresses all points of the sensor field. In this paper, as a new approach, they focus on the problem of maximizing the area covered by the sensor field with a given number of sensors. For the notation of the problem, they use the multidimensional multiple-choice knapsack problem. In contrast to [14], they base their study on n static sensors of k types and each type of sensor can cover a given area with an arbitrary fixed radius $r_1, r_2, ..., r_k$. They assume there is at least one sensor for each type of sensors. Their objective is to find locations $(x_1, y_1), (x_2, y_2), ..., (x_n, y_n)$ for all n sensors that produce the maximum coverage for a given domain. So they proposed a novel normalization method for the problem that could improve the performance of genetic algorithms they use. This method tailored to the MCSDP to associate the search by GAs with the original solution space. After presenting some GAs, (RANDOM, PGA and MGA) that are without normalization, they propose the OPTGA which, like MGA, uses the Monte Carlo method with an increasing number of random samples to evaluate solutions, and rearranges the genes of the second parent to minimize the distance sum before applying recombination. They then make a comparison between their proposal and these algorithms and show that their proposal is more stable and faster.

With a different approach, Sami Mnasri and al have worked on target coverage and presented a mathematical model to provide a deployment scheme while optimizing the target coverage of the location in an audio sensors networks [6]. They aim to optimize the placement of nodes with the most possible uniform distribution of nodes (anchors and mobile nodes) around the target to locate. To model the problem of target coverage considering the localization, they define the objective function in two sub-functions (coverage and location), which they then add. The sub-function objective for coverage is the ability of a mobile node to compute coverage based on the targets covered, while that of the location computes the ability for each target to be monitored by at least n nodes (mobile or anchor). In doing so, they aim to optimize target coverage and location. They publish thirteen constraints which constitute the rules of management of the network and evaluate the performance of the proposed genetic algorithm in terms of coverage rate, degree of coverage (k-coverage), number of iterations, and the front Pareto.

Nguyen ThiHanh and in [4] based their work on the same problem formulation as [12]. They proposed a genetic algorithm called MIGA, based on IGA, to solve the problem of maximizing area coverage in a WSN with heterogeneous sensing ranges. Their modified IGA algorithm, approaches the problem with a new individual representation, a different heuristic initialization, a combination of Laplace Crossover (LX) and Arithmetic Crossover Method (AMXO) operators, and a local search (VFA). K different types of sensors were used in the set of sensor nodes, in which one type of sensor i has a detection range r_i. The goal of their work is to find an optimal placement scheme for all the sensor nodes without overlapping so that the coverage of the area is maximized. To increase the reliability of results, they provided an exact integral area calculation for the fitness function for computing the area coverage corresponding to a given number of sensor nodes. At the end they compare the performance of their algorithm with the performance of MIGA, IGA, PSO, DPSO, ICS and CFPA.

3   Problem Definition

The problem of deploying sensors for maximum coverage (MCSDP) aims to cover as much as possible an area with a minimum number of sensor nodes. The maximum coverage problem addressed in this paper is based on the calculation of the total area covered by deployed sensor nodes. Our contribution is based on GA and focuses on to choose the optimal positions of sensors for the best coverage.

3.1   Assumptions

Let us consider n be the number of homogeneous sensors with identical sensing range r_s. We assume that each sensor has omnidirectional antena. We also assume that the sensing range and the communication range are the equal. The detection by each sensor is modeled as a circle on the two-dimensional grid. The center of the circle indicates the sensor and its radius indicates its sensing range. The cover model is Boolean. Boolean disk coverage model might be the most widely used sensor coverage model in the literature [4,7,12,14]. The coverage function used is define by:

$$f(d(s,p)) = \begin{cases} 1 & \text{if } d(s,p) \leq r_s \\ 0 & \text{otherwise.} \end{cases} \tag{1}$$

where $d(s,p)$, the Euclidean distance between the sensor s and a point p is given by:

$$d(s,p) = \sqrt{(s_x - p_x)^2 + (s_y - p_y)^2} \tag{2}$$

The observed area (A) to be covered with sensors is a rectangular region in two-dimensional Euclidean space. Ideally, having a fully covered area is given by:

$$Area(A) = n\pi r_s^2. \tag{3}$$

The parameter of sensing range is generated from the real-world sensor in formations as in literature [12–14].

3.2 Problem Formulation

The problem can be described as follows: we seek to have the maximum coverage with a given number n of sensors of the same radius r_s. We formulate that the area really covered by the sensors, is the difference between the area of the surface A that we note $Area(A)$ and the sum of the areas of the sensors $Area(S_i)$. We call this difference $Coverage(A)$ and calculate as presented in Eq. (4):

$$Coverage(S_i; A) = Area(A) - \sum_{i=1}^{n} Area(S_i) \tag{4}$$

However, due to several overlaps, calculating $\sum_{i}^{n} Area(S_i)$ is not obvious. For two or three sensors, this can be less complex. Figure 1 illustrates the case with three sensors nodes.

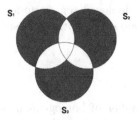

Fig. 1. Three overlapped sensors coverage

We note the area covered by the three sensors in Fig. 1, $Area(S_1 \cup S_2 \cup S_3)$. Finding this area, would mean sum up the areas of the sensors and then subtract the sum of the areas of the intersections, being careful not to remove the same intersection area several times. This gives us:

$$Area(U_{i=1}^{N} S_i) = \sum_{i=1}^{n} Area(S_i) - (\sum_{i=1}^{n} \sum_{j=i+1}^{n-1} Area(S_i \cap S_j)) \tag{5}$$

This makes the task more complex when the number of sensors exceeds three. Found the total area covered by four or more sensors is mathematically very complex (Fig. 2).

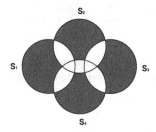

Fig. 2. Four overlapped sensors coverage

Based on the above, we formalize the problem as follows:

$$Object = \begin{cases} maximize \ Area(U_{i=1}^{N} S_i) \\ subject \ to \ (x_i, y_i) \in A, i = 1, 2, ..., n. \end{cases} \tag{6}$$

4   Genetic Algorithm

4.1   Genetic Framework

Genetic Algorithm (GA) is a widely used optimization algorithm, and it has been used frequently in different applications especially in WSN. It is an optimization method which is based on natural selection, the process that drives biological evolution. Especially, when no deterministic methods exist GA are used and it performs well. This technique is based on basic principles. First, it works on an initially random population of individuals, each representing a solution of the problem being addressed. Each solution is represented by a genotype or chromosome and is expressed as a phenotype. It is then necessary to develop an evaluation function (fitness) that distinguishes the best performing phenotype. The latter have a higher probability of bequeathing their genotype to their offspring. The rules governing this gene transfer are described in the form of genetic operators. Three operators are required:

- The selection operator which describes how the candidates are chosen for recombination;
- The crossover operator is the main operator and corresponds to the method used to order mix the parents in order to generate offspring;
- The mutation operator, used to add diversity to the population.

Algorithm 1. Genetic algorithm's steps

1: Generation of a random population;
2: Evaluation for each individual;
3: Selection of individuals for recombination;
4: Generation of offspring thanks to a crossover operator;
5: Using mutation of offspring;
6: Go back to step 2 until stop criterion is met.

In the proposed genetic algorithm, we represent a chromosome as a set of n sensors with different geographic coordinates. Figure 3 shows the structure of a chromosome.

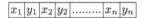

Fig. 3. Representation of our chromosome

4.2 Evaluation of a Solution

As shown in Sect. 3, calculating the total area covered by sensor nodes with mathematical formulas is very hard to apply. Sensors can have one or more overlapping zones and this calculation when the number of sensor nodes is equal or greater than four is complex.

In this paper, we evaluate the area covered by considering each sensor node as an image and proposing another and new way to find the area covered. Indeed, we simulate a two-dimensional surface as a background with a number of 0. The number of 0 depends on the resolution initially selected. For example, a surface of 10×20 with a resolution of 0.5 will provide a table of 20×40 of 0 in the matrix. The circular stamp is the square of 0 which is entirely filled by a circle formed by 1. The size of the circle is based on a radius defined in the chromosome. Each chromosome provides the x and y axes of the position of the circles within the defined rectangular area as well as the proposed radius r_s of the circles. The coordinates x and y change with each generation so that at the end we have the best positions. Finally, the program counts the number of 1 inside the rectangular field. Coverage corresponds to found by the ratio between the number of 1 and the initial number of 0. The following pseudo-code briefly summarizes how we arrive at the final result:

Algorithm 2. Evaluation function

 1: Simulated a two-dimensional surface A as a background included a numerous of 0;
 2: Counts the number of 0 inside A;
 3: $a \leftarrow number\ of\ 0$;
 4: Circular Stamp is the square of 0 which is fully filled by a circle shaped by 1;
 5: The diagonal of the circle is based on a defined radius in the chromosome;
 6: Each chromosome provides the x-axes and y-axes of circles position inside A;
 7: Put the center of a stamp on the position of defined circles in the chromosome;
 8: Count the number of 1 inside A;
 9: **if** Overlap between circle **then**
10:     Count once 1 in the overlapped area;
11: **else**
12:     Count number of 1;
13: **end if**
14: **if** Circle goes outside of A **then**
15:     Count only the 1 that are inside;
16: **else**
17:     Count number of 1;
18:     $b \leftarrow number\ of\ 1$;
19: **end if**
20: $Coverage = b/a \times 100$;

---

5   Experimental Results

In this section, we evaluate our proposal. Let N be the size of the population. In our study, a population of $N = 100$ individuals was used. The population consists of N genotype, and each genotype is a set of $n = 15$ sensors. The genetic algorithm ends after 350 iterations. As crossover and mutation operators, BLX-α and the Gaussian mutation are respectively used [3,13]. P_c and P_m are respectively the crossover and mutation rate and the sensing area A is a rectangular of dimension 100×90.

Unlike the other methods listed in the literature review, ours puts itself in a real deployment situation, and takes into account overlapping between one or more sensors to calculate and give the maximum coverage rate. Several tests have been done. Figure 4 show the initial deployment at the beginning of the algorithm of one of these tests. At each iteration the positions of the sensors are displayed and at the end, the best population (Fig. 5) and the final position of sensors (Fig. 6) are given. For each set of values coverage - crossover, coverage-number of sensor, coverage-permutation and coverage - iteration, we have conduced several experiments which show that we can reach a coverage rate of up to 88%. For the cases shown in Fig. 5 and Fig. 6 we reach 85% of coverage rate. The results showed our algorithm is efficient and the proposed approach produces better coverage than initial deployment.

Fig. 4. Initial random deployment

Fig. 5. Best population

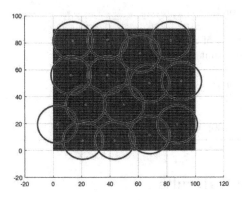

Fig. 6. Final position of sensors

Table 1 and Table 2 shows respectively the variation in the coverage rate as a function of the percentage of crossover and the variation in the coverage rate as a function of the number of sensors. For Table 1 the number of sensors is unchanged and the permutation rate is $P_m = 0.1$ and for Table 2 the permutation rate is $P_m = 0.1$ and crossover rate is $P_c = 0.7$ (Fig. 7).

Table 1. Coverage rate according to crossover

Crossover percentage	0.3	0.4	0.5	0.6	0.7	0.8	0.9
Coverage rate		75.5	77.1	73.35	76.61	79.3	75.9	80

Table 2. Coverage rate according to number of sensors

Nb sensors	10	12	14	15	17	20	22
Coverage rate	65	70.52	74.1	78.37	76.35	76.76	74.73

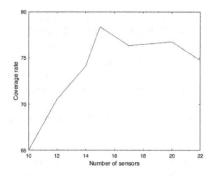

Fig. 7. Coverage rate according to number of sensors

Keeping the number of sensors and the parameters fixed, we carried out several tests, some of the numerical results of which are listed in Table 3. These tests show a coverage average of 77.36% and a standard deviation of 2.14. This significant standard deviation shows that on average the coverage rate between two tests varies slightly and hovers around 77%.

Table 3. Coverage rate according to number of test

Nb sensors	1	2	3	4	5	6	7	8	9	10	11	12	13	14
Coverage rate	73.63	77.22	79.18	73.67	75.93	75.96	76.48	76.25	79.88	77.43	78.66	79.63	79.25	79.86

6   Conclusion

Wireless sensor networks are used in several fields. Thus, covering an area as much as possible, with a limited number of sensor nodes is a major problem. This problem is called the Maximum Coverage Sensor Deployment Problem (MCSDP). Due to the increasing number of sensor nodes, and the likely overlaps of these nodes during deployment, this problem cannot be resolved using discrete mathematical formulas. This paper presents a solution to this problem. We have formally defined the problem of maximum coverage sensor deployment (MCSDP), which occurs frequently in real world applications. Given this definition, we proposed a new method for the evaluation of the areas of sensor nodes by combining this method with a genetic algorithm. Our approach, tested with homogeneous sensor nodes shows the effectiveness and the efficiency of the proposed system. In future work, it is planned to include obstacles and areas that do not require coverage in the deployment space and also use this approach for heterogeneous sensor nodes.

References

1. Aznoli, F., Navimipour, N.J.: Deployment strategies in the wireless sensor networks: systematic literature review, classification, and current trends. Wirel. Pers. Commun. **95**(2), 819–846 (2016). https://doi.org/10.1007/s11277-016-3800-0

2. Devi, G., Bal, R.S.: Node deployment coverage in large wireless sensor networks. J. Netw. Commun. Emerg. Technol. (JNCET) **6** (2016)
3. Eshelman, L.J., Schaffer, J.D.: Real-coded genetic algorithms and interval-schemata. In: Whitley, L.D. (ed.) Foundations of Genetic Algorithms, Foundations of Genetic Algorithms, vol. 2, pp. 187–202. Elsevier (1993). https://doi.org/10.1016/B978-0-08-094832-4.50018-0, http://www.sciencedirect.com/science/article/pii/B9780080948324500180
4. Hanh, N.T., Binh, H.T.T., Hoai, N.X., Palaniswami, M.S.: An efficient genetic algorithm for maximizing area coverage in wireless sensor networks. Inf. Sci. **488**, 58–75 (2019). https://doi.org/10.1016/j.ins.2019.02.059. http://www.sciencedirect.com/science/article/pii/S0020025519301823
5. Litvinchev, I., Infante, L., Ozuna, L.: Packing circular-like objects in a rectangular container. J. Comput. Syst. Sci. Int. **54**(2), 259–267 (2015). https://doi.org/10.1134/S1064230715020070
6. Mnasri, S., Thaljaoui, A., Nasri, N., Val, T.: A genetic algorithm-based approach to optimize the coverage and the localization in the wireless audio-sensors networks. In: 2015 International Symposium on Networks, Computers and Communications (ISNCC), pp. 1–6, May 2015. https://doi.org/10.1109/ISNCC.2015.7238591
7. Njoya, A.N., Abdou, W., Dipanda, A., Tonye, E.: Evolutionary-based wireless sensor deployment for target coverage. In: 2015 11th International Conference on Signal-Image Technology Internet-Based Systems (SITIS), pp. 739–745, November 2015. https://doi.org/10.1109/SITIS.2015.62
8. Orick, G.L., Stephenson, K., Collins, C.: A linearized circle packing algorithm. Comput. Geom. **64**, 13–29 (2017). https://doi.org/10.1016/j.comgeo.2017.03.002. http://www.sciencedirect.com/science/article/pii/S0925772117300172
9. Rashid, B., Rehmani, M.H.: Applications of wireless sensor networks for urban areas: a survey. J. Netw. Comput. Appl. **60**, 192–219 (2016). https://doi.org/10.1016/j.jnca.2015.09.008. http://www.sciencedirect.com/science/article/pii/S1084804515002702
10. Sangwan, A., Singh, R.P.: Survey on coverage problems in wireless sensor networks. Wirel. Pers. Commun. **80**(4), 1475–1500 (2015). https://doi.org/10.1007/s11277-014-2094-3
11. Wang, B.: Coverage problems in sensor networks: a survey. ACM Comput. Surv. **43**(4) (2011). https://doi.org/10.1145/1978802.1978811
12. Yoon, Y., Kim, Y.: An efficient genetic algorithm for maximum coverage deployment in wireless sensor networks. IEEE Trans. Cybern. **43**(5), 1473–1483 (2013). https://doi.org/10.1109/TCYB.2013.2250955
13. Yoon, Y., Kim, Y.H.: The Roles of Crossover and Mutation in Real-coded Genetic Algorithms (2012). https://doi.org/10.5772/38236
14. Zorlu, O., Sahingoz, O.K.: Increasing the coverage of homogeneous wireless sensor network by genetic algorithm based deployment. In: 2016 Sixth International Conference on Digital Information and Communication Technology and Its Applications (DICTAP), pp. 109–114, July 2016. https://doi.org/10.1109/DICTAP.2016.7544010

Object-Oriented Internet Reactive Interoperability

Mariusz Postół^(⊠) (iD)

Institute of Information Technology, Lodz University of Technology, Łódź, Poland
mariusz.postol@p.lodz.pl

Abstract. Information and Communication Technology has provided society with a vast variety of distributed applications. By design, the deployment of this kind of application has to focus primarily on communication. This article addresses research results on the systematic approach to the design of the meaningful Machine to Machine (M2M) communication targeting distributed mobile applications in the context of new emerging disciplines, i.e. Industry 4.0 and Internet of Things. This paper contributes to the design of a new architecture of mobile IoT solutions designed atop of the M2M communication and composed as multi-vendor cyber-physicals systems. The described reusable library supporting this architecture designed using the reactive interoperability archetype proves that the concept enables a systematic approach to the development and deployment of software applications against mobile IoT solutions based on international standards. Dependency injection and adaptive programming engineering techniques have been engaged to develop a full-featured reference application program and make the proposed solution scalable and robust against deployment environment continuous modifications. The article presents an executive summary of the proof of concept and describes selected conceptual and experimental results achieved as an outcome of the open-source project Object-Oriented Internet targeting multi-vendor plug-and-produce interoperability scenario.

Keywords: Industrial communication · Industry 4.0 · Internet of Things · Machine to Machine communication · OPC Unified Architecture

1 Introduction

Information and Communication Technology has provided society with a vast variety of new distributed applications. By design, the deployment of this kind of applications has to focus primarily on communication technologies. This article addresses further research on systematic design of Machine to Machine (M2M) communication [9,11,19,21] targeting distributed mobile applications in the context of new emerging disciplines, i.e. Industry 4.0 (I4.0) [20] and Internet of Things (IoT) [3,22]. New architecture is proposed for IoT solutions designed

© Springer Nature Switzerland AG 2020
V. V. Krzhizhanovskaya et al. (Eds.): ICCS 2020, LNCS 12141, pp. 409–422, 2020.
https://doi.org/10.1007/978-3-030-50426-7_31

atop of M2M communication deployed as multi-vendor cyber-physicals systems [20]. The architecture is backed by a proof of concept reference implementation.

All of the applications designed atop of network communication can be grouped as follows:

- **human-centric** - information origin or ultimate information destination is an operator,
- **machine-centric** - information production, consumption, networking, and processing are achieved entirely without human interaction.

A typical **human-centric** approach is a web-service supporting, for example, a web UI to monitor conditions, and manage millions of devices and their data in a typical cloud-based IoT approach. It is characteristic that, in this case, any uncertainty and necessity to make a decision can be relaxed by human interaction.

Coordination of robots behavior in a work-cell is a **machine-centric** example. In this case, it is essential that any human interaction is impractical or even impossible. This interconnection scenario requires the machine to machine communication (M2M) demanding multi-vendor devices integration. From the M2M communication concept, a broader concept of a smart factory can be derived. In this concept, the mentioned robots are only executive parts of an integrated supervisory control system responsible for macro optimization of any industrial process composed into one whole. This approach is called the fourth industrial revolution and coined as Industry 4.0. It is worth stressing that machines - or more general parts - interconnection is not enough, and additionally, parts interoperability has to be expected for the deployment of this concept.

Examining the information exchange over a network, the first challenge to be faced up is information reusability and security. The second challenge is how to make the mentioned machines interoperable if they are provided by a vast variety of vendors.

M2M communication requires information exchange technology. Unfortunately, information is an abstract knowledge and, hence, cannot be directly processed/transferred using technical means. Fortunately, there is a solution that overcomes this issue, i.e. the information must be represented as bitstreams called data. To make this representation useful in the context of information processing there must be defined semantic and syntax rules called semantic-context. The syntax is used to validate the correctness of the bitstream, and the semantics rules associate meaning to the correct bitstreams.

Going beyond the smart factory realm, a similar concept may be used to make any general-purpose entities interoperable. Finally, we get cyber-physical systems where a variety of entities may be aggregated into distributed information processing solutions. In this case, we are opening the public connectivity domain, which requires a globally scoped infrastructure, i.e. the Internet. Parties interconnected over any network require special precautions that must be taken against malicious users to assure the best possible level of security. It is especially crucial if public resources are used to exchange data. To address this security demand suitable protection methods may be applied to:

- **network traffic** - accomplished using intermediary devices to enforce traffic selective availability based on predetermined security rules against unauthorized access,
- **data transfer object (DTO)** - accomplished using cipher algorithms against bitstreams formatted as a message traversing the network and containing process data.

For any generic solution addressing the design of the cyber-physical system, the data holder mobility must be considered as well. Mobile data means that it may come from mobile devices or be generated in unpredictable attachment points. If the data places exposition is arbitrary it means that the data appearance must be recognized and processed as an event. A good example of this scenario is a product (e.g. hygiene goods, cosmetics, drugs, etc) global tracking system - an application domain where the IoT term has been coined [3,22]. One of the arguments for the IoT is allowing distributed yet interlinked devices, machines, and objects (data holders) to interact with each other without relying on human interaction to set-up and commission the embedded intelligence. In case any kind of mobility has to be considered, the next engineering challenge is dynamic discoverability on the network and the possibility of establishing the semantic and security contexts of the parts composing the IoT application.

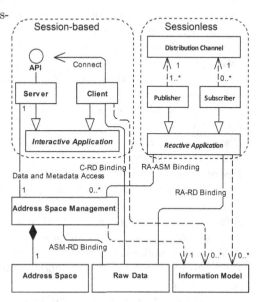

Fig. 1. Interactive/reactive communication

The remainder of this paper is structured as follows. Section 2 presents the generic architecture that is to be used as a foundation for further decisions addressing the systematic design of the multi-vendor cyber-physical systems. It focuses on the reusability and security of data processing. In Sect. 3 open and reusable software model is presented. It promotes a reactive interoperability pattern and a generic approach to establishing interoperability-context. A reference implementation of this archetype is described in Sect. 4. The most important findings and future work are summarized in Sect. 5.

2   Machine to Machine Interoperability

In this section, the most important features of the mobile applications compliant with the more general IoT concept have been abstracted to settle a foundation for a further discussion addressing systematic design, development, and deployment methods and tools. A detailed description of this model is covered by [19]. A starting point for further discussion is a generic architecture of the Machine to Machine (M2M) interoperability presented in Fig. 1. It can be used as a design foundation of the cyber-physical systems. In this approach the following classes have been distinguished:

- *Server* - the front-end component of the server software application,
- *Client* - the front-end component of the client software application,
- *Interactive Application* - a framework supporting the client-server communication pattern,
- *Publisher* - a part derived from *Reactive Application* and implementing *Publisher* role,
- *Subscriber* - a part derived from *Reactive Application* and implementing *Subscriber* role,
- *Reactive Application* - a framework supporting the publisher-subscriber communication pattern,
- *Distribution Channel* - a set of intermediary network nodes interconnected by communication links carrying the published messages to the destination delivery points,
- *Address Space Management* - maintenance services of the *Address Space*,
- *Address Space* - an instance of an address space that can be recognized as a replica of the underling process,
- *Information Model* - a formal model defining the syntax of the exchanged process data,
- *Raw Data* - data origin representing information describing the underlying process state and behavior.

It is essential, that the proposed architecture has no single point where human interaction is possible in compliance with the requirements of the M2M communication concept. In [19] it is recommended how the OPC Unified Architecture international standard may be mapped partially on this architecture to satisfy multi-vendor environment requirements.

M2M communication must assure the reusability od the process data. It could be accomplished as a result of the available communication patterns (Fig. 1):

- **session-oriented** - the data source and ultimate data destinations are tightly coupled by a connection-oriented relationship established between *Server* and *Client* entities,
- **sessionless** - the data source and ultimate data destinations are loosely coupled by *Distribution Channel* carrying the process data over the network.

The session-oriented client/server archetype is the data exchange scenario that requires establishing, in advance, a session relationship before any process data can be sent over the wire. In this case, the connection-oriented services set up a virtual link making a tight relationship between the communicating parties. The session is responsible for retaining state information about each communicating party for the duration of multiple requests. In this scenario, data sharing is supported because many data destinations may establish an independent session with the *Server* at the same time and, as a result, access the shared data.

The sessionless archetype is a message distribution scenario where senders of messages, called publishers, do not send them directly to specific receivers, called subscribers, but instead, categorize the published messages into topics without knowledge about which subscribers if any, there may be. Similarly, subscribers express interest in one or more topics and only receive messages that are of interest, without knowledge about which publishers, if any, there are. In this scenario, the publishers and subscribers are loosely coupled by *Distribution Channel* filtering, buffering and routing the messages to preselected delivery points. They are decoupled in time, space and synchronization [5].

In the session-oriented communication scenario to establish a new session, the client must send an originating request (*Connect*) over the network to the server exposing relevant data. In some circumstances, it may be difficult or even impossible because of the following (Sect. 1) restrictions related to the Internet-based cyber-physical systems:

- **traffic asymmetry** - intentional limits of the network traffic propagation for the security reasons, for example, enforced by a firewall,
- **mobility** - due to data origin mobility the network node may need to move from one attachment point to another losing its previous endpoint address.

The *Server* - hidden behind the firewall (omitted in Fig. 1 for sake of simplicity) to protect the data origin against malicious users - is an example of asymmetric behavior of the network traffic. It causes that the in front *client* instances cannot efficiently send the originating session request to the server. A similar issue is encountered if the server exposing the process data from the underlying origin is mobile, i.e. the network attachment point could change, and, as a result, the network address is not deterministic. In both scenarios, the sessionless scenario may be employed to address this connectivity limitation issue. It is the main reason for selecting a sessionless pattern to be engaged as a communication foundation to implement M2M interoperability over the Internet.

In [15] the Object-Oriented Internet concept is proposed as a systematic approach to be a foundation of M2M meaningful communication targeting session-oriented archetype. It is derived from well-known Object-Oriented Programming (OOP), which is a paradigm defining objects and their interactions to design computer programs. The goal is to provide a generic solution for publishing and updating information in a context that can be utilized to describe and discover it. Now the possibility to expand this concept addressing the sessionless scenario is researched.

To make two parties interoperable both must use the same semantic-context to assign the information (meaning) to bitstreams (data) exchanged over the wire (Sect. 1). In other words, there must be a shared understanding of the mutually processed data. In general, the semantic-context may be agreed upon at runtime or design-time, but always in advance. According to the proposed model, this process is supported by the *Address Space*, *Address Space Management*, and *Information Model* concepts, which may be partially implemented [19] in compliance with the OPC UA standard [1, 2, 18].

The runtime approach is straightforward after establishing the session because the session is a tight relationship that can be used to exchange appropriate metadata [15]. In the context of the metadata exposed by the mentioned entities, the client can select what data it is interested in and implement necessary rules to establish the semantic-context using the *Server* services. Because there is no similar tight relationship between the *Subscriber* and *Publisher*, the *Subscriber* must deal with this issue in the reactive rather than proactive way. Lack of standardization in this respect (for example in the OPC UA PubSub [2]) shall be recognized as an interoperability issue that is difficult to overcome in the multi-vendor environment.

Protecting data exchanged over the network requires shared security artifacts that have to change over time to increase the protection strength. It needs a dynamic but stateful relationship between the data source and the ultimate data destination. The session makes the communication parties tightly coupled and therefore may be used as the foundation for establishing also a stateful security-context. In case the sessionless communication pattern is considered the only option is the indirect security-context established using out-of-band communication means. Again, this procedure must be precisely described by an interoperability standard in the multi-vendor environment.

3   Reactive Interoperability Domain Model

As it was pointed out in Sect. 2 the *Subscriber* must deal with data exchange and establishing a semantic-context coupled with security-context in reactive rather than a proactive way. In Fig. 2 a generic domain model of the reactive interoperability archetype is proposed. The *Publisher* and *Subscriber* are derived from a generic *Reactive Application* class, which represents common functionality. They express the publisher and subscriber roles behavior accordingly and fulfill more specific functionality aimed at allowing message centric communication where the primary relationship between process data origin and ultimate data destination is the shared understanding of:

- semantics (meaning) of exchanged process data encoded into and carried by a *Message*,
- the syntax and semantics of *Messages* that include the process data,
- a common *Distribution Channel*.

In this relationship pattern the *Publisher* is responsible for the encoding of the process data into *Message* entities and for pushing the messages to the *Distribution Channel*. The *Distribution Channel* accomplishes a set of virtual communication routes interconnecting the *Publisher* and *Subscriber* instances.

The *Distribution Channel* is a composition of *Intermediary* network nodes interconnected by communication links. The *Intermediary* performs messages multicasting, cloning, filtering, queueing, and forwarding to carry a copy of the *Message* from the *Publisher* to all interested *Subscribers*. This way a virtual path traversed by *Messages* is created. The mentioned functionality is parametrized using *Topic* values and the *Message* entities content attributed by metadata. This asynchronous delivery scenario (colloquially speaking 'fire and forgot') may be recognized as message centric communication.

At preparation time, the subscribers express an interest in one or more topics and, by design, they receive only messages that are of interest, without knowledge about which publishers, if any, there are. As a result, the message must be self-contained and meaningful also outside of the process data origin context. To implement this functionality, the *Subscriber* instances have to parametrize the routing behavior of the *Distribution Channel* after association with *Distribution Channel*. Finally, the messages are pulled and processed by the *Subscriber*. Fulfilling this role the *Subscriber* entities are being reactive. It means that they don't

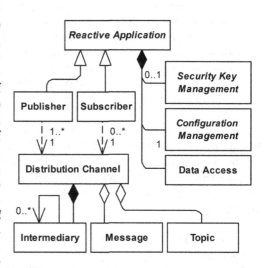

Fig. 2. Reactive interoperability model

take any initiative or make strategic decisions at runtime. Because in this scenario the *Publisher* uses push operation to forward messages and the *Subscriber* uses pull operation to recover messages, the end-to-end interconnection is limited to one way only. It is worth stressing that two independent communication channels and role coupling are required to obtain bidirectional communication.

The proposed domain model (Fig. 2) allows the implementation of all possible multiplicity relationships between the process data origin and ultimate data destination. For example, *2..* to 0..** relationships are important if data origin redundancy is required. *1..* to 2..** relationships may be considered if the same data should be processed in many locations and/or for a variety of reasons.

In the presented domain model, the *Publisher* and *Subscriber* instances are loosely coupled. It means that their interoperability is not based on a common context established directly between communicating parties similar to a session. Instead, the interoperability semantic-context must be created against the *Topic* values. It makes the *Topic* metadata a vital factor that has to contribute to the behavior of the *Intermediary* nodes and a piece of common knowledge that is required to establish interoperability.

Nevertheless, a prerequisite for establishing interoperability of the communicating parties is an underlying semantic-context that must be created at the preparation step. For the session-based communication pattern, in advance exchange of control messages is applied for this purpose. By design, any bidirectional exchange of messages is unacceptable for the publisher-subscriber communication pattern. A variety of solutions can be applied in this case but all can be derived from one of the following approaches or combination thereof:

- **static** - based on consistent or common configuration files,
- **dynamic** - based on the out-of-band communication between the *Publisher-Subscriber* and *Distribution Channel* instances,
- **remote** - based on the out-of-band communication between an independent configuration server and *Distribution Channel* instance,
- **attributed** - based on metadata (attributes) added to the messages by the forwarding mechanism.

Security between the process data origin and ultimate data destination refers to data protection measures against malicious users. The main tasks of the security mechanism aim at protecting against unauthorized data access, guaranteeing data consistency and non-repudiating while *Message* is transferred by the *Intermediary* nodes. Implementation of any security mechanism requires common knowledge of:

- **cipher algorithms** - mathematical formula necessary to protect the data concerned,
- **keys** - arbitrary streams of bits applied as actual parameters of the cipher algorithms.

If security requires common knowledge it must be deployed based on a mutually shared entity, e.g. session, link, semantic-context, etc. Security integrated with the session is a typical approach for the tightly coupled scenario. Links interconnect adjacent *Intermediary* nodes, but applying link-based security we must deal with a collection of security contexts and hop-by-hop security. This security aggregation approach is recognized as not robust because it is prone to a man-in-the-middle attack. Using semantic-context or establishing an independent context for the security purpose enables the implementation of the end-to-end security between publishers and subscribers that must be a common harmonized activity represented as the *Security Key Management* in Fig. 2. This architecture enables the implementation of the security artifacts management using practically any centralized or distributed technology including but not limited to the blockchain [12].

Because the interoperability of the *Reactive Application* is based on the semantic-context shared at the preparation step, the *Configuration Management* services have to be considered as a common activity.

4   Reactive Interoperability Implementation

This section covers an analysis of how to use the proposed domain model (Sect. 3) to make strategic design decisions and distribute functionality to reusable loosely coupled parts using the dependency injection software engineering and adaptive programming using the approach proposed in [16].

The *Reactive Interoperability* concept has been implemented as an open-source library named *SemanticData* in the project Object-Oriented Internet [17] designed to be a foundation for developing application programs that are taking part in the message-centric communication pattern. From the above discussion, we can learn that the main design decisions must concern standardization and flexibility. Standardization needs the selection of an international interoperability specification to make the library ready to be adopted by the multi-vendor environment. Flexibility requires an architecture that promotes the polymorphic independent implementation of essential functions.

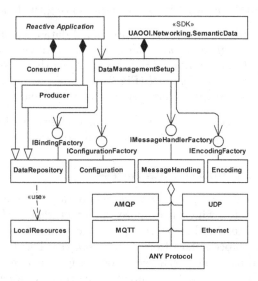

Fig. 3. Reactive interoperability implementation

Piece by piece integration of a cyber-physical system using multi-vendor products requires that M2M communication employs international standards as the interoperability foundation. Following the presented conclusions, OPC Unified Architecture Part 14 PubSub [2] is selected in this respect. By design, this standard should support the required publisher-subscriber communication pattern. Unfortunately, as it is pointed-out in Sect. 3, it covers only partially the requirements of the applications concerned. It must be stressed that by design it provides only an abstract specification. Abstract means that the standard must not limit the implementation strategy. This relationship shall be recognized as the proof of concept to verify that the implementation of the proposed model is feasible to be compliant with the selected standard as envisioned and open to support all functionality required to establish the interoperability context.

For many parts of the *Reactive Application* domain model (Sect. 3) a polymorphic approach to implementation is required. To promote the polymorphic ready solution the following concepts have been adopted:

– **separation of concerns** - to allow an independent development of the parts [8],
– **dependency injection** - to allow late binding of separately implemented parts [6].

In Fig. 3 the implementation architecture of the *Reactive Application* is proposed. The common functionality has been implemented as the Software Development Kit (SDK) available as the NuGet package [17]. To promote the polymorphic approach, it has the factory class *DataManagementSetup* that is a placeholder to gather all injection points used to compose external parts. To be injected the parts must be compliant with appropriate contracts expressed as the following interfaces representing the following functionality:

- *IBindingFactory* - bidirectional data exchange with the underlying process,
- *IConfigurationDataFactory* - the configuration data access,
- *IMessageHandlerFactory* - pushing the *Message* entities to/pulling from *Distribution Channel* (Fig. 2),
- *IEncodingFactory* - searching a dictionary containing value converters.

It is expected that the functionality implementation represented by these interfaces is provided as external composable parts.

4.1   Process Data Access

The *DataRepository* represents data holding assets in the *Reactive Application* and, following the proposed architecture, the *IBindingFactory* interface is implemented by this external part. It captures functionality responsible for accessing the process data from *LocalResources*. The *LocalResources* represents an external part that has a very broad usage purpose. For example, it may be any kind of the process data source/destination, i.e. *Raw Data* or *Address Space Management* (Fig. 1). By design, the *DataRepository* and associated entities, i.e. *Local Resources, Consumer, Producer* have been implemented as external parts, and consequently, the application scope may cover practically any concern that can be separated from the core *Reactive Application* implementation.

Depending on the expected network role the library supports the external implementation of:

- *Consumer* - entities processing data from incoming messages,
- *Producer* - entities gathering process data and populating outgoing messages.

The *Consumer* and *Producer* parts are derived from the *DataRepository* (Fig. 3). The *Consumer* uses the *IBindingFactory* to gather the data recovered from the *Message* instances pulled from the *Distribution Channel*. The received data may be processed or driven to any data destination. The *Producer* mirrors the *Consumer* functionality and, after reading data from an associated source, populates the *Message* using the gathered data.

4.2   Configuration Management

Based on the domain model analysis we can infer that the interoperability of communicating parties requires establishing a semantic-context and security-context between them in advance. The *Topic* based semantic-context is a foundation to ensure a common understanding of the process data encoded into the

Message instances. On the other hand, the security-context assures mutually shared security artifacts (i.e. cipher algorithms, keys, etc.) used to protect the *Message* as one whole. Both are necessary for handling messages. It is assumed that the security-context is established on top of the semantic-context. In that approach, the data is recognized as a principal of the security measures and the topic is regarded as a name of a collection containing the *Message* instances that are to be protected using the associated security-context independently of the endpoints that produce or consume them.

Therefore, a prerequisite for establishing interoperability of the communicating parties is an underlying semantic-context that must be created at the preparation step. In the proposed implementation, realization of this step is based on the configuration using the implementation of the *IConfigurationDataFactory* interface.

Decoupling of this functionality implementation from the functionality activation using an abstract contract and late binding mechanism allows:

- implementation of practically any configuration management scenario,
- modification of this functionality later after releasing the library or deploying the application program in the production environment.

The preparation phase concerns all the parts composed to make a running instance of the application program using the dependency injection approach to allow separate development and late binding. This approach makes any modification straightforward at any development and maintenance stage. From the end-user point of view, the composition process of the running program using independently developed parts is invisible. Because in any case a single program instance is created in a typical approach, we shall expect a commonly shared configuration mechanism providing mutually exclusive parameters to separately developed parts. This scenario has been addressed by the proposed implementation thanks to making the configuration expandable.

4.3   Distribution Channel Access

Messages preparation and the pull/push operations, which are essential to get access to *Distribution Channel* require an implementation of the interfaces *IEncodingFactory* and *IMessageHandlerFactory*.

The *IEncodingFactory* is used by the SDK to encode and decode the *Message* entities. To make the parties associated with the same *Distribution Channel* interoperable they all must use the same *Message* syntax. Implementation of the *IEncodingFactory* should address one of the options defined in the specification: JSON or binary.

The *IMessageHandlerFactory* creates object supporting operations: pull incoming messages from and push outgoing messages to adjacent *Intermediary* embedded in an abstract *Distribution Channel* (Fig. 2). To fulfill this task it has to use a concrete protocol stack. The OPC UA PubSub [2] specification lists the following protocol stacks:

- *UDP* - UDP protocol [13] that is used to transport UADP NetworkMessages,
- *Ethernet* - Ethernet-based protocol that is used to transport UADP NetworkMessages,
- *AMQP* - Advanced Message Queuing Protocol (AMQP) [7] that is used to transport JSON and UADP NetworkMessage,
- *MQTT* - Message Queue Telemetry Transport (MQTT) [4] that is used to transport JSON and UADP NetworkMessage.

The library [17] contains a reference implementation of the *IMessageHandlerFactory* for the *UDP* protocol. Because the UDP protocol is used as the *Distribution Channel*, the external filtering of massages is possible only based on the IP address. In this case, the destination port and multicast IP address combination may be recognized as the *Topic* (Sect. 3). Unfortunately, the Pub-Sub specification doesn't provide any mapping outline addressing the question of how to express *Topic* for a particular underlying communication stack. Additionally, it seems difficult or even impossible to create any directory services based on the IP addressing mechanism because it is used for nodes identification and localization on the global network, but not to express data semantics (data meaning). To overcome this limitation it is proposed to use globally unique identifiers of types defined in compliance with OPC UA Information Model [10,14] as the *Topic* entities to establish semantic-context and security- context of the reactive assets interoperability. In this approach, the filtering and multiplexing functionality must be embedded locally in the implementation of the UDP communication stack.

It is worth stressing that the proposed separation of concerns and dependency injection approach make the architecture ready to utilize *ANY Protocol* that supports transparent data transfer over the wire (Fig. 3).

5   Conclusions

Based on the architecture proposed in [19] and abstracted in Sect. 2 the session-less and session-oriented communication patterns are examined against the IoT requirements reviewed in Sect. 1. The discussion concludes that the connection-less pattern better suites issues related to the assets mobility and traffic asymmetry that is characteristic for the application domains concerned. Additionally, to promote interoperability and address the demands of the M2M communication in the context of a multi-vendor environment the implementation of products derived from the proposals must be compliant with the selected international standard. The mapping of this architecture and OPC UA is discussed in [19].

In Sect. 3 a generic domain model of the reactive interoperability archetype is introduced. It has been designed based on the above-mentioned findings. The main goal is to provide a foundation for the future development of standards, best practice rules and supporting reusable frameworks. It was pointed out that improvements of the existing interoperability standards addressing the reactive interoperability based on the publisher-subscriber archetype, e.g. OPC UA Pub-Sub [2], AMQP [7] and MQTT [4] shall be scoped on establishing the semantic and security contexts. Considering best practice rules it is worth stressing that the process of establishing an interoperability context is based on the firm but abstract rules. However, the concrete implementation must be flexible enough to deal with a variety of polymorphic algorithms. Addressing the development of a reusable framework needs the proposed model to be backed by proof of concept, i.e. a reference implementation described in Sect. 4.

The main aim of Sect. 4 is to present the experience gained during the implementation of the reactive interoperability concept outlined in Sect. 3. This concept was implemented consistently with the Object-Oriented Internet paradigm[1] [17] worked out in an open-source project. The domain model proposed in Sect. 3 is used to make significant implementation decisions. The description of a reference application program implementation proves that it is possible to design universal architecture targeting reactive interoperability as a consistent part of the Object-Oriented Internet concept compliant with the OPC UA PubSub [2] international standard. According to the presented implementation and evaluation, using the dependency injection and late binding, the application program can be seamlessly adapted to the production environment and scales well.

The results presented in the article prove that the composite nature of the disciplines concerned can be relaxed by the composite nature of the running programs and postponing parts binding up to the system deployment stage. It also improves flexibility and adaptability of the existing solutions against any modification of the production environment including but not limited to the selected interoperability standard change.

Future work is focused on Machine to Sensors connectivity based on Process-Observer concept introduced in [15]. By design, it allows access to plant floor devices using a variety of Fieldbus industrial network protocols. The main goal of this project is to prove that, based on the presented results, the fetching data from Process-Observer will make the factoring of the structural data as the composition of simple data possible.

References

1. OPC unified architecture specification part 3 - address space model (2017). https:// opcfoundation.org/developer-tools/specifications-unified-architecture/part-3-addr ess-space-model/
2. OPC unified architecture specification part 14 - PubSub (2018). https://opcfoun dation.org/developer-tools/specifications-unified-architecture/part-14-pubsub/
3. Ashton, K.: That 'internet of things' thing. RFID J. **2009**, 1 (2009). https://www.rfidjournal.com/articles/pdf?4986
4. Cohn, R.J., Coppen, R.J.: MQTT version 3.1.1 plus errata 01, 10 December 2015. http://docs.oasis-open.org/mqtt/mqtt/v3.1.1/mqtt-v3.1.1.html
5. Eugster, P.T., Felber, P.A., Guerraoui, R., Kermarrec, A.M.: The many faces of publish/subscribe. ACM Comput. Surv. **35**(2), 114–131 (2003). https://doi.org/ 10.1145/857076.857078

---

[1] https://github.com/mpostol/OPC-UA-OOI.

6. Fowler, M.: Inversion of control containers and the dependency injection pattern, 23 January 2004. https://www.martinfowler.com/articles/injection.html

7. Jeyaraman, R., Telfer, A.: OASIS advanced message queuing protocol (AMQP) version 1.0, 29 October 2012. http://docs.oasis-open.org/amqp/core/v1.0/os/amqp-core-overview-v1.0-os.html

8. Kulkarni, V., Reddy, S.: Separation of concerns in model-driven development. IEEE Softw. **20**(5), 64–69 (2003). https://doi.org/10.1109/MS.2003.1231154

9. Lawton, G.: Machine-to-machine technology gears up for growth. Computer **37**(9), 12–15 (2004). https://doi.org/10.1109/MC.2004.137

10. Mahnke, W., Leitner, S.H., Damm, M.: OPC Unified Architecture, 1st edn. Springer, Heidelberg (2009). https://doi.org/10.1007/978-3-540-68899-0

11. Meng, Z., Wu, Z., Muvianto, C., Gray, J.: A data-oriented M2M messaging mechanism for industrial IoT applications. IEEE Internet Things J. **4**(1), 236–246 (2017). https://doi.org/10.1109/JIOT.2016.2646375

12. Novo, O.: Blockchain meets IoT: an architecture for scalable access management in IoT. IEEE Internet Things J. **5**(2), 1184–1195 (2018). https://doi.org/10.1109/JIOT.2018.2812239

13. Postel, J.: User datagram protocol, 28 August 1980. http://www.ietf.org/rfc/rfc768.txt

14. Postol, M.: OPC From Data Access to Unified Architecture, sec. Information Model, 4th revised edn., pp. 111–130. VDE VERLAG GMBH (2010)

15. Postol, M.: Object oriented internet. In: 2015 Federated Conference on Computer Science and Information Systems (FedCSIS), pp. 1069–1080 (2015). https://doi.org/10.15439/2015F160

16. Postol, M.: Csharp in practice adaptive programming. Technical report, Lodz University of Technology, 26 February 2019. https://doi.org/10.5281/zenodo.2578244

17. Postol, M.: Object-oriented internet. Technical report 5.1.0 (2019). https://doi.org/10.5281/zenodo.3345043

18. Postół, M.: OPC UA information model deployment. Technical report, CAS, 18 April 2016. https://doi.org/10.5281/zenodo.2586616

19. Postół, M.: Machine to machine semantic-data based communication: comprehensive survey. In: Computer Game Innovations 2018, pp. 83–101. Lodz University of Technology Press, Łódź (2018). https://www.researchgate.net/publication/335524620_Computer_Game_Innovations_2018

20. Schlick, J.: Cyber-physical systems in factory automation-towards the 4th industrial revolution. In: 9th IEEE International Workshop on Factory Communication Systems (WFCS) (2012)

21. Verma, P.K., et al.: Machine-to-machine (M2M) communications: a survey. J. Netw. Comput. Appl. **66**, 83–105 (2016). https://doi.org/10.1016/j.jnca.2016.02.016

22. Xu, L.D., He, W., Li, S.: Internet of things in industries: a survey. IEEE Trans. Ind. Inform. **10**(4), 2233–2243 (2014). https://doi.org/10.1109/TII.2014.2300753

Impact of Long-Range Dependent Traffic in IoT Local Wireless Networks on Backhaul Link Performance

Przemyslaw Wlodarski[(✉)]

Department of Signal Processing and Multimedia Engineering,
West Pomeranian University of Technology, 70-313 Szczecin, Poland
przemyslaw.wlodarski@zut.edu.pl

Abstract. Performance evaluation in Internet of Things (IoT) networks is becoming more and more important due to the increasing demand for quality of service (QoS). In addition to basic statistical properties based on the distribution of interarrival times of packets, actual network traffic exhibits correlations over a wide range of time scales associated with long-range dependence (LRD). This article focuses on examining the impact of both LRD and number of nodes that transmit packets in a typical IoT wireless local network, on performance of the backhaul link. The analysis of latency and packet loss led to an interesting observation that the aggregation of packet streams, originating from single nodes, lowers the importance of LRD, even causing an underestimation of performance results when compared to the queueing system with Markovian input.

Keywords: IoT wireless network · Backhaul link · Long-range dependence · Traffic analysis · Performance evaluation

1 Introduction and Related Work

Internet of Things (IoT) wireless networks integrate many physical devices and sensors that send data through wire or wireless connections and have different architectures (e.g. fog, edge, cloud). Because of the heterogeneity of IoT systems it is not easy task to design reliable and efficient communication systems [8]. One of the big issue is energy consumption, which depends on: incorrect selection of the microcontroller, energy-inefficient software [3] or communication protocol parameters [12]. Another big issue is the impact of offered load on quality of service (QoS) and quality of experience (QoE), especially for interactive or streaming services. There are two main measures in terms of QoS that relate to end-to-end connection performance: packet loss and latency. This article presents the results of these measures for different scenarios, taking into account long-range dependent (LRD) feature of the traffic.

Wireless communication in IoT networks is usually based on one of the Medium Access Control (MAC) protocols: Time Division Multiple Access (TDMA) or

© Springer Nature Switzerland AG 2020
V. V. Krzhizhanovskaya et al. (Eds.): ICCS 2020, LNCS 12141, pp. 423–435, 2020.
https://doi.org/10.1007/978-3-030-50426-7_32

Carrier Sense Multiple Access with Collision Avoidance (CSMA/CA). In TDMA protocol each node is assigned a time slot for data transmission in a predetermined order. The main disadvantage of this system is that it requires accurate synchronization, which reduces the efficiency of the entire system. The second MAC protocol widely used in IoT systems is well known CSMA/CA (example use: IEEE 802.11 family, IEEE 802.15.4, etc.), that was extensively studied in many articles. Comprehensive study on the throughput, delay and stability performance of CSMA networks was presented in [6]. The problem of stable throughput and bounded mean delay was discussed in [24]. Another interesting contribution in the area of CSMA/CA protocols was a proposition of Handshake Sense Multiple Access with Collision Avoidance (HSMA/CA) protocol [19], which protects a densely deployed network from the classical hidden and exposed terminal problems. This idea was developed using Markov modeling and simulations. Recently, authors of [9] considered maximum effective throughput and suggested that the minimum mean access delay parameter of CSMA/CA system is of great practical interest.

Network traffic affects the reliability and performance of a network. Information on the statistical distribution of interarrival times of packets is not sufficient for evaluation of network performance, since actual network traffic exhibits second-order properties associated with long-range dependence (LRD) [1,20]. It is very important to consider LRD, because of the impact on queueing performance [16]. Many models were developed from the properties of fractional Gaussian noise [17] or fractional autoregressive moving average process [4]. One of the most popular model that incorporates LRD properties is on/off source, recently analyzed in [25]. There is also a modified version of the Pareto on/off model [13], which is used further in this article.

Network traffic analysis and modeling is crucial for design and implementation of efficient and reliable transmission networks. It helps explain possible problems before they occur. The simulation results can be used to identify anomalies [5,7,10] or detect Distributed Denial of Service flood attacks [11,15]. Furthermore, most of the symptoms that lead to congestion and high level of packet loss rate can be detected in the simulation process of network traffic [2,23].

Experimenting with physical devices is uneconomical in the first phase of the project, especially if the number of devices is large. Therefore, the simulation approach is optimal in developing new methods and testing new scenarios. For the purposes of this article, all simulations were carried out using OMNeT++ framework [21,22], which is a powerful open-source discrete event simulation tool. In fact, it is component-based C++ simulation library and framework, but instead of providing components specifically for computer networks, it includes generic component architecture to create any simulation.

The article is organized as follows. Next section characterizes basic properties of long-range dependence and introduces estimation methods used in further sections of this paper. Section 3 presents considered network structure as well as the model of network traffic generated by single node. In addition, it contains all the statistics for the reference queueing system, which are further studied from the point of view of performance evaluation. In Sect. 4, the results obtained for the analyzed network are then compared with the corresponding statistics of a commonly used queueing model described in Sect. 3.3.

2   Long-Range Dependence

2.1   Basic Properties

In order to explain the concept of long-range dependence (LRD), one needs to take a closer look at the stochastic process for different time scales. Let $Y(t)$ be the stationary stochastic process. The following simple equation describes the relationship for the process that is rescaled in time:

$$Y(at) \stackrel{d}{=} a^H Y(t), \quad a > 0, \tag{1}$$

where a is a stretching factor and H is the Hurst exponent and $\stackrel{d}{=}$ denotes equality in distributions. If $0.5 < H < 1$ then second-order properties associated with correlation structure are preserved regardless of scaling in time and the process becomes LRD. The higher value of H the stronger dependence. The autocorrelation function for the incremental process $X(i) = Y(i) - Y(i-1)$, $i = 1, 2, ...$, which reflects the similarity between $X(i)$ and $X(i+k)$, has the following form:

$$r_k = \frac{\sigma^2}{2} \left((k+1)^{2H} - 2k^{2H} + |k-1|^{2H} \right), \quad k = 0, 1, ... \tag{2}$$

and for $H > 0.5$ is not summable:

$$\sum_{k=0}^{\infty} r_k \to \infty. \tag{3}$$

The value of autocorrelation function decays slowly for LRD processes. In case of no-LRD processes, there is no dependency ($H = 0.5$) and $r_k = 0$ for $k \geq 1$.

2.2   Estimation

In order to evaluate Three methods of estimation of Hurst exponent were used. First one is variance-time method, which is based on the aggregated process of $X(n)$ for discrete times $n = 0, 1, ..., N$ that corresponds to fixed-length intervals:

$$X^{(m)}(n) = m^{-1} \sum_{t=mn}^{m(n+1)-1} X(t), \quad n = 0, 1, ..., \lfloor N/m \rfloor - 1, \tag{4}$$

where m denotes the level of aggregation, i.e.:

$$X^{(m)} = m^{-1} \sum_{t=1}^{m} X(t) = m^{H-1} X \tag{5}$$

The variance of the aggregated random variable in (5) is:

$$Var\left(X^{(m)}\right) = \sigma^2 m^{2H-2}, \tag{6}$$

where σ^2 is the variance of X. It can be easily seen that if one performs logarithmic operation on both sides of (6) then Hurst exponent can be estimated from the slope of linear regression $(2H - 2)$ for all aggregated samples according to (4).

The next, similar method of estimation is Index of Dispertion for Counts (IDC). It is defined as a relation of variance-to-mean ratio of the sum of random variable X for period L:

$$IDC(L) = Var\left(\sum_{n=1}^{L} X(n)\right) \bigg/ E\left(\sum_{n=1}^{L} X(n)\right) \approx cL^{2H-1}, \tag{7}$$

where c is a positive value. As with the variance-time method, one can use the linear regression to get the estimated $\tilde{H}$ from the slope $(2H - 1)$.

Another method of estimation is periodogram based on the approximated value of spectral density of LRD processes:

$$f(\lambda, H) \approx sin(\pi H)\Gamma(2H + 1)|\lambda|^{1-2H}, \tag{8}$$

where λ is the frequency value for analyzed random variable X. Although the estimation operation is done in the frequency domain, the Hurst exponent can also be calculated from linear regression. In this case, the FFT values should be taken as the regression points. LRD refers to the lowest frequencies, which is reflected in the formula (8), where most of the energy concentrates near 0. For that reason, only 10% of the lowest frequencies is considered in the periodogram estimation method.

3   Framework

3.1   Network

In order to analyze network traffic in a typical and commonly used structure for IoT devices shown in Fig. 1, a simple and efficient non-persistent CSMA/CA protocol is assumed. This protocol was chosen because it can be easily implemented even in basic and cheap microcontrollers and does not consume much power during wireless operation, which is crucial for IoT wireless sensors. In non-persistent version of CSMA protocol waiting node does not listen to the channel continuously until it becomes idle (like in 1- or p-persistent versions), which reduces energy consumption.

The network consists of $nNodes$ wireless nodes and two stations: $st0$ and $st1$. Each node can transmit packets to the station $st0$ and can sense transmission from another node to avoid collisions. All nodes can hear each other, so there is no hidden node problem [14]. When the channel is busy, because another node is

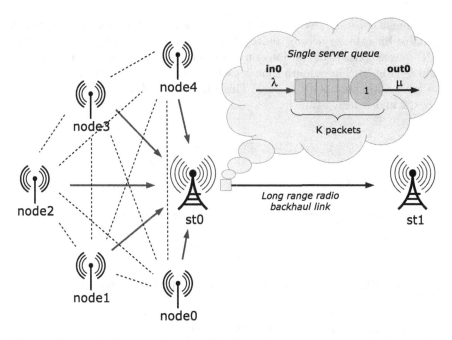

Fig. 1. Schematic diagram of analyzed IoT wireless network, example for 5 nodes.

transmitting, the node that wants to send packets must wait a period of time and then listen (sense) again. If the channel is idle, the node starts its transmission immediately.

Wireless station $st0$ receives packets from all nodes and tries to send them all across the long range radio link to the station $st1$. All packets must pass through the wireless network interface of $st0$ that connects this station to another one ($st1$). The output interface at $st0$ actually incorporates the queueing system that has K places for packets (including the one being transmitted) and the transmission circuit limited by the bandwidth of the radio link. Therefore, because all the cumulative traffic from the nodes goes there, all interesting performance statistics, associated with the impact of LRD traffic, can be found at the radio link network interface of station $st0$. Before the packet leaves $st0$, either it is immediately processed (if the queue is empty) or goes to the queue buffer. If there is not enough buffer space, packet is dropped (buffer overflow). By changing the bandwidth of this radio link channel, one can examine the impact of the local traffic on queueing performance, and thus on the level of packet loss and latency.

3.2   Source Node

Every node sends fixed length packets to the np-CSMA channel according to Pareto Modulated Poisson Process (PMPP) [13]. This model was chosen because it is versatile, efficient and introduces long-range dependence to the traffic while maintaining a constant level of packet rate. In addition, it resembles the behavior of variable-bit-rate services or protocols that transmit packets in batches. The PMPP source consists of two Poisson sources with alternating traffic intensities λ_1 and λ_2 (Fig. 2). The sojourn time in each state has Pareto distribution $P\{X \geq x\} = x^{-\alpha}$ with parameter $\alpha > 0$. This distribution has infinite variance for $1 < \alpha < 2$ and is heavy-tailed.

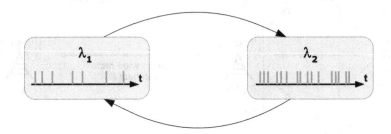

Fig. 2. Two Poisson sources of PMPP packet generator

The approximated value of IDC for PMPP source is:

$$IDC(t) \approx 1 + \frac{(\lambda_1 - \lambda_2)^2}{\lambda_1 + \lambda_2} \left(\frac{\alpha - 1}{\alpha} \right) t^{2-\alpha}, \tag{9}$$

where H can be easily obtained from:

$$H = \frac{3 - \alpha}{2} \tag{10}$$

and is compatible with (7) in terms of the same exponent $(2H - 1)$. The λ_1 and λ_2 values should be selected so that the expected value of number of packets $E(N(t)) = 0.5(\lambda_1 + \lambda_2)t$ corresponds to the desired value of the generated network traffic.

3.3   Performance Evaluation

The network performance of backhaul link between $st0$ and $st1$ depends on statistical properties of the inbound traffic as well as the service rate and packet length distribution. Since all packets have fixed size, a deterministic service is assumed. All traffic from local network goes to the input of queueing system inside the output interface of $st0$ (Fig. 1). The queueing system consists of one server and has $K - 1$ slots as a buffer space for packets. If the buffer overflows (K packets in the system) then the next incoming packet is dropped. Most common type of queueing system that meets the above assumptions is M/D/1/K, for which explicit formulas of blocking probability, stationary distribution and mean system sojourn time were derived in [18]. Both latency and packet loss can be expressed in terms of steady state probabilities of number of packets in the system:

$$D = \frac{1}{\lambda(1 - P_{LOSS})} \sum_{k=0}^{K} k \cdot p_k^{(K)} \tag{11}$$

$$P_{LOSS} = p_K^{(K)}, \tag{12}$$

where:

$$p_k^{(K)} = \begin{cases} (1 + \rho \Lambda_{K-1})^{-1} & \text{for} \quad k = 0 \\ (\Lambda_k - \Lambda_{K-1}) p_0^{(K)} & \text{for} \quad k = 1, \ldots, K-1 \\ 1 - \Lambda_{K-1} p_0^{(K)} & \text{for} \quad k = K \end{cases} \tag{13}$$

$$\Lambda_k = \sum_{i=0}^{k} \frac{(\rho(i-k))^i}{i!} exp\left((k-i)\rho\right). \tag{14}$$

These relationships are the reference for comparing them with the data received from the interface of *st0* for different scenarios, i.e. different levels of LRD as well as different number of nodes.

4 Results

All results were obtained using the OMNeT++ [22] simulation platform. The simulation framework was described in the previous section.

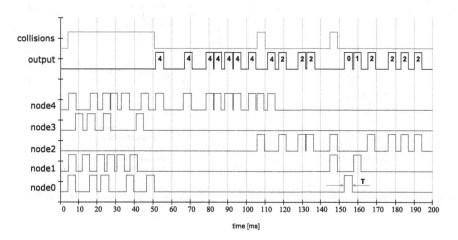

Fig. 3. Sample result for the first 200 ms of highly congested traffic, nNodes: 5, bandwidth: 1 Mbps, packet transmission time (T): 4.096 ms.

The performance results were compared to a typical M/D/1/K queuing system, presented in 3.3, where all performance measures refer to the classical Poisson model of input traffic (without LRD feature).

The purpose of the experiments was to examine the effect of different Hurst exponent values in the range of $0.5 < H < 1$ as well as different number of nodes ($nNodes$) on performance of the backhaul link (Fig. 1). The simulation time, bandwidth and packet length for each scenario were 60 min., 1 Mbps and 4096 bits (512 bytes), respectively. The packet length corresponds to the packet transmission time of 4.096 ms shown in Fig. 3.

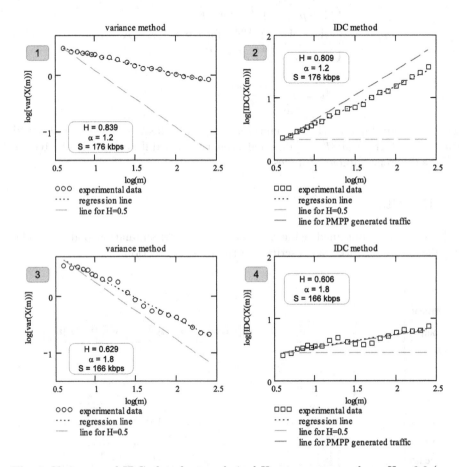

Fig. 4. Variance and IDC plots for two desired Hurst exponent values: $H = 0.9$ ($\alpha = 1.2$) and $H = 0.6$ ($\alpha = 1.8$).

Figure 4 shows the estimation results of Hurst exponent ($\tilde{H}$) for the aggregated outbound traffic from 5 nodes. Two methods of estimation were applied: variance-time and IDC plot (see Sect. 2.2 and Eqs. (5), (7)). The desired H was

0.9 and 0.6, which corresponds to the $\alpha = 1.2$ and $\alpha = 1.8$ of Pareto distribution in PMPP model.

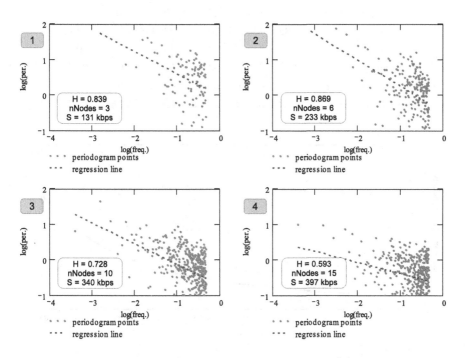

Fig. 5. Periodogram plots for different number of nodes.

In the next figure (Fig. 5) there are periodogram estimation plots for different number of nodes for desired $H = 0.9$. All estimation results of Hurst exponent were obtained based on the formula (8). It is clearly seen that the values of $\tilde{H}$ decreases as the number of nodes increases. This is due to the disappearance of the LRD structure in aggregated stream that consists of many independent network flows coming from single nodes. It can be better observed in Table 1, where the mean $\tilde{H}$ estimated for nodes stays the same (approximately), while the mean $\tilde{H}$ at the input of the queue ($in0$) decreases to the low value, suggesting that the LRD properties gradually disappears.

Table 2 shows the main statistics for different desired H values constant number of nodes ($nNodes = 5$). A slight increase of all measured values can be observed, which suggests that increasing H value raises the level of collisions and hence increases values for other statistics.

The observations from Table 1 are confirmed by the backhaul link performance results. In Fig. 6 there are curves for mean number of packets in the queueing system for different number of nodes versus offered traffic load. The theoretical curve for $M/D/1/20$ system, marked with solid black line, is calculated as the mean value of all $p_k^{(K)}$ in (13) for $K = 20$. In the next Fig. 7 one

Table 1. Main statistics for $H = 0.9$ and different number of nodes.

Number of nodes:	2	3	6	8	10	15
Throughput [kbps]	81	131	233	290	397	326
Packets at $in0$ [pkts]	70829	112264	192267	239938	291615	345220
Collision rate [%]	2	4	12	16	24	39
$\bar{\lambda}$ at $in0$ [pkts/s]	20	31	53	67	81	96
Mean $\tilde{H}$ for nodes	0.943	0.881	0.889	0.917	0.859	0.91
Mean $\tilde{H}$ at $in0$	0.873	0.813	0.861	0.841	0.758	0.574

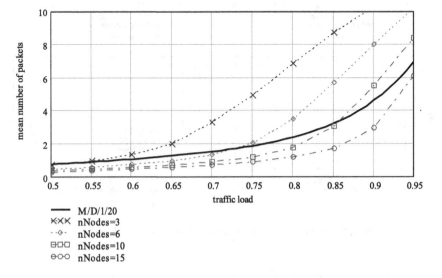

Fig. 6. Mean number of packets in the queueing system for different number of nodes, $\alpha = 1.2$.

Table 2. Main statistics for 5 nodes and different H values.

H value:	0.6	0.7	0.8	0.9
Throughput [kbps]	180.5	181.8	183.8	189.5
Packets at $in0$ [pkts]	158681	159830	161598	166602
Collision rate [%]	7.9	7.8	8.1	8.5
$\bar{\lambda}$ at $in0$ [pkts/s]	40.6	40.8	44.5	48.4

can observe the same tendency - the packet loss becomes lower as the number of nodes increases. Last Fig. 8 presents changing latency when number of nodes increases. For $nNodes = 10$ and $nNodes = 15$ the empirical curves goes below

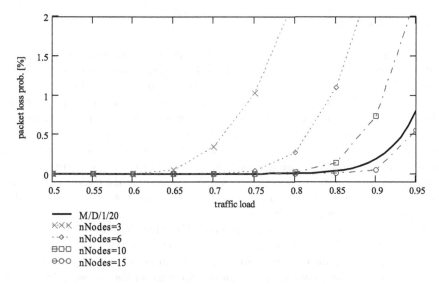

Fig. 7. Packet loss for different number of nodes, $\alpha = 1.2$.

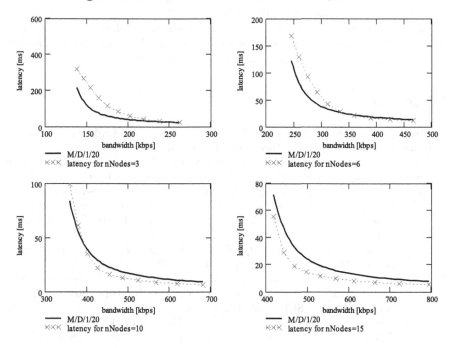

Fig. 8. Latency for different number of nodes compared to M/D/1/20 queueing system.

the levels of theoretical counterpart calculated from (11), which can be explained by the fact that the aggregation of streams from single nodes causes big change in LRD as well as in distribution.

5 Conclusions

The number of IoT devices and networks is constantly increasing, which means that congestion can occur, especially when the communication channel capacity does not increase. The main performance measures analyzed in this article were latency and packet loss. It is obvious that for higher offered traffic load the values of both measures increases causing poor performance of the connection link. However, situation becomes worse when LRD is considered. There is no doubt that this feature exists in network traffic. The question is, how strong are these long-term relationships and how they influence the performance. The simulation results of analysis of a typical IoT wireless CSMA/CA network with backhaul link provided more insights on the impact of both LRD and number of nodes on latency and packet loss.

All estimation results of Hurst exponent show that $\tilde{H}$ is stable for the same number of nodes. If number of nodes increases, then $\tilde{H}$ becomes smaller. Furthermore, all performance results show that with an increasing number of nodes, performance improves, causing even underestimation of the classical M/D/1/K model of queueing system. It implies that the aggregated stream consisting of many single node streams has changed its structure in terms of LRD feature as well as the distribution. This phenomenon can be used to determine the parameters of the IoT network system in order to reduce the value of latency and packet loss, which in turn has a positive effect on QoS and QoE.

References

1. Abry, P., Flandrin, P., Taqqu, M., Veitch, D.: Theory and Applications of Long-Range Dependence, 1st edn. Birkhäuser Basel, Cambridge (2002)
2. Jan, M.A., Jan, S.R.U., Alam, M., Akhunzada, A., Rahman, I.U.: A comprehensive analysis of congestion control protocols in wireless sensor networks. Mob. Netw. Appl. **23**(3), 456–468 (2018). https://doi.org/10.1007/s11036-018-1018-y
3. Al-Kofahi, M.M., Al-Shorman, M.Y., Al-Kofahi, O.M.: Toward energy efficient microcontrollers and Internet-of-Things systems. Comput. Electr. Eng. **79**, 106457 (2019). https://doi.org/10.1016/j.compeleceng.2019.106457
4. Bai, S., Taqqu, M.: On the validity of resampling methods under long memory. Ann. Stat. **45**, 2365–2399 (2017). https://doi.org/10.1214/16-AOS1524
5. Bhuyan, M.H., Bhattacharyya, D.K., Kalita, J.K.: Network anomaly detection: methods, systems and tools. IEEE Commun. Surv. Tutorials **16**(1), 303–336 (2014). https://doi.org/10.1109/SURV.2013.052213.00046
6. Dai, L.: Toward a coherent theory of CSMA and Aloha. IEEE Trans. Wireless Commun. **12**, 3428–3444 (2013). https://doi.org/10.1109/TWC.2013.052813.121605
7. Dymora, P., Mazurek, M.: Anomaly detection in IOT communication network based on spectral analysis and hurst exponent. Appl. Sci. **9**(24) (2019). https://doi.org/10.3390/app9245319
8. Fortino, G., Gravina, R., Russo, W., Savaglio, C.: Modeling and simulating Internet-of-Things systems: a hybrid agent-oriented approach. Comput. Sci. Eng. **19**(5), 68–76 (2017). https://doi.org/10.1109/MCSE.2017.3421541

9. Gao, Y., Dai, L.: Random access: packet-based or connection-based? IEEE Trans. Wireless Commun. **18**(5), 2664–2678 (2019). https://doi.org/10.1109/TWC.2019. 2906596
10. Iglesias, F., Zseby, T.: Analysis of network traffic features for anomaly detection. Mach. Learn. **3**, 59–84 (2014). https://doi.org/10.1007/s10994-014-5473-9
11. Jing, X., Yan, Z., Jiang, X., Pedrycz, W.: Network traffic fusion and analysis against ddos flooding attacks with a novel reversible sketch. Inf. Fusion **51**, 100–113 (2019). https://doi.org/10.1016/j.inffus.2018.10.013
12. Koseoglu, M., Karasan, E.: Energy-optimum throughput and carrier sensing rate in csma-based wireless networks. IEEE Trans. Mob. Comput. **13**(6), 1200–1212 (2014). https://doi.org/10.1109/TMC.2013.124
13. Le-Ngoc, T., Subramanian, S.N.: A pareto-modulated poisson process (PMPP) model for long-range dependent traffic. Comput. Commun. **23**, 123–132 (2000)
14. Ley-Bosch, C., Alonso-González, I., Sanchez-Rodriguez, D., Ramírez-Casañas, C.: Evaluation of the effects of hidden node problems in IEEE 802.15.7 uplink performance. Sensors **16**, 216 (2016). https://doi.org/10.3390/s16020216
15. Li, G., Yan, Z., Fu, Y., Chen, H.: Data fusion for network intrusion detection: a review. Secur. Commun. Netw. **2018**, 1–16 (2018). https://doi.org/10.1155/2018/8210614
16. Park, K., Willinger, W.: Self-Similar Network Traffic and Performance Evaluation, 1st edn. Wiley, Hoboken (2000)
17. Purczynski, J., Wlodarski, P.: On fast generation of fractional Gaussian noise. Comput. Stat. Data Anal. **50**, 2537–2551 (2006)
18. Seo, D.W.: Explicit formulae for characteristics of finite-capacity M/D/1 queues. ETRI J. **36**(4), 609–616 (2014). https://doi.org/10.4218/etrij.14.0113.0812
19. Shafiq, M., Ahmad, M., Afzal, M., Ali, A., Irshad, A., Choi, J.G.: Handshake sense multiple access control for cognitive radio-based IOT networks. Sensors **19**(2), 241 (2019). https://doi.org/10.3390/s19020241
20. Smith, R.: The dynamics of internet traffic: self-similarity, self-organization, and complex phenomena. Adv. Complex Syst. **14**(6), 905–949 (2011). https://doi.org/10.1142/S0219525911003451
21. Varga, A.: OMNeT++. In: Wehrle, K., Güneş, M., Gross, J. (eds.) Modeling and Tools for Network Simulation, pp. 35–59. Springer, Heidelberg (2010). https://doi.org/10.1007/978-3-642-12331-3_3
22. Varga, A.: OMNeT++. https://omnetpp.org/. Accessed 19 Dec 2019
23. Wang, J., Yang, X., Liu, Y., Qian, Z.: A contention-based hop-by-hop bidirectional congestion control algorithm for ad-hoc networks. Sensors **19**(16) (2019). https://doi.org/10.3390/s19163484
24. Wong, P.K., Yin, D., Lee, T.T.: Analysis of non-persistent CSMA protocols with exponential backoff scheduling. IEEE Trans. Commun. **59**(8), 2206–2214 (2011). https://doi.org/10.1109/TCOMM.2011.051811.100241
25. Zhang, Y., Huang, N., Xing, L.: A novel flux-fluctuation law for network with self-similar traffic. Phys. A Stat. Mech. Appl. **452**(C), 299–310 (2016). https://doi.org/10.1016/j.physa.2016.02.031

Computer Graphics, Image Processing and Artificial Intelligence

OpenGraphGym: A Parallel Reinforcement Learning Framework for Graph Optimization Problems

Weijian Zheng[1] , Dali Wang[2]([⊠]) , and Fengguang Song[1]

[1] Indiana University-Purdue University, Indianapolis, IN 46202, USA
zheng273@purdue.edu, fgsong@iupui.edu
[2] Oak Ridge National Laboratory, Oak Ridge, TN 37831, USA
wangd@ornl.gov

Abstract. This paper presents an open-source, parallel AI environment (named *OpenGraphGym*) to facilitate the application of reinforcement learning (RL) algorithms to address combinatorial graph optimization problems. This environment incorporates a basic deep reinforcement learning method, and several graph embeddings to capture graph features, it also allows users to rapidly plug in and test new RL algorithms and graph embeddings for graph optimization problems. This new open-source RL framework is targeted at achieving both high performance and high quality of the computed graph solutions. This RL framework forms the foundation of several ongoing research directions, including 1) benchmark works on different RL algorithms and embedding methods for classic graph problems; 2) advanced parallel strategies for extreme-scale graph computations, as well as 3) performance evaluation on real-world graph solutions.

Keywords: Reinforcement learning · Graph optimization problems · Distributed GPU computing · Open AI software environment

1 Introduction

Solving graph optimization problems effectively is critical in many important domains, including social networks, telecommunications, marketing, security, transportation, power grid, bioinformatics, traffic planning, scheduling, and emergency preparedness. However, many of the graph optimization problems are in the class of NP-hard problems, and require exponential time algorithms to search for optimal solutions. Due to the exact graph algorithms' exponential time complexity, practical approaches most often use either *approximation*

This research was funded by the U.S. Department of Energy, Office of Science, Advanced Scientific Computing Research (Interoperable Design of Extreme-scale Application Software).

V. V. Krzhizhanovskaya et al. (Eds.): ICCS 2020, LNCS 12141, pp. 439–452, 2020.
https://doi.org/10.1007/978-3-030-50426-7_33

algorithms or *heuristic algorithms* to tackle big graphs. The approximation algorithms are of polynomial time (if they do exist), but in theory can be several times worse than the optimal solutions. The heuristic algorithms are fast, but do not have the same guaranteed solution quality as that of approximation algorithms. Also, heuristic algorithms typically require experts' knowledge, insights, and repeated redesigns to create efficient heuristics.

Instead of devising different heuristics for different graph problems and distinct graph datasets, we aim to utilize machine learning techniques to "learn" effective heuristics automatically. Since 2016, a few researchers have started to design reinforcement learning and deep learning methods to solve combinatorial optimization problems [4,11–14,16,22]. The rational behind it is that graphs from the same application domain or similar types are not totally different from each other; they may have similar structures and are often solved repeatedly. Hence, it can be beneficial to use machine learning to generalize the methods or heuristics to find near optimal solutions.

To investigate different deep reinforcement learning methods, and design new domain-specific graph embeddings to capture graph features, we design and implement an open source AI environment to allow users to rapidly plug in and test new RL algorithms and graph embeddings for graph optimization problems. The new open source RL framework, named *OpenGraphGym*, is targeted at achieving both high performance and high quality of the computed graph solutions. Our work has the following contributions. 1) We design and create an extensible framework for generic graph problems. A suit of NP-hard graph problems and graph embedding methods can be added into our framework conveniently. Our framework can also be used to benchmark several RL algorithms for graph optimization problems. 2) Our distributed RL framework can utilize multiple GPUs. 3) Case study shows that our framework can help to provide better solutions for Minimum Vertex Cover problems (a classic NP-hard graph problems).

In the remainder of the paper, we will first introduce the related work, then describe how to convert (or map) conventional graph problems to RL problems in Sect. 3. In Sect. 4, we will present the *OpenGraphGym* framework design and implementation details. A case study of using *OpenGraphGym* to solve the Minimum Vertex Cover problem with different types of graphs will be shown in Sect. 5. Finally, Sect. 6 will present our conclusions and future work.

2   Related Work

Reinforcement learning (RL) was commonly used in the field of playing games [17,20,21]. Recently, researchers started to investigate if RL can be used to help solve NP-hard graph problems. Based on the observation that knowledge learned from some problem instances can be applied to a similar type of problem instances, Dai et al. [11] created an end-to-end RL model that combines graph embedding and the objective Q function to tackle NP-hard graph problems. In their work, the solution is built by incrementally adding vertices. They studied the Minimum Vertex Cover, the Maximum Cut, and the Traveling Salesman problems by applying the Q-learning algorithm. Meanwhile, their results also proved that the strategy learned by the smaller size of graphs could be applied to the larger size of graphs. Bello et al. [4] also applied RL to graph optimization problems. However, they focused on euclidean Travelling Salesman Problems (TSP), and their methods cannot be applied to other graph problems conveniently.

Besides RL methods, researchers have also employed supervised machine learning methods to solve graph problems. Li et al. [12,14] applied the Graph Convolution Network (GCN) to find multiple solutions in one step, then used a tree search model to select the best solution. The labeled SATLIB dataset [10] was used to train their GCN model. Another similar work was done by Mittal et al., who also used GCN to generate multiple solutions [16]. However, instead of using tree search, they took advantage of RL to select the best solution. In addition, Vinyals et al. applied a neural network architecture called *pointer network* to address graph combinatorial optimization problems [22]. The following study by Kool et al. modified the pointer network by introducing an attention-based encoder-decoder model and applied it to the TSP problem [13].

Compared to the existing work, our project targets creating an open AI framework that is optimized for solving big graph optimization problems. The major differences are as follows. First, our *OpenGraphGym* framework is an open environment, in which additional graph embedding methods, different RL algorithms, and new graph problems can be plugged in and tested rapidly. Second, *OpenGraphGym* is designed to be a high performance computing solution that can support distributed GPU systems. By contrast, the existing work is either constrained to a very small subset of graph problems, or works on shared-memory systems only. Third, the end-to-end learning approach realized in *OpenGraph-Gym* follows the line of research done by Dai et al. [11], but we extend it with new parallel GPU computing algorithms and a distinct software design and implementation using Tensorflow [1] and Horovod [19].

3   Methodology

In this section, we describe how to apply RL to solve graph optimization problems, which involves processing input graphs, reducing graph problems to RL problems, and executing RL training and testing.

3.1   Graph Processing for Reinforcement Learning

In conventional RL applications such as Atari games [17], input data are typically represented as matrices. For instance, pixel images may be taken as input to train deep neural network (DNN) models.

To handle graphs, an intuitive way is to feed a graph's corresponding adjacency matrix to DNN models. It is feasible. However, there are two major issues: 1) It requires a lot of memory space to train a DNN model due to graphs' large dimensions; 2) The successfully trained model only works for the graphs that have the same number of vertices as that of the training graph. To solve the issues, we use the technique of graph embedding, which is currently an active research area [5]. In brief, graph embedding can take a graph or vertex as input, then produce a p dimension vector that represents the useful information of the graph or vertex. Here, the dimension of p is predefined by users.

In our current implementation, we support two graph embedding of *structure2vec* [6] and *node2vec* [8]. Other graph embedding methods can be added to *OpenGraphGym* by extending certain classes. In Sect. 4.3, we explain how to add a new graph embedding method to *OpenGraphGym*.

3.2 Reinforcement Learning Formulation

In reinforcement learning, an agent and an environment interact with each other repeatedly in every *step*. For each *step*, the agent will take an *action*, then the environment will provide the agent with a *reward* and the old and new *states*. Eventually, the RL process will stop at a special "finished" state, which is called the *terminal state*. The above sequence of *steps* until *terminal state* is called an *episode*.

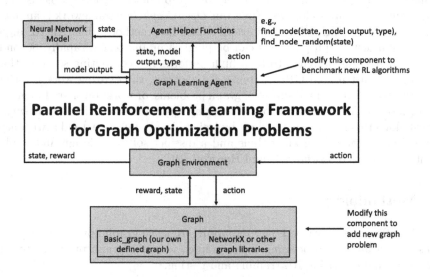

Fig. 1. The OpenGraphGym framework architecture.

Figure 1 shows the architecture of our *OpenGraphGym* framework. In the framework, the *Graph Learning Agent* takes an *action* by selecting and adding the "best" node to the graph problem's partial solution. Then, the *Graph Environment* returns the *reward*. The *reward* is used to justify the quality of a solution. It varies for different graph problems. More details of the framework will be introduced in Sect. 4.1.

For different types of graph problems, there are different formulations for the graph problem's RL algorithm. For instance, an RL algorithm for a distinct

Table 1. Examples of NP-hard graph problems that are defined in RL algorithms

Problem	State	Action	Reward	Termination
MVC	Subset of nodes selected as the partial solution	Add a new node to the partial solution	Number of nodes used to cover all edges at the end of the episode	All edges are covered
MAX	Subset of nodes selected as the partial solution	Add a new node to the partial solution	Cut set weight at the end of the episode	Cut set weight cannot be improved

graph problem may have a new representation of *state*, a problem-specific *action*, and a redefined *reward*.

As an example, Table 1 shows two graph problems' *states*, *actions*, *termination states*, and *rewards*. The graph problems of the Minimum Vertex Cover (MVC) and the Maximum Cut (MAX) are defined briefly as follows:

- **Minimum Vertex Cover (MVC):** Given an undirected graph, find the smallest subset of nodes to cover all the edges.
- **Maximum Cut (MAX):** Given an undirected graph, a subset of nodes S, assume the cut set is the set of edges that only has one end in S, find S with the largest weight of the cut set.

In Table 1, we can observe that the *state* and the *action* for MVC and MAX are same. However, the *reward* and the *termination* varies. As to MVC, the *reward* and the *termination* are related to the number of edges. As to MAX, the *reward* and *termination* are related to the cut set weight. Note that although we pick two NP-hard graph problems, our framework can be extended to solve more graph optimization problems.

3.3   Graph-RL Training and Testing Algorithms

In the previous Sect. 3.2, we have introduced how to formulate a graph RL algorithm. The next step is to train and test the model. In this section, we will summarize the algorithm of training and testing.

As shown in Algorithm 1, we will first initialize the experience replay memory buffer and the objective Q function (lines 2–4). Then, for each episode, we will select a random graph from the distribution D (line 6). One distribution of graphs include graphs generated using the same model and parameters. Next, we will initialize three sets of vertices (lines 8 and 9). Two of them (S_{new} and S_{old}) are for the solution. S_{new} is a set of vertices that includes all the nodes which have been selected as the solution in the current step. S_{old} is a set of vertices that includes all the nodes which have been selected as the solution in the previous step. Another one (C) is for the candidate nodes. A temporary replay buffer is also initialized (line 10). At each step, the agent will either randomly or according to a policy to select a node v_t from the candidate nodes set C (line 12). Then,

---

Algorithm 1. Q-learning greedy algorithm training

1: /* Q-learning algorithm training for the Minimum Vertex Cover problem */
2: Initialize experience replay memory buffer R
3: Initialize the function Q as the objective function
4: L: number of episodes used for training
5: **for episode** $e = 1$ **to** L **do**
6:     Sample a random graph in distribution D, G
7:     Vertices of G, V; Edges of G, E; $T = |V|$
8:     Initialize two sets of vertices S_{old} and S_{new} to empty
9:     Initialize a set of vertices $C = V$ as candidate vertices
10:    Initialize a temp experience replay memory buffer R_temp
11:    **for step** $s = 1$ **to** T **do**
12:        $v_t = \begin{cases} \text{Random node } v \in C \text{ w.p. } \epsilon \\ argmax_{v \in C} Q(embed(v, S_{old})) \end{cases}$
13:        Add v_t to S_{old} as S_{new}; Remove v_t from C
14:        Mark all edges linked to v_t as covered
15:        Add tuple(S_{new} ; v_t ; S_{old}) to R_temp
16:        Sample a batch of tuples $B\_samples$ from the R
17:        Update Q using $B\_samples$; $S_{old} = S_{new}$
18:        **if** All edges in E are covered **then**
19:            Assign rewards for all tuples in R_temp
20:            Add tuples in R_temp to R
21:            break
22:        **end if**
23:    **end for**
24: **end for**

---

we will update the solution sets S_{old} and S_{new} (lines 13–16). Meanwhile, v_t will also be removed from the candidate nodes set C. S_{old}, S_{new} and the selected node v_t will be combined and be added to the temporary replay buffer R_temp. Tuples in the temporary buffer will be pushed to the replay buffer R when we finish one episode (lines 19–20). At each step, we will also update the objective function Q by sampling a batch of tuples from the replay buffer R.

After we have trained an RL agent successfully, we can utilize the trained agent to find solutions to a set of new unseen graphs afterwards. Such an algorithm is called an RL Testing algorithm. The RL Testing algorithm is nearly the same as the training algorithm Algorithm 1 except for two differences: 1) Only the best candidate node will be selected every step (in line 12), and 2) the RL agent will not update the objective function (in line 17).

4   Design and Implementation of OpenGraphGym

This section presents 1) the main components of our framework, 2) how we design the framework to support parallel computing on multiple GPUs, and 3) how to extend the framework to support new graph optimization problems and graph embedding methods. Our code can be found at https://github.com/zwj3652011/OpenGraphGym.git.

4.1   Main Software Components

As shown in Fig. 1, *OpenGraphGym* has five main components, which are described as follows:

- **Graph Learning Agent:** It is the agent that is responsible for reinforcement learning for graph problems. It constantly receives input from three other components: the *Neural Network Model*, the *Agent Helper Functions*, and the *Graph Environment*. The *Neural Network Model* provides DNN model output (e.g., Q values), the *Agent Helper Functions* provides actions, and the *Graph Environment* provides states and rewards. On the other hand, the *Graph Learning Agent* also sends information to the three components constantly.
- **Neural Network Model:** It defines the graph embedding function and the RL agent's DNN model. During RL training, the *Neural Network Model* takes a graph state as input and produces a *Q value*. The *Q* value will be sent to the *Agent* for making decisions. In the current implementation of *OpenGraphGym*, we use the deep network Q-learning (DQN) method. Our next work will add the support of other RL methods such as A2C and A3C.
- **Agent Helper Functions:** It is a set of functions that are used by the RL agent to compute its appropriate *action*. By receiving the parameters of *states*, *models outputs* and *graph problem types*, the helper functions computes the *action* needs to be taken by checking the *model output* and the *graph problem type*. Please note that the *Agent Helper Functions* varies for different graph problems. In our framework, we include the *Agent Helper Functions* for some graph problems.
- **Graph Environment:** The *Graph Environment* is the interface between the *Graph* and the *Graph Learning Agent*. *Action*, *reward* and *state* will be transferred through it. For example, when the *action* is received from the *Graph Learning Agent*, the *Graph Environment* will call the function *step* to send the *action* to the component *Graph*. Then, it will receive the *state* and the *reward* from the component *Graph*. Finally, it will push the *state* and the *reward* to the *Graph Learning Agent*.
- **Graph:** It is a graph object implemented by our framework. Each graph object stores a set of graph-related information (e.g., number of nodes, number of vertices). Currently, our framework supports two types of graph objects. In the *Basic Graph*, we store the node lists, edge lists, number of nodes, and other basic information of a graph. Another one is defined by the networkX graph library [9]. NetworkX will read graph objects from edgelist files. The *Basic Graph* object is more flexible and can be extended by the user. If users cannot find a metric of graphs in networkX or other graph libraries, they can define and add their metrics to the *Basic Graph* object.

In general, the above components can be classified into two categories. The first category is designed to support for the agent part in RL, which includes the first three components. The *Graph Learning Agent* works as the interface between them. The second category is designed to support the environment part, for which the *Graph Environment* is the interface of them.

Furthermore, our framework is designed to be modular. By modifying a couple of components, it can be extended to support new graph embedding methods, RL algorithms, and graph optimization problems. For example, if a user desires to study another graph problem, the user needs to modify the *Graph* component to do it. In Sect. 4.3, we will show more details about it.

4.2   Parallel Implementation Using Multiple GPUs

The *OpenGraphGym* framework is able to support RL training on multiple GPUs. The following content describes how we distribute the workload and compute the graph RL in parallel among multiple GPUs.

- **Parallel Setup and Initialization:** Our framework will launch n processes given a number of n available GPUs. One CPU and one GPU will be mapped to one process. Inside each process, we create an instance of *Graph Learning Agent* and an instance of *Graph Environment*. Each *Graph Learning Agent* has its own copy of the global DNN model (i.e., a single model but duplicated multiple times on multiple GPUs), as well as a private RL replay buffer. At the beginning of the parallel execution, we use the distributed deep learning framework Horovod [19] to ensure each agent's DNN model will be initialized with the same weights.
- **Exploring Graphs in Parallel:** We use an asynchronous algorithm to let each process explore training on different graphs in parallel. At the start of each *episode*, every process will select a random graph from all the training graphs based on their unique random seeds. At the end of the *episode*, each process will then push its experience tuples to its own replay buffer. Note that all the training graphs are generated automatically by our framework.
- **Computing Gradients:** In the previous step, each process has started to explore graphs asynchronously. Then, at the end of every step, as shown in Algorithm 1, line 16, each process needs to sample some tuples from their replay memory buffer and compute the gradients. Assume the batch size is b and we have n processes, each process will sample b/n tuples and compute the averaged gradients of b/n tuples. Thus, each process will have one gradient. Finally, all processes' gradients will be averaged. We use the distributed deep learning framework Horovod [19] to finish the gradient computing. Horovod will accomplish two major tasks in this step: 1) add a barrier to wait for all processes to finish the gradient computing, and 2) average all processes' gradients and broadcasts it to them.
- **Updating Model:** In the previous step, all processes have received the averaged gradients. Then, each process needs to update their DNN model using the new gradient, as shown in Algorithm 1, line 17. Please note that all processes' DNN models are still the same after updating for the following two reasons: 1) DNN models are initialized to be the same, and 2) the gradients used to update the DNN models are identical for all processes.

Please note that the above operations are not executed in the same frequency. As shown in the Algorithm 1, the agent's DNN model will be updated in each *step*. Hence, the operations of **Computing Gradients** and **Updating Model** will be called every *step*. Moreover, in each *episode*, a new graph will be explored. Therefore, the operation of **Exploring Graphs** will be called every episode.

4.3 Framework Extensibility

In this section, we demonstrate the extensibility of our framework from two perspectives: 1) how to support other graph optimization problems, and 2) how to add new graph embedding methods.

Now, we use an example to show to extend *OpenGraphGym* to support other graph optimization problems. In the case of adding the Max Cut problem (MAX), we need to modify the *Graph* component as shown in Fig. 1. More specifically, two functions will be modified, which are: 1) the environment constructor (function _init_ in graph.py), and 2) the step function (function *update_state* in graph.py). In the environment constructor, we need to add a new local variable to represent the cut set weight inside the environment initialization function. In the step function, we then calculate the MAX-specific cut set weight to decide the termination state and the corresponding reward. Table 1 shows the definitions of the reward and termination state for the MAX graph problem.

To demonstrate how to add the new graph embedding method, we will use an example of adding the node2vec embedding method. Node2vec is a graph embedding method aiming at preserving each node's neighborhood information [8]. We use the open-source node2vec library Node2Vec in our framework [18]. Two functions needs to be modified in the *Graph* component are 1) the environment constructor (function _init_ in graph.py), and 2) the step function (function *update_state* in graph.py). As to the environment constructor, we need to set up the node2vec model using the APIs provided by the Node2Vec library. As to the step function, we need to reset and update the node2vec model when the graph is modified.

5 A Case Study on Minimum Vertex Cover (MVC)

To evaluate the performance and accuracy of the *OpenGraphGym* framework, we compare the solution found by our RL framework with that of other classical solvers on the MVC problem. In addition, we did experiments to show the improved convergence rate by utilizing multiple GPUs.

5.1 Experimental Setting

The software and hardware configurations for all our experiments are provided as follows.

Software: To implement *OpenGraphGym*, we use Horovod version 0.16.4 to compute DNN gradients in a distributed setting, as mentioned in Sect. 4.2.

Horovod can help with our distributed training by doing the following three tasks: 1) initialize each agent's DNN models to be the same, 2) add a barrier for multiple processes gradient computing, and 3) average the gradients from all processes and broadcast it to them. We also use Tensorflow version 1.12.0 [1], graph library networkX version 2.4 [9], and the HPC environment toolkit Docker version 18.09.2 [15]. Tensorflow is responsible for the agent's neural network training and inference. As for networkX, we use it to generate, manipulate, and evaluate different graphs. As to Docker, we create an environment using Docker and install all required libraries of our framework. Then, we can deploy the environment conveniently to a new HPC system. Docker helps us to manage the software and libraries used in our framework.

Graph Datasets Used: Our graph datasets contain two types of graphs, which are generated by two different distributions or random graph generation models. In the following content, we will present how graphs are generated using them. We will also present the parameter values we set for each model.

- *Erdős-Rényi (ER) Graphs:* The Erdős-Rényi model will generate a random graph with the graph size m and the edge possibility r [7]. We use the function *erdos_renyi_graph* in networkX to generate different sizes of ER graphs. The edge probability is set to 0.15, which means every possible edge has the possibility of 0.15 to exist.
- *Barabási-Albert (BA) Graphs:* The Barabási-Albert model generate the random graph based on the graph size m and the edge density d [2]. Edge density d is equal to the number of edges from a new node to the existing nodes. We use the function *barabasi_albert_graph* in networkX with the edge density of 4 to generate BA graphs.

Computer System: We use an Nvidia DGX workstation to do all experiments, which consists of 40 CPU cores and 4 Volta V100 GPUs. More details of the system is provided in Table 2.

Others: In addition, we use the learning rate 1.0×10^{-5} and batch size 128 to train our DNN model. The RL parameter exploration rate ϵ is set to 0.1. The size of the RL replay buffer is set to store up to 50,000 experience tuples.

5.2 Quality of Graph Problem Solutions

To evaluate the quality of our graph solutions, we compare the results generated by five different MVC solvers:

- **Graph-RL:** This is our RL framework of *OpenGraphGym*.
- **Random:** The Random solver finds a solution by randomly taking a node from the graph in each step.
- **2-OPT:** The 2-OPT approximation algorithm takes both endpoints of an edge from the graph in each step. Its solution is guaranteed to be less than twice of the optimal solution [3].

Table 2. An Nvidia DGX System.

CPU	Intel Xeon CPU E5-2698 v4 2.20 GHz
GPU	Nvidia Volta V100
Number of CPU Cores	40
Number of GPUs	4
Memory per GPU	16 GB
Operating System	Ubuntu
OS Version	16.04.6

- **Greedy:** The Greedy algorithm builds the solution by simply adding the node with the largest degree in each step.
- **Exhaustive Search:** For the verification purpose, we also implement a "brute-force" random search MVC solver. We let this program continuously search for large numbers of solutions until no better solution could be found in one hour.

In our validation experiment, we use 40 graphs to train the RL agent, and use 10 graphs to test the agent. The 10 test graphs has never been seen by the agent during training. Note that the training graphs and test graphs belong to the same type of graphs: either ER or BA type.

Table 3 shows four different datasets used in our experiments (one dataset per row), each with different average number of nodes and edges in the ER or BA type. In the table, the graph type, average number of vertices and edges are shown in the first three columns. The best two solutions are also highlighted in a bold font.

As to the ER graph dataset with 20 nodes, the exhaustive search algorithm obtains the best solution with 9.4. Our Graph-RL algorithm is the second-best with 10.1. As to the BA graph dataset with 20 nodes, our solver is the best one with MVC size 11.4. The second-best is the exhaustive search algorithm with the best MVC size 12. For the rest of the solvers, the 2-OPT and the Greedy are similar, whose solutions are around 15 and 16. Finally, the random method produces the worst solutions for both ER and BA graphs.

We also use ER and BA graph datasets that have 50 nodes for training and testing. As shown in the two rows at the bottom of Table 3, our Graph-RL solver always found the best MVC solutions.

Based on the above comparison, we can say that our Graph-RL solver and the exhaustive search solver are constantly better than the other three solvers. In addition, the quality of the Graph-RL solutions is comparable to that of the long-time exhaustive searching algorithm.

Table 3. Experiments with MVC on both ER and BA graphs. All MVC solutions shown here are averaged over 10 testing graphs. Five solvers (Graph-RL, Random, 2-OPT, Greedy, Exhaustive Search) are compared. The best two solutions are highlighted in bold for each dataset. Graph-RL is the solution obtained by our OpenGraphGym framework.

Graphs	Avg#nodes	Avg#edges	Graph-RL	Random	2-OPT	Greedy	Exhaustive search
ER	20	30.1	**10.1**	17.8	15.2	16.3	**9.4**
BA	20	64	**11.4**	18	16.2	16	**12**
ER	50	190.4	**33.2**	48	45	47.4	**36.4**
BA	50	184	**28.8**	48	41.8	46	**34.3**

5.3 Deploying an Agent Trained by Small Graphs to Test Bigger Graphs

The DNN model implemented by our framework can support graphs with various sizes. This feature enables us to train and test graphs with distinct sizes. To test the generalization performance of our model, we train the model on smaller size of graphs first. Then, we test the learned model using larger size of graphs.

The new experimental results are shown in Fig. 2. For this experiment, we use a dataset with 40 ER graphs that have an average number of 20 nodes for training. However, we use a dataset of 10 ER graphs that have a number of 50 nodes for testing. For every 20 episodes of training, we use the 50-node testing dataset to test the model's solution quality. From Fig. 2, we can observe that after around 75 episodes, the average number of nodes to cover the test graph dataset reaches 34.8, which is close to the exhaustive search algorithm's solution. This result demonstrates that an RL agent trained from small graphs can be generalized to solve larger size graphs.

5.4 Effect of Using Multiple GPUs

Finally, we use multiple GPUs to accelerate the RL training process with our *OpenGraphGym* framework. In the experiment, the training dataset has 40 ER graphs and the test dataset has 10 ER graphs. All the training and testing graphs have an average number of 20 nodes. Also, we evaluate the trained RL model's solution with the test graph dataset in every episode.

As shown in Fig. 3, the blue line represents the MVC solutions computed by a single GPU. The orange line represents the solutions computed by four GPUs. From the figure, we can observe that when we use one GPU, our framework can find the best solutions after 80 episodes. By contrast, the framework takes only 62 episodes to find the best solutions when using four GPUs. This experiment shows that our RL framework is able to find the best solution of a problem by taking fewer episodes (i.e., converging faster) when more GPUs are used.

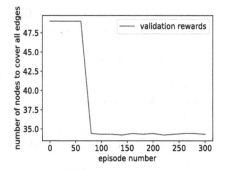
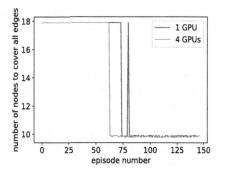
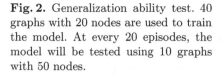

Fig. 2. Generalization ability test. 40 graphs with 20 nodes are used to train the model. At every 20 episodes, the model will be tested using 10 graphs with 50 nodes.

Fig. 3. In this set of experiments, we use multiple GPUs for training on ER graphs with 20 nodes. The orange line is for the testing results with four GPUs. The blue line is for the results with one GPU. (Color figure online)

6   Conclusion

In this work, we design and implement a parallel reinforcement learning framework OpenGraphGym for graph optimization problems. Then, we use the MVC as the test case to demonstrate that the solution provided by our framework is better than some classical MVC solvers. This work focuses on three research directions: 1) We aim to use the open framework to benchmark various new RL algorithms and embedding methods. 2) Many real-world graphs are extreme-scales. We will add the support of extreme-scale graphs to our framework. 3) Currently, we only support a few basic parallel strategies. To better utilize the high performance computing resources, we will extend the *OpenGraphGym* framework to design more advanced and efficient parallel strategies.

References

1. Abadi, M., et al.: TensorFlow: a system for large-scale machine learning. In: 12th USENIX Symposium on Operating Systems Design and Implementation (OSDI 16), pp. 265–283 (2016)
2. Albert, R., Barabási, A.L.: Statistical mechanics of complex networks. Rev. Mod. Phys. **74**(1), 47 (2002)
3. Bar-Yehuda, R., Even, S.: A local-ratio theorem for approximating the weighted vertex cover problem. Technical report, Computer Science Department, Technion (1983)
4. Bello, I., Pham, H., Le, Q.V., Norouzi, M., Bengio, S.: Neural combinatorial optimization with reinforcement learning. arXiv preprint arXiv:1611.09940 (2016)
5. Cai, H., Zheng, V.W., Chang, K.C.C.: A comprehensive survey of graph embedding: problems, techniques, and applications. IEEE Trans. Knowl. Data Eng. **30**(9), 1616–1637 (2018)

6. Dai, H., Dai, B., Song, L.: Discriminative embeddings of latent variable models for structured data. In: International Conference on Machine Learning, pp. 2702–2711 (2016)
7. Erdős, P., Rényi, A.: On the evolution of random graphs. Publ. Math. Inst. Hung. Acad. Sci. **5**(1), 17–60 (1960)
8. Grover, A., Leskovec, J.: node2vec: scalable feature learning for networks. In: Proceedings of the 22nd ACM SIGKDD International Conference on Knowledge Discovery and Data Mining, pp. 855–864. ACM (2016)
9. Hagberg, A., Swart, P., Chult, D.S.: Exploring network structure, dynamics, and function using NetworkX. Technical report, Los Alamos National Lab. (LANL), Los Alamos, NM (United States) (2008)
10. Hoos, H.H., Stützle, T.: SATLIB: an online resource for research on SAT. In: SAT 2000, pp. 283–292 (2000)
11. Khalil, E., Dai, H., Zhang, Y., Dilkina, B., Song, L.: Learning combinatorial optimization algorithms over graphs. In: Advances in Neural Information Processing Systems, pp. 6348–6358 (2017)
12. Kipf, T.N., Welling, M.: Semi-supervised classification with graph convolutional networks. arXiv preprint arXiv:1609.02907 (2016)
13. Kool, W., van Hoof, H., Welling, M.: Attention solves your TSP, approximately. Statistics **1050**, 22 (2018)
14. Li, Z., Chen, Q., Koltun, V.: Combinatorial optimization with graph convolutional networks and guided tree search. In: Advances in Neural Information Processing Systems, pp. 539–548 (2018)
15. Merkel, D.: Docker: lightweight linux containers for consistent development and deployment. Linux J. **2014**(239), 2 (2014)
16. Mittal, A., Dhawan, A., Medya, S., Ranu, S., Singh, A.: Learning heuristics over large graphs via deep reinforcement learning. arXiv preprint arXiv:1903.03332 (2019)
17. Mnih, V., et al.: Human-level control through deep reinforcement learning. Nature **518**(7540), 529 (2015)
18. Node2Vec (2019). https://github.com/eliorc/node2vec
19. Sergeev, A., Del Balso, M.: Horovod: fast and easy distributed deep learning in TensorFlow. arXiv preprint arXiv:1802.05799 (2018)
20. Silver, D., et al.: Mastering the game of go with deep neural networks and tree search. Nature **529**(7587), 484 (2016)
21. Silver, D., et al.: Mastering the game of go without human knowledge. Nature **550**(7676), 354 (2017)
22. Vinyals, O., Fortunato, M., Jaitly, N.: Pointer networks. In: Advances in Neural Information Processing Systems, pp. 2692–2700 (2015)

Weighted Clustering for Bees Detection
on Video Images

Jerzy Dembski and Julian Szymański[(⊠)]

Faculty of Electronic Telecommunications and Informatics,
Gdańsk University of Technology, Gdańsk, Poland
{jerzy.dembski,julian.szymanski}@eti.pg.edu.pl

Abstract. This work describes a bee detection system to monitor bee
colony conditions. The detection process on video images has been
divided into 3 stages: determining the regions of interest (ROI) for a
given frame, scanning the frame in ROI areas using the DNN-CNN classifier, in order to obtain a confidence of bee occurrence in each window
in any position and any scale, and form one detection window from a
cloud of windows provided by a positive classification. The process has
been performed by a method of weighted cluster analysis, which is the
main contribution of this work. The paper also describes a process of
building the detector, during which the main challenge was the selection
of clustering parameters that gives the smallest generalization error.

The results of the experiments show the advantage of the cluster analysis method over the greedy method and the advantage of the optimization of cluster analysis parameters over standard-heuristic parameter values, provided that a sufficiently long learning fragment of the movie is
used to optimize the parameters.

Keywords: Automatic bee's image detection · Convolutional deep
neural networks · Weighted clustering · Bee monitoring

1 Introduction

In this paper, we present the approach used for building a bee detection system that is part of a larger project on apiary monitoring with the usage of IT
Technologies [1,2]. The main goal of the research presented here is to build a
system that allows us for non-invasive, real-time monitoring of the bee family
using video analysis. Usage of cameras and algorithms allows us to quantify the
amount of the bees coming out and coming into the hive that is an important
factor for a beekeeper indicating how the bee colony develops during the season.

Bees tracking is a challenging task, which is related to the specificity of this
problem. There are at least four causes for this: bees move fast, they are small,
on a single video frame can be a large number of items and they are very similar
one each other.

The bees tracing and hive entrance monitoring can be done in different ways.
For example paper [3] shows how to use for that purpose RFID Tags. In our

© Springer Nature Switzerland AG 2020
V. V. Krzhizhanovskaya et al. (Eds.): ICCS 2020, LNCS 12141, pp. 453–466, 2020.
https://doi.org/10.1007/978-3-030-50426-7_34

research, we assume a usage non-invasive methods and we decided to focus on image analysis. In that domain, some works have been already done.

One of the first systems aiming at monitoring bees traffic from images has been presented in [4]. The study has been based on SVM classifier and it allows to identify the individual honeybee. The analysis of bees flight activity at the beehive entrance using tracking of flight paths has been shown in [5]. Honeybee hive health monitoring by image processing has been presented and analyzed [6] by the usage of two different approaches based on the illumination-invariant change detection algorithm and signal-to-noise ratio to estimate the number of bees at the entrance of the hives. Tracking of honeybees has been also done using an integrated Kalman Filter and Hungarian algorithm [7]. The problem of bees detection has been also tackled in 3D space. Paper [8] presents a stereo vision approach that allows detecting bees at the beehive entrance and is sufficiently reliable for tracking.

The solutions mentioned above have been constructed and most of all tested on data prepared for the particular model. The images have been taken for a fixed scene and do not show the applicability of the solutions while cameras, unlike during the training, are put in different angels to the hive as well as while light conditions are changing. Also, the images used in the above-mentioned research have been made in low resolutions that is obvious simplification due to the efficiency of processing online date in real-life applications.

Our paper is constructed as follows: In the next section we describe the data acquisition and processing used in our research. Then we describe the architecture and algorithms used in our system. In Sect. 4 the experiments and results are described. The paper finalizes with conclusions.

2 Data Preparation

For the requirements of the project to tag the data, we have implemented own software, for fast and easy selection of windows containing bees on video images. It allows preparing examples for a window classifier training. The contours of the bees touch at least two edges of the window, which results from the fact that for simplicity the windows are square – it is easier to mark bees with square rather than rectangular windows. The selection of one individual on subsequent frames is facilitated by indicating its center and the length of the side of the square because the size of a bee changes at a lower pace than its location, so a human can mark bees with single clicks on many successive frames. Also, areas, where there are no bees, are marked on each video to generate negative examples for the initial stage of classifier learning.

Additionally, special areas of exclusion were marked. These are, for example, around the outlet of the hive or places where dead individuals are found. It was done to avoid excessive effort and controversial marking decisions. Exclusion areas are ignored when negative examples are generated for training the classifier (Sect. 3.2) and when detection is evaluated (see Sects. 3.3 and 4).

According to the idea of adapting the system to the environment described in Sect. 1, the system was built and tested for three different camera positions and settings. For each of these, a Full-HD (1920 × 1080) pixel movie was recorded at a frequency of 50 frames/sec. Each of the movies was manually marked with windows containing bee images, including all bee images except for exclusion areas. Large windows containing no bees were also marked, from which negative examples were generated for training purposes and selected small windows containing objects resembling bees (e.g. bee shadows, knots, inflorescences of some plants, e.g. of the genus Burnet (Sanguisorba L.), potentially be false positive examples.

Each of the original movies was divided into two almost equal parts. The first part was used for training the system, the other was intended only for the final tests of the tuned system and was not used at any stage of learning.

Statistics of the data for particular movies used for training and for test are summarized in Table 1.

The data and software used in this research we have made publicly available and they are accessible in the web url: https://goo.gl/KNV7sd.

Table 1. Statistics of sample movies

Movie	DSC_0559_A	DSC_0559_B	DSC_0562_A	DSC_0562_B	01091_A	01091_B
num.frames	1149	630	2097	2631	1270	1634
num.pos.windows	2726	2728	6755	6750	6004	6009
num.individuals	94	92	121	224	81	79

3   System Architecture

The components of our system, that aims at precisely extract bees from the background on particular movie frames has been shown in Fig. 1. It consists of three parts: procedure for determining ROI where a bee can potentially be found, the procedure for window classification by scanning particular frames within the ROI, and procedure for extracting windows that match bee images.

Our preliminary research on bee detection algorithms have been presented in [2] where the study of different color models for image representation have been given. We extend the system introducing three significant modifications that contribute to this paper:

– new models of artificial neural networks were trained separately for three selected environments (camera settings for different hives) by two-stage method of false positive error reduction described in Sect. 3.2,

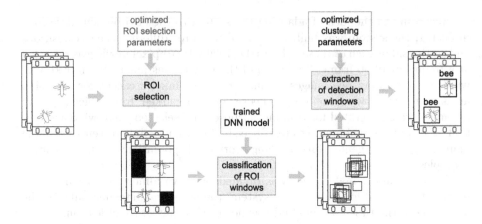

Fig. 1. Diagram of the bee detection system

- changed the greedy method of window selection to a method based on a weighted clustering algorithm with parameter tuning using the simulated annealing method described in details in Sect. 3.3,
- the function of evaluating detection results has been developed (described in Sect. 3.3).

3.1   Procedure for Determining Regions of Interest

Due to two assumptions of the system: the camera is stationary and only bees in motion are considered, it is possible to extract regions of interest by relatively low computational cost by accepting only those windows in which motion was detected for further processing. The easiest way to detect motion is to compare the contents of the window in two adjacent movie frames. If the difference exceeds a certain threshold θ, the window is qualified as potentially containing a bee. Then it is only necessary to determine how to calculate the difference based on image features. In the simplest case, the set of image features may consist of the intensities of particular channels in the RGB model or the intensities of pixels in the B&W model. They can also consist of histograms of intensity levels or histograms of gradient directions [9]. After determining the type of feature vector, the next dilemma may be due to the function of the distance between vectors. In an extension of the simple Euclidean distance, the cosine distance can be used as less sensitive to general changes in brightness, e.g. due to obscuring the sun by clouds or Pearson distance. Additionally, the window can be divided into smaller blocks for thresholding, which reduces the system's sensitivity to local noise and lighting changes, and increases the sensitivity to displacement of objects which only occupy a part of the window. For example, a bee with open wings covers about 30% of the window surface.

After the preliminary experiments, the simplest feature vector variant was adopted in the form of a pixel intensities vector divided into R, G and B channels, Euclidean distance between vectors and a division into blocks of 16×16 pixels appeared to be sufficient. For each of these blocks, the distance is zeroed when the actual distance value does not exceed the threshold of η.

For each movie, optimization of the parameters θ and η was performed using a training fragment of the movie e.g. with A ending. We used the "brute-force" searching approach consisting of calculation the criterion for each point of the grid of 100×100 pairs of parameter values, which gives about 10000 cases. Both parameters were in the $(0, 1)$ range of values. As a criterion for optimization we used the weighted error sum $E_w = w_{fn} * E_{fn} + w_{fp}E_{fp}$, where w_{fn}, w_{fp} - partial error weights, E_{fn} - false negative error (the ratio of the sum of manually marked windows representing bees recognized as the background to the sum of all manually marked windows representing bees), E_{fp} - false positive error (the ratio of the sum of windows containing the background considered ROI to the sum of all windows containing the background). These windows are generated from larger areas that certainly do not contain bees. The weight values were set at $w_{fn} = 0.95, w_{fp} = 0.05$, which means that it is more important to leave bees in the ROI than skipping the background windows, which only results in increased calculations. Additionally, the maximum error of false negative $E_{fn}^{max} = 0.005$ was assumed. It is not zero due to the occurrence of isolated cases when the bee is almost motionless in two successive frames.

3.2 Windows Classification Procedure by Sliding Window Method

After determining the ROIs, these regions are scanned with the usage of the square sliding window at different scales, shifted in horizontal and vertical directions. Each window was classified whether it contains a bee or not. By default, scanning begins with windows that are 64 pixels wide and tall, and then the window width and height is multiplied by 1.2 factor up to 440 pixels. The window shift step is 20% horizontal and vertical of frame width. In total, for a Full-HD format frame, this gives 51644 of windows for evaluation. Each window is transformed to the standard resolution of 48×48 pixels and is fed as an input image to the deep artificial neural network with convolutional layers (DNN-CNN). The last softmax layer of DNN-CNN returns the probability that there is a bee in the window.

The two-step DNN-CNN model training was proposed to match the model to the specific camera and environment settings. In the first stage, positive examples are used in the form of bee images extracted from manually selected windows in the training movie. Negative examples are extracted randomly from manually selected areas in various places in a training movie where there were definitely no bees. After training, the first version of the model it was used as a window classifier on all frames of the training movie with non-ROI skipped windows and outside the exclusion areas. As a result of this scanning process, some background windows are classified as containing bee e.g. with false positive decision. The images from these windows are used as negative examples in the second step

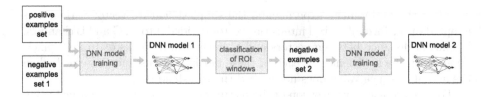

Fig. 2. Diagram of the two-step learning process of the DNN-CNN model as a window classifier

of DNN-CNN model training process. The final result of the two-step learning process was the DNN model with a significantly reduced number of false positive classifications for specific camera settings and the environment. The scheme of the learning system is shown in the Fig. 2. The 6-layer artificial neural network with three convolutional layers and three fully connected layers was used for building window classifier. As an input representation, the RGB three-channel model was chosen from 4 different color models, which was evaluated in the experiments described in [2]. The network was trained using a dropout technique with a keep probability 0.5 in all fully connected layers apart from the last one. We used cross entropy as a loss function and ADAM optimization. All learning parameters were selected experimentally as part of the work [2]. The output layer with softmax activation function returns the probabilities whether an input contains bee and background. To generate negative examples, as well as to calculate the classification error, it was assumed that the classification of the window in terms of bee content occurs when the probability of a bee occurrence is greater than or equal to 0.5. The network diagram is shown in Fig. 3. The training process has been done using the TensorFlow library. The classification error of the final model version for particular test movies is given in Table 2.

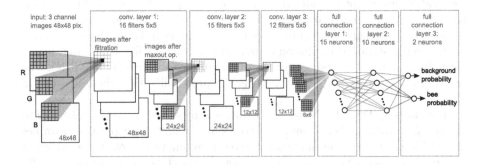

Fig. 3. Diagram of the DNN-CNN model - artificial neural network with convolutional layers determining the probability value that a bee is in the window

Table 2. DNN-CNN generalization error calculated for test movies

DSC_0559_B			DSC_0562_B			01091_B		
error	fp	fn	error	fp	fn	error	fp	fn
0.0044	0.0070	0.0018	0.0115	0.0018	0.0212	0.0022	0.0000	0.0043

3.3 Window Extraction Form Classification Windows Using Weighted Clustering Algorithm

In many detection systems, both classic [10] and compact [11–13] the non-maximum suppression (NMS) method is used for window extraction. NMS is based on greedy removal of windows with a lower probability of object occurrence and more similar to other windows in the sense of covering. [14] describes the improved version of Soft-NMS, but still concerns a simple elimination of windows. There are more advanced methods for window extraction using different grouping methods – clustering. In our work we propose an original grouping method with the possibility of adapting its parameters to specific environmental conditions and camera settings.

We consider the window classification as positive if the probability returned by DNN-CNN network that it contains a bee image is greater or equal 0.5. However, during scanning the frame by sliding window the classifier positively classifies not only the window perfectly coinciding with the bee image, but also the windows slightly shifted and scaled relative to the true bee image. The reason for this is not the classifier inaccuracy, but rather the inaccuracy associated with manual selection of windows containing bees. For this reason, after the scanning process, we are obtaining a cloud of positive detection windows, instead of one window, perfectly matching the image of a bee. It became necessary to use an algorithm for determining windows that coincide with bee images.

The idea of the algorithm is to bring each dense window cluster to a single window with average parameters that should match the image of a bee. The algorithm should break the clusters into two windows, thus allowing detection of the case while two bees are very close together. It should also ignore clusters with a small number of windows that may arise due to false positive classification. Our algorithm was designed as an extension of the K-means clustering, adopting the following assumptions:

- Each square window in the movie frame is a point in the 3-dimensional space $p \in \mathcal{P}$ with the parameters x, y, d, where x, y - the coordinates of the center of the window, d - the length of its side,
- each cluster center $c \in \mathcal{C}$ is also a point in 3-dimensional space,
- the p_i window is assigned to the cluster with center c_j, when $n(p_i) = j$,
- the IOU function decides about the window assignment to the cluster, given by the formula $IOU(p, q) = \text{area}(p \cap q)/\text{area}(p \cup q)$ for two windows p and q, item the probability that i-th window contains a bee - $Pr_{bee}(p_i)$ is used to calculate the window weight from the formula $w_i = Pr_{bee}(p_i)^\beta$, where β - one of the clustering parameters, item the window weight w_i is used for calculation of the cluster center parameters, as well as decides if the cluster

is sufficiently represented by the weighted sum of the windows that belong to it.

The algorithm is described by a pseudo-code:

Require:
$\mathcal{P}$ – window set with positive classification
$\mathcal{C}$ – initial set of cluster centers
$\alpha, \beta, \gamma, \delta$ – clustering parameters
while the specified number of cycles has not been reached **do**
    for all $p_i \in \mathcal{P}$ **do** $\triangleright$ 1. Assignment of windows to clusters
      if $\max\limits_{c_j \in \mathcal{C}} \text{IOU}(p_i, c_j) > \alpha$ **then**
        $n(p_i) \leftarrow \underset{j, c_j \in \mathcal{C}}{\text{argmax}}\, \text{IOU}(p_i, c_j)$
      else
        $n(p_i) \leftarrow$ null
      end if $\triangleright$ $n(p_i)$ – number of cluster which contain window p_i
    end for
    for all $c_j \in \mathcal{C}$ **do** $\triangleright$ 2. Modification of location of cluster centers

$$x(c_j) \leftarrow \sum_{i, n(p_i)=j} x(p_i)w_i \Big/ \sum_{i, n(p_i)=j} w_i$$

$$y(c_j) \leftarrow \sum_{i, n(p_i)=j} y(p_i)w_i \Big/ \sum_{i, n(p_i)=j} w_i$$

$$d(c_j) \leftarrow \sum_{i, n(p_i)=j} d(p_i)w_i \Big/ \sum_{i, n(p_i)=j} w_i$$

$$\sigma_x(c_j) \leftarrow \sum_{i, n(p_i)=j} x(p_i)^2 w_i \Big/ \sum_{i, n(p_i)=j} w_i - x(c_j)^2$$

$$\sigma_y(c_j) \leftarrow \sum_{i, n(p_i)=j} y(p_i)^2 w_i \Big/ \sum_{i, n(p_i)=j} w_i - y(c_j)^2$$

    end for
    for all $c_j \in \mathcal{C}$ **do** $\triangleright$ 3. Removing clusters from a set $\mathcal{C}$
      if $\sum\limits_{i, n(p_i)=j} w_i < \gamma$ **then**
        $\mathcal{C} \leftarrow \mathcal{C} \setminus \{c_j\}$
      end if
    end for
    for all $c_j \in \mathcal{C}$ **do** $\triangleright$ 4. Adding new cluster with high std.dev.
      $\sigma_{max} \leftarrow \max(\sigma_x(c_j), \sigma_y(c_j))$
      if $\sigma_{max}/d(c_j)^2 > \delta$ **then**
        $coord \leftarrow \underset{x,y}{\text{argmax}}(\sigma_x(c_j), \sigma_y(c_j))$ $\triangleright$ selection of coordinate
        $c_{new} \leftarrow c_j$
        $\mathcal{C} \leftarrow \mathcal{C} \cup \{c_{new}\}$ $\triangleright$ adding a new cluster
        $coord(c_j) \leftarrow coord(c_j) - \sqrt{\sigma_{max}}/3$ $\triangleright$ shifting of new clusters centers
        $coord(c_{new}) \leftarrow coord(c_j) + \sqrt{\sigma_{max}}/3$
      end if
    end for
end while
return $\mathcal{C}$

To the input of the weighted clustering algorithm is given a set of positively classified $\mathcal{P}$ windows by DNN-CNN network, initial cluster centers, and clustering parameters. The initial locations of the cluster centers are determined using the greedy algorithm described in [2] employing the gradual averaging of parameter pairs of overlapping windows. The main algorithm loop consists of four nested "for" loops. The first and second of them are analogous to the *K-means* and allow the assignment of windows to clusters alternately with correction of a location of cluster centers based on their windows. However, due to the specifics of the application, there are two main differences. The first is the additional condition of window assignment to the cluster, which is based on a test if the maximum match in the sense of IOU is less than the α threshold. If so, then the window is significantly outside the bee images represented by the cluster centers and should therefore not have any assignment as probably false positive classification. The second difference is the usage of weight for each i-th window. The next two nested "for" loops allow us to remove and break the clusters. The first option allows for removing false positive detections, which are usually represented by a small number of windows. Breaking the cluster into two allows extracting images of two bees that are very close together or even partially overlapping. In the algorithm, this is solved by adding a copy of the primary cluster and then moving these two clusters towards the coordinate with the maximum standard deviation. This procedure was carried out when the maximum coordinate standard deviation of σ_{max} normalized with the window area exceeds the given threshold value δ.

3.4   Clustering Parameter Optimization

The optimization of $\alpha, \beta, \gamma, \delta$ parameters, similarly as in the case of optimization of ROI determination parameters and DNN-CNN model learning, takes place each time for a given camera setting on the training movie. We used simulated annealing as optimization algorithm due to its simplicity, although of course other optimization methods such as genetic algorithms or PSO could be used. The applied algorithm of simulated annealing, after adopting the initial parameter values, works periodically, randomly shifting the parameter vector in each cycle, and then after calculating the evaluation, can reject or accept new solution with probability $Pr_{accept} = (1 + \exp(\Delta E/bT))^{-1}$. This probability value depends on the change in the evaluation value of the solution $\Delta E = E(t) - E(t-1)$ in step t and the randomness factor T – the temperature that initially allows greater exploration of solutions (acceptance of worse solutions). The b constant differentiate the effect of temperature on the acceptance probability and the magnitude of the parameter vector shift. Determining the assessment of the solution in a given step of the annealing algorithm t: $E(t)$ is based on running a weighted clustering algorithm with new parameters for a given number of cycles, and then calculating the coverage error of the obtained detection windows in relation to the windows determined manually in the training movie. Unfortunately, the detection error measures known from the literature such as Mean Average Precision (mAP) do not change for small window shifts, which is necessary to direct the search in optimization. The second reason why we do not use mAP is the difficulty in determining the probability that the window contains a bee because these windows are not determined directly by the DNN network, but are obtained as a result of extraction - averaging of many windows. Due to the requirements of the learning system, we used its own measure of coverage error represented by the Eq. 1 together with the algorithm to calculate it.

$$ E = \frac{\sum_{k=1}^{K} |\mathcal{O}_k| + |\mathcal{C}_k| - m_{\mathcal{O}_k} - m_{\mathcal{C}_k}}{\sum_{k=1}^{K} |\mathcal{O}_k| + |\mathcal{C}_k|} \tag{1} $$

where K – the number of movie frames, $|\mathcal{O}_k|, |\mathcal{C}_k|$ – the number of window sets determined manually and by means of the weighted clustering algorithm on the frame k, $m_{\mathcal{O}_k}$ - the sum of the coverage degrees of windows determined manually with windows determined algorithmically, $m_{\mathcal{C}_k}$ - the sum of the coverage degrees of the windows determined algorithmically with windows determined manually on the frame k. If the manually selected windows perfectly coincide with the windows determined by the clustering algorithm than $E = 0$. The error will be non-zero when the bees are not detected and in false positive cases when the system classifies the background image as containing the bee.

The sums of coverage degrees on the frame are calculated by the algorithm:

Require:
$\quad \mathcal{O}$ – a set of windows containing manually selected bee images
$\quad \mathcal{C}$ – set of windows obtained by the weighted clustering algorithm
$\quad \mathcal{R} : \mathcal{O} \times \mathcal{C} \equiv \{(o, c) | o \in \mathcal{O}, c \in \mathcal{C}\}$ – set of window pairs
$\quad m_{\mathcal{O}} \leftarrow 0$
$\quad m_{\mathcal{C}} \leftarrow 0$
$\quad$**while** $|\mathcal{R}| > 0$ **do**
$\qquad (a, b) \leftarrow \underset{i,j}{\operatorname{argmax}} \operatorname{IOU}(o_i, c_j)$ $\qquad$ ▷ pair with the highest matching degree
$\qquad m_{\mathcal{O}} \leftarrow m_{\mathcal{O}} + \operatorname{IOW}(o_a, c_b)$
$\qquad m_{\mathcal{C}} \leftarrow m_{\mathcal{C}} + \operatorname{IOW}(c_b, o_a)$
$\qquad$**for all** $(o_i \in \mathcal{O}, c_j \in \mathcal{C})$ **do** $\qquad$ ▷ for all window pairs (o_i, c_j)
$\qquad\quad$**if** $i = a \lor j = b$ **then**
$\qquad\qquad \mathcal{R} \leftarrow \mathcal{R} \setminus \{(o_i, c_j)\}$ $\qquad$ ▷ removing a pair $\{(o_i, c_j)\}$
$\qquad\quad$**end if**
$\qquad$**end for**
$\quad$**end while**
$\quad$**return** $m_{\mathcal{O}}, m_{\mathcal{C}}$

The IOW function determines the coverage degree of the first window by the second and can be represented by the formula: $\operatorname{IOW}(p, q) = \operatorname{area}(p \cap q) / \operatorname{area}(p)$. Standard parameter values and values optimized for three training movies are shown in the Table 3.

Table 3. Cluster analysis parameter values obtained as a result of optimization for particular training movies

Symbol	Interpretation	Standard par.	DSC_0559_A	DSC_0562_A	01091_A 01091_A
α	IOU threshold	0	0	0.044748	0.02259
β	Power at probability that bee	2	-1.1917	2.1359	-2.6423
γ	Sum of window weights threshold	5	4.8284	4.7435	4.7063
δ	Normalized std. dev. threshold	0.5	2.1140	39.1152	18.5411

4 The Experiments and Results

The experiments were carried out for the same ROI regions and DNN-CNN models trained separately for each movie at earlier stages. The results of the experiments presented in the Table 4 allow us to compare movies registered in different conditions and three window selection methods: cluster analysis with optimized parameter values, cluster analysis with heuristically accepted – standard parameter values and the greedy method. Each experiment consisted of usage of a fragment of the movie to learn, for example DSC_0559_A to determine ROI areas, DNN classifier training and in the case of parameter optimization – searching for optimal clustering parameters with the criterion given by the Eq. 1. Then for the test movie, for example DSC_0559_B, the error was calculated according to the Eq. 2:

$$E_{0.5} = \frac{\sum_{k=1}^{K} |\mathcal{O}_k| + |\mathcal{C}_k| - 2N_k}{\sum_{k=1}^{K} |\mathcal{O}_k| + |\mathcal{C}_k|}, \tag{2}$$

where N_k - the sum of pairs of windows determined manually and algorithmically (o_i, c_j) in the frame k, such that $\text{IOU}(o_i, c_j) \geq 0.5$, assuming that each window determined manually could only coincide with one window determined algorithmically and vice versa. The 0.5 threshold value is the value most commonly used in other works related to image detection. For the method with the optimization of cluster analysis parameters, the table also provides the detection error of false positives $E_{0.5}^{fp} = \sum_{k=1}^{K}(|\mathcal{C}_k| - N_k)/\sum_{k=1}^{K}|\mathcal{C}_k|$, which can be interpreted as the proportion of algorithmically determined windows that do not cover any manually designated windows. False negative error $E_{0.5}^{fn} = \sum_{k=1}^{K}(|\mathcal{O}_k| - N_k)/\sum_{k=1}^{K}|\mathcal{O}_k|$ is the proportion of manually marked windows that were not covered after window extraction process. As it can be seen, the error when using cluster analysis is always smaller than while using the greedy method. The error in two out of three setting cases was smaller for standard cluster analysis parameters, which is because tests were done on a different fragment of the movie from the fragment for parameters optimization. Such weak generalization may be related to short length of the movies, which may be confirmed by a fact that in both cases with weak generalization: DSC_0559_A and 01091_A the movies were much shorter than in the case of DSC_562_A where results after optimization is slightly better than with standard parameters. This may lead to the conclusion that the movies used for parameter optimization should be longer. The significantly greater error for the movie 01091 is probably due to greater consideration of the outlet area combined with a large number of mutually obscuring bees.

Table 4. Detection error of windows containing bees for test movies

Movie name	Methods of extraction windows with bee images				
	Parameters optimized on training movie			Standard parameters	Greedy method from [2]
	$E_{0.5}$	$E_{0.5}^{fp}$	$E_{0.5}^{fn}$	$E_{0.5}$	$E_{0.5}$
DSC_0559_B	0.0688	0.0504	0.0865	0.0619	0.0926
DSC_0562_B	0.0626	0.0546	0.0705	0.0646	0.0686
01091_B	0.427	0.494	0.340	0.417	0.524

Figure 4 shows the dependence of the test error on the assumed IOU threshold at which positive detection was considered. As can be seen in the case of DSC_0559 and DSC_562 for $x > 0.5$, the error increases very quickly, but in some applications, such as bee counting at low density, high detection precision is not required and even a lower threshold at the level of $0.1 \div 0.2$ can be used. In the case of further image analysis, e.g. to determine if a bee carries pollen, precision should be increased. Sample images after subsequent stages of the detection process are shown in Fig. 5.

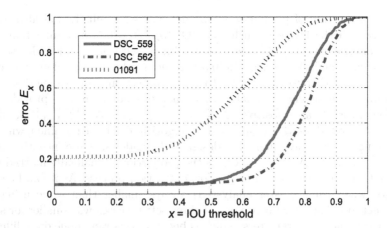

Fig. 4. Test error E_x depending on x - IOU threshold, above which detection is considered as positive

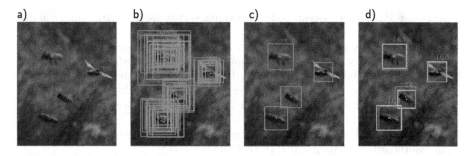

Fig. 5. Sample images from frame 1203 from the movie DSC_0562_B: a) original image, b) window clouds after window classification stage, c) windows after extraction by weighted clustering algorithm, d) comparison of positive detections (green) with windows marked manually (yellow) (Color figure online)

5 Conclusions

The results of the experiments indicates that it is possible to build an effective bee detection system that can adapt to the specific camera settings and environmental conditions. This system can be implemented using the classical method in the form of three subsystems: the ROI area subsystem, the window classification subsystem along with the procedure of scanning individual frames of the video stream, and the subsystem for determining positive detection windows by the weighted clustering method proposed in this paper. The disadvantage of the current system is a longtime scanning and window classification does not yet allow the system to be operated on-line on high resolution images. Two-step training process of the window classifier allows the elimination of false positive windows in specific camera setting and in a specific environment, so that after the determination of positive detections there is almost no false positive detection despite the simple, only 6-layer architecture of DNN-CNN network. Weighted clustering is always better than the greedy method of windows selection, and the additional optimization of its parameters allows to achieve better results in the case of sufficiently long training movies.

Acknowledgements. The work has been supported by founds of Faculty of Electronics, Telecommunications and Informatics, Gdańsk University of Technology.

References

1. Cejrowski, T., Szymański, J., Mora, H., Gil, D.: Detection of the bee queen presence using sound analysis. In: Nguyen, N.T., Hoang, D.H., Hong, T.-P., Pham, H., Trawiński, B. (eds.) ACIIDS 2018. LNCS (LNAI), vol. 10752, pp. 297–306. Springer, Cham (2018). https://doi.org/10.1007/978-3-319-75420-8_28
2. Dembski, J., Szymański, J.: Bees detection on images: study of different color models for neural networks. In: Fahrnberger, G., Gopinathan, S., Parida, L. (eds.) ICDCIT 2019. LNCS, vol. 11319, pp. 295–308. Springer, Cham (2019). https://doi.org/10.1007/978-3-030-05366-6_25

3. de Souza, P., et al.: Low-cost electronic tagging system for bee monitoring. Sensors **18**, 2124 (2018)
4. Chen, C., Yang, E.C., Jiang, J.A., Lin, T.T.: An imaging system for monitoring the in-and-out activity of honey bees. Comput. Electron. Agric. **89**, 100–109 (2012)
5. Magnier, B., Gabbay, E., Bougamale, F., Moradi, B., Pfister, F., Slangen, P.: Multiple honey bees tracking and trajectory modeling. In: Multimodal Sensing: Technologies and Applications, vol. 11059. International Society for Optics and Photonics (2019). 110590Z
6. Tashakkori, R., Ghadiri, A.: Image processing for honey bee hive health monitoring. In: SoutheastCon 2015, pp. 1–7. IEEE (2015)
7. Ngo, T.N., Wu, K.C., Yang, E.C., Lin, T.T.: A real-time imaging system for multiple honey bee tracking and activity monitoring. Comput. Electron. Agric. **163**, 104841 (2019)
8. Chiron, G., Gomez-Krämer, P., Ménard, M.: Detecting and tracking honeybees in 3D at the beehive entrance using stereo vision. EURASIP J. Image Video Process. **2013**(1), 1–17 (2013). https://doi.org/10.1186/1687-5281-2013-59
9. Dalal, N., Triggs, B.: Histograms of oriented gradients for human detection. In: 2005 IEEE Computer Society Conference on Computer Vision and Pattern Recognition (CVPR 2005), vol. 1, pp. 886–893 (2005)
10. Viola, P., Jones, M.J.: Robust real-time face detection. Int. J. Comput. Vision **57**, 137–154 (2004). https://doi.org/10.1023/B:VISI.0000013087.49260.fb
11. Girshick, R., Donahue, J., Darrell, T., Malik, J.: Rich feature hierarchies for accurate object detection and semantic segmentation. In: 2014 IEEE Conference on Computer Vision and Pattern Recognition (2014)
12. Redmon, J., Divvala, S., Girshick, R., Farhadi, A.: You only look once: unified, real-time object detection. In: 2016 IEEE Conference on Computer Vision and Pattern Recognition (CVPR) (2016)
13. Liu, W., et al.: SSD: single shot multibox detector. In: Leibe, B., Matas, J., Sebe, N., Welling, M. (eds.) ECCV 2016. LNCS, vol. 9905, pp. 21–37. Springer, Cham (2016). https://doi.org/10.1007/978-3-319-46448-0_2
14. Bodla, N., Singh, B., Chellappa, R., Davis, L.S.: Soft-NMS—improving object detection with one line of code. In: 2017 IEEE International Conference on Computer Vision (ICCV) (2017)

Improved Two-Step Binarization of Degraded Document Images Based on Gaussian Mixture Model

Robert Krupiński⬤, Piotr Lech⬤, and Krzysztof Okarma$^{(\boxtimes)}$⬤

Department of Signal Processing and Multimedia Engineering,
Faculty of Electrical Engineering, West Pomeranian University
of Technology in Szczecin, Sikorskiego 37, 70-313 Szczecin, Poland
{rkrupinski,piotr.lech,okarma}@zut.edu.pl

Abstract. Image binarization is one of the most relevant preprocessing operations influencing the results of further image analysis conducted for many purposes. During this step a significant loss of information occurs and the use of inappropriate thresholding methods may cause difficulties in further shape analysis or even make it impossible to recognize different shapes of objects or characters. Some of the most typical applications utilizing the analysis of binary images are Optical Character Recognition (OCR) and Optical Mark Recognition (OMR), which may also be applied for unevenly illuminated natural images, as well as for challenging degraded historical document images, considered as typical benchmarking tools for image binarization algorithms.

To face the still valid challenge of relatively fast and simple, but robust binarization of degraded document images, a novel two-step algorithm utilizing initial thresholding, based on the modelling of the simplified image histogram using Gaussian Mixture Model (GMM) and the Monte Carlo method, is proposed in the paper. This approach can be considered as the extension of recently developed image preprocessing method utilizing Generalized Gaussian Distribution (GGD), based on the assumption of its similarity to the histograms of ground truth binary images distorted by Gaussian noise. The processing time of the first step, producing the intermediate images with partially removed background information, may be significantly reduced due to the use of the Monte Carlo method.

The proposed improved approach leads to even better results, not only for well-known DIBCO benchmarking databases, but also for more demanding Bickley Diary dataset, allowing the use of some well-known classical binarization methods, including the global ones, in the second step of the algorithm.

Keywords: Document images · Image binarization · Gaussian Mixture Model · Monte Carlo method · Thresholding

1 Introduction

Analysis of binary images still belongs to the most popular applications of machine vision both in industry and some other computer vision tasks, where

© Springer Nature Switzerland AG 2020
V. V. Krzhizhanovskaya et al. (Eds.): ICCS 2020, LNCS 12141, pp. 467–480, 2020.
https://doi.org/10.1007/978-3-030-50426-7_35

the shape of objects plays the dominant role. Although in some industrial applications with controlled lighting conditions, as well as in the analysis of high quality scanned documents, some classical global thresholding algorithms, such as e.g. well-known Otsu [24] method, may be sufficient, for unevenly illuminated objects or degraded document images, even the use of more advanced adaptive methods might be challenging in some cases. For some of the popular adaptive methods, e.g. proposed by Niblack [18] or Sauvola [29], the obtained results may be far from expectations, especially in outdoor scenarios. On the other hand, some more sophisticated methods may be troublesome to implement in some embedded systems and devices with low computing performance.

Some typical areas of applications, where the quality of binary images obtained from natural images is important, are Optical Text Recognition (OCR), Optical Mark Recognition (OMR), recognition of QR codes, self-localization, terrain exploration and path following in autonomous navigation of vehicles and mobile robots, video monitoring and inspection, etc. Nevertheless, due to the lack of image and video datasets, containing both natural and ground truth images other than document images, a widely accepted approach to performance evaluation of image binarization methods is the use of the datasets provided yearly by the organizers of Document Image Binarization COmpetitions (DIBCO), taking place during two major conferences, namely International Conference on Document Analysis and Recognition (ICDAR) and International Conference on Frontiers in Handwriting Recognition (ICFHR).

Although these datasets contain images with more and more challenging image distortions each year, another interesting possibility is the additional verification of the proposed methods for the images included in Bickley Diary dataset [5], containing 92 photocopies of individual pages from a diary written ca. 100 years ago by the wife of one of the first missionaries in Malaysia – Bishop George H. Bickley. Since the distortions in this dataset are related not only to overall noise caused by photocopying, but also discolorization and water stains, as well as differences in ink contrast for different years, it may be considered as even more challenging in comparison to DIBCO datasets [26]. To ensure a reliable verification of the advantages of the method proposed in this paper, all currently available DIBCO datasets together with Bickley Diary database have been used.

Although many various approaches to image binarization have been presented over the years, including adaptive methods e.g. proposed by Bradley [2], Feng [6], Niblack [18], Sauvola [29] or Wolf [35], and their modifications [28,30], for each newly developed algorithm its required computational effort usually increases. Good examples may be the applications of local features with the use of Gaussian Mixture Models [17] or the use of deep neural networks [32], where multiple processing stages are necessary. Some comparisons of popular methods and their overviews can be found in recent survey papers or books [3,31]. In many methods the additional background removal, median filtering or morphological processing are required, as well as time-consuming training process for recently popular deep convolutional neural networks. Therefore, our motivation is the increase of performance of some classical methods due to efficient image preprocessing rather than comparison with sophisticated state-of-the-art methods and solutions based on deep learning, considering also the time-quality efficiency challenges [14].

2   Modelling the Histograms of Distorted Images

2.1   Generalized Gaussian Distribution and Gaussian Mixture Model

The application areas of the Generalized Gaussian Distribution (GGD) cover a wide range of signal and image processing methods utilizing the designation of various models, including e.g. tangential wavelet coefficients used to compress three-dimensional triangular mesh data [12] or generation of augmented quaternion random variables with the GGD [7]. Some other popular applications are related to no-reference image quality assessment (IQA) based on natural scene statistics (NSS) model, used to describe certain regular statistical properties of natural images [38], as well as image segmentation [33] and approximation of an atmosphere point spread function (APSF) kernel [34].

One of the main advantages of the GGD is the coverage of the other popular distributions, namely Gaussian distribution, Laplacian distribution, a uniform one, an impulse function, as well as some other special cases [9,10]. Estimation of its parameters is possible using various methods [37]. Its extension into multidimensional case [25] and covering the complex variables [19] is also possible.

The probability density function of the GGD can be expressed as [4]:

$$f(x) = \frac{\lambda \cdot p}{2 \cdot \Gamma\left(\frac{1}{p}\right)} e^{-[\lambda \cdot |x|]^p}, \tag{1}$$

where p denotes the shape parameter, $\Gamma(z) = \int_0^\infty t^{z-1} e^{-t} dt, z > 0$ [23] and λ is the parameter based on the standard deviation σ of the distribution. Their relation is given by the equation $\lambda(p, \sigma) = \frac{1}{\sigma} \left[\frac{\Gamma(\frac{3}{p})}{\Gamma(\frac{1}{p})} \right]^{\frac{1}{2}}$. Choice of the parameter $p = 1$ corresponds to Laplacian distribution, whereas $p = 2$ is typical for Gaussian distribution. When $p \to \infty$, the GGD density function goes to a uniform distribution and for $p \to 0$, $f(x)$ becomes an impulse function.

A Gaussian Mixture Model (GMM) consists of Gaussian distribution components defined by their locations μ and standard deviations σ and additionally, a vector of mixing proportions. Due to the main purpose of investigation, related to image binarization, the application of two Gaussian distribution components is considered, since only two classes of pixels are assumed. In the other words, it is assumed that only two clusters are present in the image, consisting of pixels representing text and background respectively, and – in the ideal case – each cluster is represented by a single Gaussian distribution component. Having computed the parameters of the GMM with two Gaussian distribution components,

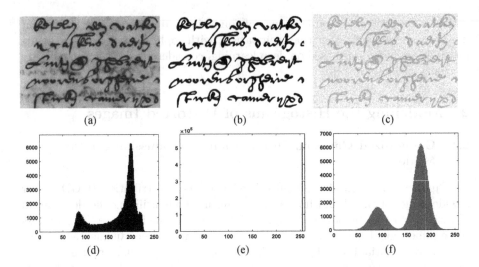

Fig. 1. Illustration of the general assumption of the proposed approach based on the similarity of histograms: (a) - sample greyscale document image, (b) - ground truth binary image, (c) - binary image corrupted by Gaussian noise, (d)–(f) - histograms of respective images.

the initial threshold should be located between the locations μ of both distributions and may be calculated in several ways e.g. as the intersection point of two determined curves.

The GMM parameters can be determined using the iterative Expectation-Maximization (EM) algorithm. The algorithm iterates over two steps until the convergence is achieved. The first step would estimate the expected value for each observation and the maximization step would optimize the parameters of the probability distributions using the maximum likelihood.

2.2 General Assumptions for Natural Images

Natural images, representing old handwritten or machine-printed documents, contain some specific distortions, being the result of gradual degradation of original manuscripts or printings during years. Some visible imperfections, such as faded and low contrast ink, as well as the presence of noisy distortions and some stains, influence the histogram of the image. Hence, assuming the analysis of greyscale images, more intermediate grey levels may be observed, similarly as for binary images corrupted by Gaussian noise, as illustrated in Fig. 1 for the sample image no. 7 from DIBCO2017 dataset. This similarity is especially well visible assuming the use of two Gaussian distributions modelling the histogram of the ground truth (GT) binary image corrupted by Gaussian noise.

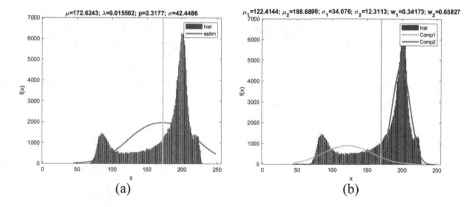

Fig. 2. Exemplary results of the approximation of the histogram of the sample image from DIBCO2017 dataset with obtained thresholds: (a) – using the GGD, (b) – for two Gaussian distribution components of the GMM.

Therefore, the approximation of histograms by the GMM with two Gaussian distribution components should be useful for the initial thresholding step, eliminating the most of the background information. It can be conducted by choosing a threshold between two peaks of the approximated histogram defined by their location parameters μ. Another possible approach, investigated in one of the previous papers [11], is the use of a single Gaussian distribution or the GGD, being its extended version. Nevertheless, the choice of its location parameter μ as the initial threshold leads to elimination of less background information. The comparison of parameters of the GGD and GMM with two Gaussian distribution components (further referred as GMM2), obtained for the sample image no. 7 from DIBCO2017 with a typical bimodal histogram, is shown in Fig. 2, where the threshold selected for the GMM is marked as the intersection point of both Gaussian curves.

For some images the GMM2 components may be located closer to each other and therefore the choice of an appropriate threshold may be more troublesome. In such situations the solution proposed in the paper [11] may be insufficient and the application of the GMM2 makes it possible to remove the background information better. Some exemplary results, obtained for sample image no. 8 from more challenging DIBCO2018 dataset, are presented in Fig. 3, where the greater ability to remove unnecessary background can be clearly observed for the GMM2. Such obtained images may be subjected to further binarization steps.

3 Proposed Method

3.1 Improved Two-Step Binarization Algorithm

Taking the advantage of similarity of histograms of degraded document images converted to greyscale and binary GT images corrupted by Gaussian noise, chosen as the most widespread type of noise in practical applications, the first step

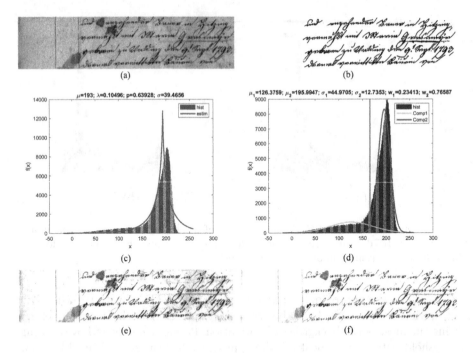

Fig. 3. Exemplary results obtained for image no. 8 from DIBCO2018 dataset: (a) – original image, (b) – ground truth image, (c) – histogram and obtained using the GGD, (d) – histogram and threshold for the GMM2, (e) and (f) – respective thresholding results obtained for two methods.

of the proposed algorithm is the calculation of parameters of the Gaussian distributions. Due to a great importance of the universality of the proposed approach, three possible models are used: a single Gaussian distribution, GGD and GMM2. Using the most relevant parameters: μ and σ, several variants of possible thresholds X_{thr} have been tested, including:

- location parameter μ_{GGD} of the GGD (originally proposed in [11]),
- location parameter μ_G of the single Gaussian distribution,
- location parameter μ_G lowered by Gaussian standard deviation σ_G,
- intersection of two GMM2 curves (thr), as shown in Fig. 2b,
- upper location parameter of two GMM2 curves μ_{GMMmax},
- weighted average of two GMM2 locations: $\mu_{GMM01} \cdot w_{01} + \mu_{GMM02} \cdot w_{02}$,
- weighted average of two GMM2 locations lowered by the respective standard deviations: $(\mu_{GMM01} - \sigma_{01}) \cdot w_{01} + (\mu_{GMM02} - \sigma_{02}) \cdot w_{02}$,
- minimum values of the above thresholds.

The weighting coefficients w_{01} and w_{02} have been determined during the calculation of the GMM and normalized so that $w_{01} + w_{02} = 1$. To avoid the necessity of using sophisticated estimators based on maximum likelihood, moments, entropy matching or global convergence [27], the values of the four GGD parameters: shape parameter p, location parameter μ, variance of the distribution λ, and standard deviation σ, as well as parameters of the GMM2, have been determined using the fast approximated method based on the standardized moment, described in the paper [8].

Additionally, all the above parameters have also been calculated after filtration of the 256-bin image histograms using 5-element median filter used to remove peaks. However, the results obtained using this approach have been worse for all databases and slightly higher binarization accuracy has been observed only for a few images. Therefore, all further experiments have been conducted using the original histograms without the additional time-consuming filtering.

To improve text readability, the determined thresholds (X_{thr}) are used instead of the maximum intensity values in the classical normalization of pixel intensity levels, applied only for the intensity levels not exceeding X_{thr}, as

$$Y(i,j) = \left| \frac{(X(i,j) - X_{min}) \cdot 255}{X_{thr} - X_{min}} \right|, \tag{2}$$

where $0 \le X_{min} < X_{thr} < X_{max} \le 255$ and

- X_{thr} is the upper threshold determined during the proposed preprocessing,
- X_{min} is the minimum intensity of all image pixels,
- X_{max} is the maximum intensity of all image pixels,
- $X(i,j)$ is the intensity level of the input pixel at (i,j) coordinates,
- $Y(i,j)$ is the intensity level of the output pixel at (i,j) coordinates.

Assuming the presence of a dark text on a brighter background, to remove partially the bright background data, usually containing some distortions not influencing the text information, intensity values for all pixels with brightness higher than X_{thr} are set to 255 independently on the formula (2).

As the result, the limitation of the brightness range from $\langle X_{min} ; X_{max} \rangle$ to $\langle X_{min} ; X_{thr} \rangle$ with additional normalization to the range $\langle 0 ; 255 \rangle$ is obtained, where the increase of dynamic range for images with overexposure or visible low ink contrast is achieved regardless of the selected upper threshold. Finally, the intermediate image with partially eliminated background is obtained, which is the input for some other classical global or adaptive binarization methods. Since such obtained images are better balanced in terms of text and background information, is it assumed that the finally obtained thresholds should be closer to expectations in comparison with those achieved by the same methods without the proposed preprocessing.

3.2 Acceleration of Calculations Using the Monte Carlo Method

The idea of the Monte Carlo method is based on the significant decrease of the number of analysed pixels, preserving the statistical properties of the image histogram. According to the law of large numbers and the central limit theorem, for a statistical experiment the sequence of successive approximations of the estimated value is convergent to the sought solution. Therefore, using the pseudo-random number generator with a uniform distribution, a limitation of the number of analysed pixels, decreasing the computational burden, is possible [21].

To prevent the necessity of using two independent generators to draw the coordinates of pixels, the image is initially reshaped into one-dimensional vector V containing the intensities of all $M \times N$ pixels. Applying the pseudo-random number generator with possibly good statistical properties and a uniform distribution, n independent draws of the positions in the vector V are conducted. To build an estimate of the simplified histogram, the total number of the randomly drawn pixels (k) for each intensity level is calculated according to

$$\hat{L}_{MC} = \frac{k}{n} \cdot M \cdot N, \tag{3}$$

where k denotes the number of randomly chosen pixels of the given intensity, n is the total number of draws and $M \times N$ determines the image size. In some applications a random choice of pixels can also be made in parallel to increase the computational speed.

Analysing the convergence of the method [11], the estimation error can be determined as

$$\varepsilon_\alpha = \frac{u_\alpha}{\sqrt{n}} \cdot \sqrt{\frac{K}{M \cdot N} \cdot \left(1 - \frac{K}{M \cdot N}\right)}, \tag{4}$$

where K is the total number of pixels for a given intensity and u_α represents the two-sided critical range. Nevertheless, the influence of even relatively high values of the above estimation error (calculated for the histogram) on the determined binarization thresholds is marginal.

Such obtained estimated simplified histogram may be successfully used as the input data for histogram based global thresholding methods [13,22], however its use for adaptive thresholding would be possible assuming the division of images into regions. Nevertheless, the direct application of this approach for typical adaptive methods based on the analysis of the local neighbourhood of each pixel, such as Bradley [2], Niblack [18] or Sauvola [29], would be troublesome.

In the proposed approach the Monte Carlo method is applied to reduce the computational effort of the first step of the algorithm. Due to the use of the simplified histogram it is possible to estimate the initial upper threshold X_{thr} using a significantly reduced number of samples. To reduce the possibility of the influence of potentially imbalanced intensities of the randomly chosen pixels for a small number of draws, the Monte Carlo experiment may be repeated and then the median from the determined thresholds would be selected as the result. To verify the stability of this approach, some experiments have been conducted with the use of 2.5%, 5%, 7.5%, 10%, 12.5% and 15% of the total numbers of pixels in consecutive images from all available DIBCO datasets, as well as for all 92 images from Bickley Diary database. For the lower percentages median values from 3, 5, 7 and 9 Monte Carlo experiments have been chosen, although it should be noted that from computational point of view e.g. the use of 9 draws for 5% of pixels can be treated as equivalent to a random choice of 45% of all pixels (not considering the time necessary for selection of the median value).

The second stage of the proposed approach, assuming the use of some previously proposed binarization methods, is not based on the use of the Monte Carlo method, although for some of the global methods, it might be possible as well and should be considered in further research. Nevertheless, in the first experiment it has been assumed that n is equal to the total number of pixels ($M \times N$), however – as described further – it may be significantly reduced applying the Monte Carlo method, without affecting the accuracy of binarization.

4   Experimental Results

The verification of the proposed approach has been made using 208 images: 116 images from 9 available DIBCO datasets (2009 to 2018), converted to greyscale according to popular ITU-R Recommendation BT.601, and 92 monochrome images from Bickley Diary dataset. All the calculations have been made for full images, as well as for the limited number of samples, applying the Monte Carlo method with repetitions and median choice, as stated above. During the first stage of the algorithm all threshold variants listed in Sect. 3.1 have been examined. Such obtained images with partially eliminated background information have been subjected to further binarization in the second stage, using popular thresholding methods, such as: fixed threshold (0.5 of the intensity range), Bernsen [1] (also with the local Gaussian window), Bradley [2], Otsu [24], Sauvola [29] and Wolf [35].

Finally, the obtained results have been compared with the direct use of the above mentioned methods without the proposed preprocessing. To make a reliable comparison of the final binarization results, according to widely accepted methodologies [20], some typical metrics based on the counting of true positive (TP) pixels, true negatives (TN), false positives (FP) and false negatives (FN), such as Precision, Recall, F-Measure, Specificity and Accuracy, have been calculated, assuming the pixels representing text as "ones" and background pixels as "zeros". Additionally, some other metrics, such as PSNR, Distance Reciprocal Distortion (DRD) [15] and Misclassification Penalty Metric (MPM) [36], have also been computed.

Although the values of the estimated parameters may be slightly different for each independent execution of the Monte Carlo method, especially for a low number of drawn samples (n), the overall influence of the number of randomly drawn pixels on the final binarization accuracy is unnoticeable, even for the use of 2.5% of the pixels assuming the 3-fold drawing and the choice of the median threshold in the first stage. Hence, only the results obtained for full images, considered as easier for potential recalculation, are presented in this paper, although the same results have been achieved applying the Monte Carlo method almost for all images. To avoid the presentation of all metrics based on the number of TP, TN, FP and FN pixels, we have focused on accuracy and PSNR, as well as some alternative metrics, such as DPD and MPM.

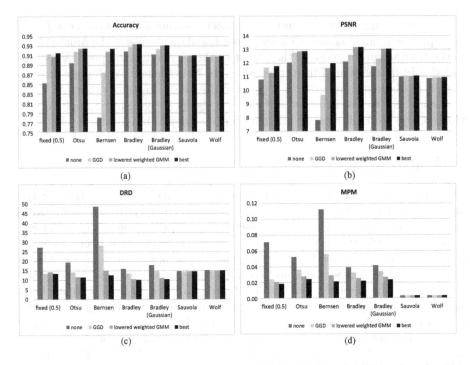

Fig. 4. Comparison of the average accuracy, PSNR, DRD and MPM values for 208 images using popular binarization methods with and without proposed preprocessing.

The results of experiments conducted for all DIBCO datasets and Bickley Diary database are illustrated in Fig. 4. Better results are represented by higher accuracy and PSNR, but lower DRD and MPM values.

As it may be observed, the influence of preprocessing for Sauvola and Wolf methods is marginal, however its application for some other methods, including the simplest fixed thresholding and the classical global Otsu method, leads to a significant improvement of all average metrics presented in Fig. 4. Analysing the accuracy, PSNR and DRD values, the use of the proposed preprocessing for Otsu method, as well as for adaptive Bradley and Bernsen algorithms, leads to better results than achieved by Sauvola and Wolf algorithms, even though these methods applied directly have led to much worse binarization results. In almost all cases the application of the proposed preprocessing method, based on the weighted average of two GMM2 locations lowered by the respective standard deviations $(\mu_{GMM01} - \sigma_{01}) \cdot w_{01} + (\mu_{GMM02} - \sigma_{02}) \cdot w_{02}$, leads to better results, also in comparison with the previously proposed method [11] using the location parameter μ_{GGD} of the GGD.

Nevertheless, considering the results indicated as "best", some minor exceptions occur, especially for the MPM results, as the results obtained for some other variants of possible thresholds X_{thr} are slightly better. For the fixed threshold (0.5) the best accuracy, PSNR and DRD may be obtained applying the

Fig. 5. Comparison of exemplary final binarization results: (a) and (b) – without preprocessing, (c) and (d) – with the use of $X_{thr} = \mu_{GGD}$ [11], (e) and (f) – with the use of the proposed preprocessing with $X_{thr} = (\mu_{GMM01} - \sigma_{01}) \cdot w_{01} + (\mu_{GMM02} - \sigma_{02}) \cdot w_{02}$, for Bernsen (left images) and Bradley (right images) thresholding.

$X_{thr} = max(\mu_{GMM01}, \mu_{GMM02})$, whereas the use of $X_{thr} = \mu_{GGD} - \sigma_{GGD}$ leads to the best DRD and MPM values for Otsu and Bradley methods, as well as the DRD for Wolf and all metrics for Bernsen thresholding. The use of weighted average of two GMM2 locations (without lowering) slightly improves the MPM results for Wolf method and DRD for Sauvola, as well as PSNR and accuracy for both of them. Nonetheless, as can be seen in Fig. 4, the differences between the results for "best" variants and the most universal method, utilizing the formula $X_{thr} = (\mu_{GMM01} - \sigma_{01}) \cdot w_{01} + (\mu_{GMM02} - \sigma_{02}) \cdot w_{02}$, are relatively small.

The best overall accuracy equal to 0.9336 and PSNR = 13.1614 has been achieved applying the proposed method followed by Bradley thresholding. The same popular adaptive method, implemented e.g. as the *adaptthresh* function in MATLAB environment, with the GGD based preprocessing [11], leads to noticeably worse results (ACC = 0.9279 and PSNR = 12.5857), whereas its direct application gives the accuracy equal to 0.9187 and PSNR = 12.1072 (without preprocessing).

A visual comparison of the final binarization results for a sample image no. 8 from the challenging DIBCO2018 dataset is shown in Fig. 5, where the advantages of the proposed approach, also over the GGD based preprocessing [11], are clearly visible, especially for Bernsen method shown in the left part. Nevertheless, the improvements of results can also be noticed for Bradley thresholding.

5   Summary and Future Work

The experimental results presented in the paper confirm the usefulness of the preprocessing of degraded document images based on the histogram modelling using the GGD and GMM based methods. A combination of the proposed approach with some well-known thresholding algorithms makes it possible to enhance the binarization results significantly, making them comparable with the use of more sophisticated methods. Even though the final results may be outperformed by some other methods, e.g. utilizing deep learning [32], the presented approach is relatively fast, also due to the use of the Monte Carlo method, and does not require the long training process with many images. The proposed approach may be easily combined with some other methods proposed by some other researchers, although – considering the results achieved for Sauvola and Wolf methods – achieved improvements may be smaller.

Some of the directions of our future research will be an attempt to a further simplification of the histogram modelling step, as well as the combination of the proposed preprocessing with statistical methods [13] and some region based thresholding methods [16], being usually much faster in comparison with typical adaptive methods, which require the analysis of the local neighbourhood of each pixel. Considering potential applications in robotics, related to the real-time analysis of natural images, our efforts will be oriented towards a further acceleration of image processing operations preceding the final binarization step.

References

1. Bernsen, J.: Dynamic thresholding of grey-level images. In: Proceedings 8th International Conference on Pattern Recognition (ICPR), pp. 1251–1255 (1986)
2. Bradley, D., Roth, G.: Adaptive thresholding using the integral image. J. Graph. Tools **12**(2), 13–21 (2007). https://doi.org/10.1080/2151237X.2007.10129236
3. Chaki, N., Shaikh, S.H., Saeed, K.: Exploring Image Binarization Techniques. SCI, vol. 560. Springer, New Delhi (2014). https://doi.org/10.1007/978-81-322-1907-1
4. Clarke, R.J.: Transform Coding of Images. Academic press, New York (1985)
5. Deng, F., Wu, Z., Lu, Z., Brown, M.S.: Binarization shop: a user assisted software suite for converting old documents to black-and-white. In: Proceedings of Annual Joint Conference on Digital Libraries, pp. 255–258 (2010)
6. Feng, M.L., Tan, Y.P.: Adaptive binarization method for document image analysis. In: Proceedings of 2004 IEEE International Conference on Multimedia and Expo (ICME), vol. 1, pp. 339–342 (2004). https://doi.org/10.1109/ICME.2004.1394198
7. Krupiński, R.: Generating augmented quaternion random variable with generalized Gaussian distribution. IEEE Access **6**, 34608–34615 (2018). https://doi.org/10.1109/ACCESS.2018.2848202
8. Krupiński, R.: Approximated fast estimator for the shape parameter of generalized Gaussian distribution for a small sample size. Bull. Pol. Acad. Sci. Tech. Sci. **63**(2), 405–411 (2015). https://doi.org/10.1515/bpasts-2015-0046
9. Krupiński, R.: Reconstructed quantized coefficients modeled with generalized Gaussian distribution with exponent 1/3. Image Process. Commun. **21**(4), 5–12 (2016)
10. Krupiński, R.: Modeling quantized coefficients with generalized Gaussian distribution with Exponent 1 / m, m = 2, 3, ... In: Gruca, A., Czachórski, T., Harezlak, K., Kozielski, S., Piotrowska, A. (eds.) ICMMI 2017. AISC, vol. 659, pp. 228–237. Springer, Cham (2018). https://doi.org/10.1007/978-3-319-67792-7_23

11. Krupiński, R., Lech, P., Tecław, M., Okarma, K.: Binarization of degraded document images with generalized Gaussian distribution. In: Rodrigues, J.M.F., et al. (eds.) ICCS 2019. LNCS, vol. 11540, pp. 177–190. Springer, Cham (2019). https://doi.org/10.1007/978-3-030-22750-0_14

12. Lavu, S., Choi, H., Baraniuk, R.: Estimation-quantization geometry coding using normal meshes. In: Proceedings of the Data Compression Conference (DCC 2003), p. 362, March 2003. https://doi.org/10.1109/DCC.2003.1194027

13. Lech, P., Okarma, K.: Optimization of the fast image binarization method based on the Monte Carlo approach. Elektronika Ir Elektrotechnika $20(4)$, 63–66 (2014). https://doi.org/10.5755/j01.eee.20.4.6887

14. Lins, R.D., Bernardino, R.B., de Jesus: D.M.: A quality and time assessment of binarization algorithms. In: Proceedings of the 15th IAPR International Conference on Document Analysis and Recognition, ICDAR 2019, Sydney, Australia, 20–25 September 2019, pp. 1444–1450. IEEE (2019). https://doi.org/10.1109/ICDAR.2019.00232

15. Lu, H., Kot, A.C., Shi, Y.Q.: Distance-reciprocal distortion measure for binary document images. IEEE Signal Process. Lett. $11(2)$, 228–231 (2004). https://doi.org/10.1109/LSP.2003.821748

16. Michalak, H., Okarma, K.: Adaptive image binarization based on multi-layered stack of regions. In: Vento, M., Percannella, G. (eds.) CAIP 2019. LNCS, vol. 11679, pp. 281–293. Springer, Cham (2019). https://doi.org/10.1007/978-3-030-29891-3_25

17. Mitianoudis, N., Papamarkos, N.: Document image binarization using local features and Gaussian mixture modeling. Image Vis. Comput. 38, 33–51 (2015). https://doi.org/10.1016/j.imavis.2015.04.003

18. Niblack, W.: An introduction to Digital Image Processing. Prentice Hall, Englewood Cliffs (1986)

19. Novey, M., Adali, T., Roy, A.: A complex generalized Gaussian distribution - characterization, generation, and estimation. IEEE Trans. Signal Process. $58(3)$, 1427–1433 (2010). https://doi.org/10.1109/TSP.2009.2036049

20. Ntirogiannis, K., Gatos, B., Pratikakis, I.: Performance evaluation methodology for historical document image binarization. IEEE Trans. Image Process. $22(2)$, 595–609 (2013). https://doi.org/10.1109/TIP.2012.2219550

21. Okarma, K., Lech, P.: Monte Carlo based algorithm for fast preliminary video analysis. In: Bubak, M., van Albada, G.D., Dongarra, J., Sloot, P.M.A. (eds.) ICCS 2008. LNCS, vol. 5101, pp. 790–799. Springer, Heidelberg (2008). https://doi.org/10.1007/978-3-540-69384-0_84

22. Okarma, K., Lech, P.: Fast statistical image binarization of colour images for the recognition of the QR codes. Elektronika Ir Elektrotechnika $21(3)$, 58–61 (2015). https://doi.org/10.5755/j01.eee.21.3.10397

23. Olver, F.W.J.: Asymptotics and Special Functions. Academic Press, New York (1974)

24. Otsu, N.: A threshold selection method from gray-level histograms. IEEE Trans. Syst. Man Cybern. $9(1)$, 62–66 (1979). https://doi.org/10.1109/TSMC.1979.4310076

25. Pascal, F., Bombrun, L., Tourneret, J.Y., Berthoumieu, Y.: Parameter estimation for multivariate generalized Gaussian distributions. IEEE Trans. Signal Process. $61(23)$, 5960–5971 (2013). https://doi.org/10.1109/TSP.2013.2282909

26. Pratikakis, I., Zagoris, K., Kaddas, P., Gatos, B.: ICFHR 2018 competition on handwritten document image binarization (H-DIBCO 2018). In: 2018 16th International Conference on Frontiers in Handwriting Recognition (ICFHR), pp. 489–493, August 2018. https://doi.org/10.1109/ICFHR-2018.2018.00091

27. Roenko, A.A., Lukin, V.V., Djurović, I., Simeunović, M.: Estimation of parameters for generalized Gaussian distribution. In: 2014 6th International Symposium on Communications, Control and Signal Processing (ISCCSP), pp. 376–379, May 2014. https://doi.org/10.1109/ISCCSP.2014.6877892

28. Samorodova, O.A., Samorodov, A.V.: Fast implementation of the Niblack binarization algorithm for microscope image segmentation. Pattern Recogn. Image Anal. **26**(3), 548–551 (2016). https://doi.org/10.1134/S1054661816030020

29. Sauvola, J., Pietikäinen, M.: Adaptive document image binarization. Pattern Recogn. **33**(2), 225–236 (2000). https://doi.org/10.1016/S0031-3203(99)00055-2

30. Saxena, L.P.: Niblack's binarization method and its modifications to real-time applications: a review. Artif. Intell. Rev. **51**(4), 673–705 (2017). https://doi.org/10.1007/s10462-017-9574-2

31. Shrivastava, A., Srivastava, D.K.: A review on pixel-based binarization of gray images. In: Satapathy, S.C., Bhatt, Y.C., Joshi, A., Mishra, D.K. (eds.) Proceedings of the International Congress on Information and Communication Technology. AISC, vol. 439, pp. 357–364. Springer, Singapore (2016). https://doi.org/10.1007/978-981-10-0755-2_38

32. Tensmeyer, C., Martinez, T.: Document image binarization with fully convolutional neural networks. In: 14th IAPR International Conference on Document Analysis and Recognition, ICDAR 2017, Kyoto, Japan, 9–15 November 2017, pp. 99–104. IEEE (2017). https://doi.org/10.1109/ICDAR.2017.25

33. Wang, C.: Research of image segmentation algorithm based on wavelet transform. In: 2015 IEEE International Conference on Computer and Communications (ICCC), pp. 156–160, October 2015. https://doi.org/10.1109/CompComm.2015.7387559

34. Wang, R., Li, R., Sun, H.: Haze removal based on multiple scattering model with superpixel algorithm. Signal Process. **127**, 24–36 (2016). https://doi.org/10.1016/j.sigpro.2016.02.003

35. Wolf, C., Jolion, J.M.: Extraction and recognition of artificial text in multimedia documents. Formal Pattern Anal. Appl. **6**(4), 309–326 (2004). https://doi.org/10.1007/s10044-003-0197-7

36. Young, D.P., Ferryman, J.M.: PETS metrics: on-line performance evaluation service. In: Proceedings of 2005 IEEE International Workshop on Visual Surveillance and Performance Evaluation of Tracking and Surveillance, pp. 317–324 (2005). https://doi.org/10.1109/VSPETS.2005.1570931

37. Yu, S., Zhang, A., Li, H.: A review of estimating the shape parameter of generalized Gaussian distribution. J. Comput. Inf. Syst. **21**(8), 9055–9064 (2012)

38. Zhang, Y., Wu, J., Xie, X., Li, L., Shi, G.: Blind image quality assessment with improved natural scene statistics model. Digital Signal Process. **57**, 56–65 (2016). https://doi.org/10.1016/j.dsp.2016.05.012

Cast Shadow Generation Using Generative Adversarial Networks

Khasrouf Taif⊙, Hassan Ugail(⊠)⊙, and Irfan Mehmood⊙

Centre for Visual Computing, University of Bradford, Bradford, UK
h.ugail@bradford.ac.uk

Abstract. We propose a computer graphics pipeline for 3D rendered cast shadow generation using generative adversarial networks (GANs). This work is inspired by the existing regression models as well as other convolutional neural networks such as the U-Net architectures which can be geared to produce believable global illumination effects. Here, we use a semi-supervised GANs model comprising of a PatchGAN and a conditional GAN which is then complemented by a U-Net structure. We have adopted this structure because of its training ability and the quality of the results that come forth. Unlike other forms of GANs, the chosen implementation utilises colour labels to generate believable visual coherence. We carried forth a series of experiments, through laboratory generated image sets, to explore the extent at which colour can create the correct shadows for a variety of 3D shadowed and un-shadowed images. Once an optimised model is achieved, we then apply high resolution image mappings to enhance the quality of the final render. As a result, we have established that the chosen GANs model can produce believable outputs with the correct cast shadows with plausible scores on PSNR and SSIM similarity index metrices.

Keywords: Cast shadows · Generative Adversarial Networks · Visualisation · Semi-supervised learning

1 Introduction

Shadow generation is a popular computer graphics topic. Depending on the level of realism required, the algorithms can be real-time such as Shadow mapping [23] and Shadow projection [3], or precomputed such as ray marching techniques, which are an expensive way to generate realistic shadows. Thus, pre-computing often become a compelling route to take, such as [28], [2] and [6]. Generative Adversarial Networks have been implemented widely to perform graphical tasks, as it requires minimum to no human interaction, which gives GANs a great advantage over conventional deep learning methods, such as image-to-image translation with single D, G semi-supervised model [7] or unsupervised dual learning [26].

We apply image-to-image translation to our own image set to generate correct cast shadows for 3D rendered images in a semi-supervised manner using

© Springer Nature Switzerland AG 2020
V. V. Krzhizhanovskaya et al. (Eds.): ICCS 2020, LNCS 12141, pp. 481–495, 2020.
https://doi.org/10.1007/978-3-030-50426-7_36

colour labels. We then augment a high-resolution image to enhance the overall quality. This approach can be useful in real-time scenarios, such as in games and Augmented Reality applications, since recalling a pre-trained model is less costly in time and quality compared to 3D real-time rendering, which often sacrifices realism to enhance performance. Our approach eliminates the need to constantly render shadows within a 3D environment and only recalls a trained model at the image plane using colour maps, which could easily be generated in any 3D software. The model we use utilises a combination of PatchGAN and Conditional GAN, because of their ability to compensate for missing training data, and tailor the output image to the desired task.

There are many benefits of applying GANs to perform computer graphic tasks. GANs can interpret predictions from missing training data, which means a smaller training data set compared to classical deep learning models. GANs can operate in multi modal scenarios, and a single input can generalise to multiple correct answers that are acceptable. Also, the output images are sharper due to how GANs learn the cost function, which is based on real-fake basis rather than traditional deep learning models as they minimise the Euclidean distance by averaging all plausible outputs, which usually produces blurry results. Finally, GAN models do not require label annotations nor classifications.

The rest of this paper is structured as follows. Section 1 reviews related work in terms of traditional shadow algorithms, machine learning and GANs, then in Sect. 2 we explain the construction of the generative model we used. In Sect. 3 we present our experiments from general GAN to cGAN and DCGAN and ending with Pix2pix. In Sect. 4 we discuss our results. Then, in Sects. 5 and 6 we present the conclusion and future work, respectively.

1.1   Generative Networks

Recent advancements in machine learning has benefited computer graphics applications immensely. In terms of 3D representation, modelling chairs of different styles from large public domain 3D CAD models was proposed by [1]. [25] applied Deep Belief Network to create representation of volumetric shapes. Similarly, supervised learning can be used to generate chairs, tables, and cars using up-convolutional networks [5],[4]. Furthermore, the application of CNN extended into rendering techniques, such as enabling global illumination fast rendering in a single scene [19], and Image based relighting from small number of images [18]. Another application filters Monte Carlo noise using a non-linear regression model [8].

Deep convolutional inverse graphics network (DC-IGN) enabled producing variations of lighting and pose of the same object from a single image [10]. Such algorithms can provide full deep shading, by training end to end to produce dense per-pixel output [16]. One of the recent methods applies multiple algorithms to achieve real-time outputs, which is based on a recommendation system that learns the user's preference with the help of a CNN, and then allow the user to make adjustments using a latent space variant system [30].

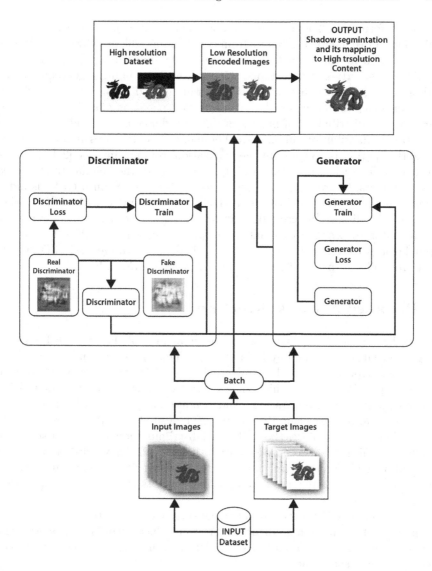

Fig. 1. The proposed framework emphasises on the combination of image pairs tailored to a specific function using pix2pix, producing high resolution content.

GANs have been implemented widely to perform graphical tasks. In cGANs, for instance, feeding a condition into both D, G networks is essential to control output images [14]. Training the domain-discriminator to maintain relevancy between input image and generated image, to transfer input domain into a target domain in semantic level and generate target images at pixel level [27]. Semi-supervised and unsupervised models has been approached in various ways such as trading mutual information between observed and predicted categorical class information [20]. This can be done by enabling image translators to be trained from two unlabelled images from two domains [26], or by translating both the image and its corresponding attributes while maintaining the permutation invariance property of the instance [15], or even by training a generator along with a mask generator that models the missing data distribution [11]. [12] proposed an image-to-image translation framework based on Coupled GANs that learns a joint distribution of images in different domains by using images from the marginal distributions in individual domains. Also, [13] uses an unsupervised model that performs image-to-image translation from few images. [21] made use of other factors such as structure and style. Another method eliminated the need for image pairs, by training the distribution of $G : X \rightarrow Y$ until $G(X)$ is indistinguishable from the Y distribution which is demonstrated with [29]. Another example is by discovering cross-domain relations [9]. Another way is forcing the discriminator to produce class labels by predicting which of $N + 1$ class the input belongs to during training [17]. Another model can generate 3D objects from probabilistic space with volumetric CNN and GANs [24].

2   Proposed Network Structure

The model we use in this paper is a Tensorflow port of the Pytorch Image-to-Image translation presented by Isola et al. [7]. This approach can be generalised to any semi-supervised model. However, this model serves us better for its two network types; PatchGAN, which allows better learning for interpreting missing data and partial data generation. And conditional GAN, which allows semi-supervised learning to facilitate control over desired output images by using colour labels. It also replaces the traditional discriminator with a U-Net structure with a step, this serves two purposes. First, it solves the known drop model issue in traditional GAN structure. Second, it helps transfer more features across the bottle neck which reduces blurriness and outputs larger and higher quality images.

The objective of GAN is to train two networks (see Fig. 1) to learn the correct mapping function through gradient descent to produce outputs believable to the human eye y. The conditional GAN here learns from observed image x and the random noise z, such that,

$$y, G : \{x, z\} \rightarrow y. \tag{1}$$

Where x is a random noise vector, z is an observed image, y is the output image. The Generator G, and a Discriminator D operate on "real" or "fake" basis. This is achieved by training both networks simultaneously with different objectives, G is trained to produce as realistic images as possible, while D is trained to distinguish which are fake, thus conditional GAN ($cGAN$) can be expressed as,

$$\mathcal{L}_{cGAN}(G, D) = \mathbb{E}_{x,y}[logD(x, y)] + \mathbb{E}_{x,z}[log(1 - D(x, G(x, z)))]. \tag{2}$$

The aim here for G to minimise the objective against the discriminator D which aims to maximise it such that,

$$G^* = arg \min_G \max_D \mathcal{L}_{cGAN}(G, D). \tag{3}$$

By comparing it to an unconditional variant where the discriminator does not observe x it becomes

$$\mathcal{L}_{cGAN}(G, D) = \mathbb{E}_y[log D(y)] + \mathbb{E}_{x,z}[log(1 - D(G(x, z)))]. \tag{4}$$

Here the distance of $\mathcal{L}_{L1}$ is used instead of $\mathcal{L}_{L2}$ to reduce blurring such that,

$$\mathcal{L}_{L1}(G) = \mathbb{E}_{x,y,z}[||y - G(x, z)||1]. \tag{5}$$

The final objective becomes,

$$G^* = arg \min_G \max_D \mathcal{L}_{cGAN}(G, D) + \lambda \mathcal{L}_{L1}(G). \tag{6}$$

Both networks follow the convolution-BatchNorm-ReLu structure. However, the generator differs by following the general U-Net structure, and the discriminator is based on Markov random fields. The application of a U-Net model allows better information flow across the network than the encoder-decoder model by adding skip connections over bottle necks between layer i and layer $n - i$, where n is the total number of layers, by concatenating all channels at layer i with the ones in $n - i$, thus, producing sharper images.

For discriminator D, a PatchGAN of $N \times N$ is applied to minimize blurry results by treating the image in small patches that are classified as real-fake across the image, then averaged to produce the accumulative results of D, such that the image is modelled as a Markov random field, Isola et al. [7] refers to this PatchGAN as a form of texture/style loss.

The optimisation process alternates descent steps between D and G, by training the model to maximize $log D(x, G(x, z))$ and dividing the objective by 2 to slow the learning rate of D, minibatch Stochastic Gradient Descend with Adam solver is applied at the rate of $0 : 0002$ for learning, and its momentum parameters are set to $\beta_1 = 0.5, \beta_2 = 0 : 999$. This allows the discriminator to compare minibatch samples of both generated and real samples. The G network runs at the same setting as the training phase at inference time. Dropout and batch normalization to the test batch is applied at test time with batch size of 1. Finally, random jitter is applied by extending the 256×256 input image size to 286×286 and then crop back to its original size of 256×256. Further processing using Photoshop, is applied manually to enhance the quality of output image, by mapping a higher resolution render of the model over the output model image, thus delivering a more realistic final image.

3   Experiments

Here we report our initial experiments for shadow generation as well as minor shading functions to support it. Our approach is data driven; it focuses on adjusting the image set in every iteration to achieve the correct output. For that we manually created the conditions we intended to test. Also, our image set is created using Maya with Arnold renderer. All of our experiments are conducted on an HP Pavilion laptop with 2.60 GHz Intel core i7 processor, 8 Ghz of RAM and Nvidia Geforce GTX 960M graphics card.

3.1 Image-to-Image Approach

We start with the assumption that GANs can generate both soft and hard shadows on demand, using colour labels and given a relatively small training image set. Our evaluation is based on both real-fake basis as well as similarity index matrices. Real-fake implies that the images can be clearly evaluated visually, for the network itself does not allow poor quality images by design. The similarity index matrices applied here are proposed by [22], namely, PSNR which measure the peak signal to noise ratio, which is scored between 1 and 100, and SSIM which computes the ratio between the strength of the maximum achievable power of the reconstructed signal and the strength of the corrupted noisy signal, which is scored between 0 and 1.

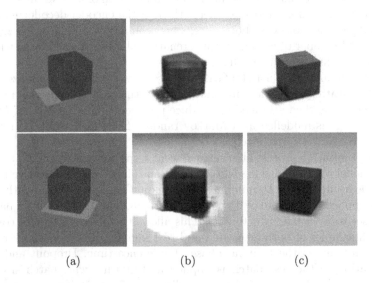

(a)   (b)   (c)

Fig. 2. Problematic output samples from the initial Image-to-image approach given a limited training image set, showing colour map (a), generated output (b) and target (c). Top (b) shows correct translation with some tiling effects due to the small training set, while bottom (b) is more problematic with incorrect environment translation such as large white patches. (Color figure online)

For our image-to-image translation approach, we started with a small image set of 100 images: 80 for training, 20 for testing and validation of random cube renders, with different lighting intensities and views. The results showed correct translations as shown in Fig. 2 (b) top section. However, the output colours of the background sometimes differed from the target images. Also, some of the environments contained patches and tiling artefacts as shown in Fig. 2 bottom section. Which is understandable given the small number of training images.

Next, we trained the model with Stonehenge images. It is also lit with a single light minus the variable intensity. The camera rotates 360 degrees around the Y axis. The total image number is 500 images, 300 for training and 200 for testing and validation.

We started with the question; can we generate shadows for non-shadowed images that are not seen during training? We worked around it by designing the colour label to our specific need. In the validation step we fed images with no shadows as in Fig. 3 (c), paired with colour labels that contains the correct shadows (a). As the results show in Fig. 3 (b), the model translated the correct shadow with the appropriate amount of drop off to the input image.

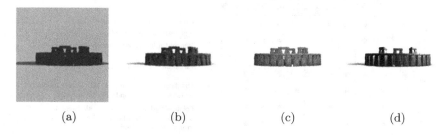

(a)   (b)   (c)   (d)

Fig. 3. Stonehenge samples showing the colour map (a), and correct translation of shadows (b) from non-shadowed images (c) and high resolution mapping (d). (Color figure online)

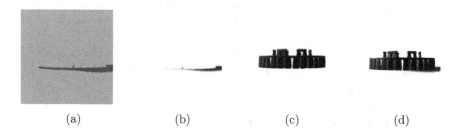

(a)   (b)   (c)   (d)

Fig. 4. Generating shadows with correct translation (b) for non-shadowed image (c) using only a shadow colour map (a) of the input image (c), then the output (b) is mapped to a high resolution image (d). (Color figure online)

Next we explore generating only accurate shadows Fig. 4 (b) for non-shadowed images (c), which is accomplished by constructing the colour map to only contain shadows (a), while training the network with full shadowed images as previously, paired with shadowed labels. The results show accurate translation of shadow direction and intensity (see Fig. 4 (b) and (d)).

For the third and final set of experiments, we used two render setups of the Stanford Dragon, one for training, the second for testing and validation. The camera setup that rotate 360 degrees around the dragon's Y axis with different step angle from training and testing. Also, the camera elevates 30 degrees across the X axis to show a more complex and elevated view of the dragon model rather than an orthographic one. The image set is composed of 1600 training images, 600 testing images and 800 for validation.

Table 1. Showing all the various categories applied in the final set of experiments.

Categories	Object		Colour map			Shadow			Position	
1	●	●	●	●		●	●		●	
2		●		●		●	●		●	
3	●	●	●			●	●			●
4		●	●			●			●	
5	●		●			●			●	
6	●			●		●	●		●	
7	●		●			●	●		●	
8		●		●		●			●	
9	●	●		●		●				●
10	●	●		●		●				●
11	●	●		●		●			●	
12	●	●	●	●		●			●	
13	●	●	●	●		●				●
14	●	●	●			●				●
15	●	●	●		●	●	●			●
16	●	●			●	●				●
17	●	●			●	●				●
18		●	●			●	●		●	
19		●	●			●			●	
20		●	●			●			●	
21	●	●		●		●			●	

Legend:

- Categories that are represented in both training and testing phases.
- Non-Partial: Images with models completely visible in them.
- Partial: Images with models cut in some areas.
- Standard Colour Map: Images with original model and shadow colours, and white backgrounds.
- Coloured Background: Images with different background colours of which there are six, in addition to the standard white.
- Switched colour maps: images that are paired with colour maps that belongs to another category.
- Soft Shadows: models that are rendered with soft cast shadows.
- Hard Shadows: models that are rendered with hard cast shadows.
- Non-Shadowed: Models with cast shadows removed from them.
- Syncronized: Image pairs with the models and colour maps has the same position and rotation.
- Non-Syncronized: Image pairs with the models and colour maps has different position and rotation.

The training set (shown in Table 1) is broken into multiple categories, with each one represented within 200 images with overlapping features sometimes. These images help understand how input image/label affects the behaviour of the output images. For example, if we fed a standard colour map to a coloured image, will it be able to translate the colours across or will the output be of standard colour, this will better inform the training process for future applications. All networks are trained from scratch using our image sets, and the weights are initialized from a Gaussian distribution with mean 0 and standard deviation 0.02.

3.2 Testing and Validation

The test image set consisted of 600 images that are not seen in the training phase. During this, some of the features learnt from the training set are tested and the weights are adjusted accordingly.

For validation, a set of 800 images that are not seen in the training set were used, they are derived from the testing image set, but have been modified heavily. The objective here is to test the ability to generate soft and hard shadows from unseen label images, as well as colour shadows, partial shadows, shadow generation for images with no shadows. Also, in some cases we have taken the image set to extremes in order to generate shadows for images that has different orientation and colour maps than their original labels.

From here, the experiments progressed in three phases. First, we trained the model with the focus on standard labels to produce soft and hard shadows, using an image set of 1000 images, 400 of them are dedicated for standard colours and shadows. The remaining 600 images are an arbitrary collection of cases mentioned in the training section (Table 1), with similar arbitrary testing and validation image sets of 300 images each. In this experiment, we had the standard colours and shadows provide comprehensive 360-degree view of the 3D model. While purposefully choosing arbitrary coloured samples of 10–25 images, we created 6 colour variations for both images and their respective colour labels. Our first objective was to observe whether the model can generalise colour information into texture, meaning to fill the details of the image from what has been learnt from the 360 view, and overlaying colour information on top of it. Even though the general details can be seen on the models, there were heavy artefacts such as blurring and tiling in some of the coloured images with fewer training images.

With that knowledge in mind, a second experiment was carried out. We adjusted the image set to reduce the tiling effect, which is mainly due to the lack of sufficient training images for specific cases. Hence, the number of training images was increased to 200 per case, to increase the training set to 1600 images.

In the validation image set, we pushed the code to extreme cases, such as pairing images of different colour maps and different directions, as well as partial and non-shadowed images. Thus, accumulating the validation set to 800 images, assuming beforehand that we will get the same tiling effect from the previous experiment in cases where we have different angles or different colours.

4 Discussion

The significant training time is one of the main challenges that we face, as the training time for our set of experiments ranged between 3 days to one week using

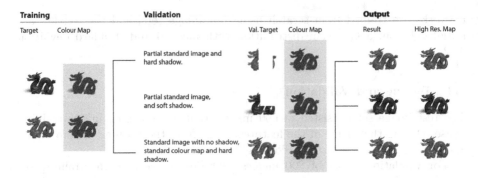

Fig. 5. The initial phase of the third set, which shows direct translation tasks. Our focus here is mainly the ability to generate believable soft and hard shadows, as well as inpainting for missing patches.

our laptop, which is considered long for training 1600 images. This why our image set was limited to 256×256 pixels. For this work, we overcome the issue with augmenting the output image with a high-resolution render. Once trained, however, with some optimisation the model should be capable of real-time execution, but this issue has not been tested by us. The two biggest limitations for this method are, it is a still semi-supervised model that needs a pair of colour maps and a target image. The second limitation is that the colour maps and image pairs are manually created, and the process is labour intensive. These issues should be considered for future work.

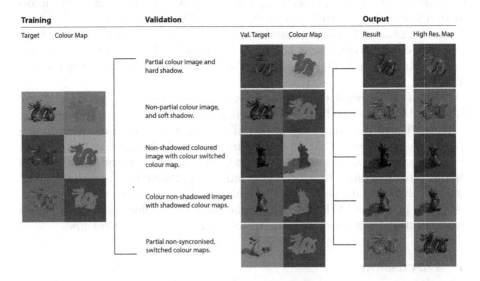

Fig. 6. Phase two of the third set focused on the interpretation of coloured shadows, since shadows are darker values of colours and not gray values. (Color figure online)

Our method performed well in almost all cases with minimal to no errors and sharp image reproduction, especially when faced with direct translation tasks, such as Fig. 5, and colour variations Fig. 6. Even with partial and non-shadowed images, the colours remained consistent and translated correctly across most outputs. This is promising, given a relatively small training set (approximately 200 images per case) we have used.

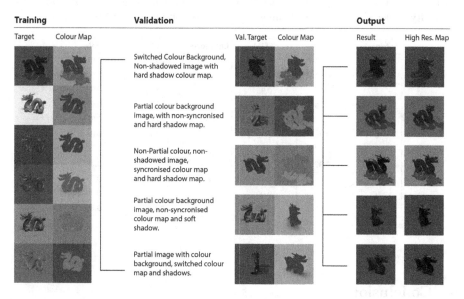

Fig. 7. In the third and final phase of the third set, we pushed the code to its limits by switching targets, colour background and colour maps. (Color figure online)

By examining Fig. 7, we notice that the model generalised correctly in most cases even though the colour maps are non-synchronised. This means our method has the breadth to interpret correctly when training set fails. However, it tends to take a more liberal approach in translating colour difference between the label and target with bias towards the colour map. This was also visible in partial images, non-shadowed images, as well as soft and hard shadows. The network struggled mostly when more than two parameters are changed, for example, a partial image and non-shadowed model will translate well. However, partial image, non-synchronised shadow and position will start to show tiling in the output image. The model seems to struggle the most with position switching than any other change, especially when paired with non-synchronised colour map as well. This is usually manifested in the form of noise, blurring and tiling (see Fig. 7), while the colours remain consistent and true to training images, and the shadows are correct in shape and intensity but are produced with noise and tiling artefacts. We conducted our quantitative assessment by applying the similarity matrices PSNR and SSIM [22] and we can confirm the previous observations.

When looking at the Table 2, the lowest score were in the categories with non-synchronized image pairs such as categories 3 and 19, while the image pairs that were approximately present in both training and testing performed the highest, which are categories 4 and 5, with overall performance leaning towards a higher scores spectrum.

Table 2. This table shows how each category performed in both PSNR and SSIM similarity indices between output images and corresponding ground truth images.

Categories	PSNR	SSIM	Categories	PSNR	SSIM
1	29.93	0.93	12	29.85	0.92
2	28.14	0.89	13	25.59	0.90
3	10.02	0.35	14	27.82	0.92
4	30.19	0.94	15	28.19	0.91
5	31.33	0.95	16	27.33	0.89
6	28.25	0.90	17	24.71	0.88
7	16.87	0.81	18	16.87	0.81
8	27.09	0.88	19	14.23	0.77
9	27.59	0.89	20	30.74	0.93
10	25.66	0.90	21	25.68	0.90
11	23.50	0.88	**Average**	**25.80**	**0.88**

5   Conclusion

This paper explored a framework based on conditional GANs using a pix2pix Tensorflow port to perform computer graphic functions, by instructing the network to successfully generate shadows for 3D rendered images given training images paired with conditional colour labels. To achieve this, a variety of image sets were created using an off-the-shelf 3D program.

The first set targeted soft and hard shadows under standard conditions and coloured labels and backgrounds, using 6 colour variations in the training set to test different variations, such as partial and non-shadowed images. The image set consisted of 1000 training images and 600 images for testing and validation. The results were plausible in most cases but showed clear blurring and tiling with coloured samples that did not have enough training images paired with it.

Next, we updated the image set to 3000 images, with 1600 training images, providing an equal number of training images for each of the 8 cases. We used 600 images for testing, and 800 for validation, which included more variations such as partial and non-shadowed images. In the validation set, the images included extreme cases such as non-sync pairing of position and colour. The results were believable in all cases, except the extreme cases, which resulted in tiling and blurring.

The results are promising for shadow generation especially when challenged to produce accurate partial shadows from training image set. The model is reliable to interpret successful output for images not seen during the training phase, except when paired with different colours viewpoints. However, there are still challenges to resolve. For example, the model requires a relatively long time to be trained and the output images still suffer from minor blurriness.

6   Future Work

This is only a proof of concept. The next logical step is to optimise the process by training the model to create highly detailed renders from lower poly-count models. This also can be tested with video-based models such as HDGAN. It is expected to output flickering results due to its learning nature and current state of the art. Another direction of interest may be to automate the generation of colour maps from video or live feed such as the work in [29]. The main challenge, however, is the computation complexity, especially for higher resolution training.

References

1. Aubry, M., Maturana, D., Efros, A.A., Russell, B.C., Sivic, J.: Seeing 3D chairs: exemplar part-based 2D–3D alignment using a large dataset of cad models. In: Proceedings of the IEEE conference on computer vision and pattern recognition, pp. 3762–3769 (2014)
2. Baran, I., Chen, J., Ragan-Kelley, J., Durand, F., Lehtinen, J.: A hierarchical volumetric shadow algorithm for single scattering. In: ACM Transactions on Graphics (TOG), vol. 29, p. 178. ACM (2010)
3. Blinn, J.: Me and my (fake) shadow. IEEE Comput. Graph. Appl. 8(1), 82–86 (1988)
4. Dosovitskiy, A., Springenberg, J.T., Tatarchenko, M., Brox, T.: Learning to generate chairs, tables and cars with convolutional networks. IEEE Trans. Pattern Anal. Mach. Intell. 39(4), 692–705 (2017)
5. Dosovitskiy, A., Tobias Springenberg, J., Brox, T.: Learning to generate chairs with convolutional neural networks. In: Proceedings of the IEEE Conference on Computer Vision and Pattern Recognition, pp. 1538–1546 (2015)
6. Herzog, R., Eisemann, E., Myszkowski, K., Seidel, H.P.: Spatio-temporal upsampling on the GPU. In: Proceedings of the 2010 ACM SIGGRAPH Symposium on Interactive 3D Graphics and Games, pp. 91–98. ACM (2010)
7. Isola, P., Zhu, J.Y., Zhou, T., Efros, A.A.: Image-to-image translation with conditional adversarial networks. In: Proceedings of the IEEE Conference on Computer Vision and Pattern Recognition, pp. 1125–1134 (2017)
8. Kalantari, N.K., Bako, S., Sen, P.: A machine learning approach for filtering Monte Carlo noise. ACM Trans. Graph. 34(4), 122 (2015)
9. Kim, T., Cha, M., Kim, H., Lee, J.K., Kim, J.: Learning to discover cross-domain relations with generative adversarial networks. In: Proceedings of the 34th International Conference on Machine Learning-Volume 70, pp. 1857–1865. JMLR. org (2017)

10. Kulkarni, T.D., Whitney, W.F., Kohli, P., Tenenbaum, J.: Deep convolutional inverse graphics network. In: Advances in Neural Information Processing Systems, pp. 2539–2547 (2015)

11. Li, S.C.X., Jiang, B., Marlin, B.: MisGAN: learning from incomplete data with generative adversarial networks. arXiv preprint arXiv:1902.09599 (2019)

12. Liu, M.Y., Breuel, T., Kautz, J.: Unsupervised image-to-image translation networks. In: Advances in Neural Information Processing Systems, pp. 700–708 (2017)

13. Liu, M.Y., et al.: Few-shot unsupervised image-to-image translation. In: Proceedings of the IEEE International Conference on Computer Vision, pp. 10551–10560 (2019)

14. Mirza, M., Osindero, S.: Conditional generative adversarial nets. CoRR abs/1411.1784 (2014). http://arxiv.org/abs/1411.1784

15. Mo, S., Cho, M., Shin, J.: InstaGAN: instance-aware image-to-image translation. arXiv preprint arXiv:1812.10889 (2018)

16. Nalbach, O., Arabadzhiyska, E., Mehta, D., Seidel, H.P., Ritschel, T.: Deep shading: convolutional neural networks for screen-space shading. Comput. Graph. **36**(4), 65–78 (2017)

17. Odena, A., Olah, C., Shlens, J.: Conditional image synthesis with auxiliary classifier GANs. In: Proceedings of the 34th International Conference on Machine Learning-Volume 70, pp. 2642–2651. JMLR. org (2017)

18. Ren, P., Dong, Y., Lin, S., Tong, X., Guo, B.: Image based relighting using neural networks. ACM Trans. Graph. **34**(4), 111:1–111:12 (2015)

19. Ren, P., Wang, J., Gong, M., Lin, S., Tong, X., Guo, B.: Global illumination with radiance regression functions. ACM Trans. Graph. **32**(4), 130:1–130:12 (2013)

20. Springenberg, J.T.: Unsupervised and semi-supervised learning with categorical generative adversarial networks. arXiv preprint arXiv:1511.06390 (2015)

21. Wang, X., Gupta, A.: Generative image modeling using style and structure adversarial networks. In: Leibe, B., Matas, J., Sebe, N., Welling, M. (eds.) ECCV 2016. LNCS, vol. 9908, pp. 318–335. Springer, Cham (2016). https://doi.org/10.1007/978-3-319-46493-0_20

22. Wang, Z., Bovik, A.C., Sheikh, H.R., Simoncelli, E.P., et al.: Image quality assessment: from error visibility to structural similarity. IEEE Trans. Image Process. **13**(4), 600–612 (2004)

23. Williams, L.: Casting curved shadows on curved surfaces. SIGGRAPH Comput. Graph. **12**(3), 270–274 (1978)

24. Wu, J., Zhang, C., Xue, T., Freeman, B., Tenenbaum, J.: Learning a probabilistic latent space of object shapes via 3D generative-adversarial modeling. In: Advances in neural information processing systems, pp. 82–90 (2016)

25. Wu, Z., et al.: 3D ShapeNets: a deep representation for volumetric shapes. In: 2015 IEEE Conference on Computer Vision and Pattern Recognition (CVPR), pp. 1912–1920 (2015)

26. Yi, Z., Zhang, H., Tan, P., Gong, M.: DualGAN: unsupervised dual learning for image-to-image translation. In: 2017 IEEE International Conference on Computer Vision (ICCV), pp. 2868–2876 (2017)

27. Yoo, D., Kim, N., Park, S., Paek, A.S., Kweon, I.S.: Pixel-level domain transfer. In: Leibe, B., Matas, J., Sebe, N., Welling, M. (eds.) ECCV 2016. LNCS, vol. 9912, pp. 517–532. Springer, Cham (2016). https://doi.org/10.1007/978-3-319-46484-8_31

28. Zhou, K., Hu, Y., Lin, S., Guo, B., Shum, H.Y.: Precomputed shadow fields for dynamic scenes. In: ACM Transactions on Graphics (TOG), vol. 24, pp. 1196–1201. ACM (2005)

29. Zhu, J.Y., Park, T., Isola, P., Efros, A.A.: Unpaired image-to-image translation using cycle-consistent adversarial networks. In: Proceedings of the IEEE international conference on computer vision, pp. 2223–2232 (2017)
30. Zsolnai-Fehér, K., Wonka, P., Wimmer, M.: Gaussian material synthesis. ACM Trans. Graph. (TOG) **37**(4), 76 (2018)

Medical Image Enhancement Using Super Resolution Methods

Koki Yamashita and Konstantin Markov[⊠][iD]

University of Aizu, Aizuwakamatsu, Fukushima 965-8580, Japan
{m5231120,markov}@u-aizu.ac.jp

Abstract. Deep Learning image processing methods are gradually gaining popularity in a number of areas including medical imaging. Classification, segmentation, and denoising of images are some of the most demanded tasks. In this study, we aim at enhancing optic nerve head images obtained by Optical Coherence Tomography (OCT). However, instead of directly applying noise reduction techniques, we use multiple state-of-the-art image Super-Resolution (SR) methods. In SR, the low-resolution (LR) image is upsampled to match the size of the high-resolution (HR) image. With respect to image enhancement, the upsampled LR image can be considered as low quality, noisy image, and the HR image would be the desired enhanced version of it. We experimented with several image SR architectures, such as super-resolution Convolutional Neural Network (SRCNN), very deep Convolutional Network (VDSR), deeply recursive Convolutional Network (DRCN), and enhanced super-resolution Generative Adversarial Network (ESRGAN). Quantitatively, in terms of peak signal-to-noise ratio (PSNR) and structural similarity index (SSIM), the SRCNN, VDSR, and DRCN significantly improved the test images. Although the ERSGAN showed the worst PSNR and SSIM, qualitatively, it was the best one.

Keywords: Medical image processing · OCT image enhancement · Image super resolution

1 Introduction

In recent years, Deep Neural Networks (DNN) have shown great success in image processing and analysis, outperforming humans in some tasks such as image classification [20]. It has been a matter of time, when DNNs would find their way in the area of medical image processing. The enhancement of medical images is a task of high practical value since many of the current MRI or CT images are of low quality. Classical image enhancement methods are mostly based on histogram equalization techniques [19] which don't work well with medical images. Lately, there have been some studies where the DNN are used for image enhancement [15] and MRI scans denoising [8].

In this work, we focus on enhancing or rather denoising images obtained by Optical Coherence Tomography (OCT) [21]. The OCT technology has become

© Springer Nature Switzerland AG 2020
V. V. Krzhizhanovskaya et al. (Eds.): ICCS 2020, LNCS 12141, pp. 496–508, 2020.
https://doi.org/10.1007/978-3-030-50426-7_37

a widely used tool for assessing optic nerve head tissues and monitoring many ocular pathologies. However, the quality of OCT scans is hampered by mainly speckle noise [7] as well as some other artifacts [1]. There exist some methods, both hardware and software based, to denoise OCT scans. For example, the multi-frame averaging [10] is a hardware technique which greatly improves the image quality, but requires long scanning time. This inflicts discomfort and strain in many patients. Software based image denoising approaches include filtering [16] or some numerical methods [6].

So far, with respect to the OCT image processing, the usage of deep learning has been limited to image segmentation [22] and classification [14]. The only other work on OCT denoising we are aware of is [4].

The goal of the OCT image enhancement task is to improve the quality of a single OCT scan to match the quality of multi-frame averaged image produced by the OCT device. This would greatly reduce the time needed to obtain high-quality image, because one multi-frame scan can takes about 3 min while a single scan - only few seconds. From machine learning point of view, this is a supervised multiple regression task as depicted in Fig. 1, where the input is the low quality (LQ) single scan and the output is an enhanced high quality (HQ) image resembling the multi-frame OCT scan.

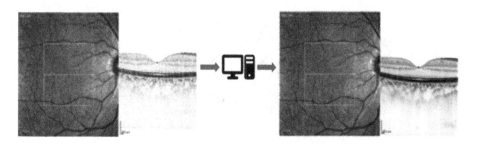

Fig. 1. The task of OCT scan enhancement. Low quality single scans are processed to obtain high quality images resembling the multi-frame scans as closely as possible.

In [4], researchers try to solve this task by adding Gaussian noise to the HQ multi-frame scans and use them as input to their denoising network based on the popular U-net [17]. This approach avoids problems with the image registration, because often there is a misalignment between single scans and their multi-frame counterparts. However, it ignores the actual speckle noise distribution which could be far from Gaussian and is OCT device dependent as well. Our approach differs in two main ways. First, we don't add artificial noise to the HQ multi-frame scans, but use the original LQ single scans. This apparently requires image registration which we performed using the excellent SimpleITK toolkit [2]. Second, we don't use DNN architectures targeted at image denoising, but adapt several state-of-the-art single images super resolution (SR) networks for the purposes of our task. They include super-resolution Convolutional Neural Network

(SRCNN), very deep Convolutional Network (VDSR), deeply recursive Convolutional Network (DRCN), and enhanced super-resolution Generative Adversarial Network (ESRGAN). The way we use the SR networks for image enhancement and some details for each of them are given in the next section. Later, we describe our data, experimental conditions and results we obtained.

2   Single Image Super Resolution

Single image super resolution (SR) is a classical problem in computer vision where the aim is to recover high-resolution (HR) image from a single low-resolution (LR) image. With the rise of deep convolutional networks, the number of proposed solutions and network architectures has increased dramatically [24, 26]. In practice, since the HR image size is bigger, during processing, the input LR image has to be upsampled to match the size. There are different strategies where and how to do this in the processing pipeline. Two widely used approaches are shown in Fig. 2. In the first one, the LR image is upsampled in advance using some form of interpolation and then is passed to the SR model as in Fig. 2(a). The other way is to keep the LR image size and perform upsampling at the last processing step as in Fig. 2(b).

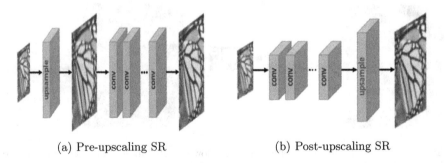

(a) Pre-upscaling SR   (b) Post-upscaling SR

Fig. 2. Two widely used SR architectures where image upsampling is done either before a) or after b) the processing.

Since in our task, the size of the image should not change, we cannot use those SR architectures directly. However, if we remove the upsampling step in the case of Fig. 2(a), we end up with a system that essentially enhances the input image without changing its size. This is illustrated in Fig. 3(a). Unfortunately, this approach does not work with the architecture of Fig. 2(b). In this case, the upsampling step is part of the processing pipeline and its parameters are trainable. We solve this problem by first downsampling the input image and then passing it to the system as shown in Fig. 3(b).

In the next four subsections we describe briefly each of the SR networks we used in this study.

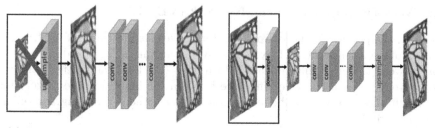

(a) In pre-upscaling SR, the first up-sampling block is deleted

(b) In post-upscaling SR, a new down-sampling block is added

Fig. 3. Changes made to accommodate the two SR architectures for image enhancement purposes.

2.1 Super Resolution Convolutional Neural Network (SRCNN)

The SRCNN [5] is a simple network consisting of two hidden convolutional layers as can be seen in Fig. 4. The input is supposed to be the upscaled version of the LR image, so the architecture corresponds to the pre-upsampling SR from Fig. 2(a).

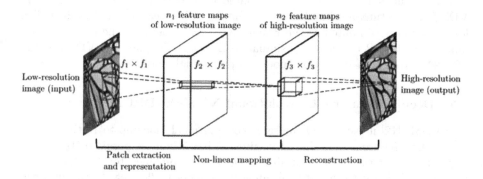

Fig. 4. SRCNN architecture.

Each hidden layer performs standard convolutional operation with output clipped to be positive. The loss function is the mean squared error (MSE) between the output image $\tilde{Y}_i$ and the target HR image Y_i averaged over the training set:

$$L(\Theta) = \frac{1}{n} \sum_{i=1}^{n} \parallel \mathbf{Y}_i - \hat{\mathbf{Y}}_i \parallel^2 \tag{1}$$

The MSE loss function favors a high peak signal-to-noise ratio (PSNR) which is a widely-used metric for quantitative evaluation of SR quality. However, the PSNR is only partially related to the perceptual quality and in practice, sometimes images with high PSNR don't look perceptually very good.

2.2   Very Deep Convolutional Network (VDSR)

Based on the popular VGG network [18] for image classification, the VDSR [11] consists of many convolutional layers with ReLU activation. The residual connection between the input and the last hidden layer (the long line in Fig. 5), forces the network to learn only the difference between the input and the target and as a result allows network to be much more deeper without vanishing/exploding gradients problem.

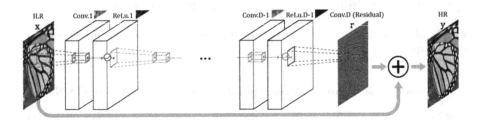

Fig. 5. VDSR architecture.

The input is an upsampled interpolated low-resolution (ILR) image, so the VDSR architecture falls into the pre-upsampling SR category as in Fig. 2(a). The loss function is computed as the Euclidean distance between the reconstructed image and the HR target image similar to Eq. (1). Therefore, the VDSR as the SRCNN favors high PSNR, but not high perceptual quality.

2.3   Deeply Recursive Convolutional Network (DRCN)

The VDSR [12] makes use of the same convolutional block up to 16 times. The main difference from the other structures is that a multi-supervised strategy is applied, so that the outputs of all the blocks are combined together as shown in Fig. 6. This approach not only allows gradients to flow easily through the network, but also encourages all the intermediate representations to reconstruct the HR image. In such multi-supervised approach, there are multiple objectives to minimize. The loss for the intermediate outputs is defined as:

$$l_1(\theta) = \frac{1}{2DN} \sum_{d=1}^{D} \sum_{i=1}^{N} \| \mathbf{y}_i - \hat{\mathbf{y}}_i^d \|^2 \tag{2}$$

where D is the number of recursions. For the final output with is a weighted sum of all intermediate outputs the loss is:

$$l_2(\theta) = \frac{1}{2N} \sum_{i=1}^{N} \| \mathbf{y}_i - \sum_{d=1}^{D} w_d \hat{\mathbf{y}}_i^d \|^2 \tag{3}$$

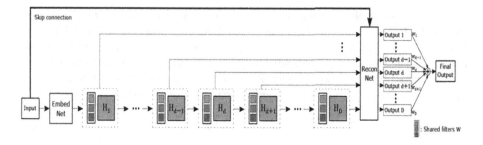

Fig. 6. DRCN architecture.

The final loss function includes both the l_1 and l_2 as well as a regularization term:

$$L(\theta) = \alpha l_1(\theta) + (1 - \alpha)l_2(\theta) + \beta \parallel \theta \parallel^2 \tag{4}$$

where α controls the trade-off between the intermediate and final losses and β - the amount of regularization. Note that all losses use the MSE criterion, so the DRCN also favors high PSNR images.

2.4 Enhanced Super Resolution Generative Adversarial Network (ESRGAN)

The ESRGAN [23] is an improved version of the super resolution generative adversarial network (SRGAN) [13]. It consists of two networks - Generator and Discriminator working together. The structure of each of them is shown in Fig. 7. The Generator includes multiple blocks called residual in residual dense block (RRDB) which combine multi-level residual network and dense connections. The upsampling block is located at the end of the pipeline, so the ESRGAN architecture is of the type shown in Fig. 2(b). The Discriminator has a simpler structure

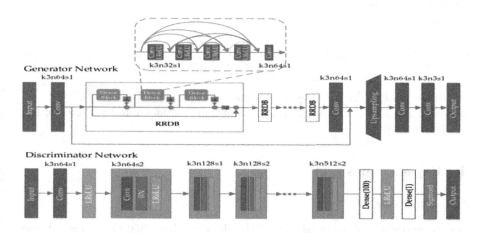

Fig. 7. ESRGAN architecture.

consisting of multiple convolution layers each followed by a batch normalization and Leaky ReLU activation. One important difference between the ESRGAN and other SR networks described above is that the Generator utilizes an improved version of the so called perceptual loss [9]. Originally, it is defined on the activation layers of a pre-trained network where the distance between two activated features is minimized. Thus, the Generator total loss is expressed as:

$$L_G^{tot} = L_{percep} + \lambda L_G + \eta L_1 \tag{5}$$

where $L_1 = \mathbb{E}_x \parallel G(x) - y \parallel_1$ is the 1-norm difference between the Generator output $G(x)$ given input image x and the target HR image y. Using such loss makes the ESRGAN to produce images of higher perceptual quality than the PSNR oriented networks.

3   Performance Evaluation

There exist various quantitative performance metrics adopted in image processing among which the peak noise-to-signal ratio (PSNR) and the structural similarity index measure (SSIM) [25] are the most widely used. In [4], authors used pure SNR and SSID metrics, while we utilize the PSNR and SSID.

The MSE and PSNR between ground truth image I and reconstructed image $\hat{I}$ both of which have N pixels as defined as:

$$MSE = \frac{1}{N} \sum_{i=1}^{N} (I(i) - \hat{I}(i))^2 \tag{6}$$

$$PSNR = 10 \log(\frac{L^2}{MSE}) \tag{7}$$

where $L = 255$ for 8-bit pixel encoding. Typical PSNR values vary from 20 to 40, higher is better.

On the other hand, the SSID is defined as:

$$SSIM(I, \hat{I}) = \frac{(2\mu_I \mu_{\hat{I}} + C_1)(2\sigma_{I\hat{I}} + C_2)}{(\mu_I^2 + \mu_{\hat{I}}^2 + C_1)(\sigma_I^2 + \sigma_{\hat{I}}^2 + C_2)} \tag{8}$$

where $C_1 = (k_1 L)^2, C_2 = (k_2 L)^2$ are constants for avoiding instability, $k_1 \ll 1, k_2 \ll 1$ are small constants, and μ and σ^2 are the mean and variance of the pixels intensity.

4   Experiments

4.1   Database

For the experiments, we used a small database of about 350 OCT scans. Some of the HQ multi-frame scans had several corresponding LQ single scans, so the same targets were used for those LQ images. Most of the HQ/LQ pairs required alignment and for this purpose we used the SimpleITK image registration toolkit [2]. Six HQ/LQ pairs were selected for testing, and the remaining data were split into training and validation sets by 9:1 ratio.

Since the number of scans is quite small, we did exhaustive data augmentation which includes horizontal and vertical flips, rotation by several different degrees, etc., commonly used in image processing practice. In addition, each scan was cropped into non-overlapping sub-images of size 224×224. Thus, we managed to increase the number of training data roughly 100 fold.

4.2   Results

Here, we present the results in terms of PSNR and SSIM metrics for each of the network architectures described in Sect. 2. In each case, we tried to tune the network hyper-parameters to achieve the best possible result. The results shown in the tables below reflect the performance dependence on the two most impactful parameters we found for each network.

All the networks were trained with up to 100 epochs and for testing we used the model obtained from the epoch where the PSNR of the validation data was the highest.

SRCNN Results. The SRCNN is trained using small patches of size 33×33 taken from the input image with stride 14. This network in known to take many training iterations to achieve good performance, so we chose a small learning rate of 5.0e-6. We found that the batch size and the size of the filter of the first convolutional layer have the biggest influence on the SRCNN performance. The obtained PSNR and SSIM values are given in Table 1.

Table 1. SRCNN performance in terms of PSNR (dB) and SSIM.

Metric	Batch size	f_1 size		
		7×7	9×9	11×11
PSNR	64	**25.23**	25.15	24.81
	128	25.18	24.93	23.76
	256	24.96	25.10	24.38
SSIM	64	0.794	**0.798**	0.797
	128	0.795	0.798	0.790
	256	0.791	0.796	0.792

VDSR Results. The patch size during the VDSR training was set to 41×41 with no overlap. We experimented with the number of convolutional blocks and the batch size. The learning rate was set to 0.001 and the other hyper-parameters were used as recommended by the VDSR developers. Table 2 shows the PSNR and SSID values obtained during the experiment.

Table 2. VDSR performance in terms of PSNR (dB) and SSIM.

Metric	Batch size	Number of blocks		
		8	16	32
PSNR	32	25.10	25.22	24.18
	64	25.12	25.38	25.30
	128	24.18	25.45	**25.51**
SSIM	32	0.791	0.785	0.543
	64	0.791	0.789	0.779
	128	0.791	**0.795**	0.778

DRCN Results. With the DRCN, we used the same patch size as for the VDSR, but with stride 21 [11]. Initially, the learning rate was set to 0.01 and during training was decreased 10 times every time validation performance plateaus. The main architectural hyper-parameters of the DRCN are the number of blocks and the number of filters in each block. We varied those parameters and the results with batch size of 128 are presented in Table 3.

Table 3. DRCN performance in terms of PSNR (dB) and SSIM.

Metric	Filter number	Number of blocks		
		4	8	16
PSNR	16	24.81	25.02	25.12
	32	24.48	24.26	23.02
	48	18.07	22.37	**25.77**
SSIM	16	0.768	0.774	**0.778**
	32	0.761	0.762	0.535
	48	0.723	0.535	0.687

We have to note that we could not find a good trade-off between the intermediate loss l_1 and final loss l_2 functions given in Eq. (2) and Eq. (3) respectively. The best results we obtained when the combination parameter λ from Eq. (4) was set to 0.

ESRGAN Results. In terms of parameters, this is the biggest network among all the networks we experimented with, and so is the number of possible hyper-parameters. Structurally, for the generator, important are the RDDB number, the RDB number in each RDDB as well as the number of convolutional layers and the number of filters. The discriminator's structure has no big influence on the performance. As can be seen from Table 4, in our case, the RDDB number and the filter number were the most sensitive to the ESRGAN performance. We couldn't obtain results for the case of RDDB number = 7 and filter number = 16 since the model was so big and did not fit in our GPU memory. The other parameters were as follows: number of RDBs inside each RRDB = 6, number of convolutional layers inside a RDB = 4, learning rate = 4.0e-4 with decay factor of 2. For training and evaluation of the ESRGAN, we used the ISR toolkit [3] and all the other parameters we left at their default values.

Table 4. ESRGAN performance in terms of PSNR (dB) and SSIM.

Metric	RDDB number	RDB filter number		
		4	8	16
PSNR	3	19.56	19.01	18.98
	5	19.53	**21.25**	18.92
	7	19.64	18.69	NA
SSIM	3	0.670	0.639	0.725
	5	0.432	0.722	**0.730**
	7	0.658	0.377	NA

Networks Comparison. Here, we compare the best obtained performance from all the networks we evaluated in terms of PSNR and SSIM. Figure 8 shows bar plots for each metric together with the case when no enhancement is applied. In terms of PSNR, the DRCN achieved the best result, while the best SSIM was achieved by the SRCNN and VDSR. In both cases, the obtained metrics values are much better than the baseline, i.e. the case of unprocessed single scan images.

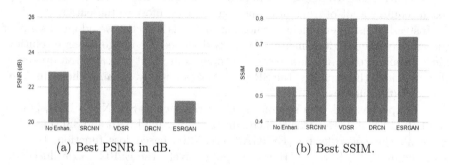

(a) Best PSNR in dB.   (b) Best SSIM.

Fig. 8. Comparison of the networks best performances in terms of PSNR and SSIM with the baseline ("No Enhan.")

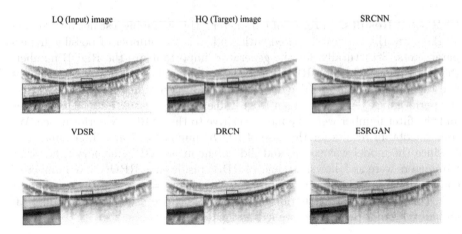

Fig. 9. Example test single scan (first row, left), the corresponding multi-frame averages scan (first row, center), and the results from each network.

The ERSGAN, however, showed PSNR even lower than the baseline. This can be explained with the fact that the ESRGAN is trained to improve the perceptual loss more than the mean absolute error (MAE) which is the L_1 in Eq. (5) and is related to the PSNR. To verify this hypothesis, we looked at all the test images enhanced by each of the networks and visually compared them. Indeed, the ESRGAN has produced the best looking images with sharper edges and higher contrast. As an example, we show one of the test single scans and its corresponding multi-frame scan as well as its enhanced versions by all the networks in Fig. 9.

5   Conclusion

In this study, we focused on enhancing single scans obtained from Optical Coherence tomography. They all contain speckle noise as well as some other artifacts making the interpretation of the OCT data cumbersome. Many OCT devices apply multi-frame averaging techniques to alleviate this problem, but this approach requires a lot of time and causes great discomfort to the patients.

Instead of using enhancing/denoising methods directly, we adopted some of the state-of-the-art deep neural networks designed for image super resolution. Since in many cases the low resolution images are first upscaled, an operation that degrades their quality, the SR networks essentially enhance those upscaled low resolution images.

We experimented with several SR networks such as SRCNN, VDSR, DRCN and ERSGAN and evaluated them quantitatively using PSNR and SSIM metrics. Since all the networks but ESRGAN use MSE based loss function, they all achieved high PSNR values. However, qualitatively, the ESRGAN produced the best looking images which we attribute to the use of a perceptual loss function.

Our results are still preliminary, because the amount of training data was clearly insufficient to reliably train big networks such as DRCN or ESRGAN. Also, the OCT scans come from healthy patients only and many pathological artifacts haven not been learned. In addition, we expect scans from different OCT devices to have different noise distributions. All these problems we intend to address in the future.

Acknowledgment. We are grateful to Prof. Sekiryu from Fukushima Medical University for providing the OCT scan data.

References

1. Asrani, S., Essaid, L., Alder, B.D., Santiago-Turla, C.: Artifacts in spectral-domain optical coherence tomography measurements in glaucoma. JAMA Ophthalmol. **132**(4), 396–402 (2014)
2. Beare, R., Lowekamp, B., Yaniv, Z.: Image segmentation, registration and characterization in R with simpleITK. J. Stat. Softw. **86**(8), 1–35 (2018). https://doi.org/10.18637/jss.v086.i08
3. Cardinale, F., John, Z., Tran, D.: ISR (2018). https://github.com/idealo/image-super-resolution
4. Devalla, S.K., et al.: A deep learning approach to denoise optical coherence tomography images of the optic nerve head. Sci. Rep. **9**(1), 1–13 (2019)
5. Dong, C., Loy, C.C., He, K., Tang, X.: Image super-resolution using deep convolutional networks. IEEE Trans. Pattern Anal. Mach. Intell. **38**(2), 295–307 (2016)
6. Du, Y., Liu, G., Feng, G., Chen, Z.: Speckle reduction in optical coherence tomography images based on wave atoms. J. Biomed. Opt. **19**(5), 056009 (2014)
7. Esmaeili, M., Dehnavi, A.M., Rabbani, H., Hajizadeh, F.: Speckle noise reduction in optical coherence tomography using two-dimensional curvelet-based dictionary learning. J. Med. Signals Sensors **7**(2), 86 (2017)
8. Jiang, D., Dou, W., Vosters, L., Xu, X., Sun, Y., Tan, T.: Denoising of 3D magnetic resonance images with multi-channel residual learning of convolutional neural network. Japan. J. Radiol. **36**(9), 566–574 (2018)
9. Johnson, J., Alahi, A., Fei-Fei, L.: Perceptual losses for real-time style transfer and super-resolution. In: Leibe, B., Matas, J., Sebe, N., Welling, M. (eds.) ECCV 2016. LNCS, vol. 9906, pp. 694–711. Springer, Cham (2016). https://doi.org/10.1007/978-3-319-46475-6_43
10. Kennedy, B.F., Hillman, T.R., Curatolo, A., Sampson, D.D.: Speckle reduction in optical coherence tomography by strain compounding. Opt. Lett. **35**(14), 2445–2447 (2010)
11. Kim, J., Kwon Lee, J., Mu Lee, K.: Accurate image super-resolution using very deep convolutional networks. In: Proceedings of the IEEE Conference on Computer Vision and Pattern Recognition, pp. 1646–1654 (2016)
12. Kim, J., Kwon Lee, J., Mu Lee, K.: Deeply-recursive convolutional network for image super-resolution. In: Proceedings of the IEEE Conference on Computer Vision and Pattern Recognition, pp. 1637–1645 (2016)
13. Ledig, C., et al.: Photo-realistic single image super-resolution using a generative adversarial network. In: Proceedings of the IEEE Conference on Computer Vision and Pattern Recognition, pp. 4681–4690 (2017)

14. Lee, C.S., Baughman, D.M., Lee, A.Y.: Deep learning is effective for classifying normal versus age-related macular degeneration OCT images. Ophthalmol. Retin. **1**(4), 322–327 (2017)
15. Lu, L., Zheng, Y., Carneiro, G., Yang, L. (eds.): Deep Learning and Convolutional Neural Networks for Medical Image Computing. ACVPR. Springer, Cham (2017). https://doi.org/10.1007/978-3-319-42999-1
16. Ozcan, A., Bilenca, A., Desjardins, A.E., Bouma, B.E., Tearney, G.J.: Speckle reduction in optical coherence tomography images using digital filtering. JOSA A **24**(7), 1901–1910 (2007)
17. Ronneberger, O., Fischer, P., Brox, T.: U-Net: convolutional networks for biomedical image segmentation. In: Navab, N., Hornegger, J., Wells, W.M., Frangi, A.F. (eds.) MICCAI 2015. LNCS, vol. 9351, pp. 234–241. Springer, Cham (2015). https://doi.org/10.1007/978-3-319-24574-4_28
18. Simonyan, K., Zisserman, A.: Very deep convolutional networks for large-scale image recognition. arXiv preprint arXiv:1409.1556 (2014)
19. Suganya, P., Gayathri, S., Mohanapriya, N.: Survey on image enhancement techniques. Int. J. Comput. Appl. Technol. Res. **2**(5), 623–627 (2013)
20. Szegedy, C., Ioffe, S., Vanhoucke, V., Alemi, A.A.: Inception-v4, inception-resnet and the impact of residual connections on learning. In: Thirty-First AAAI Conference on Artificial Intelligence (2017)
21. van Velthoven, M.E., Faber, D.J., Verbraak, F.D., van Leeuwen, T.G., de Smet, M.D.: Recent developments in optical coherence tomography for imaging the retina. Prog. Retin. Eye Res. **26**(1), 57–77 (2007)
22. Venhuizen, F.G., et al.: Robust total retina thickness segmentation in optical coherence tomography images using convolutional neural networks. Biomed. Opt. Express **8**(7), 3292–3316 (2017)
23. Wang, X., et al.: ESRGAN: enhanced super-resolution generative adversarial networks. In: Proceedings of the European Conference on Computer Vision (ECCV) (2018)
24. Wang, Z., Chen, J., Hoi, S.C.: Deep learning for image super-resolution: a survey. arXiv preprint arXiv:1902.06068 (2019)
25. Wang, Z., Bovik, A.C., Sheikh, H.R., Simoncelli, E.P., et al.: Image quality assessment: from error visibility to structural similarity. IEEE Trans. Image Process. **13**(4), 600–612 (2004)
26. Yang, W., Zhang, X., Tian, Y., Wang, W., Xue, J.H., Liao, Q.: Deep learning for single image super-resolution: a brief review. IEEE Trans. Multimed. **21**(12), 3106–3121 (2019)

Plane Space Representation in Context of Mode-Based Symmetry Plane Detection

Lukáš Hruda[1]($\boxtimes$), Ivana Kolingerová[1], and Miroslav Lávička[2]

[1] Faculty of Applied Sciences, Department of Computer Science and Engineering,
University of West Bohemia, Pilsen, Czech Republic
hrudalu@kiv.zcu.cz
[2] Faculty of Applied Sciences, Department of Mathematics and NTIS,
University of West Bohemia, Pilsen, Czech Republic

Abstract. This paper describes various representations of the space of planes. The main focus is on the plane space representation in the symmetry plane detection in E^3 where many candidate planes for many pairs of points of the given object are created and then the most often candidate is found as a mode in the candidate space, so-called Mode-based approach. The result depends on the representation used in the mode-seeking process. The most important aspect is how well distances in the space correspond to similarities of the actual planes with respect to the input object. So, we describe various usable distance functions and compare them both theoretically and practically. The results suggest that, when using the Mode-based approach, representing planes by reflection transformations is the best way but other simpler representations are applicable as well. On the other hand, representations using 3D dual spaces are not very appropriate. Furthermore, we introduce a novel way of representing the reflection transformations using dual quaternions.

Keywords: Space of planes · Symmetry · Distance function · Mode

1 Introduction and Related Work

Symmetry detection in 3D objects is a very large and progressive field with many possible applications, mainly in computer graphics or computer vision, such as object alignment, compression or reconstruction of incomplete objects. One very popular approach to symmetry detection is to create a number of candidate transformations by matching different points or parts of the input object and then finding those transformations that occur most often in the transformation space (see e.g. [13]). This can also be described as seeking modes in the transformation space, so the approach is called Mode-based symmetry detection. It can be applied to detect symmetries of various types, however, in this paper we only focus on the detection of the planes of symmetry (reflectional symmetries) of 3D objects.

© Springer Nature Switzerland AG 2020
V. V. Krzhizhanovskaya et al. (Eds.): ICCS 2020, LNCS 12141, pp. 509–523, 2020.
https://doi.org/10.1007/978-3-030-50426-7_38

Mitra et al. [13] and Shi et al. [17] used the Mode-based approach to find quite general partial (or local) symmetries – the transformations can contain rotation with or without reflection, translation and uniform scaling or any subgroup of these transformations including pure reflections (symmetry planes). To find the modes these methods employ Mean shift clustering algorithm [3]. Li et al. [11] used the same approach to detect symmetry planes of damaged skulls. Other uses for symmetry plane detection are in [9,12]. Similar approach has recently been used by Hruda et al. in rigid surface registration [8] which can be understood as symmetry detection between two objects. The candidate space contained rigid transformations and a mode was found by a density peak estimation algorithm. This could also be utilized to find the global plane of symmetry. A related approach was used by Podolak et al. [16] and Caillière et al. [2] where Hough transform and voting are employed to find the symmetry planes. The space of planes was divided into non-uniform discrete bins to count plane occurrences. Regardless of the specific algorithm, any Mode-based method for symmetry plane detection requires to define some representation of the space of planes, the definition influences the result. The important aspect is to have distances between points in the space well corresponding to the actual similarity/dissimilarity of the planes in E^3. The mode(s) can be found in an arbitrary non-Euclidean space only using distances between the points, their coordinates are not needed [8,18]. Proximity queries in non-Euclidean spaces can be accelerated using the Vantage Point Tree data structure [19] as done in [8]. In context of symmetry detection, planes can be understood as transformations reflecting points over the given plane. The problem of computing distances between rigid transformations has been analyzed in [8], however, the same problem does not seem to be sufficiently addressed in the literature for reflection transformations or planes, in spite of the popularity of the Mode-based approach.

This paper describes and analyzes several different representations of the space of planes. For each representation, reasonable distance functions, suitable in Mode-based symmetry plane detection, are discussed. The distance functions are compared to the distance function closest to the ground truth but useless in practice due to its large computation cost. The information about the described representations can be useful in other applications where a plane representation is needed, although the presented distance functions might require some application-based adjustment. We thus believe that researchers from various fields, not restricted to symmetry plane detection, could benefit from this paper.

2   Background

A general plane P can be defined by its implicit equation as $P : ax + by + cz + d = 0$ where $\mathbf{n} = [a, b, c]^T$ is the normal vector of the plane. We always consider the coefficients to be normalized such that $\|\mathbf{n}\| = 1$ in which case d represents the signed distance of the plane from the origin. A function $\mathbf{r}_P(\mathbf{x}) \in E^3$ that reflects an arbitrary point $\mathbf{x} \in E^3$ over the plane P can be defined as shown in Eq. (1).

$$\mathbf{r}_P(\mathbf{x}) = \mathbf{x} - 2(\mathbf{n}^T\mathbf{x} + d)\mathbf{n} \tag{1}$$

2.1   Candidate Creation Algorithm

Let us have a set of input points representing the object for which the set of candidate symmetry planes is to be found: $X = \{\mathbf{x}_1, \mathbf{x}_2, \ldots, \mathbf{x}_N\}$, $\mathbf{x}_i \in E^3$, $i = 1, 2, \ldots, N$. We use the point set representation due to its generality. We first create a 3D uniform grid with cell size $\frac{l_{avrg}}{\delta} \times \frac{l_{avrg}}{\delta} \times \frac{l_{avrg}}{\delta}$ where l_{avrg} is the estimated size of the object computed as the average distance of the points of X from their centroid. We mark each cell as either occupied if any point from X falls into it, otherwise unoccupied. Then we start randomly selecting pairs of points $\mathbf{x}_i, \mathbf{x}_j$ from X and construct a plane P such that $\mathbf{r}_P(\mathbf{x}_i) = \mathbf{x}_j$. To avoid clutter in the candidate space we perform a quick check to determine whether P is a plausible candidate by randomly selecting another five points from X, reflecting them over P and checking whether all of them end up in an occupied cell of the previously created grid. If they do P is accepted as a candidate. If at least one of them reflects into an unoccupied cell then P is rejected. We keep iterating this process until we have k accepted candidates and if not stated otherwise we set $\delta = 5$, $k = 2000$. The key idea behind the Mode-based approach is that now there should be significant modes in the candidate space of planes corresponding to the strongest symmetries of the input point set X.

2.2   Dependence on Scale and Position

The a, b, c coefficients are bounded on finite interval $\langle -1; 1 \rangle$. The value of the d coefficient of any candidate plane depends on the position and overall scale of the input object because d represents the distance of the given candidate plane from the origin and if the size of the object changes, the span of the d coefficient will change as well. However, the a, b, c coefficients will stay the same.

The dependence on the position is less obvious. If we translate the input object (all points in X) by some arbitrary vector $\mathbf{t}$, then for an arbitrary candidate plane P, d will change by $\mathbf{t}^T\mathbf{n}$ against the original position. As the change of d depends also on the orientation of the given plane, the change of d is inconsistent throughout the candidate planes and this inconsistency grows with distance of the input object from the origin. Therefore, the position of the input object influences the span of d but does not influence the span of a, b, c.

Generally, the distance functions for planes are negatively influenced by significantly different span of d and a, b, c, therefore, we always translate the input object's centroid into the origin and, if necessary, we also normalize d by l_{avrg} to make the spans of d and of a, b, c similar. For those distance functions where the translation to origin is not necessary, this fact will be pointed out explicitly.

2.3   Ground Truth

As pointed out in [8], distance between transformations cannot be well defined without the context (the object on which the transformations are applied), which is consistent with Sect. 2.2. Therefore, the most meaningful distance function for planes is the one used for error evaluation of registration results in [8], only with reflection transformations instead of rigid ones. Given two arbitrary planes P_1 and P_2, the distance function measures the exact difference between the effects of two reflections defined by P_1 and P_2 on the input object. It will be considered the ground truth distance function, denoted $D_{GT}(P_1, P_2)$ and defined

$$D_{GT}(P_1, P_2) = \sum_{i=1}^{N} \|\mathbf{r}_{P_1}(\mathbf{x}_i) - \mathbf{r}_{P_2}(\mathbf{x}_i)\|, \quad \text{where} \quad \mathbf{x}_i \in X, i = 1, \dots, N. \quad (2)$$

The D_{GT} distance function is not affected by the position of the input object, so it does not require the translation to origin, and the object size only effects its overall scale. Unfortunately, the time complexity of computing D_{GT} is $O(N)$ where N is the point count of the input object, which makes it too computationally expensive and, therefore, virtually unusable in any Mode-based symmetry detection algorithm. However, we can use it to compare other distance functions.

3   Plane Space Representations

In this section we describe and visualize various representations of the space of planes in E^3 plus possible distance functions usable in an arbitrary Mode-based symmetry plane detection algorithm. We use the algorithm from Sect. 2.1 to create a set of candidate symmetry planes of the object shown in Fig. 1. The black line in the figure represents the correct symmetry plane, the object is rotated to have this plane perpendicular to the projection. We purposely selected a slightly asymmetrical object. Although the object is visualized as a triangle mesh, only its vertices are used to compute the candidates.

Fig. 1. Model object with its correct symmetry plane

3.1   Dual Representation in E^3

The implicit equation of a plane has four coefficients but there are only three degrees of freedom when defining a plane because the space of planes is a 3-dimensional manifold embedded in 4-dimensional space. We can thus use a dual representation of any plane as a point in E^3. We denote $\rho(P) \in E^3$ a dual

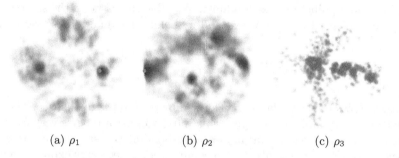

(a) ρ_1 (b) ρ_2 (c) ρ_3

Fig. 2. Dual representations of the candidate symmetry planes. The colors represent density (the darker, the larger density), the red spot shows the correct symmetry plane. (Color figure online)

representation of a plane P. Euclidean metric can then be used to compute the distance between two planes P_1, P_2 as $D_\rho(P_1, P_2) = \|\rho(P_1) - \rho(P_2)\|$.

One possible representation is to encode the plane orientation into a vector in E^3 with the same direction as the plane normal vector, and d into the length of this vector. Such dual representation can be defined as $\rho_1(P) = d\mathbf{n}$. Obviously, for $d \to 0$ there is ambiguity because such planes are shrunk into a single point. To solve this problem, the value of d is shifted by a constant μ, then these planes get spread on the surface of a sphere with radius μ instead of being all at the origin. We set $\mu = \frac{1}{2}l_{avrg}$ so that rotating the normal by π and changing d by l_{avrg} make approximately similar change in position of the point in the dual space. The dual representation is therefore finally defined as

$$\rho_1(P) = \begin{cases} (d + \frac{1}{2}l_{avrg})\mathbf{n} & d \geq 0 \\ (d - \frac{1}{2}l_{avrg})\mathbf{n} & d < 0 \end{cases}.$$

Distances in such dual space still do not very well correspond to similarities of the actual planes. Mainly, two planes with d close to 0 and similar normal vectors can be on the other sides of the sphere, and therefore more than 2μ apart, although they are actually very similar. However, such representation can be very good for visualization as each point in the dual space represents the plane quite intuitively. Figure 2a shows the generated candidates on the given model in the dual E^3 space transformed with ρ_1. The darker spots correspond to larger density of the points in the space, the red spot to the correct plane from Fig. 1. The viewpoint was selected manually to maximize the information in the image. It can be seen that the correct plane is in a noticeable mode (dense spot) but this mode is split on the sphere surface corresponding to $d = 0$ and its non-negligible part is on the other side. This is undesirable because it makes the mode much less significant than it would be if the two parts were together.

Another duality, also called polar duality (described e.g. in [6]), uses normalization of the plane coefficients such that $d = 1$, then the a, b, c coefficients are used as coordinates in E^3 i.e $\frac{1}{d}\mathbf{n}$. This again poses a problem for $d \to 0$ which makes the dual points approach infinity. We solve this issue in the same way as with ρ_1 by shifting the d coefficient and we define this dual representation as

$$\rho_2(P) = \begin{cases} \frac{1}{(d+\frac{1}{2}l_{avrg})}\mathbf{n} & d \geq 0 \\ \frac{1}{(d-\frac{1}{2}l_{avrg})}\mathbf{n} & d < 0 \end{cases}.$$

Figure 2b shows the candidates transformed by ρ_2 into the dual space. The right plane is in a noticeable mode, split again into two separate parts very far apart. There are also other significant modes corresponding to very different planes.

Another duality commonly used in computational geometry expresses a plane using its coefficients in explicit representation [1]. There are three possible explicit representations of a plane in E^3:

$$x = -\frac{b}{a}y - \frac{c}{a}z - \frac{d}{a}, \quad y = -\frac{a}{b}x - \frac{c}{b}z - \frac{d}{b}, \quad z = -\frac{a}{c}x - \frac{b}{c}y - \frac{d}{c}.$$

For demonstration, we select the first one, the dual representation is then defined as $\rho_3(P) = [\frac{b}{a}, \frac{c}{a}, \frac{d}{l_{avrg} \cdot a}]$. The division of d by l_{avrg} is necessary to normalize the span of d. Such duality obviously cannot represent planes parallel to the x-axis and planes with $a \rightarrow 0$ approach infinity in the dual space. We could solve this by shifting a but this time, we do not include l_{avrg} into the shift because the span of a does not depend on the size of the input object, so we get

$$\rho_3(P) = \begin{cases} [\frac{b}{a+\frac{1}{2}}, \frac{c}{a+\frac{1}{2}}, \frac{d}{l_{avrg}(a+\frac{1}{2})}] & a \geq 0 \\ [\frac{b}{a-\frac{1}{2}}, \frac{c}{a-\frac{1}{2}}, \frac{d}{l_{avrg}(a-\frac{1}{2})}] & a < 0 \end{cases}.$$

Figure 2c shows the candidates in the dual space transformed by ρ_3 and in this case there do not seem to be any significant modes.

In general, the dual representations appear not very appropriate for representing planes in Mode-based symmetry detection due to their singularities. This problem can be solved by shifting the value of some coefficient by a constant but the choice of this constant is rather arbitrary and even then the distances between points in the dual space might not well correspond to similarities of the planes. However, the dual representations can be useful to visualize the candidates as the dual points are 3-dimensional.

3.2 4D Vector Representation

Probably the most intuitive way of representing a plane is by a 4D vector of the plane coefficients. Given a plane P we represent it by a vector $\mathbf{p} = [a, b, c, \frac{d}{l_{avrg}}]^T$. In such a space we can easily define a distance function as the Euclidean distance of the two 4D vectors. However, $\mathbf{p}$ and $-\mathbf{p}$ represent the same plane so we need to take this into account. The Euclidean distance function is then defined as

$$D_{ED}(P_1, P_2) = \begin{cases} \|\mathbf{p}_1 - \mathbf{p}_2\| & \mathbf{p}_1^T \mathbf{p}_2 \geq 0 \\ \|\mathbf{p}_1 + \mathbf{p}_2\| & \mathbf{p}_1^T \mathbf{p}_2 < 0 \end{cases}.$$

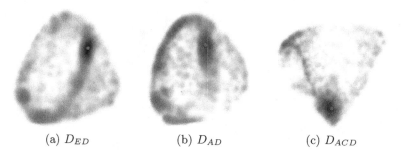

(a) D_{ED} (b) D_{AD} (c) D_{ACD}

Fig. 3. The candidates represented by 4D vectors projected into E^3 with MDS using different distance functions.

In this case the points cannot be visualized directly, so we use the multidimensional scaling (MDS) technique to transform the points into E^3 while maintaining their distances, w.r.t. the given distance function. However, the projection into E^3 might cause some imprecision in the visualization. Figure 3a shows the candidate planes projected into E^3 with MDS using the D_{ED} distance function and there is a very significant mode visible around the correct symmetry plane.

The distances in 4D vector space of planes can also be measured as angles between the vectors because the length of the vector $\mathbf{p}$ does not influence the plane P it represents. The angle distance function can be defined as

$$D_{AD}(P_1, P_2) = \arccos\left(\frac{|\mathbf{p}_1^T \mathbf{p}_2|}{\|\mathbf{p}_1\|\|\mathbf{p}_2\|}\right).$$

Figure 3b shows the candidates after using MDS with the D_{AD} distance function and the correct plane is again placed inside a noticeable mode.

We can also use only the cosine of the angle and measure its deviation from 1. The angle cosine distance function can be defined as

$$D_{ACD}(P_1, P_2) = 1 - \frac{|\mathbf{p}_1^T \mathbf{p}_2|}{\|\mathbf{p}_1\|\|\mathbf{p}_2\|}$$

and its visualization using MDS is shown in Fig. 3c. There is again a noticeable mode around the correct plane.

The 4D representation of the plane space is more appropriate for any Mode-based symmetry detection algorithm than the dual representations in E^3. However, they are not as convenient for visualization as the points first need to be projected into a lower dimensional space which causes a loss of information.

3.3   Transformation Representation

As already mentioned, the distance between arbitrary two planes P_1 and P_2 can be defined as the distance between the two reflection transformations $\mathbf{r}_{P_1}$ and $\mathbf{r}_{P_2}$ defined according to Eq. (1). One way of doing this is using the compound

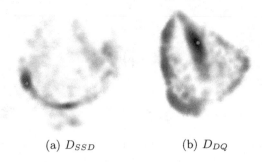

(a) D_{SSD} (b) D_{DQ}

Fig. 4. The candidates represented as transformations projected into E^3 with MDS using different distance functions.

metric evaluated as the most suitable for rigid transformations in [8]. It is based on sum of squared distances between the transformed points and is defined as

$$D_{SSD}(P_1, P_2) = \sqrt{\sum_{i=1}^{N} \|\mathbf{r}_{P_1}(\mathbf{x}_i) - \mathbf{r}_{P_2}(\mathbf{x}_i)\|^2}$$

where $\mathbf{x}_i \in X, i = 1, \ldots, N$. Certain similarity between D_{SSD} and the ground truth distance function D_{GT} (see Eq. (2)) can be noticed with two major differences. First, D_{SSD} uses squared distances, favouring smaller displacements over larger ones, which leads to different distances. Second, unlike D_{GT}, D_{SSD} can be computed in $O(1)$ with $O(N)$ preprocessing [8]. The transformations must be expressed as $\mathbf{Mx} + \mathbf{t}$ where $\mathbf{M}$ is an orthogonal transformation matrix, $\mathbf{t}$ is an arbitrary translation vector and $\mathbf{x}$ is the transformed point. From Eq. (1) we get

$$\mathbf{r}_P(\mathbf{x}) = \mathbf{x} - 2\mathbf{nn}^T\mathbf{x} - 2d\mathbf{n} = (\mathbf{I} - 2\mathbf{nn}^T)\mathbf{x} - 2d\mathbf{n}$$

where $\mathbf{I}$ is the identity matrix. If we denote $\mathbf{M} = (\mathbf{I} - 2\mathbf{nn}^T)$ and $\mathbf{t} = -2d\mathbf{n}$, then the reflection is $\mathbf{r}_P(\mathbf{x}) = \mathbf{Mx} + \mathbf{t}$. As $\mathbf{M}$ is orthogonal (and symmetric), we can use the same approach as in [8] to compute D_{SSD} in $O(1)$ with $O(N)$ preprocessing. The D_{SSD} distance function, as well as D_{GT}, is not affected by the position of the input object, so the translation to the origin is not required, and the object size only effects the overall scale of the distance function.

Figure 4a shows the candidates projected into E^3 using MDS with the D_{SSD} distance function and the correct plane is again in a significant mode. There is another smaller significant mode visible in the figure, however, this can very likely be caused by the distortion of the MDS projection.

Dual Quaternions. Dual quaternions combine the concepts of quaternions and dual numbers. Let us show how to use them to represent a reflection over an arbitrary plane. A general quaternion is defined as $Q = q_0 + q_1 i + q_2 j + q_3 k$ where the i, j, k units multiply according to the following rules

$$i^2 = j^2 = k^2 = ijk = -1, ij = k = -ji, jk = i = -kj, ki = j = -ik.$$

A conjugate Q^* of a quaternion Q is defined as $Q^* = q_0 - q_1 i - q_2 j - q_3 k$. We denote $v(\mathbf{x}) = xi + yj + zk$ a quaternion that represents an arbitrary point $\mathbf{x} = [x, y, z]^T \in E^3$. If $\mathbf{u}$ is an arbitrary unit vector and we set $Q = \cos \alpha + v(\mathbf{u}) \sin \alpha$, then $Qv(\mathbf{x})Q^*$ represents the point $\mathbf{x}$ rotated by angle 2α around the axis that passes through the origin and has the direction of $\mathbf{u}$. Similarly, if we set $Q = v(\mathbf{u})$ then $Qv(\mathbf{x})Q$ represents the point $\mathbf{x}$ reflected over the plane with normal $\mathbf{u}$ that passes through the origin. For details about quaternions we refer to [7].

A dual quaternion, for more detail see e.g. [15], is defined as

$$Q_d = Q + \epsilon Q_\epsilon = q_0 + q_1 i + q_2 j + q_3 k + \epsilon(q_{\epsilon 0} + q_{\epsilon 1} i + q_{\epsilon 2} j + q_{\epsilon 3} k)$$

where Q and Q_ϵ are quaternions and ϵ is the dual unit which commutes with the quaternion units i, j, k and it is that $\epsilon^2 = 0$. A quaternion conjugate of Q_d is defined as $Q_d^* = Q^* + \epsilon Q_\epsilon^*$, a dual conjugate of Q_d is defined as $\overline{Q_d} = Q - \epsilon Q_\epsilon$. These conjugations can be combined into $\overline{Q_d^*} = Q^* - \epsilon Q_\epsilon^*$.

We denote $v_d(\mathbf{x}) = 1 + \epsilon(xi + yj + zk)$ a dual quaternion representing an arbitrary point $\mathbf{x} = [x, y, z]^T \in E^3$. If Q is a quaternion that represents rotation and $Q_\epsilon = \frac{v(\mathbf{t})Q}{2}$ where $\mathbf{t} = [t_x, t_y, t_z]^T$ is an arbitrary translation vector then for $Q_d = Q + \epsilon Q_\epsilon$, using the rules of dual quaternion algebra, it can be shown that

$$Q_d v_d(\mathbf{x})\overline{Q_d^*} = 1 + \epsilon(Qv(\mathbf{x})Q^* + v(\mathbf{t}))$$

which represents the point $\mathbf{x}$ rotated via Q and then translated by $\mathbf{t}$. This shows how to use dual quaternions in connection with rigid transformations. Note that Q_d represents the same transformation as $-Q_d$ with the identity being represented by either 1 or -1. The transformations can be concatenated by multiplying the corresponding dual quaternions and if Q_d represents a rigid transformation, Q_d^* represents its inverse. Next, for Q_{d1}, Q_{d2} representing rigid transformations, these transformations are the same only if $Q_{d1}Q_{d2}^* = 1$ or $Q_{d1}Q_{d2}^* = -1$.

Consider now a plane P and a dual quaternion $Q_d = Q + \epsilon Q_\epsilon$ defined such that $Q = v(\mathbf{n})$ and $Q_\epsilon = \frac{v(\mathbf{t})Q}{2}$ where $\mathbf{t} = -2d\mathbf{n}$. Now Q_d represents a transformation that first rotates by π around the axis that passes through the origin and has the direction of $\mathbf{n}$, and then translates by $-2d\mathbf{n}$. However, if we apply the transformation on $-\mathbf{x}$ instead of $\mathbf{x}$, it can be shown that we will get

$$Q_d v_d(-\mathbf{x})\overline{Q_d^*} = 1 + \epsilon(Qv(\mathbf{x})Q + v(\mathbf{t})) = 1 + \epsilon(v(\mathbf{n})v(\mathbf{x})v(\mathbf{n}) - v(2d\mathbf{n})) = v_d(\mathbf{r}_P(\mathbf{x}))$$

which exactly represents $\mathbf{r}_P(\mathbf{x})$. This shows that a dual quaternion can also represent a reflection transformation by representing a rigid transformation that transforms $-\mathbf{x}$ to $\mathbf{r}_P(\mathbf{x})$. Therefore, to measure distances between reflection transformations we can use a distance function for dual quaternions.

We denote $vec(Q_d) = [q_0, q_1, q_2, q_3, q_{\epsilon 0}, q_{\epsilon 1}, q_{\epsilon 2}, q_{\epsilon 3}]^T \in E^8$ an 8-dimensional vector that is equivalent to Q_d. Given a plane P, we create the corresponding dual quaternion Q_d such that $Q = v(\mathbf{n})$ and $Q_\epsilon = \frac{v(-2d\mathbf{n})Q}{2l_{avrg}}$, i.e. $Q_d = v(\mathbf{n}) + \epsilon \frac{v(-2d\mathbf{n})v(\mathbf{n})}{2l_{avrg}}$. The division by l_{avrg} is to normalize the translation part. Using the algebra of dual quaternions we can actually get that $Q_\epsilon = \frac{d}{l_{avrg}}$, so Q_d can be finally expressed as shown in Eq. (3).

$$Q_d = v(\mathbf{n}) + \epsilon \frac{d}{l_{avrg}} = ai + bj + ck + \epsilon \frac{d}{l_{avrg}} \tag{3}$$

There are two common distance functions for dual quaternions. The first one uses differences between the equivalent 8-dimensional vectors [15]. Suppose two arbitrary planes P_1 and P_2 represented by dual quaternions Q_{d1} and Q_{d2} respectively. Such distance function can be defined as

$$\min\{\|vec(Q_{d1}) - vec(Q_{d2})\|, \|vec(Q_{d1}) + vec(Q_{d2})\|\}$$

but given Eq. (3) this is exactly the same as D_{ED}. The second distance function [5] uses a difference transformation $Q_{d1}Q_{d2}^*$ and computes its distance from the identity, i.e. from 1 or -1. It is defined as

$$D_{DQ}(P_1, P_2) = \min\{\|vec(1 - Q_{d1}Q_{d2}^*)\|, \|vec(1 + Q_{d1}Q_{d2}^*)\|\}.$$

Figure 4b shows the candidates projected into E^3 using MDS with D_{DQ} and the correct plane is in an obvious mode.

4   Results

We compared the distance functions by generating the candidate symmetry planes of a given object (using the model algorithm from Sect. 2.1), comparing values of the given distance function and of the ground truth one. We did this for the six test objects from Fig. 5, taken from datasets [4,10]. The objects are represented by triangle meshes for easier visualization, but we again only used their vertices as the input points for the candidate creation process.

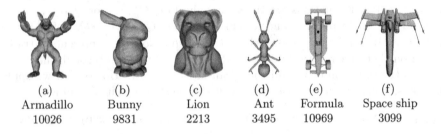

(a)	(b)	(c)	(d)	(e)	(f)
Armadillo	Bunny	Lion	Ant	Formula	Space ship
10026	9831	2213	3495	10969	3099

Fig. 5. The test objects used to generate the candidate sets for comparing the distance functions. The number under the name of each object expresses its point count.

Let $C = \{P_1, P_2, ..., P_k\}, k = 2000$ be the set of candidate planes created for a given input object. The error of a given distance function D against the ground truth is defined as

$$Err(D) = \frac{1}{Count(k)} \sum_{i=1}^{k} \sum_{j=i+1}^{k} \left| \frac{D_{GT}(P_i, P_j)}{Avrg(D_{GT})} - \frac{D(P_i, P_j)}{Avrg(D)} \right|$$

where

$$Avrg(D) = \frac{1}{Count(k)} \sum_{i=1}^{k} \sum_{j=i+1}^{k} D(P_i, P_j)$$

is the average distance between candidates in C and $Count(k) = \frac{1}{2}(k^2 - k)$ is the total number of candidate pairs used for the computation. The normalization by $Avrg$ is used because the overall scales of the distance functions do not matter so the differences are computed after both D_{GT} and D are divided by their mean values. Table 1 shows the errors of all the distance functions described above for all the test objects. We include the dual representations in the comparison. The smallest error is obviously achieved using D_{SSD} probably due to the same principle of D_{SSD} and D_{GT}. It is still rather surprising that the D_{SSD} function which uses squared distances is so similar to D_{GT} that uses absolute distances. The D_{ED}, D_{AD} and D_{DQ} all have very similar errors (D_{DQ} usually has the lowest error) which are overall lower than those of D_{ACD} and the distances in the dual spaces, but in case of D_{ACD} this can be explained by its resemblance to the cosine function ($D_{ACD} = 1 - \cos(D_{AD})$). The function D_{ρ_3} exhibits similar or lower error than D_{ACD} on some objects (Arm, Ant) but also considerably larger error on different ones (For, Shi) which suggests that D_{ρ_3} is quite unpredictable.

Table 1. Errors of the distance functions for the candidate sets for different objects.

	Arm	Bun	Ant	For	Lio	Shi	Average
D_{ED}	0.120	0.277	0.163	0.093	0.130	0.234	0.169
D_{AD}	0.133	0.281	0.157	0.098	0.144	0.236	0.174
D_{ACD}	0.299	0.388	0.264	0.250	0.306	0.352	0.309
D_{SSD}	0.012	0.023	0.009	0.014	0.011	0.012	0.013
D_{DQ}	0.118	0.277	0.162	0.093	0.129	0.232	0.168
D_{ρ_1}	0.382	0.399	0.503	0.596	0.326	0.425	0.438
D_{ρ_2}	0.401	0.408	0.488	0.563	0.360	0.489	0.451
D_{ρ_3}	0.280	0.446	0.269	0.730	0.362	0.447	0.422

The graphs in Fig. 6 show the relation between D_{GT} and the other distance functions. We generated 50 candidates on the Armadillo and for each pair of the candidates we put its distance computed by D_{GT} on the horizontal axis and the distance computed by a given different distance function on the vertical axis. We normalize each value by $Avrg$. If some distance function D was the same as D_{GT} (apart from overall scale) there would be a perfect linear dependency and the points would lie on a line in the graph. Figure 6a shows the relations of D_{SSD}, D_{ED}, D_{ACD} to D_{GT}. The similarity of D_{AD}, D_{DQ}, D_{ED} is shown in Fig. 6b, Fig. 6c shows the dual representations. For different objects the graphs are slightly different but very similar. There is an obvious almost linear dependency between D_{GT} and D_{SSD} (see Fig. 6a), however, D_{ED}, D_{AD}, D_{DQ} exhibit

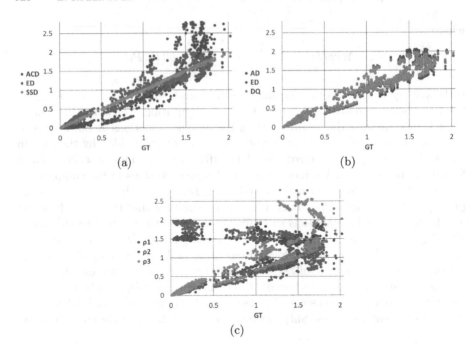

Fig. 6. Relations between a) $D_{SSD}/D_{ED}/D_{ACD}$ and D_{GT}; b) $D_{DQ}/D_{ED}/D_{AD}$ and D_{GT}; c) $D_{\rho_1}/D_{\rho_2}/D_{\rho_3}$ and D_{GT} for the Armadillo object.

relation to D_{GT} that is also nearly linear (see Fig. 6b). For D_{ACD} the resemblance to cosine is visible. The dual representations show rather unstable behavior (see Fig. 6c). This is mostly caused by the shift in some of their coordinates but without it the dual representations would suffer from singularities which would make them unusable. Different shifting constant could lead to better results, at least for some objects, but there is no general way how to choose it.

Table 2 shows the Pearson correlations [14] between all pairs of the distance functions for the data from Fig. 6. Value of 1 indicates perfect linear dependency and the closer to 0 the weaker the dependency. Expectedly, D_{SSD} shows the best linear correlation with D_{GT}. The correlations of D_{ED}, D_{AD}, D_{DQ} and even D_{ACD} with D_{GT} are all rather high. On the other hand, the distances in the dual spaces have mostly low correlation with D_{GT}. The high correlations among D_{ED}, D_{AD} and D_{DQ} confirm the similarity of these three distance functions.

Based on the results, the most appropriate representation of the space of planes in any Mode-based symmetry detection method is the transformation representation with the D_{SSD} distance function. But, except for the dual representations, all the distance functions are similar and none of them deviates significantly from D_{GT} making all of them well applicable. However, all the distance functions except D_{SSD} require translating the input object to the origin, otherwise the normalization of the d coefficient would have to be done differently.

Table 2. Pearson correlations of the distance functions for the Armadillo object.

	GT	ED	AD	ACD	SSD	DQ	ρ_1	ρ_2	ρ_3
GT	1.0000	0.9723	0.9644	0.9120	0.9998	0.9738	0.5361	0.3265	0.7238
ED	0.9723	1.0000	0.9989	0.9679	0.9713	0.9998	0.5537	0.3460	0.7621
AD	0.9644	0.9989	1.0000	0.9767	0.9635	0.9983	0.5499	0.3474	0.7548
ACD	0.9120	0.9679	0.9767	1.0000	0.9105	0.9664	0.5196	0.3111	0.7262
SSD	0.9998	0.9713	0.9635	0.9105	1.0000	0.9728	0.5351	0.3286	0.7214
DQ	0.9738	0.9998	0.9983	0.9664	0.9728	1.0000	0.5516	0.3423	0.7637
ρ_1	0.5361	0.5537	0.5499	0.5196	0.5351	0.5516	1.0000	0.9248	0.4217
ρ_2	0.3265	0.3460	0.3474	0.3111	0.3286	0.3423	0.9248	1.0000	0.1934
ρ_3	0.7238	0.7621	0.7548	0.7262	0.7214	0.7637	0.4217	0.1934	1.0000

4.1 Theoretical Comparison

There are some theoretical differences between the various representations. The dual and the 4D vector representations are basically Euclidean and the candidates can easily be stored in a data structure such as a KD-tree or a grid. In case of the D_{AD} and D_{ACD} distance functions some structure can be built using the polar coordinates in 4D. Also, there are quite many possible algorithms for mode-seeking in Euclidean data. The transformation representations and the D_{SSD} and D_{DQ} distance functions are non-Euclidean, with smaller choice of data structures and possible mode seeking algorithms [8,18,19]. Also, the implementation of the D_{SSD} and D_{DQ} is more complex since they require implementing the matrix and the dual quaternion algebras.

Although the D_{DQ} distance function does not bring any considerable improvement over simpler distance functions, the idea of representing reflections by dual quaternions seems novel and can possibly find its use in other applications or new symmetry detection methods created in the future.

5 Conclusion

We described several representations of the space of planes suitable for any Mode-based algorithm for symmetry plane detection and computation of the distances in these representations. We showed that the 3D dual space representations are not very appropriate for this purpose but usable for visualizations. In order to represent the space of planes suitably in context of the Mode-based symmetry detection, spaces of higher dimensions must be used and the most appropriate representation is even non-Euclidean. However, the results suggest that apart from the 3D dual spaces all the plane space representations are well applicable in this context, although there are some theoretical differences between them.

Acknowledgement. This work was supported by Ministry of Education, Youth and Sports of the Czech Republic, project PUNTIS (LO1506) under the program NPU I and

University specific research project SGS-2019-016 Synthesis and Analysis of Geometric and Computing Models.

References

1. de Berg, M., Cheong, O., van Kreveld, M., Overmars, M.: Computational Geometry: Algorithms and Applications, 3rd edn. Springer, Heidelberg (2008). https://doi.org/10.1007/978-3-540-77974-2
2. Caillière, D., Denis, F., Pele, D., Baskurt, A.: 3D mirror symmetry detection using Hough transform. In: 2008 15th IEEE International Conference on Image Processing. ICIP 2008, pp. 1772–1775 (2008)
3. Comaniciu, D., Meer, P.: Mean shift: a robust approach toward feature space analysis. IEEE Trans. Pattern Anal. Mach. Intell. **24**(5), 603–619 (2002)
4. Fang, R., Godil, A., Li, X., Wagan, A.: A new shape benchmark for 3D object retrieval. In: Bebis, G., et al. (eds.) ISVC 2008. LNCS, vol. 5358, pp. 381–392. Springer, Heidelberg (2008). https://doi.org/10.1007/978-3-540-89639-5_37
5. Figueredo, L., Adorno, B.V., Ishihara, J.Y., Borges, G.A.: Robust kinematic control of manipulator robots using dual quaternion representation. In: 2013 IEEE International Conference on Robotics and Automation, pp. 1949–1955 (2013)
6. Gallier, J.: Notes on convex sets, polytopes, polyhedra, combinatorial topology, voronoi diagrams and delaunay triangulations. arXiv:0805.0292 (2008)
7. Goldman, R.: Rethinking Quaternions: Theory and Computation. Morgan and Claypool Publishers, San Rafael (2010)
8. Hruda, L., Dvořák, J., Váša, L.: On evaluating consensus in RANSAC surface registration. In: Computer Graphics Forum, vol. 38, pp. 175–186 (2019)
9. Kroemer, O., Amor, H.B., Ewerton, M., Peters, J.: Point cloud completion using extrusions. In: 2012 12th IEEE-RAS International Conference on Humanoid Robots (Humanoids 2012), pp. 680–685 (2012)
10. Levoy, M., Gerth, J., Curless, B., Pull, K.: The stanford 3D scanning repository (2005). http://www.graphics.stanford.edu/data/3Dscanrep/
11. Li, X., Yin, Z., Wei, L., Wan, S., Yu, W., Li, M.: Symmetry and template guided completion of damaged skulls. Comput. Graph. **35**(4), 885–893 (2011)
12. McCrae, J., Singh, K., Mitra, N.J.: Slices: a shape-proxy based on planar sections. ACM Trans. Graph. **30**(6), 168 (2011)
13. Mitra, N.J., Guibas, L.J., Pauly, M.: Partial and approximate symmetry detection for 3D geometry. In: ACM Transactions on Graphics (TOG), vol. 25, pp. 560–568. ACM (2006)
14. Mudelsee, M.: Estimating pearson's correlation coefficient with bootstrap confidence interval from serially dependent time series. Math. Geol. **35**(6), 651–665 (2003). https://doi.org/10.1023/B:MATG.0000002982.52104.02
15. Pham, H.L., Perdereau, V., Adorno, B.V., Fraisse, P.: Position and orientation control of robot manipulators using dual quaternion feedback. In: 2010 IEEE/RSJ International Conference on Intelligent Robots and Systems, pp. 658–663 (2010)
16. Podolak, J., Shilane, P., Golovinskiy, A., Rusinkiewicz, S., Funkhouser, T.: A planar-reflective symmetry transform for 3D shapes. ACM Trans. Graph. (TOG) **25**(3), 549–559 (2006)
17. Shi, Z., Alliez, P., Desbrun, M., Bao, H., Huang, J.: Symmetry and orbit detection via lie-algebra voting. In: Computer Graphics Forum, vol. 35, pp. 217–227 (2016)

18. Vedaldi, A., Soatto, S.: Quick shift and kernel methods for mode seeking. In: Forsyth, D., Torr, P., Zisserman, A. (eds.) ECCV 2008. LNCS, vol. 5305, pp. 705–718. Springer, Heidelberg (2008). https://doi.org/10.1007/978-3-540-88693-8_52
19. Yianilos, P.N.: Data structures and algorithms for nearest neighbor search in general metric spaces. In: Soda 1993, pp. 311–21 (1993)

Impression Curve as a New Tool in the Study of Visual Diversity of Computer Game Levels for Individual Phases of the Design Process

Jarosław Andrzejczak[(✉)] [iD], Marta Osowicz, and Rafał Szrajber [iD]

Institute of Information Technology, Lodz University of Technology,
215 Wólczańska Street, 90-924 Lodz, Poland
{jaroslaw.andrzejczak,rafal.szrajber}@p.lodz.pl
http://it.p.lodz.pl

Abstract. Impression curve is a widely used method in urban and landscape design to assess visual diversity of the space. In these studies, the method is applied for game level design. The goal of conducted research was the analysis of space perception in successive design phases related to the process of game environment formation. Next steps of the design process define the space burdened with more and more information. It aims to evaluate if initial assumptions, made by a designer at the beginning of the designing process, are maintained with the increase in the number of details and the content of locations. These studies are also a background for research in automation of visual diversity assessment. This, in turn, is related to making a player focused and interested during a gameplay, by the means of space defining an action scene. By applying a method from domain of urban planning and architecture in human-computer interaction (HCI) studies related to virtual space, we show that both - defining the surroundings and its impact on recipient - are subject to the same rules in either case.

Keywords: Virtual environment · Level design · Game design · Virtual architecture · Impression curve · Experience design

1 Introduction

Game design became a pretty complex process, where experts from many domains are engaged in order to produce a user-oriented product. Games, being under continuous development, represent a very broad area of research. In classic media as movies, space is thought to be a background of events defining only the location and the time of action. In interactive environments, in turn, it is an interface dedicated to navigational and narrative purposes, which highly influences an observer because of its spatial composition. This kind of interaction determines if a system is perceived as an attractive one, or not. Experience of

© Springer Nature Switzerland AG 2020
V. V. Krzhizhanovskaya et al. (Eds.): ICCS 2020, LNCS 12141, pp. 524–537, 2020.
https://doi.org/10.1007/978-3-030-50426-7_39

space, triggered by movement, requires spaces to be formed in such a way the impression of monotony and repeatability are diminished so that a system seems attractive to an observer.

Creation of diversified attractive, and engaging virtual environments led to strategies applied in other areas of studies, like architecture or urban design. Furthermore, digital characteristic of both description and exploration manners made these strategies enable to study and verify the theses stated in this paper for virtual environments.

The contributions to virtual worlds (Virtual Reality as well as video games levels) design and evaluation presented in this article are:

- The impression curve idea adaptation for virtual world creation (especially for video games level design process).
- Usability tests with 112 participants confirming the impact of usage of the impression curve in virtual worlds creation and evaluation.
- Identification of parameters affecting impression in virtual world.

We start with a impression curve definition and history. Next the impression curve adaptation for virtual world creation process idea detailed description in given. This is followed by a evaluation methodology and its stages construction as well as the some details of research environment design. Next, tests results and their discussion are presented. Finally, ideas for further development and final conclusions will be given.

2   Impression Curve

The strategy of impression curve was firstly elaborated on by Wejcherts in [8]. The method relies on relation among space, time, and velocity of impression forming. The author defined space as an interior or an interior layout being a component part of the structure of space. He emphasized that time and space, in successive interior layouts (e.g. street sequence), are inseparable. An observer who moves (motion enables three-dimensional perception) perceives spatial images, which are bonded to the shape of space where they are and which change over time. For impression curve no measure can be established, since it is a way how particular elements of space influence an observer.

Impression curve is depicted as a chart, where the horizontal axis is a time scale and the vertical one describes how a particular element of a scene influences an observer on the scale from 1 up to 10. The maximal value of 10 is a conventional value being the result of the commonly used decimal system. Each value (from 1 to 10) was described by the Wejchert: value 1 describes monotonous system devoid of architectural merits. On the other hand, value 10 describes/presents a system of meaningful strong points, dominating as an element of city's structure. The studies showed that there is a clear group which react in the similar manner despite the subjective assessment of observers. Thanks to this, preparing an average chart on the basis of individual charts is possible. Based on the average chart, impression curve-based appraisal of the spatial system may be done and some basic conclusions about space perception may be drawn.

Wejchert suggests that street sequence should provide impression changes every 2 or 3 min to avoid the feeling of monotony. If so, it is clear that the level of diversity is dependent on the velocity of an observer and how much attention may he or she pay to the particular fragment of a route. Having analysed charts of impression curve for different streets, one may notice the decrease of value for long-lasting motion along repetitive fragments, even though they seemed interesting at the beginning. As a corollary, space changes are said to be as essential as visual quality of a scene.

Impression curve may be implemented both for new spaces design and for assessment of already existing areas. Thanks to it, valuable elements and elements requiring improvements might be easily identified [12]. Impression curve was widely used as a method for studying space diversity: identification of the most precious places at the Piotrkowska street, evaluation of landscape of countryside Panieńszczyzna in vicinity of Lublin [10] or organizing agrarian and landscape structures in Pojałowice countryside [18] are just a few examples to the point [1].

When analyzing the state of knowledge, it is worth referring to the search for various methods of exploring the reception of space in games and the player's experience. Research shown in [9,11,16,19] affected the choice of our method and its application domain. Additionally [17] authors employs surveys to appraise emotional state of an entity. The key element is that those questions are introduced to the world of game so a survey may be carried out while playing. The goal of our search was first to create a system embedded in a virtual environment.

Other domains, like urban planning, interior or landscape design may also be helpful to figure comprehend how to build a functional and visually attractive game level. Dan Cox proves in his presentations that virtual spaces design and interior design have a lot in common. He also presents well known techniques of interior design, which are supportive in game level design [2,3,5].

To sum up, impression curve may be thought of as a method, based on subjective grade, of valorisation of a landscape in either local or regional scale. A chart of impression curve depicts an average observers' judgement and a chart of deviation from the average reaction tends to form Gaussian distribution [7]. Although this method was introduced for architecture domain and it is widely applied therein, it may be considered in game level design as well. The first application of impression curve in design and analysis of virtual environments was proposed by Rafał Szrajber during [13] and [14]. Those speeches initiated the studies presented in this paper.

3   Method

The idea of the test method is to use an impression curve for the analysis of the visual diversity and attractiveness of game level. It assesses subjective attraction of a given space. An entity, being a subject of tests, moves along prepared route (in virtual space- controlling an avatar) and in predefined, equally spaced location, he or she evaluates how appealing the given fragment of environment is. The entity assigns grades from 1 to 10, where 1 means a boring scene, and 10- an exciting scene of high aesthetic merits. Impression curve seems to enable virtual spaces assessment even though it comes from architectural domain. Thanks to it, interesting elements of surroundings may be identified. An average chart of impression curve may be used also to localize and enhance poorly graded level fragments.

Sinusoidal shape is said to be the proper distribution of the computer game level impression curve. In uniformed long-lasting environment, the values of impression curves fall because a brain receives still the same set of stimuli in similar intensity [8]. In order to keep the constant or increasing trend, impressions should grow continuously. In case of game development industry it often wreaks much more resources to produce better set of models and more interesting environment. This may be difficult to achieve. If so, impression level may periodically grow and fall to reach successive increases without any extra expenses with respect to the previous ones. In addition, impression level should be always kept above some predefined threshold. Otherwise, it means that some parts of the game level observed should be definitively improved.

The proposed method is constrained to assessing quality of a single route. Without further extensions, it may appraise only linear levels, where a player cannot select an alternative direction. For more complex levels it should be assured that all players follow the predefined testing routes.

The method of impression curve was introduced to test different versions of computer game levels during successive phases of virtual space design. It aimed to answer if impression curve changes in successive design stages and by which factors it is affected. The goal was also to indicate from which stage of design process it is possible to evaluate virtual space diversity.

There is a variety of methods in the literature to evaluate quality of game levels. A few of them rely on questionnaires whereas others suggest monitoring biological changes in a player's organism. E.g. in [6,15], the authors monitored biological changes in a player's organism in order to adjust the level of difficulty during gameplay. Based on the results of EEG, the system may adjust requirements of a game to maintain constant level of player's focus. Thanks to this solution, players do not feel neither bored nor frustrated.

4 Experiment

4.1 Research Environment

Research environment is constituted by a set of trail applications done with the Unreal Engine 4, each of which contains different version of the prepared game level. The environment was designed in such a way that no extra explanations or help were required. Each application consists of the following elements: start screen (defining the goal a player has to achieve and informing about a test);

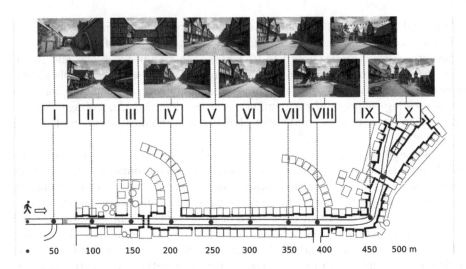

Fig. 1. Figure shows successive points where a player assess a level, as well as the visual representation of each place (for final stage of design process - main models with final materials with extra fine detailed models (called final level version).

actual game level; questionnaire to be filled with impression curve (appears every time a player has to grade a scene); and end screen.

The game level used in experiment depicts a fragment of a main road of a medieval town. The level is linear and a player, in first-person perspective, is allowed to move only along predefined path (Fig. 1). The graphics of the final level version is realistic and no stylised elements were added so as to make the virtual environment as close to the real space as possible. Thanks to this, disturbing variables influencing the level perception could be diminished. Velocity of the movement is constant (walk) and does not change in individual variants. Every 50 meters, a player stops and grades the surroundings. There are 10 such places in the entire level, hence impression curves have 10 points on time axis.

Following level variants showing successive design stages were arranged:

- Simple blockout – 3D game level sketch. Only essential elements were places on a scene, mainly buildings. Each element was constructed from one or more cuboid (A)
- Advanced blockout - detailed 3D sketch. Main level elements are presented in forms of simplified blocks (B)
- Main models without materials (C)
- Main models with monochromatic materials (D)
- Main models with final materials (E)
- Main models with final materials with extra fine detailed models (called final level version) (F)

All variants represents successive stages of level design (Fig. 2). There is a clear dependency that each next stage introduces another details to the presented

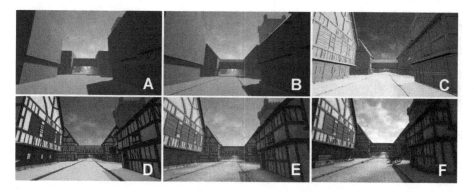

Fig. 2. The stages of computer game level design. A - simple blockout; B - advanced blockout; C - main models without materials; D- main models with monochromatic material; E - main models with final materials; F - main models with final materials with extra fine detailed models.

location. This process resembles the way a human perceives visually according to the theory of vision of David Marr [4].

4.2   Variants of the Level with Changes of Selected Environment Elements

In order to assess influence of other factors on the final shape of impression curve, additional trials were conducted. Following potentially influential factors, split into two groups:

1. Factors whose impact is uniform across the level - Lightening condition changes (L), Weather changes (W) (Fig. 3)
2. Factors which impact on small fragments of the level - Geometrical changes (G), Material changes (M), Extra models and objects in the environment (O), Adding expression (X) (Fig. 4).

   In order to conduct these extra tests, another variants of the level were created. Each of them let a researcher assert if a particular factor impacts on the shape of impression curve or not. Introduced changes are natural yet conspicuous. Thanks to this, it was possible to screen out a particular factor and see if the trials show it does not change the shape of impression curve. In case of level variants with lightening condition and weather changes, the general appearance of the environment was changed. The subject of those modification was the final level version.

   The level variant considering changes in lightening conditions presents a medieval town by night brightened with torches. The lightening varies from uniform to the spotlight, which causes changes in distributions of fair and dark areas and increase of contrast.

Fig. 3. Figure shows Screenshots of selected fragments of modified levels with global changes - factors whose impact is uniform across the level - Lightening condition changes (L), Weather changes (W).

The level version with weather changes depicts a town in the rain. Following extra elements were added: pools, cloudy sky, rain (done with particles), and rain drops on the screen (postprocess) (Fig. 3).

In those versions of the level which contain spot factors, the elements requiring modification were selected in advance. In each variant changes concern fifth and sixth point on the impression curve. They lie on the relatively monotonous fragment of curve where possible influence may be explicitly visible. Changes in geometry were realized as modifications of advanced blockout whereas the others – of the final version of the level.

In the case of changes in geometry, extra tall distinctive elements were added, like guard towers. Also, two houses facing each other were connected with a passage above the street.

In versions with material changes, some of the houses were modified so as to make them distinctive among other buildings and to increase the contrast. Those houses look like painted in flashy and contrasted colours.

In variant with extra models, some tiny items were placed, like a set of boxes, barrels, and a wagon, in front of a house. On the other hand, a variant with added expression was enriched by elements which could have some impact on the observer's emotions so that the level appears to be inhabited and some particular events took place there. As emotion fear was opted, since it is easy to trigger and it affects significantly. The first of the modified impression curve points contains a ruined building. Another one, blood stains leading an observer to the broken scratched doors so they appear to be attacked by an animal or a monster.

4.3   Test Group

Trials were conducted on 112 people. All of them were accustomed with computer games and control system (first-person perspective). Most of the people tested were students of computer science. Subjects are from 19 to 35 years old. It was assumed that one person cannot test more than one version of the level. Each variant of the level was examined by 7 to 15 people.

Fig. 4. Figure shows Screenshots of selected fragments of modified levels with local changes - (G) geometry, (M) material, (O) objects, (X) expression

4.4 Experiment and Results Storing

Each trial began with the start screen containing the description of the studies. Afterwards, participants played the levels fulfilling grading questionnaires until end screen was reached. To make trials objective and to make trials reliable, following constraints were put on experiments: no one (including a researcher) should look at the screen during a trial and a participant should not be able to see values provided by the other players.

Having done all trials, CSV file was generated. It contains the following information: an unique name of a variant, Values of impression curve for a player, time of the level accomplishment.

4.5 Results

For each variant of the level, the average impression curve was estimated. The values of impression curves as well as associated standard deviations are presented in Table 1. (variants with successive stages of level design) and Table 3 (variants with level modifications).

The results suggest that despite the similarity of individual charts, values used by players differ. Some of them used virtually all available values from 1 to 10 whereas the others did not distinguish so much. This is the reason for relatively high values of standard deviations for individual average impression curves.

5 Discussion

Prior to comparing the impression curves charts, let us focus on grades given by players. Although individual shapes seem to be similar, scale used by the players differ. Some of them made use of the entire scale, unlike the others who were satisfied with values from 1 to 5 or from 5 to 10. In order to evaluate which elements of the game level are visually appealing, local behaviour as well as local and global extrema are taken as indicators rather than particular values of impression curve.

The crucial factor which ought to be considered while assessing successive level variants is the shape of impression curve. There are several strategies to appraise similarity of the two impression curves:

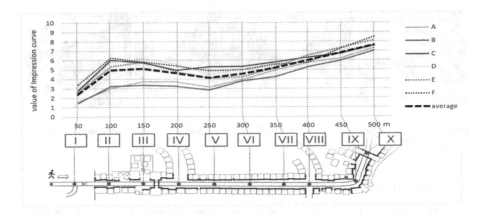

Fig. 5. The average values of impression curve for variants of levels concerning successive stages of virtual space design and the average value for all stages.

- Global minima and maxima lie in the same points or their close neighbourhood on both charts
- Local minima and maxima lie in the same points or their close neighbourhood on the two charts
- Both charts feature increasing trend in the same ranges
- Both charts feature decreasing trend in the same ranges.

5.1   Comparison of Impression Curve Charts for Next Stages of Creating a Computer Game Level

In Fig. 6 visual similarity of average impression curves may be seen for variants representing successive stages of level design (exact numbers with standard deviations are presented on Table 1). They are said to be similar due to the following reasons:

- Local minimum of each chart is placed in the first point of impression curve (narrow passage surrounded with wall)
- Local maximum of all charts are in second and third point (crossing a marketplace)
- Local minimum of charts are in fourth and fifth point of impression curve (beginning with the street surrounded with small houses)
- From fifth and sixth point on the impression curve, values increase up to the global maximum in the tenth (or eleventh in case of variant with monochromatic materials) point of impression curve.

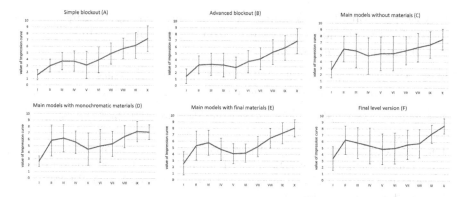

Fig. 6. Average value of impression curve - stages of the design process

Table 1. The average values of impression curve for variants of levels concerning successive stages of virtual space design and the average value for all stages. AV - average value of the impression curve for each point; SD - standard deviation of the impression curve average value for each point.

Curve point	I	II	III	IV	V	VI	VII	VIII	IX	X
Simple blockout (A)										
AV	1.57	3.00	3.71	3.71	3.14	3.93	4.93	5.71	6.21	7.29
SD	0.79	1.00	1.35	1.58	2.12	2.01	1.64	1.60	2.00	1.98
Advanced blockout (B)										
AV	1.44	3.22	3.33	3.22	2.83	3.78	4.22	5.28	5.94	7.00
SD	1.13	1.39	1.73	1.86	1.62	1.72	1.70	1.92	1.91	1.94
Main model without material (C)										
AV	2.79	6.00	5.71	4.93	5.29	5.31	5.79	6.29	6.71	7.57
SD	1.29	2.00	2.63	2.83	2.63	2.66	2.20	2.21	2.06	1.62
Main model with momochromatic material (D)										
AV	2.61	5.84	6.18	5.58	4.50	5.02	5.34	6.48	7.26	7.18
SD	0.82	2.41	2.11	1.72	2.51	2.54	1.94	1.74	1.61	1.16
Main model with final material (E)										
AV	2.57	5.33	5.78	4.78	4.11	4.22	5.22	6.56	7.33	8.11
SD	1.80	2.24	1.92	1.72	1.54	1.39	1.56	1.51	1.66	1.36
Main model with final material with extra fined detailed model - final level version (F)										
AV	3.40	6.25	5.81	5.38	4.88	5.00	5.56	5.75	7.25	8.50
SD	1.83	2.19	2.39	2.77	2.36	2.39	2.26	2.19	1.39	1.20

5.2   Comparison of the Highest, Lowest and Average Values of the Impression Curves for Next Stages of Creating a Computer Game Level

Table 2 contains the lowest, the highest and the average values of impression curve for variants of levels concerning successive stages of virtual space design:

- For "Simple blockout"(A) and "Advanced blockout"(B) those values are the lowest ones
- For variant "Main models without materials"(C), "Main models with monochromatic materials"(D) and "Main models with final materials"(E) those values are higher than in case of "Simple blockout"(A) and lower than for the "final level version"(E).
- Value of impression curve are highest for "Final level version - Main models with final materials with extra fine detailed models"(F).

The above data indicates that the more advanced stage of level design, the higher values of impression curve occur. Both the average and extreme values grow.

Table 2. The highest, lowest and average values of the impression curve for individual level variants, showing the next stages of work on the virtual environment.

Stage of design process	Value of impression curve		
	Lowest value	Highest value	Average value
Simple blockout (A)	1.57	7.29	4.32
Advanced blockout (B)	1.44	7.00	4.03
Main model without material (C)	2.79	7.57	5.64
Main model with momochromatic material (D)	2.61	7.26	5.60
Main model with final material (E)	2.57	8.11	5.40
Final level version (F)	3.40	8.50	5.78

5.3   Comparison of Impression Curve Charts for Modified Variants of the Computer Game Level

Having confirmed that impression curves for successive stages of level design are alike, curves for modified versions of the level might be compared. This should provide the information which modifications have the effect on impression curve.

In Fig. 7 and Fig. 8 visual similarity of average impression curves for modified levels may be juxtaposed with the curve of any stage of level design process (exact numbers with standard deviations are presented on Table 3). It may be noted that:

- Changes of weather, lightening conditions and materials do not impact on the shape of impression curve

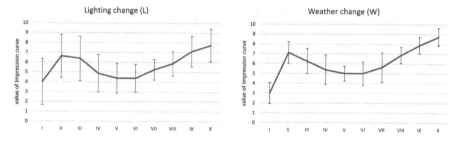

Fig. 7. Average value of impression curve - variants with global changes - (W) weather, (L) Lighting

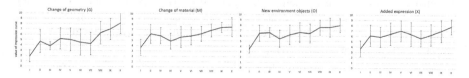

Fig. 8. Average value of impression curve - variants with level modifications - local changes - (G) geometry, (M) material, (O) objects, (X) expression

Table 3. The average values of impression curve for variants of levels concerning successive stages of virtual space design and the average value for all stages. AV - average value of the impression curve for each point; SD - standard deviation of the impression curve average value for each point.

Curve point	I	II	III	IV	V	VI	VII	VIII	IX	X
Weather change (rain)										
AV	3.00	7.13	6.25	5.38	5.00	5.00	5.63	6.88	7.88	8.75
SD	1.07	1.13	1.28	1.51	0.76	1.20	1.51	0.83	0.83	0.89
Change of geometry - advanced blockout										
AV	1.91	4.68	3.82	5.23	5.05	4.55	4.32	6.36	7.09	8.18
SD	1.14	2.17	0.78	1.99	2.15	2.45	2.12	2.11	1.97	1.94
Adding expression										
AV	3.65	6.15	5.88	6.46	7.05	6.42	5.65	6.31	7.04	7.85
SD	1.52	1.99	1.87	2.11	1.96	2.25	1.52	2.02	2.05	1.39
Change of materials										
AV	3.64	6.21	5.86	4.86	5.64	5.86	6.29	6.93	7.50	7.57
SD	1.55	1.76	1.51	1.51	1.95	1.66	1.38	1.21	0.94	1.79
New environment objects										
AV	3.50	6.50	6.63	5.50	6.25	6.69	6.50	7.63	7.63	8.00
SD	0.76	1.51	0.92	1.60	1.83	1.83	1.77	1.41	1.77	1.31
Lighting change (night)										
AV	4.00	6.63	6.38	4.88	4.38	4.38	5.25	5.88	7.13	7.75
SD	2.33	2.20	2.26	1.89	1.51	1.41	1.04	1.25	1.55	1.67

- For variants enriched by expression elements or tiny surrounding items, the curve grew and reached a local maximum in the modified locations (fifth and sixth point on impression curve)
- Changes in geometry dramatically influence the impression curve. The values of curve decrease for points from fourth to seventh on the impression curve.

6   Summary and Conclusions

The goal of this paper was to examine, by means of impression curve, how visual perception varies throughout successive stages of level design process as well as to identify factors which influence on scene perception by a player. To conduct studies, several variants of the same level were prepared. Each version was tested by players. All presented charts contain average impression curves which were compared to one another (Fig. 5).

Based on conducted experiments, following conclusions might be drawn:

- The method of impression curve could be a useful tool to assess virtual worlds, especially in case of game level design. It is supposed to support the design of virtual worlds being visually appealing and diversified, where the gameplay and appearance complement each other
- Charts of average impression curves for successive stages of level design are alike. This means that the virtual environment may be tested with impression curve from the very beginning
- The more advanced the design phase, the higher average impression curve values are
- Disturbances of buildings blocks (geometry changes, extra tiny items, added expression) have impact on the shape of impression curve
- Neither lightening condition changes, nor weather or material changes influence impression curve

There are possible following directions of further research: investigation of dependency between player's reaction and forming of the impression curve (eye-tracking [20], EEG [21]), examination if graphical style or perspective (used camera type) have any impact on the impression curve or check if the strategy of impression curve can be used for 2D.

References

1. Senetra, A., Szczepańska, A., Bajerowski, T., Wycena krajobrazu. In: Rynkowe aspekty oceny i waloryzacji krajobrazu, Educaterra, Olsztyn, PL (2000)
2. Cox, D.J.: Interior design and environment art: mastering space. In: Mastering Place on Game Developers Conference 2015 (2015). Accessed 21 May 2019
3. Cox, D.J.: What Modern Interior Design Teaches Us About Environment Art on Game Developers Converence 2014 (2014). Accessed 21 May 2019
4. Marr, D.: Vision. The MIT Press, Cambridge (2010)
5. Jenkins, H.: Game design as narrative architecture. Computer **44**, 118–30 (2004)

6. Mikami, K., Kondo, K.: Adaptable game experience based on player's performance and EEG. In: 2017 Nicograph International (NicoInt), Kyoto, JP, pp. 1-8 (2017). https://doi.org/10.1109/NICOInt.2017.11
7. Solecka, I.: Polish experience in landscape identyfication and valorisation. Inżynieria Ekologiczna **2016**(50), 223–231 (2016)
8. Wejchert, K.: Elementy kompozycji urbanistycznej. Arkady, Warszawa (1984)
9. Norman, K.: GEQ (Game Engagement/Experience Questionnaire): a review of two papers. Interact. Comput. **25**(4), 278–283 (2013)
10. Zając, M., Bałaga, K., Janicki, G.: Waloryzacja krajobrazu podmiejskiej wsi (okolice Lublina) dla potrzeb gospodarki przestrzennej. Problemy Ekologii Krajobrazu **2014**(37), 77–86 (2014)
11. Ølsted, P.T., Ma, B., Risi, S.: Interactive evolution of levels for a competitive multiplayer FPS. In: IEEE Congress on Evolutionary Computation (CEC), Sendai, JP, pp. 1527–1534 (2015). https://doi.org/10.1109/CEC.2015.7257069
12. Szrajber, R.: Architecture in virtual worlds as a field of research. In: Fross, Klaudiusz (ed.) book: Badania Interdyscyplinarne w Architekturze 2, vol. 1, pp. 55–66. Wydział Architektury Politechniki Śląskiej, Gliwice (2017)
13. Szrajber, R.: Impression Curve and the Gameplay Curve as an Attempt to Record the Player's Experience during International Conference for Young Researchers - Computer Game Cultures 2: Game as an Experience, 26–28 June. Lodz, PL (2014)
14. Szrajber, R.: Architecture in Video Games - Seeing or Experiencing during International Conference for Young Researchers - Computer Game Cultures 4: Criticism and Methodology, 30 June, Lodz, PL (2017)
15. Berta, R., Bellotti, F., De Gloria, A., Pranantha, D., Schatten, C.: Electroencephalogram and physiological signal analysis for assessing flow in games. IEEE Trans. Comput. Intell. AI Games **5**(2), 164–175 (2013). https://doi.org/10.1109/TCIAIG.2013.2260340
16. Almeida, S., Veloso, A., Roque, L., Mealha, O., Moura, A.: A video game level analysis model proposal. In: 16th International Conference on Information Visualisation, pp. 474–479 (2012). https://doi.org/10.1109/IV.2012.82
17. Freytag, S.-C., Wienrich, C.: Evaluation of a virtual gaming environment designed to access emotional reactions while playing. In: 9th International Conference on Virtual Worlds and Games for Serious Applications (VS-Games), Athens, 2017, pp. 145–148 (2017). https://doi.org/10.1109/VS-GAMES.2017.8056585
18. Litwin, U., Bacior, S., Piech, I.: Metodyka waloryzacji i oceny krajobrazu. Geodezija, kratografija i aerofotoznimanie **70**, 14–25 (2009)
19. IJsselsteijn, W.A., de Kort, Y.A.W., Poels, K., Jurgelionis, A., Bellotti, F.: Characterising and measuring user experiences in digital games. In: Bernhaupt, R., Tscheligi, M. (eds.) Proceedings of the International Conference on Advances in Computer Entertainment Technology (ACE 2007), 13–15 June 2007, pp. 1–4. Salzburg, Austria (2007)
20. Rynkiewicz, F., Daszuta, M., Napieralski, P.: Pupil detection methods for eye tracking. J. Appl. Comput. Sci. **26**(2) (2018). pp. 201–211. Technical University Press, Łódź, Poland, ISSN 1507–0360
21. Opałka, S., Stasiak, B., Szajerman, D., Wojciechowski, A.: Multi channel convolutional neural networks architecture feeding for effective EEG mental tasks classification. Sensors **18**(10), 3451 (2018). https://doi.org/10.3390/s18103451

Visual Analysis of Computer Game Output Video Stream for Gameplay Metrics

Kamil Kozłowski[1], Marcin Korytkowski[2] (ID), and Dominik Szajerman[1]([✉]) (ID)

[1] Institute of Information Technology, Lodz University of Technology,
ul. Wólczańska 215, 90-924 Łódź, Poland
`dominik.szajerman@p.lodz.pl`
[2] Częstochowa University of Technology, Al. Armii Krajowej 36,
42-200 Częstochowa, Poland
`marcin.korytkowski@pcz.pl`
`http://kisi.pcz.pl, http://it.p.lodz.pl`

Abstract. This work contains a solution for game metrics analysis based on a visual data stream dedicated for the player. The solution does not require interference in the programming code of the analyzed game and it is only based on image processing. It is possible to analyze several aspects of the game simultaneously, for example health/energy bars, current weapon used, number of objects worn (aid kits, ammunition). There have been presented methods using cascading classifiers and their training to detect the desired objects on the screen and to prepare data for other stages of processing, e.g. OCR. The effect of the methods is a gameplay chart that allows a thorough analysis of the player's actions in the game world and his or her advancement. The solution is fast enough that it can be used not only in previously recorded gameplay analysis, but also in real time during simultaneous gameplay.

Keywords: Gameplay metrics · HUD · Image processing

1 Introduction

Gameplay metrics are important parts of testing and analyzing computer games. They rely on the interpretation of raw data from the game, which may contain information about the player's current activities [10] – including, but not limited to, moving around the world, performing defined operations, or interacting with the graphical interface [4]. Acquiring data for the gameplay metrics process may require some access to the game's source code, whether by modding or compiling open source projects. Another way may be to use specialized equipment such as e.g. eyetracker to analyze the player's behavior, but for obvious reasons it is limited to laboratory conditions [9]. However, a lot of information could be obtained from the game result streams alone – image and sound. This is very

V. V. Krzhizhanovskaya et al. (Eds.): ICCS 2020, LNCS 12141, pp. 538–552, 2020.
https://doi.org/10.1007/978-3-030-50426-7_40

important, because it allows to analyze games at which access to the source code is impossible, such as those recently released – and do it very effectively [6].

The methods presented here can also be used in a broader scope, such as testing player behavior (including attempts to cheat in online games), or elements of automated tests, including tests of games produced by third-parties. The data obtained here can be a subset of the inputs of a bot's program, which tests the game by building a map and navigating within it [3] or acting in other way [8].

The paper presents a solution consisting of three methods. The first combines known HUD[1] analysis with shape detection based on a cascade classifier to obtain a robust method independent of user screen configuration. The second expands the game analysis presented in the literature with a new algorithm. It involves recognizing shapes using cascade classifiers based on Haar-like [12] and Local Binary Patterns (LBP) [2] features to recognize moving technical objects – in this case, weapons in first-person shooter game (FPS). The third method allows to analyse the HUD detected by the cascade classifiers to recognize the text it contains, and interpret the strings obtained in this way. This allows to read secondary information about the game, such as the number of first aid kits or ammunition.

2   Related Work

Usually game metrics are understood as telemetry, i.e. remote collection of game-play data for many players for analysis and statistics [5]. Only game producers can do this because they have access to data collected on their game servers. The data can be various: equipped weapon, play time, difficulty level, current mission, avatar location, enemies'/NPCs' parameters, save/load actions, player drop-off per mission, custom use of game mechanics. Applications other than play testing are, for example, bug tracking and QA. Game metrics results are often presented as graphs of measured value versus time or heat-maps showing the intensity of various measurements collected in 3D game space [5].

Audiovisual analysis of the game through postprocessing does not need to access the game code. Thus can be an effective variation of gameplay metrics for third-parties. It is based on audiovisual analysis of the recorded gameplay. It was presented in [6]. Key elements of communication with the player in the analyzed game, i.e. "Bioshock 2" [1] (e.g. health bar status, occurrence of the HUD on the screen, sounds of shots) can be precisely extracted from the game result streams. Then, these data can be presented as a graph illustrating several dozen minutes of gameplay. Such graph is a clear way to distinguish between the way a beginner player plays and an advanced one. The work [6] presents algorithms that allow to segment and index the gameplay – they are referred to as "Algorithm 1" and "Algorithm 2". Our work references to them the same way.

---

[1] Heads-Up Display is the graphics part of the game user interface by which information about the state of the game is visually presented to the player.

"Algorithm 1" checks the occurrence of HUD on the screen (during non-interactive scenes the HUD is hidden, which excludes them from performance analysis) through the static logo detection algorithm on the screen, depending on the mask created earlier. "Algorithm 2" checks the player's health and energy level as the appropriate colors constituting the health and energy bars in the areas of the screen determined by static masks.

As a field for further analysis [6] proposes to include a method of detecting objects on the screen, extracting text, and recognizing enemies by face detection.

3   Method

The solution is based on visual analysis of gameplay recordings in the game "Bioshock 2" [1], using the tools built into the OpenCV 3.3.1, and cascade classifiers that have been prepared by hand. In addition, the Tesseract text recognition system was used.

In order to get data for the analysis the gameplay recordings were made. To reduce the calculation time, the frames of the recordings for analysis should be averages of several seconds of gameplay [6]. Each image analyzed is an average of 30 frames. Averaging frames blur all the moving elements on the screen. The most static element of the world – a weapon held by the player – is also blurred, but not enough to prevent its analysis.

A certain problem in creating classifiers [7, 13] is the preparation of appropriate image samples: several hundred positive, containing the sought shape, and several thousand negative, not containing it. Although a thousand image samples can mean a huge amount of data and a long time to obtain them, it is also the number of frames from just 33 s of gameplay recording with a frequency of 30 frames per second. Samples were therefore obtained by recording a short gameplay and appropriate processing, described below.

Negative samples for learning the classifier were the entire screenshots of gameplay. To ensure the right variety of screenshots two rules were applied. The first was to quickly change the position and rotation of the camera throughout the entire recording. The second involves making recordings in several places – at least one dark and at least one bright. All classifiers were created on the basis of several uninterrupted gameplays recorded in two different chapters of "Bioshock 2" – no more than 30 s each – made separately for each weapon.

Positive samples were obtained in the same way, however, additional cropping was necessary so that they contained only the shape sought, with only small background areas at the edges. To increase the diversity of samples, a random jitter was introduced – for each frame a position offset and frame size are randomly choosen, equal to at most 5% of its width or height. To save calculation time when training the classifier, the n-th frame, instead of each were used.

The classifiers were created with the help of the traincascade program included in the OpenCV package. Each classifier for this project was based on 200–900 positive samples, and 1000–2000 negative ones. The maximum false alarm rate[2] as well as the specific number of samples were manually choosen to get the best results while keeping the training time below an hour. The exact values are presented in Table 1. For all classifiers the Gentle AdaBoost training algorithm was used, 19 was the maximum number of training epochs achieved. The classifiers were created in two variants, based on the same data sets – for Haar-like and LBP features.

Table 1. Training configurations of the different classifiers used in the method.

	Haar1/LBP1	Haar2/LBP2	Haar3	LBP3	Haar4/LBP4	Haar HUD
Negative samples	479	263	397	397	246	903
Positive samples	1501	990	1921	1921	1098	826
Max false alarm	40%	40%	40%	33%	40%	40%

3.1   Algorithm A. HUD Detection and Analysis

HUD detection is performed using a properly trained cascade classifier based on Haar-like features. The classifier reliably detects a screen section with the same shape and the same HUD fragment (the cropping differences between subsequent detections are small). It is therefore possible to train the classifier so that it detects one very characteristic fragment of the HUD and then enlarges the marked area (e.g. in percentage) so that it contains the entire HUD (Fig. 1).

The detected screen area – if the detection occurred – is intended for further processing. As in "Algorithm 2" [6], the intensity of colors was measured in two areas of the examined image (health bar and energy bar). It was based on masking the image and counting pixels with an appropriate shade relative to all non-zero pixels in the mask. In our approach, on the other hand, pixel filtering for shade and brightness in the HSV[3] color space was used, and detection of rectangular contours describing the pixels detected in this way (bar filling). The width of the bright red rectangle in the HUD area determines the amount of health of the player, and the width of the bright blue rectangle means the amount of energy of the player.

The difference from "Algorithm 2" [6] is that the operation is local for the HUD, not for the entire screen, so the transformations made on the HUD are not relevant to it. As a result, it is possible to detect the presence of HUD on the screen and advanced analysis of its content regardless of its current location. In addition, it is not necessary to prepare an appropriate mask before conducting the tests: it is enough to prepare information about the desired colors. Optionally, additional cropping of the detected area can be used to reduce the chance of incorrect color detection in the background game world.

---

[2] False positive detection percentage allowed at each stage of training the cascade classifier.

[3] hue, saturation, value.

Fig. 1. The cascade classifier detects and marks the screen area containing the HUD (processed frame from a screencast recorded from "Bioshock 2" [1]).

An intermediate "Algorithm A" has also been implemented, detecting the HUD area with a cascade classifier, and then analyzing it locally using masks, as described by "Algorithm 2".

In addition, a separate implementation was made that works exactly according to Algorithms "1" and "2" of [6] to compare the effectiveness of HUD detection.

3.2   Algorithm B. Identification of Held Weapons

In "Bioshock 2", as in the vast majority of FPS, the weapon held by the player's character is always visible in the lower right corner of the screen as a model in the game world, making small movements both when moving the camera and the character, as well as when idle. In the averaged frame, the model is therefore blurred, but still retains a reproducible set of characteristics (Fig. 2).

The image analysis was carried out in the initial stages of the plot of the game "Bioshock 2", where four weapons are available. Detections are carried out on each analyzed frame using four different classifiers, each trained to detect a different weapon: "Drill" (1), "Rivet gun" (2), "Hack tool" (3) and "Machine gun" (4). In order to optimize detection and reduce false positive errors, the analysis is performed only on a fragment of the image – specifically its right half. What counts is the fact of detecting the shape; whether the area found by the classifier contains all or only part of the weapon does not matter (Fig. 2). There is also no reason to create a general classifier (whether any weapons are kept) as part of the optimization, since the weapons are kept only when the HUD is displayed. And this is detected by the "Algorithm A".

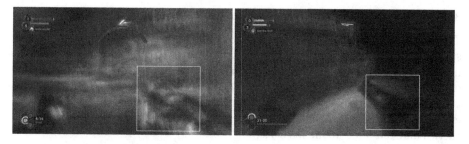

Fig. 2. The effects of the "Algorithm B". Two different weapons are effectively detected and marked on the screen (processed frames from a screencast recorded from "Bioshock 2" [1]).

Based on which classifier in the given frame detected the weapon on the screen, it is determined which weapon is currently visible.

3.3   Algorithm C. Eliminating Weapon Identification Error

Due to problems such as invisibility of the weapon at some moments, caused by environmental factors ("Bioshock 2" takes place largely in very dark locations, so a large part of the features can be visible to a very limited extent), or the possibility of incorrect detection of weapons by "Algorithm B" (no weapons, or several weapons at the same time), it is necessary to use the method of mitigating the effects of errors. Our solution to the problem is to use fuzzy logic by creating a vector of floating point values that are stored throughout the analysis. Each value in the vector corresponds to the "truth" of the occurrence of a given weapon in a given frame based on the results of detection by cascade classifiers in the current frame and previous frame. Thus, with four detected weapons, the vector has four components. Each vector field has a specific maximum (here 3.0) and minimum (0.0) value. If any weapon is recognized in a given frame, the fields corresponding to each of the detected weapons are incremented by a certain value (here 1.0), and the non-detected weapon fields are decremented by the same amount. If no weapon is detected in the frame, the values of all fields are reduced by some other value (here 0.3). The highest value in the corresponding field of the vector gives current weapon. In this way, a single incorrect detection in one frame (as well as the detection of more than one weapon in a single frame), when the weapon is detected correctly in the previous and next frame, does not affect the result. If no weapon is detected, the stored values will be accepted as current for the next few frames (maximum of 10 frames for example values). However, the detection of a correct weapon change is delayed by only two frames.

This method can be effectively applied in the vast majority of modern FPS games – it is enough to prepare other classifiers for them, for weapons present in a given game.

3.4   Algorithm D. Reading Text in the Graphical Interface

The main HUD in the game "Bioshock 2", shown in Fig. 3, in addition to two status bars also contains two numbers. The first of them means the number of first aid kits (supplementing the character's health), and the second – the number of portions of "Eve", supplementing energy. The lower HUD of the game, in turn, presents information about the quantity and type of ammunition for the currently selected weapon. These values have a huge impact on the player's tactics and tell a lot about his style.

Fig. 3. Algorithm for detecting the value of the energy bar. From the left: The input area of the image detected by the classifier (additionally automatically cropped to improve processing); the image with detected pixels of proper shade and brightness (light blue); the output image with the outline of the energy bar marked, whose width is information about the energy possessed (fragments of processed frames from a screen-cast recorded from "Bioshock 2" [1]). (Color figure online)

After detecting these HUDs, the cropped image fragments are processed using the method described in "Algorithm A". It is based on masking the interface so that only fragments that may contain text are visible, and on thresholding the image according to brightness to make the text clearer and remove unwanted shapes and gradients from the background. In the image prepared in this way, the text recognition process is carried out using the Tesseract-OCR system. The detection is based on default text recognition results for English, made available by the creators in a public repository [11].

If the pre-processing was successful, processing the detection results is very simple. Tesseract-OCR provides a text string of several lines whose analysis strictly depends on the analyzed HUD. For example, upper HUD analysis provides three lines of text: the first two are integers, the third is a simple string with the name of the plasmid[4] used.

To support text recognition, the range of characters recognized by Tesseract can be limited, e.g. to numbers only. Particularly undesirable are simple punctuation marks, such as a period and comma, which may be misdetected on the basis of individual pixels.

4   Results

A series of tests of the methods was carried out on four different recordings of the game from the "Bioshock 2" game. It was done with the use of frame averaging

---

[4] Characteristic of "Bioshock 2" gameplay object.

(30 in one) and without it. The recordings contained sequences of exploration, fighting, non-interactive story scenes (without HUD) and an open pause menu, recorded in various game locations, including well-lit and darker places, additionally differing in the colors of the environment. In total, over 11 min of various gameplay were tested, averaged into 699 samples. The recordings were made at a resolution of 1280×720 pixels, with a 30 frames per second (thus during the analysis one average frame contained data from exactly one second of the recording). Weapon classifiers operated with a scale factor of 1.02^5, and searched for areas between 250×200 and 600×600 pixels.

For averaged frames, the average accuracy (ACC) of HUD detection was 92.13%, with a precision of 99.35% (PPV). Details of the effectiveness of the HUD detection classifier are presented in Table 2 (where it was designated "Haar HUD"). The standard deviation of the reading of the values presented in the HUD at a constant level of health and energy, using "Algorithm A" was 1.07 for both bars (with values given on a scale of 0–100), respectively. Standard deviations for the values read by the implemented "Algorithm 2" were 4.03 and 2.25, respectively, for health and energy. In the case of the indirect algorithm ("Algorithm 2" operating locally in the area cut out using the cascade classifier) 4.38 and 2.22, respectively. In addition, the implemented "Algorithm 1" confirmed its 100% HUD detection efficiency mentioned in [6].

Table 2. Table of statistics on the operation of various classifiers in the application. The rows "Total Haar" and "Total LBP" contain the sums of the TP, FP, TN and FN columns of all Haar and LBP classifiers, respectively, and the average of the other columns. Numbering 1–4 next to the classifier names defines the weapons as in the Subsect. 3.2. Haar HUD is a classifier used to detect HUD in the "Algorithm A".

Classifier	TP	TN	FP	FN	Precision (PPV)	Sensitivity (TPR)	Specificity (SPC)	Accuracy (ACC)
Haar 1	124	445	107	24	53.68%	83.78%	80.62%	81.29%
Haar 2	92	495	8	105	92.00%	46.70 %	98.41%	83.86%
Haar 3	115	493	71	21	61.83%	84.56%	87.41%	86.86%
Haar 4	105	464	47	84	69.08%	55.56%	90.80%	81.29%
Total Haar	436	1897	233	234	69.15%	67.65%	89.31%	83.32%
LBP 1	132	532	20	16	86.84%	89.19%	96.38%	94.86%
LBP 2	161	466	25	37	86.56%	81.31%	95.02%	91.14%
LBP 3	94	562	2	42	97.92%	69.12%	99.65%	93.71%
LBP 4	121	431	79	69	60.50%	63.69%	84.51%	78.86 %
Total LBP	508	2002	126	164	82.95%	75.83%	93.89%	89.64%
Haar HUD	618	26	4	51	99.35%	92.37%	86.66%	92.13%

---

[5] Defines the step size when the classifier scales the searched shape. The smaller the scale factor, the more calculations the classifier makes and the greater its accuracy.

When testing the "Algorithm B", the average effectiveness of identification of the weapon held (compliance of the identified weapon with the actual value) was 58.08% for "Algorithm B" alone using LBP classifiers, 55.36% for Algorithm B" alone using Haar-like classifiers, and 66.95% for "Algorithm B" using LBP features and using "Algorithm C". Details are presented in Table 2 and Table 3.

Table 3. Weapon identification efficiency for three different detection systems, for four different gameplay recordings.

Recording number	1	2	3	4
RAW LBP	75.00%	44.64%	74.83%	76.32%
RAW Haar-like	69.57%	46.43%	73.78%	70.30%
LBP with "Algorithm C"	77.17%	62.50%	82.17%	92.48%

For non-averaged frames, both the detection of weapons and HUDs increased the occurrence of false positive error, which reduced the effectiveness of both "Algorithm B" and "Algorithm A", due to much greater information noise.

Using the presented algorithms, it was possible to create charts showing the states of health, energy and weapons used (Detected Weapon) in subsequent time samples for two (of four) choosen recordings – visible in Fig. 4. The algorithms "B" and "C" were tested independently of the "Algorithm A".

On average, full analysis took 19.692 ms, of which 9.118 ms was HUD and 10.573 ms was weapon analysis. On this scale, it should also be mentioned that loading and averaging 30 frames of the recording took an average of 188.416 ms using an HDD. Tests were carried out on a computer with an Intel Core i7-4720HQ @ 2.60 GHz processor. The created system did not contain multi-threaded elements.

Training of classifiers with Haar-like features took 20–40 minutes on the above described equipment with 5 threads. Training of LBP classifiers on the same data in all cases lasted from 50 s to 18 min (where only LBP 3 was created for more than 5 min).

HUD analysis using the "Algorithm D" in "Bioshock 2" was performed on the same recordings as the detection of weapons. The upper HUD (first aid kit, eve) and lower HUD (ammunition) analysis were tested. The average effectiveness of text analysis in frames with correctly detected areas of text was 99.36% and 98.79%, respectively. The effectiveness of text analysis in all frames in which it should be carried out (the frames in which detection of screen areas was ineffective were included) was 93.25% and 95.90%, respectively. There was no false positive error in any of the 701 frames tested (there was never an accidental successful analysis of an area that did not contain the appropriate text). Details are presented in Table 4. The numbers in the upper part of the table indicate the number of frames meeting the condition shown in the left column.

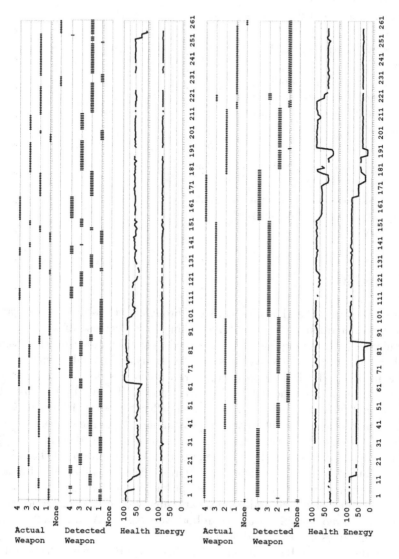

Fig. 4. Left: the first choosen gameplay recording chart. Right: the second one.

Table 4. Statistics on the analysis of two different HUD parts.

	Top HUD (aid kits, eve)	Bottom HUD (ammunition)
Correct detection of the area and its correct analysis result (A)	622	492
Correct detection of the area and incorrect analysis result (B)	4	6
Incorrect detection of visible area (FN)	41	15
Incorrect detection and full analysis of the invisible area (FP)	0	0
Correctly not detecting an invisible area (TN)	34	188
Total analysed frames	701	
Effectiveness of the analysis in the detected areas ($\frac{A}{A+B}$)	99.36%	98.80%
The effectiveness of the analysis in all frames that should be analyzed ($\frac{A}{A+B+FN}$)	93.25%	95.91%

5   Discussion

HUD detection using the Haar-like classifier brought noticeably less accuracy than the logo detection method of "Algorithm 1" (a decrease of about 8%). This, combined with the high accuracy achieved (a small number of FP errors) means that "Algorithm A" receives less data for analysis, but only a small part of them is incorrect (HUD analysis occurs rarely when it is not on the screen).

Random deviations of the detected position and HUD area by the cascade classifier should have a negative impact on the effectiveness of HUD analysis. However, there are negligible differences in the standard deviation values between the implementation of "Algorithm 2" in the full-screen version (described in [6]) and the indirect, classifier-based. This suggests that the deviations of the values are mainly caused by errors in detecting colored pixels in subsequent frames, rather than shifting areas of the masks. At the same time, the new method of analyzing the value of health and energy bars, finally described in "Algorithm A" brought much more stable results than both versions of "Algorithm 2" (with more than four times less standard deviation).

Weapon identification based on both Haar-like and LBP features has similar effectiveness – in some cases Haar-like gives better results, in others worse than LBP. The difference is always at most few percent (at most 6% points of difference). This is despite the fact that the LBP classifiers have on average clearly higher efficiency and precision than those based on Haar-like features.

For each set of samples, the use of the "Algorithm C" increased the effectiveness of weapon identification. The effectiveness of the "Algorithm C" decreases with frequent weapon changes, reaching the lowest value at first recording (Fig. 4 left). In the second recording (Fig. 4 right), where the effectiveness of identification with the "Algorithm C" is the highest of all cases, weapon changes occur least often (only 11 times in 4 min).

The number of samples that should be forwarded to the classifier learning process (and its maximum false alarm) to obtain satisfactory results is strictly dependent on the appearance of the specific weapon. For example, the most difficult classifiers to create were Haar 3 and LBP 3, detecting the "Hack Tool" (shown in Fig. 5). The reason was the distinctive appearance of the weapon, which contained two circles. This caused a false positive to occur frequently.

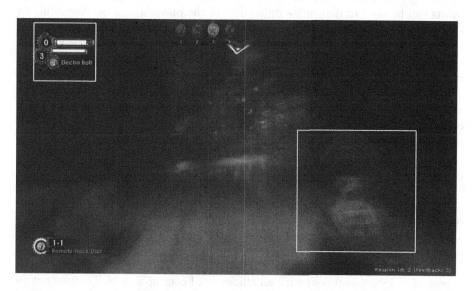

Fig. 5. Visual effect of the methods result: image from the average frame, the detected HUD area subjected to further analysis, and the identifier of the detected weapon in the lower right corner (processed frame from a screencast recorded from "Bioshock 2" [1]).

The analysis of the HUD content in terms of text content ("Algorithm D") proved to be very effective for all recordings from the game, with a particularly low (below 1.3%) percentage of erroneous full analyzes, and a complete lack of False Positive errors. Their higher frequency could lead to undetectable analysis errors, distorting the generated picture of the course of the game.

Figure 4 shows the results of analyzing two different game recordings. They presented types of weapons appearing in subsequent samples, marked manually (Actual Weapon chart line), and types detected by "Algorithm B", using LBP classifiers, supported by "Algorithm C" (Detected Weapon chart line). The results of Actual Weapon and Detected Weapon are similar to each other, which means that the identification of weapons carried out by the application is quite effective. It is clearly seen that in Fig. 4 on the left side there is a greater discrepancy between the actual weapons and weapons detected by the algorithms than on the right. The reason is the player frequently changing weapons, which reduces the effectiveness of the "Algorithm C".

Health and energy values presented in Fig. 4 were detected using the "Algorithm A". It can be seen many breaks in them, caused by the algorithm not detecting the GUI due to an error or because the GUI was hidden in the game. For example, the pause in Fig. 4 on the left side, near the sample (seconds) 246, is the result of the player entering the shopping menu. However, the gap in Fig. 4 on the right side, at sample 26, is caused by a false negative error of the classifier, extending over a dozen samples. Despite the missing data at some moments, the charts can be used to determine differences in the player's actions. According to Fig. 4 on the left, the player lost health during the fight around 11 s (samples) and healed 60 s later. At the same time he at no time used special skills that would cause loss of energy – its state was constant all the time. In Fig. 4 on the right, the player repeatedly used his skill, and in the 157th second he started a 40-s battle that caused health levels to wave over that period. Thanks to the weapon chart, it can be said that most of the aforementioned fight took place using a "Machine gun" (weapon 4), which at the end (about 181 s) was changed to a "Rivet gun" (weapon 2).

6   Conclusions

The use of cascade classifiers in the presented methods has significantly reduced its creation time. This is due to the possibility of using a small number of samples and the use of a small number of learning stages. Despite this, the classifiers achieved a precision of 75%.

Using the presented methods, it was possible to distinguish a set of four classes of objects (weapons) with an efficiency of over 66%, which turned out to be sufficient for analyzing the game. The presented algorithm for minimalization identification errors turned out to be positive in all test cases.

LBP classifiers present in this application – i.e. detecting different types of weapons – show noticeably higher efficiency than Haar-like classifiers. However, this does not translate into clear gains in the effectiveness of weapon identification without the use of an algorithm that eliminates identification errors. Much shorter learning time makes the process of improving and testing LBP classifiers in an iterative way much more efficient than with Haar-like classifiers.

The time results of the algorithms clearly show that the method can be effectively used in real time – not only as postprocessing, but also as an analysis of ongoing gameplay.

The presented method of HUD detection and analysis is effective to a degree similar to the literature method [6], however, it allows you to analyze the interface regardless of its location on the screen, so it is suitable for analyzing moving interface elements.

Further work may include the preparation of more precise classifiers, which should significantly increase the effectiveness of weapon identification. The system has the potential to significantly parallelize calculations, because of all the methods presented only the "Algorithm C" has to be executed sequentially. The presented graphical interface analysis method can be used in other games to analyze static and moving interface elements. Weapon identification algorithms can be used in most FPS games.

The algorithms presented here allowed to make an effective analysis of the player's state of change over time in three aspects: energy level, health level and the type of weapon held. The obtained statistics allow to determine the style of playing and to detect and analyze the most important moments from the recorded game (e.g. the course of the fight) based on the generated graph itself.

Of course, the methods presented have their limitations. While the analysis of GUI elements can also be useful for other game genres, the specific weapon detection algorithm will only be useful for FPS games. The method based on postprocessing and classification of graphic elements would always need to be adjusted if the type of game analyzed changes. But other game metrics methods are not without this drawback, where the collected data itself is closely related to a specific genre or even a specific game.

References

1. 2K Games: Bioshock 2 (2010). https://store.steampowered.com/app/8850/
2. Ahonen, T., Hadid, A., Pietikäinen, M.: Face recognition with local binary patterns. In: Pajdla, T., Matas, J. (eds.) ECCV 2004. LNCS, vol. 3021, pp. 469–481. Springer, Heidelberg (2004). https://doi.org/10.1007/978-3-540-24670-1_36
3. Daszuta, M., Wróbel, F., Rynkiewicz, F., Szajerman, D., Napieralski, P.: Affective pathfinding in video games. J. Appl. Comput. Sci. **26**(2), 23–29 (2018)
4. El-Nasr, M.S., Drachen, A., Canossa, A. (eds.): Game Analytics. Springer, Heidelberg (2013). https://www.ebook.de/de/product/20778057/game_analytics.html
5. El-Nasr, M.S., Drachen, A., Canossa, A. (eds.): Game Analytics. Springer, London (2013). https://doi.org/10.1007/978-1-4471-4769-5
6. Marczak, R., Schott, G., Hanna, P.: Postprocessing gameplay metrics for gameplay performance segmentation based on audiovisual analysis. IEEE Trans. Comput. Intell. AI Games **7**(3), 279–291 (2015). https://doi.org/10.1109/tciaig.2014.2382718
7. Opałka, S., Stasiak, B., Szajerman, D., Wojciechowski, A.: Multi-channel convolutional neural networks architecture feeding for effective EEG mental tasks classification. Sensors **18**(10), 3451 (2018)
8. Rogalski, J., Szajerman, D.: A memory model for emotional decision-making agent in a game. J. Appl. Comput. Sci. **26**(2), 161–186 (2018)
9. Szajerman, D., Napieralski, P., Lecointe, J.P.: Joint analysis of simultaneous EEG and eye tracking data for video images. COMPEL Int. J. Comput. Math. Electr. Electron. Eng. **37**(5), 1870–1884 (2018). https://doi.org/10.1108/compel-07-2018-0281

10. Szajerman, D., Warycha, M., Antonik, A., Wojciechowski, A.: Popular brain computer interfaces for game mechanics control. In: Zgrzywa, A., Choroś, K., Siemiński, A. (eds.) Multimedia and Network Information Systems. AISC, vol. 506, pp. 123–134. Springer, Cham (2017). https://doi.org/10.1007/978-3-319-43982-2_11
11. Tesseract Data: tesseract-ocr/tessdata. https://github.com/tesseract-ocr/tessdata/
12. Viola, P., Jones, M.: Rapid object detection using a boosted cascade of simple features. In: Proceedings of the 2001 IEEE Computer Society Conference on Computer Vision and Pattern Recognition, CVPR 2001. IEEE Computer Society (2001). https://doi.org/10.1109/cvpr.2001.990517
13. Walczak, J., Poreda, T., Wojciechowski, A.: Effective planar cluster detection in point clouds using histogram-driven KD-like partition and shifted mahalanobis distance based regression. Remote Sens. 11(21), 2465 (2019)

Depth Map Estimation with Consistent Normals from Stereo Images

Alexander Malyshev[✉]

University of Bergen, Bergen, Norway
alexander.malyshev@uib.no

Abstract. The total variation regularization of non-convex data terms in continuous variational models can be convexified by the so called functional lifting, which may be considered as a continuous counterpart of Ishikawa's method for multi-label discrete variational problems. We solve the resulting convex continuous variational problem by the augmented Lagrangian method. Application of this method to the dense depth map estimation allows us to obtain a consistent normal field to the depth surface as a byproduct. We illustrate the method with numerical examples of the depth map estimation for rectified stereo image pairs.

Keywords: Consistent normal vector field · Point cloud · Augmented Lagrangian method

1 Introduction

Estimation of the depth map for a three-dimensional (3D) scene is a major task of computer vision [17,18]. For example, depth maps can be generated from several two-dimensional (2D) color or grayscale images of a 3D scene taken by one or more cameras positioned at different space locations. More specifically, given a set of 2D images, which consists of rectified stereo pairs of images taken by a stereo pair of cameras, we have to estimate depth maps for each stereo pair of images and merge them into one 3D point cloud. The point cloud is used afterwards to reconstruct a 2D surface of the scene.

Efficient surface reconstruction methods [4,8,10,11], which produce 2D surfaces from 3D point clouds, most often use approximate normal vectors to the reconstructed surfaces. Good surface reconstructions are obtained by means of consistent normal maps [2,7,9], which respect edges and other feature points on the reconstructed surface, i.e. the normal vectors are smooth within the smooth parts of the surface but discontinuous across the edges and corners on the surface.

We estimate the depth map in the form of the global minimum of suitable functionals, which are non-convex in general. The global variational approaches to depth estimation can be more attractive than the faster local procedures

Supported by the Research Executive Agency of the European Commission, grant 778035 - PDE-GIR - H2020-MSCA-RISE-2017.

V. V. Krzhizhanovskaya et al. (Eds.): ICCS 2020, LNCS 12141, pp. 553–565, 2020.
https://doi.org/10.1007/978-3-030-50426-7_41

because they are more robust to various image corruptions and provide wider regularization conditions. However, finding global minima of non-convex functionals may be very difficult or even not feasible in practice.

Our estimation of the depth map uses the continuous variational model

$$\arg\min_u \int_\Omega \alpha |u_x(x)| dx + \int_\Omega \rho(u(x), x) dx, \tag{1}$$

where the first term is called the total variation (TV) regularization introduced in [16]. The value $|u_x(x)| = \sqrt{u_{x_1}^2 + \ldots + u_{x_n}^2}$ denotes the Euclidean length of the gradient vector $u_x = [u_{x_1}, \ldots, u_{x_n}]$. The function $\rho(t, x)$ is supposed to be not convex in variable t. Since the point-wise global minima of $\rho(u(x), x)$ are very noisy functions, the total variation times a well-chosen parameter $\alpha > 0$ enforces necessary smoothness in the solution $u(x)$ of (1).

The authors of [13] have proposed an elegant convexification procedure called the functional lifting, which reduces the non-convex model (1) to a convex variational model but with an extra dimension. Further developments and extensions are found in [12,14]. A similar method for discrete variational models has been earlier proposed by H. Ishikawa [5,6]. The paper [13] also contains a detailed comparison of the continuous functional lifting with Ishikawa's method.

The present paper contributes to the estimation of the depth maps by extending the results from [13]. The numerical method of [13] is replaced by a faster method, which is a variant of the augmented Lagrangian method (ALM); cf. [19]. We point out that the convexification from [13] is not directly suitable for ALM and must be refined as in Theorem 1. A consistent normal field to the depth map surface is obtained as a byproduct of ALM; see formula (20).

2   Convex Relaxation for Continuous Variational Models

Similar to [13], the theoretical arguments below are not fully rigorous but rather informal. For instance, the expression of the total variation, formally valid for functions from the Sobolev space H^1, is applied to functions, which are not necessarily in H^1, etc.

The variational model (1) defines an unknown real-valued function $u: \Omega \to [a, b]$ of bounded variation in the rectangular domain $\Omega = [0, L_1] \times [0, L_2] \times \cdots \times [0, L_n] \subset \mathbb{R}^n$. Note that the regularization parameter $\alpha > 0$ is introduced for convenience only.

The function $\rho(t, x)$ can be non-convex in variable t, which creates serious difficulties when developing reliable numerical methods for solving (1). Fortunately, the non-convex variational model (1) can be reformulated into a convex form by adding an extra dimension, say t, to the available dimensions $x_1, x_2, \ldots, x_n$. Such convexification uses special binary functions called the indicator functions of superlevel sets. Namely, for a given function $u(x)$, the indicator function of superlevel sets $\phi: [a, b] \times \Omega \to \{0, 1\}$ is defined as

$$\phi(t, x) = \begin{cases} 1, \text{if } u(x) \geq t, \\ 0, \text{if } u(x) < t. \end{cases} \tag{2}$$

The function $\phi(x, y)$ is binary and monotonically non-increasing with respect to the variable t. Owing to monotonicity, the original function $u(x)$ is reconstructed from $\phi(t, x)$ via the formula

$$u(x) = a + \int_a^b \phi(t, x)dt. \tag{3}$$

Theorem 1 (Functional uplifting). *If $u(x)$ is a solution of the minimization problem (1), then the indicator function of superlevel sets $\phi(t, x)$, constructed in (2), is a solution to the following minimization problem:*

$$\arg\min_{\phi \in \Phi_{\{0,1\}}} \int_\Omega \int_a^b \alpha|\phi_x(t, x)| - \rho(t, x)\phi_t(t, x)dxdt, \tag{4}$$

$$\Phi_{\{0,1\}} = \{\phi \mid \phi(t, x): [a, b] \times \Omega \to \{0, 1\}; \ \phi(a, x) = 1 \ and \ \phi(b, x) = 0 \ \forall x; \tag{5}$$
$$\phi(t_1, x) \geq \phi(t_2, x)) \ whenever \ t_1 < t_2\}.$$

The converse is also true: if ϕ is the indicator function of superlevel sets associated with u and solves the minimization problem (4)–(5), then u solves (1).

Proof. Following [13], the co-area formula from [3] gives

$$\int_\Omega |u_x(x)|dx = \int_\Omega \int_a^b |\phi_x(t, x)|dtdx,$$

where $|\phi_x(t, x)| = \sqrt{\phi_{x_1}^2 + \ldots + \phi_{x_n}^2}$. The identity $\phi_t(t, x) = -\delta(u(x) - t)$, where $\delta(\cdot)$ is the Dirac delta function, implies the equalities

$$\int_\Omega \rho(u(x), x)dx = \int_\Omega \int_a^b \rho(t, x)\delta(u(x) - t)dtdx = \int_\Omega \int_a^b -\rho(t, x)\phi_t(t, x)dtdx.$$

Theorem 2 (Convex relaxation). *If $\phi(t, x)$ is a solution of the relaxed (from $\Phi_{\{0,1\}}$ to Φ) minimization problem*

$$\arg\min_{\phi \in \Phi} \int_a^b \int_\Omega \alpha|\phi_x(t, x)| - \rho(t, x)\phi_t(t, x)dtdx, \tag{6}$$

$$\Phi = \{\phi \mid \phi(t, x): [a, b] \times \Omega \to [0, 1]; \ \phi(a, x) = 1 \ and \ \phi(b, x) = 0 \ \forall x; \tag{7}$$
$$\phi(t_1, x) \geq \phi(t_2, x)) \ whenever \ t_1 < t_2\},$$

then the binary function

$$\phi^\theta(t, x) = \begin{cases} 1, \ if \ \phi(t, x) \geq \theta, \\ 0, \ if \ \phi(t, x) < \theta, \end{cases} \tag{8}$$

is a solution of (4)–(5) for all threshold values $\theta \in (0, 1)$.

Proof. The co-area formula

$$\int_\Omega |\phi_x(t, x)|dx = \int_\Omega \int_0^1 |\phi_x^\theta(t, x)|d\theta dx.$$

and differentiation of the identity $\phi(t,x) = \int_0^1 \phi^\theta(t,x)d\theta$ with respect to t as

$$\phi_t(t,x) = \int_0^1 \phi_t^\theta(t,x)d\theta$$

allow us to derive that the energy functional

$$E(\phi) = \int_\Omega \int_a^b \alpha|\phi_x(t,x)| - \rho(t,x)\phi_t(t,x)dtdx$$

satisfies the following identity for all $\phi \in \Phi$:

$$E(\phi) = \int_0^1 \left\{ \int_\Omega \int_a^b \alpha|\phi_x^\theta(t,x)| - \rho(t,x)\phi_t^\theta(t,x)dtdx \right\} d\theta$$

$$= \int_0^1 E(\phi^\theta)d\theta.$$

When ϕ is a minimizer of the functional E in Φ, the inequality $E(\phi^\theta) \geq E(\phi)$ holds for all $\theta \in (0,1)$ because $\phi^\theta \in \Phi$. If the measure of the set of all θ for which $E(\phi^\theta) > E(\phi)$ is larger than 0, then $E(\phi) = \int_0^1 E(\phi^\theta)d\theta > E(\phi)$. Hence the measure equals zero.

An argument similar to those in [1,14] can be used to include θ belonging to the exceptional set of zero measure.

3   The Augmented Lagrangian Method

Let us introduce a dual function $p(t,x)$ such that $p = (p_0, p_1) = \nabla\phi$, where $\nabla\phi = (\phi_t, \phi_x)$ is the full gradient of $\phi(t,x)$ and $|p_1| = \|p_1\|_2$ is the Euclidean norm of p_1. Then the problem (6)–(7) is equivalent to the following constrained convex minimization problem

$$\min_{\phi,p} \int_{[a,b]\times\Omega} \alpha|p_1(t,x)| - \rho(t,x)p_0 \, dtdx, \tag{9}$$

$$p_0 = \phi_t, \; p_1 = \phi_x; \; p_0 \leq 0, \; \phi(a,x) = 1, \; \phi(b,x) = 0. \tag{10}$$

The convex variational problem (9)–(10) can be solved by means of the augmented Lagrangian method (cf. [19]) with the augmented Lagrangian

$$\mathcal{L}(\phi,p,\lambda) = \int_{[a,b]\times\Omega} \left[\alpha|p_1| - \rho p_0 + \langle \lambda, p - \nabla\phi \rangle + \frac{c}{2}\|p - \nabla\phi\|_2^2 \right] dtdx, \tag{11}$$

where $p = (p_0, p_1)$ and $\lambda = (\lambda_0, \lambda_1)$. The inner product $\langle \lambda, p - \nabla\phi \rangle$ is Euclidean. The constant $c > 0$ is sufficiently large but is not required to tend to infinity.

Each iteration of the ALM method consists of alternative minimizations with respect to ϕ and p and special updates of λ.

Algorithm ALM

1. Set $k = 0$ and initialize $p^0 = 0$ and $\lambda^0 = 0$.
2. Find the solution ϕ^{k+1} of the minimization problem

$$\phi^{k+1} = \arg \min_{\substack{\phi \\ \phi(a,x)=1 \\ \phi(b,x)=0}} \mathcal{L}(\phi, p^k, \lambda^k)$$

3. Find the dual variable p^{k+1} by solving the minimization problem

$$p^{k+1} = \arg \min_{\substack{p=(p_0,p_1) \\ p_0 \leq 0}} \mathcal{L}(\phi^{k+1}, p, \lambda^k)$$

4. Update multiplier λ in accordance with the augmented Lagrangian method

$$\lambda^{k+1} = \lambda^k + c(p^{k+1} - \nabla \phi^{k+1})$$

5. Set $k = k + 1$ and go to step 2.

Let us consider Steps 2 and 3 in detail.

Minimization with Respect to ϕ. Step 2 of the ALM algorithm concerns the optimization problem

$$\min_{\phi} \int_{[a,b] \times \Omega} \left[-\langle \lambda, \nabla \phi \rangle + \frac{c}{2} \|p - \nabla \phi\|_2^2 \right] dt dx$$

where $\phi(t, x)$ satisfies the boundary conditions $\phi(a, x) = 1$, $\phi(b, x) = 0$. Standard arguments from the variational calculus yield the Poisson equation

$$\phi_{tt} + \sum_{i=1}^{n} \phi_{x_i x_i} = (p_0 + \lambda_0/c)_t + \sum_{i=1}^{n} ((p_1)_i + \lambda_i/c)_{x_i} \tag{12}$$

with the Dirichlet and Neumann boundary conditions

$$\phi(a, x) = 1, \quad \phi(b, x) = 0, \tag{13}$$
$$\phi_{x_i}(t, x) = ((p_1)_i + \lambda_i/c)(t, x) \quad \text{if } x_i = 0 \text{ or } x_i = L_i. \tag{14}$$

Minimization with Respect to p. Step 3 of the ALM algorithm requires solution to the pointwise minimization problem

$$\arg \min_{p_0, p_1 : p_0 \leq 0} \alpha \|p_1\|_2 - \rho p_0 + \langle \lambda, p - \nabla \phi \rangle + \frac{c}{2} \|p - \nabla \phi\|_2^2 \tag{15}$$

$$= \arg \min_{p_0, p_1 : p_0 \leq 0} \frac{2\alpha}{c} \|p_1\|_2 - \frac{2\rho}{c} p_0 + \|p - (\nabla \phi - \lambda/c)\|_2^2.$$

Let us denote

$$q_0 = \phi_t - \lambda_0/c \text{ and } q_1 = \phi_x - \lambda_1/c. \tag{16}$$

Solution to $\min_{p_0 \leq 0} -\frac{2\rho}{c} p_0 + |p_0 - q_0|^2$ is $p_0 = \min(q_0 + \rho/c, 0)$.

A simple geometric argument reveals that the minimum point of the function $\frac{2\alpha}{c} \|p_1\|_2 + \|p_1 - q_1\|_2^2$ must have the form $p_1 = rq_1$ with $0 \leq r \leq 1$. For $r \in (0, 1)$ the necessary condition for extremal points is satisfied when

$$\frac{d}{dr}\left[\frac{2\alpha}{c} r + (r-1)^2 \|q_1\|_2 \right] = \frac{2\alpha}{c} + 2(r-1)\|q_1\|_2 = 0.$$

Hence $r = 1 - \alpha/(c\|q_1\|_2)$ if $\|q_1\|_2 - \alpha/c > 0$. Otherwise, $r = 0$. As a result, solution to (15) is given by the pointwise formulas

$$p_0 = \min(q_0 + \rho/c, 0), \tag{17}$$

$$(p_1)_i = \begin{cases} [1 - \alpha/(c\|q_1\|_2)](q_1)_i, & \|q_1\|_2 - \alpha/c > 0, \\ 0, & \text{otherwise.} \end{cases} \tag{18}$$

By Theorem 2, an approximation to $u(x)$ can be obtained in the form

$$u(x) = a + \int_a^b \phi^{1/2}(t, x)dt. \tag{19}$$

The normal field to the surface $t - u(x) = 0$ in the space with coordinates (t, x) is the set of the gradient vectors $n(t, x) = [1, -u_{x_1}, \ldots, -u_{x_n}]^T$. We use formula (19) to express the derivatives of $u(x)$ in terms of $\phi^{1/2}(t, x)$ as $u_{x_i} = \int_a^b \phi_{x_i}^{1/2}(t, x)dt$. Since the derivatives $\phi_{x_i}(t, x)$ in the AL method are approximated by $(p_1)_i$, the normal field can be approximated as

$$n(t, x) = \frac{v}{\|v\|_2}, \text{ where } v(t, x) = \left[1, -\int_a^b (p_1)_1(t, x)dt, \ldots, -\int_a^b (p_1)_n(t, x)dt\right]^T. \tag{20}$$

Formula (20) is one of the main contributions of the paper.

4   Numerical Experiments

The meaning of the variable x is changed in Sect. 4. Instead of the points $x = (x_1, \ldots, x_n) \in \mathbb{R}^n$, we deal with the pixels $(x, y) \in \mathbb{R}^2$.

4.1   Depth Map from Stereo Pairs of Images

Our numerical experiments deal with estimation of the dense depth map; see, e.g., [17,18] for more detail about depth estimation. Suppose that two functions $L(x, y)$ and $R(x, y)$ represent a pair of left and right 2-dimensional grayscale images with the horizontal coordinates $0 \leq x \leq L_x$ and vertical coordinates $0 \leq y \leq L_y$. We assume that the images L and R are rectified so that one can define the disparity map $t = t(x, y)$ as the parallel translation along the horizontal direction such that a point (x, y) in the left image L matches the point $(x - t(x,y), y)$ in the right image R. The disparity map is converted into the depth map via the simple formula $depth = focus \cdot baseline/disparity$, where $baseline$ is the distance between the two camera positions, from which the images L and R have been taken. We assume that both cameras are identical, have focal length $focus$ and equally oriented in 3D space.

In order to estimate the disparity map for grayscale images, we use the simplest dissimilarity function

$$\rho(t,x,y) = |L(x,y) - R(x - t(x,y), y)|. \tag{21}$$

The range of the variable t in (21) for fixed variables x and y is the interval $\max(x - L_x, a) \leq t \leq \min(x, b)$ because the first argument $x - t$ of R must satisfy $0 \leq x - t \leq L_x$. The function $\rho(t,x,y)$ equals zero outside this range. More sophisticated dissimilarity functions can be found in [17,18].

The dissimilarity function for RGB images can be constructed as the sum of absolute differences of intensities for all three color channels

$$\rho(t,x,y) = \sum_{C=\{R,G,B\}} |L_C(x,y) - R_C(x - t(x,y), y)|.$$

4.2 About Solving the Poisson Equation in Rectangular Domains

The Poisson Eq. (12) with the Dirichlet boundary conditions (13) and the Neumann boundary conditions (14) can be solved by several numerical methods.

When the size $\mathcal{N}$ of the discretized function $\phi(t,x,y)$ is not very large, the Poisson equation can be solved by means of the fast discrete sine and cosine transforms. The arithmetic complexity of the fast Poisson solver is $O(\mathcal{N} \log \mathcal{N})$.

For very large $\mathcal{N}$, the Poisson equation can be efficiently solved by the multigrid method, which has the linear arithmetic complexity $O(\mathcal{N})$.

4.3 Synthetic Dataset

In order to demonstrate behaviour of the proposed method, we use a synthetic dataset, where all functions are constant along the axis y. This dataset is convenient for visualization because it is enough to plot only sections of such functions for a fixed y. These sections are the horizontal lines of the images displaying the functions. The depth function $d(x,y) = 10 + \sin(\text{mod}(x, \pi))$ of the synthetic dataset is defined for $|x| \leq 5\pi$ and defines the surface $z = d(x,y)$. Two identical cameras with focus equal to 1 are located at the points $(x,y) = (-2,0)$ and $(x,y) = (2,0)$. The coordinate axes x, y and z of both cameras are oriented in parallel to the axes x, y and z of the surface $z = d(x,y)$. The left and right digital images $L(x,y)$ and $R(x,y)$ have size 128×10. The horizontal lines of L and R for a fixed y are shown in Fig. 1. Note that the most left 15 pixels of L and the most right 15 pixels of R on each horizontal line are outside the scene. The horizontal lines of the true disparity are shown on the right side of Fig. 2.

The augmented Lagrangian method has been implemented with the fast Poisson solver, which is based on the fast Fourier transform. The following parameters

Fig. 1. Horizontal lines of the rectified stereo images L and R

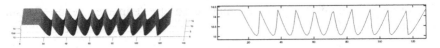

Fig. 2. The true disparity and its horizontal lines

have been used for computation by this method: $a = 13$, $b = 14.35$, $\alpha = 0.7$, $c = 0.1$. The number of nodes along the coordinate directions t, x and y are respectively 129, 128, and 16. The grid steps are $h_t = (b - a)/128$, $h_x = h_y = 1$. We have run 100 iterations of the augmented Lagrangian method. However, a sufficiently good convergence is achieved much earlier, say after 30 iterations.

Figure 3 displays the computed disparity versus the true disparity. The computed disparity is plotted using the blue solid line, and the true disparity using the red dash-dot line. We recall that the displayed disparity is defined in the coordinate system of the left image L. Note also that the most left 15 pixels lie outside the scene, i.e. the parts of both lines for $x = 1, 2, \ldots, 15$ are fictitious.

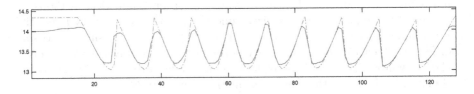

Fig. 3. Disparity computed by the AL method vs. true disparity

This section of the paper aims to justify numerically that the augmented Lagrangian method produces consistent normals to the surfaces. The surface of the synthetic example is the disparity map $d(x, y)$, which is constant along y. The computed solution $u(x, y)$ turns out to be also constant along y. The computed function $(p_1)_2(x, y)$ is zero everywhere. We recall that the normal field $n(x, y)$ is computed by the formulas

$$n(x, y) = v(x, y)/\|v(x, y)\|_2, \text{ where } v(x, y) = [1, -h_t(p_1)_1, -h_t(p_1)_2].$$

Figure 4 displays the normals along the horizontal line of the disparity map computed by the augmented Lagrangian method.

4.4 Comparison with the Numerical Method of [13]

Apart from introducing the functional lifting for convexification, the authors of [13] propose an efficient numerical method for solving the resulting convex

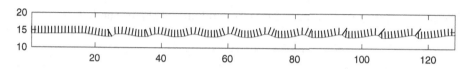

Fig. 4. Normals to the disparity surface computed by the AL method

variational problem. Provided that $\rho(t, x, y) \geq 0$, they consider the convex model

$$\arg\min_{\phi \in D} \int_a^b \int_\Omega \left[\alpha \sqrt{\phi_x(t, x, y)^2 + \phi_y(t, x, y)^2} + \rho(t, x, y)|\phi_t(t, x, y)| \right] dt\,dx\,dy,$$
(22)

$$D = \{\phi \mid \phi(t, x, y) \colon [a, b] \times \Omega \to [0, 1]; \ \phi(a, x, y) = 1 \text{ and } \phi(b, x, y) = 0\}, \quad (23)$$

which is equivalently transformed into the primal-dual formulation

$$\min_{\phi \in D} \max_{p \in C} \int_a^b \int_\Omega \left(\phi_x p_1 + \phi_y p_2 + \frac{\rho}{\alpha} \phi_t p_0 \right) dx\,dy\,dt, \quad (24)$$

where

$$C = \{p(t, x, y) \mid p(t, x, y) = (p_0, p_1, p_2)(t, x, y) \colon [a, b] \times \Omega \to [0, 1]; \quad (25)$$

$$|p_0(t, x, y)| \leq \frac{\rho(t, x, y)}{\alpha}; \ \sqrt{p_1^2(t, x, y) + p_2^2(t, x, y)} \leq 1\}.$$

Such a transformation is obtained by using the vectors $[p_1(t, x, y), p_2(t, x, y)]$ coinciding with the normalized vectors $[\phi_x, \phi_y]/\sqrt{\phi_x^2 + \phi_y^2}$. Apparently, this representation does not lead to formula (20) for a consistent normal field.

The numerical algorithm for solving (24), proposed in [13], is called a primal-dual proximal point (PDPP) method and iteratively minimizes the functional in (24) with respect to the primal variable ϕ and then maximizes the same functional with respect to the dual variable p. Each iteration consists of the following two steps:

primal step $\phi^{k+1} = \mathcal{P}_D(\phi^k + \tau_p \mathrm{div} p^k)$,
dual step $p^{k+1} = \mathcal{P}_C(p^k + \tau_d \nabla \phi^{k+1})$.

The operator $\mathcal{P}_D$ denotes the projection onto the set D, which can be computed by a simple truncation of ϕ^{k+1} to the interval $[0, 1]$ and setting $\phi(a, x, y) = 1$ and $\phi(b, x, y) = 0$. The operator $\mathcal{P}_C$ denotes the projection onto the set C, which can be computed via the formulas

$$p_1^{k+1} = p_1^{k+1} / \max\left(1, \sqrt{(p_1^{k+1})^2 + (p_2^{k+1})^2} \right)$$

$$p_2^{k+1} = p_2^{k+1} / \max\left(1, \sqrt{(p_1^{k+1})^2 + (p_2^{k+1})^2} \right)$$

$$p_0^{k+1} = p_0^{k+1} / \max\left(1, \frac{\alpha|p_0^{k+1}|}{\rho} \right).$$

The parameters τ_p and τ_d must guarantee stability of the method. The authors of [13] propose the choice $\tau_p = \tau_d = 1\sqrt{3}$. For the synthetic dataset, sufficient convergence is observed after 1000 iterations (Fig. 5).

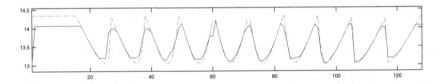

Fig. 5. Disparity computed by the PDPP method vs. true disparity

To get the plot on the upper level of Fig. 6, the PDPP method was iterated 2000 times, which is much longer than necessary for convergence. The normals have been computed by the formula $n(t, x, y) = [1, -u_x, -u_y]^T$, where $u(x, y)$ is the computed solution of (22).

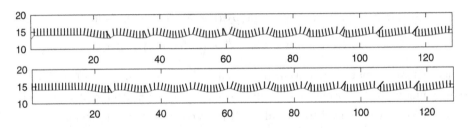

Fig. 6. Normals computed by the PDPP method are shown on on the upper level; normals computed by the AL method are shown on the lower level

The comparison of the normal fields in Fig. 6 demonstrates that both the augmented Lagrangian method and the PDPP method produce normal fields of similar visual quality.

4.5   Tsukuba Dataset

Let us use the dataset Tsukuba from http://vision.middlebury.edu/stereo in order to demonstrate the working capacity of the AL method for calculation of a consistent normal field. The rectified stereo image pair Tsukuba of the smallest size 384×288 consists of two RGB images. The image from the left camera is shown as the left image in Fig. 7.

The corresponding image of the true disparity is also available at the Middlebury repository [15]; see the right image in Fig. 7. The disparity map has 256 gray levels but it can be scaled to the interval $[0, 15]$ of integer numbers without any loss. The scaled image has only the following gray values (or labels):

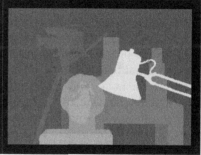

Fig. 7. The left image and the true disparity for the Tsukuba dataset

$0, 5, 6, 7, 8, 10, 11, 14$. The interval (measured in pixels) containing all disparity values is $[a, b] = [0, 16]$.

The suitable parameters in the augmented Lagrangian (11) are $\alpha = 0.1$ and $c = 0.1$. The label space $[a, b]$ is divided into 32 subintervals of the length $h_t = (b - a)/32$. We have run only 100 iterations of the AL method. Figure 8 shows the computed disparity map and consistent normals for the Tsukuba rectified stereo pair.

Fig. 8. Computed disparity and consistent normals for the Tsukuba dataset

5   Conclusion

We have developed a variant of the augmented Lagrangian method for a convex relaxation of the continuous variational problem with the total variation regularization. The most time consuming part of this method is solving a boundary value problem for the 3D Poisson equation. We solve it by the fast Poisson solver. The augmented Lagrangian method has a significantly faster convergence than the primal-dual proximity point method from [13].

An additional benefit of the augmented Lagrangian method consists in producing a consistent normal vector field to the solution surface as a byproduct. Numerical differentiation of the solution computed by the method from [13] seems to produce a normal vector field of similar quality. However, thorough verification of these properties requires further investigation.

References

1. Berkels, B.: An unconstrained multiphase thresholding approach for image segmentation. In: Tai, X.-C., Mørken, K., Lysaker, M., Lie, K.-A. (eds.) SSVM 2009. LNCS, vol. 5567, pp. 26–37. Springer, Heidelberg (2009). https://doi.org/10.1007/978-3-642-02256-2_3
2. Castillo, E., Liang, J., Zhao, H.-K.: Point cloud segmentation via constrained nonlinear least squares surface normal estimates. In: Breuß, M., Bruckstein, A., Maragos, P. (eds.) Innovations for Shape Analysis: Models and Algorithms. MATHVISUAL, pp. 283–299. Springer, New York (2013). https://doi.org/10.1007/978-3-642-34141-0_13
3. Fleming, W., Rishel, R.: An integral formula for total gradient variation. Arch. Math. **11**, 218–222 (1960)
4. Hoppe, H., Derose, T., Duchamp, T., McDonald, J., Stuetzle, W.: Surface reconstruction from unorganized points. In: Proceedings of the 19th Annual Conference on Computer Graphics and Interactive Techniques, ACM SIGGRAPH Computer Graphics, vol. 26, pp. 71–78 (1992)
5. Ishikawa, H.: Exact optimization for Markov random fields with convex priors. IEEE Trans. Pattern Anal. Mach. Intell. **25**(10), 1333–1336 (2003)
6. Ishikawa, H.: Graph cuts-combinatorial optimization in vision. In: Lézoray, O., Grady, L. (eds.) Image Processing and Analysis with Graphs, pp. 25–63. CRC Press (2012)
7. Jakob, J., Buchenau, C., Guthe, M.: Parallel globally consistent normal orientation of raw unorganized point clouds. Comput. Graph. Forum **38**(5), 163–173 (2019)
8. Kazhdan, M., Bolitho, M., Hoppe, H.: Poisson surface reconstruction. In: Proceedings of the Fourth Eurographics Symposium on Geometry Processing, Eurographics Association, pp. 61–70 (2006)
9. König, S., Gumhold, S.: Consistent propagation of normal orientations in point clouds. In: Proceedings of the Vision, Modeling, and Visualization Workshop, Braunschweig, Germany, p. 9 (2009)
10. Lempitsky, V., Boykov, Y.: Global optimization for shape fitting. In: IEEE Conference on Computer Vision and Pattern Recognition (CVPR), pp. 1–8. IEEE Xplore (2007)
11. Liang, J., Park, F., Zhao, H.: Robust and efficient implicit surface reconstruction for point clouds based on convexified image segmentation. J. Sci. Comput. **54**(2–3), 577–602 (2013)
12. Möllenhoff, T., Laude, E., Möller, M., Lellmann, J., Cremers, D.: Sublabel-accurate relaxation of nonconvex energies. In: IEEE Conference on Computer Vision and Pattern Recognition (CVPR 2016), pp. 3948–3956. IEEE Xplore (2016)

13. Pock, T., Schoenemann, T., Graber, G., Bischof, H., Cremers, D.: A convex formulation of continuous multi-label problems. In: Forsyth, D., Torr, P., Zisserman, A. (eds.) ECCV 2008. LNCS, vol. 5304, pp. 792–805. Springer, Heidelberg (2008). https://doi.org/10.1007/978-3-540-88690-7_59
14. Pock, T., Cremers, D., Bischof, H., Chambolle, A.: Global solutions of variational models with convex regularization. SIAM J. Imaging Sci. **3**(4), 1122–1145 (2010)
15. http://vision.middlebury.edu/stereo
16. Rudin, L.I., Osher, S., Fatemi, E.: Nonlinear total variation based noise removal algorithms. IEEE Trans. Pattern Anal. Mach. Intell. **60**(1), 259–268 (1992)
17. Scharstein, D., Szeliski, R.: A taxonomy and evaluation of dense two-frame stereo correspondence algorithms. Int. J. Comput. Vis. **47**(1–3), 7–42 (2002)
18. Szeliski, R.: Computer Vision: Algorithms and Applications. Springer, Heidelberg (2010). https://doi.org/10.1007/978-1-84882-935-0
19. Wu, C., Tai, X.-C.: Augmented lagrangian method, dual methods, and split bregman iteration for ROF, vectorial TV, and high order models. SIAM J. Imaging Sci. **3**(3), 300–339 (2010)

Parametric Learning of Associative Functional Networks Through a Modified Memetic Self-adaptive Firefly Algorithm

Akemi Gálvez[1,2] , Andrés Iglesias[1,2]([⊠]) , Eneko Osaba[3] ,
and Javier Del Ser[3,4]

[1] Department of Information Science, Faculty of Sciences, Toho University,
2-2-1, Miyama, Funabashi 274-8510, Japan
[2] Department of Applied Mathematics and Computational Sciences,
University of Cantabria, 39005 Santander, Spain
{galveza,iglesias}@unican.es
[3] TECNALIA, Basque Research and Technology Alliance (BRTA),
48160 Derio, Spain
{eneko.osaba,javier.delser}@tecnalia.com
[4] University of the Basque Country (UPV/EHU), 48013 Bilbao, Spain

Abstract. Functional networks are a powerful extension of neural networks where the scalar weights are replaced by neural functions. This paper concerns the problem of parametric learning of the associative model, a functional network that represents the associativity operator. This problem can be formulated as a nonlinear continuous least-squares minimization problem, solved by applying a swarm intelligence approach based on a modified memetic self-adaptive version of the firefly algorithm. The performance of our approach is discussed through an illustrative example. It shows that our method can be successfully applied to solve the parametric learning of functional networks with unknown functions.

Keywords: Artificial intelligence · Functional networks · Associative model · Parametric learning · Swarm intelligence · Firefly algorithm

1 Introduction

Models in science and engineering are usually expressed in the form of mathematical equations representing the reality with a given quality level. In addition to the free variables determining the degrees of freedom of the system, such equations usually involve some particular parameters accounting for the conditions of the problem. Learning such parameters is of paramount importance for an accurate and realistic description of the observed behavior of the system. This is also a very challenging task, typically demanding a lot of expertise, time, and effort from a human expert in order to get reliable results. Seeking to overcome this limitation, several approaches and methodologies have been devised to address this issue automatically. A classical example arises in neural networks, where

© Springer Nature Switzerland AG 2020
V. V. Krzhizhanovskaya et al. (Eds.): ICCS 2020, LNCS 12141, pp. 566–579, 2020.
https://doi.org/10.1007/978-3-030-50426-7_42

several methods for parametric learning have been reported in the literature. In this work, we focus our attention on the functional networks, which are a powerful extension of the standard artificial neural networks (see our discussion in Sect. 2 for further details).

In this paper, we consider the problem of parametric learning of a classical model of functional networks, the so-called associative model, which is used to represent the associativity operator. This problem can be formulated as a nonlinear continuous least-squares minimization problem. We solve it by applying a swarm intelligence approach based on a modified memetic self-adaptive version of the firefly algorithm. The paper is organized as follows: functional networks and their main components are described in Sect. 2. Sections 3 and 4 discuss the problem to be solved, and the firefly algorithm and its variants, respectively. Section 5 describes the method used to solve this optimization problem. Section 6 discuss an illustrative example. The paper closes with the main conclusions and a brief discussion about future work.

2 Functional Networks

In short, *functional networks* can be regarded as a generalization of the standard artificial neural network in which the classical scalar weights of the neural networks are replaced by neural functions. This methodology was firstly described in 1998 by E. Castillo in [2] as a way to extend neural networks with new capabilities. Since then, they have been successfully applied to several problems in science and engineering. The interested reader is referred to [3] for a detailed explanation about functional networks, several examples and applications. Next paragraphs describe the main components of a functional network as well as the differences between neural networks and functional networks.

2.1 Components of a Functional Network

As the functional networks generalize the neural networks, they share several common features, including a close graphical representation. Figure 1(a) shows a functional network called *associative functional network* (discussed in detail in Sect. 3.1), which represents the associativity operator. Following this figure, we can identify the main components of a functional network. They are:

1. Some layers of storing units: in Fig. 1(a), we can see a first layer of input units, which contains the input information. In our example, this input layer consists of the units ξ, ζ and v. We also have some intermediate layers of storing units. We point out that they are not neurons, but units storing some intermediate information. This set is optional and it is used to allow more than one neuron output to be connected to the same unit. For instance, in Fig. 1(a), we can see one layer with 4 intermediate units, represented by small circles in black. Finally, we have a layer of output units. In Fig. 1(a), it consists only of the unit μ.

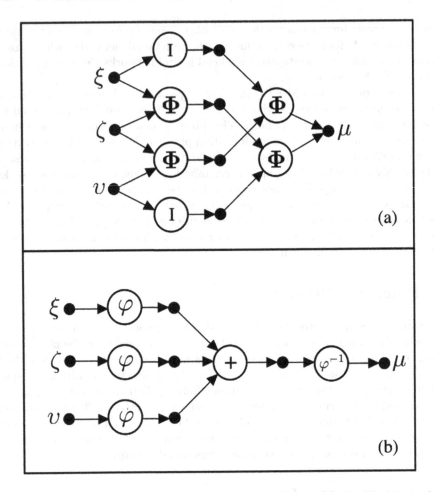

Fig. 1. Associative functional network: (a) original network; (b) simplified network.

2. One or more layers of computing units called neurons. Each neuron receives a set of input values, coming from the previous layer, makes some computations with them and returns a set of output values to the next layer. Neurons are represented graphically by circles, with the name of the corresponding neural function written inside. For example, in Fig. 1(a), we have 6 neurons arranged in two layers, where Φ and I represent the associative operator and the identity function, respectively.

3. A set of directed links, represented graphically by directional arrows. These arrows connect the input layer (or any intermediate layer, in general) to its adjacent layer of neurons, and neurons of one layer to its adjacent intermediate layers, or to the output layer. Note that the information flow is exclusively unidirectional: it always flows from the input layer to the output layer.

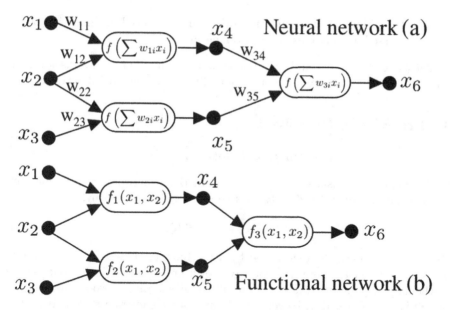

Fig. 2. Graphical differences between: (top) neural networks; (bottom) functional networks.

The collection of all these features define the so-called *network architecture* or *topology* of the functional network, which determines the functional capabilities of the network.

2.2   Differences Between Neural Networks and Functional Networks

In spite of their similarities, there is a number of differences between neural networks and functional networks. The most important ones are:

1. Each neuron of a standard neural network returns an output value $y = f(\sum w_{ik}x_k)$ that depends only on the value $\sum w_{ik}x_k$, where $x_1, x_2, \ldots, x_n$ are the received inputs (see Fig. 2 (top)). This means that each neural function is always univariate, as opposed to the case of functional networks, in which the neural functions are multivariate, as shown in Fig. 2 (bottom).
2. The neural functions of the functional networks can be *different* (such as functions f_1, f_2 and f_3 in Fig. 2 (bottom)). In contrast, the neural functions in neural networks are generally all *identical*.
3. The neural networks contain scalar values called weights, which must be learned. This component is not part of the functional networks, where neural functions have to be learned instead.
4. The neuron outputs of the neural networks are usually different. On the contrary, the neuron outputs of the functional networks can be coincident. In such cases, we obtain a set of functional equations, which have to be solved through specialized techniques, such as those reported in [2,3]. As a consequence, the

neural functions in functional networks can be reduced in dimension or can be expressed as functions of lower dimension.

All these differences and features show that functional networks are more general and exhibit more interesting capabilities than the neural networks.

3   Problem to Be Solved

3.1   The Associative Functional Network

In this paper we consider the associative functional network, which represents the associativity operator Φ between two real numbers, given by:

$$\Phi(\Phi(\xi,\zeta),v) = \Phi(\xi,\Phi(\zeta,v)). \tag{1}$$

Our goal is to learn the function Φ by using functional networks. To this purpose, we consider the network topology shown in Fig. 1(a), which replicates the mathematical structure of Eq. (1). Initially, it seems that a two-argument function Φ is to be learned. However, Eq. (1) puts some constraints on it. In fact, it can be proved that the general solution of the functional Eq. (1) takes the form (see [3] for details):

$$\Phi(\xi,\zeta) = \varphi^{-1}[\varphi(\xi) + \varphi(\zeta)] \tag{2}$$

where $\varphi(\xi)$ is a continuous and strictly monotonic (but otherwise arbitrary) function, which can only be replaced by $\eta\,\varphi(\xi)$, with η being an arbitrary constant. Replacing now Eq. (2) in Eq. (1), the two sides of Eq. (1) can be written as:

$$\varphi^{-1}[\varphi(\xi) + \varphi(\zeta) + \varphi(v)] \tag{3}$$

which means that the functional network in Fig. 1(a) is equivalent to the functional network in Fig. 1(b), where only a one-argument function φ has to be learned. This observation leads to two important conclusions:

1. This is the unique functional form for Φ that satisfies Eq. (1). So, the neurons Φ cannot be replaced by any other neurons.
2. The initial two-dimensional function Φ is fully determined by a univariate function φ. This means that the effect of the functional Eq. (1) is to reduce the initial degrees of freedom of Φ from a bivariate function to a univariate function φ.

3.2   Parametric Learning of the Associative Functional Network

Suppose now that we are provided with a set of α data points $\{(\xi_j,\zeta_j,v_j)\}_{j=1,\dots,\alpha}$, obtained from a certain unknown function $v = \Phi(\xi,\zeta)$. Let us also assume that no information is available about the form of the function, but we still know that it is associative, i.e., it follows Eq. (1). To learn this associative operator, we can take pairs of numbers and their operated values as triplets $(\xi_j, \zeta_j, \upsilon_j)$ such that $\upsilon_j = \boldsymbol{\Phi}(\xi_j, \zeta_j) = \xi_j \oplus \zeta_j$, for $j = 1, \ldots, \alpha$. From Eq. (2) we get:

$$\omega = \boldsymbol{\Phi}(\xi_j, \zeta_j) \iff \varphi(\omega) = \varphi(\xi) + \varphi(\zeta) \tag{4}$$

an interesting relation to be exploited for learning $\varphi(\xi)$. Therefore, learning the associative functional network is equivalent to learning the function $\varphi(\xi)$. To this end, we can approximate $\varphi(\xi)$ by a function:

$$\psi(\xi) = \sum_{i=1}^{\beta} \delta_i \psi_i(\xi), \tag{5}$$

where the $\{\psi_i(\xi)\}_{i_1, \ldots, \beta}$ is a given set of linearly independent functions, with the ability to approximate $\varphi(\xi)$ to the desired accuracy, and the coefficients δ_i are the parameters of the functional network. Note that this means that they assume the role played by the weights in a neural network.

In order to estimate the coefficients $\{\delta_i\}_{i=1, \ldots, \beta}$, we use the available data in the form of triplets $(\xi_j, \zeta_j, \upsilon_j)$. According to Eq. (4) we must have

$$\varphi(\upsilon_j) = \varphi(\xi_j) + \varphi(\zeta_j) \tag{6}$$

for $j = 1, \ldots, \alpha$. Then, the error of the approximation can be measured as:

$$\chi_j = \psi(\xi_j) + \psi(\zeta_j) - \psi(\upsilon_j) \tag{7}$$

To estimate the coefficients $\{\delta_i\}_{i=1, \ldots, \beta}$, we minimize the sum of squared errors:

$$\sum_{j=1}^{\alpha} \chi_j^2 = \sum_{j=1}^{\alpha} \left(\sum_{i=1}^{\beta} \delta_i \left[\psi_i(\xi_j) + \psi_i(\zeta_j) - \psi_i(\upsilon_j) \right] \right)^2 \tag{8}$$

subject to $\varphi(\xi_\kappa) \equiv \sum_{i=1}^{\alpha} \delta_i \psi_i(\xi_\kappa) = \lambda$, where λ is an arbitrary but given real constant, required to identify the constant η discussed above.

To summarize, learning the associative functional network finally reduces to perform parametric learning on this approximation function $\psi(\xi)$. This requires to solve a least-squares minimization problem:

$$\Lambda = \underset{\{\delta_i\}_i, \lambda}{\text{minimize}} \left[\sum_{j=1}^{\alpha} \left(\sum_{i=1}^{\beta} \delta_i \left[\psi_i(\xi_j) + \psi_i(\zeta_j) - \psi_i(\upsilon_j) \right] \right)^2 + \left(\sum_{i=1}^{\alpha} \delta_i \psi_i(\xi_\kappa) - \lambda \right)^2 \right] \tag{9}$$

Unfortunately, this is a difficult multivariate nonlinear continuous optimization problem. In this paper, we address this issue by applying a swarm intelligence approach based on a modified memetic self-adaptive version of the firefly algorithm. The original firefly algorithm and the modifications introduced in this paper are briefly explained in next section.

4 The Firefly Algorithm

4.1 Original Algorithm

In this work we rely on the firefly algorithm, a bio-inspired computational algorithm for optimization [14,15]. The basic inspiration for the algorithm is the observed flashing behavior of fireflies in nature; in particular, the variation of the intensity of light and the concept of attractiveness, which is assumed to be related with the encoded target function. The interested reader is referred to [16] for further details on the firefly algorithm. See also [4] for an updated review on bio-inspired computation at large.

The firefly algorithm is a population- based method in which the individuals (fireflies) are randomly distributed over the search space and perform exploration searching for the best location, related to the quality of the solution. The motion at iteration $t+1$ of a firefly i which is attracted by a more attractive (i.e., brighter) firefly j is governed by the following evolution equation:

$$\mathbf{X}_i^{t+1} = \mathbf{X}_i^t + \beta_0 e^{-\gamma r_{ij}^\mu}(\mathbf{X}_j^t - \mathbf{X}_i^t) + \alpha \left(\sigma - \frac{1}{2} \right) \tag{10}$$

where the three terms on the right-hand side of the equation account respectively for the current position of the firefly, the attractiveness of the firefly to light intensity seen by neighbor fireflies, and a random movement of the firefly if it is the brightest one. Coefficients α and σ are random numbers uniformly distributed on the interval $[0, 1]$.

Since its appearance, the firefly algorithm has been successfully applied to several problems in many different fields (see, for instance, [1,6–9] for some illustrative applications). Also, several modifications and enhanced versions on the original algorithm have been developed [10,11,13]. We refer the reader to the paper in [5] for a review and taxonomic classification of several firefly algorithms and its variants and applications.

4.2 Modified Memetic Self-adaptive Firefly Algorithm

A promising line of research nowadays is given by the so-called *memetic algorithms*. Basically, they consist of the hybridization of a global search method with a local search procedure. In agreement with this, we consider here a modified version of the firefly algorithm that enhances the original algorithm with three additional features: the use of self-adaptation schemes on some control parameters, a new elitist population model, and the hybridization with a heuristics for local search.

The first modification consists of the application of self-adaptation schemes on some control parameters; in our particular case, this strategy is applied on the randomization parameter α, the attractiveness β, and the light absorption coefficient γ. For parameter α it is convenient to consider relatively large values at initial stages, thus promoting the explorative ability at the beginning of the simulation. Consequently, we apply a self-adaptive perturbation driven by:

$$\alpha^{t+1} = \alpha^t \left(1 - \frac{t-1}{T_{max}}\right)^2$$

with $\alpha^0 = 0.9$. This means that we start with a high randomization parameter value, e.g. $\alpha = 0.9$, which corresponds to a system where fireflies are affected by a relatively large perturbation leading to a wide-range exploration, and slowly reduce it to lower values near to 0, where the system intensifies the exploitation around the local optima.

At their turn, parameters β and γ undergo a process of uncorrelated mutation where each control parameter is perturbed additively according to the normal distribution modulated by the mutation strength of that parameter, as:

$$\beta^{t+1} = \beta^t + \zeta_\beta^t N(0,1) \quad ; \quad \gamma^{t+1} = \gamma^t + \zeta_\gamma^t N(0,1) \tag{11}$$

where the mutation strength ζ_i^t, $i = \beta, \gamma$, also undergoes mutation determined by a characteristic time called the learning rate:

$$\zeta_i^{t+1} = \zeta_i^t \, exp\left[\tau' N(0,1) + \tau N_i(0,1)\right] \tag{12}$$

where $\tau' \propto 1/\sqrt{2d}$ and $\tau \propto 1/\sqrt{2\sqrt{d}}$.

The second modification concerns the population model. In the original firefly algorithm, the population of N_P individuals is entirely replaced in each generation. Consequently, it misses some valuable features typically found in the evolutionary algorithms, such as the selection pressure for survival of individuals. As a matter of fact, even the best firefly in each generation is not preserved for the next generation. In our approach, at each generation we select a percentage p of the best fireflies to be preserved unaltered to the next generation. Similarly, we select a percentage q of the worst fireflies of the swarm and split it up into two subgroups of the same size, formed respectively by fireflies that are replaced by random solutions to increase the exploratory capacity of the swarm, and by fireflies that are copies (clones) of the best members of the swarm but then undergo mutation through an additive single-point, inductive uniform mutation at a single coordinate while all other coordinates remain unaltered.

Finally, this modified firefly algorithm is enhanced by its hybridization with a local search heuristics. In this work we apply the Luus-Jaakola local search method, a heuristic proposed in 1973 to solve nonlinear programming problems [12]. The method begins with an initialization step, in which random uniform values are chosen within the search space. This is achieved by computing the upper and lower bounds for each dimension. Then, a random uniform value within these bounds is sampled for each component. This value is additively added to the current position of the firefly location to generate a new candidate solution. This new solution replaces the current one if and only if this leads to a improvement of the fitness at the new position. Otherwise, the sampling space is multiplicatively decreased by a factor, freely chosen by the user. This workflow is repeated iteratively. With each iteration, the size of the neighborhood of the point is reduced, until eventually collapsing to a point.

5   Our Method

The modified firefly algorithm described in previous section has been applied to solve the parametric learning optimization problem described in Sect. 3.2. To this purpose, we need to determine some important choices. First of all, we need an adequate representation of the unknowns of the problem. The fireflies in our method correspond to real-coded vectors of length $\beta + 1$ corresponding to the free variables, $\{\delta_i\}_{i=1,\ldots,\beta}$ and λ, of the least-squares minimization problem in Eq. (9). All individuals (fireflies) are initialized with uniformly distributed random numbers on the parametric domain for each coordinate. On the other hand, the fitness function corresponds to the evaluation of the least-squares function, given by Eq. (9).

Regarding the modified memetic self-adaptive firefly algorithm, some control parameters should be determined. As is usual in the field of metaheuristic techniques, the choice of suitable values for the control parameters becomes an important issue, as it affects the performance of the method at large extent. It is also a challenging problem, since it is strongly problem-dependent. In this work, our choice is mainly based on a large collection of empirical results. These control parameters and their values for this work are:

- the number of fireflies, n_f: we set this value to $n_f = 100$ fireflies in this paper. We also tested larger populations of fireflies (up to 300 individuals) at the expense of higher computation times, without any significant improvement, so we found this value to be appropriate in our simulations.
- the number of iterations, n_{iter}. Through numerical simulations, we found that $n_{iter} = 5000$ is a suitable value, as convergence is achieved in all our simulations and higher values for n_{iter} do not lead to any improvement in our results.
- the initial attractiveness, β_0: some theoretical results indicate that $\beta_0 = 1$ is a suitable value for many optimization problems. Accordingly, we consider this value in this paper, with positive results, as shown in next section.
- the absorption coefficient, γ: its value is set up to $\gamma = 0.5$ in this work, since it provides a quick convergence of the algorithm to the optimal solution.
- the potential coefficient, μ: in principle, any positive value can be used for this parameter. However, it is noticed that the intensity of light varies according to the inverse square law. Therefore, we decided to choose $\mu = 2$ accordingly.
- the randomization parameter, α. This parameter, which varies on the interval $[0, 1]$, is used to decide the degree of randomization introduced in the algorithm, which, in turn, is used in order to generate new solutions and avoid to getting stuck in a local minimum. However, it has been noticed that larger values introduce strong perturbations on the evolution of the firefly and, eventually, delay convergence to the global optima. Consequently, it is preferable to select values between these extreme ends of the spectrum. In this work, we select $\alpha = 0.5$.

– the percentage p of best solutions selected for elitism: it is set to $p = 0.1$, with the meaning that 10% of the best solutions are preserved to the next generation unaltered.
– the percentage q of worst solutions for replacement: it is taken as $q = 0.2$, meaning that 20% of the worst solutions are replaced in our population for each generation. Among them, 10% are replaced by random fireflies to promote exploration, while the rest are replaced by copies of the best individuals selected for elitism and then further mutated, as explained in Sect. 4.2.

After the selection of suitable values for its parameters, the modified firefly algorithm is executed iteratively for the given number of iterations. With the purpose to remove the stochastic effects, and also to avoid premature convergence, 30 independent executions have been exectued for each experiment. Then, the firefly with the best (minimum) fitness value is taken as the best solution to the problem.

6   Computational Simulations and Experimental Results

6.1   Computational Simulations

To check the performance of our approach, it has been applied to a practical example of an associative model with functional networks. The corresponding benchmark is given by the collection of data points shown in Table 1. The table displays a collection of 100 training points applied to learn the functional network. The parametric learning is achieved by solving the minimization problem in Eq. (9) through the method described in Sect. 5.

For the learning process in Eq. (5), we consider the family of Bernstein polynomials of degree ρ (which are linearly independent), given by:

$$\psi_i(\xi) = B_i^\rho(\xi) = \binom{\rho}{i} \xi^i (1 - \xi)^{\rho - i} \quad ; \quad i = 0, \ldots, \rho$$

used to approximate the neuron function φ in Eq. (2), and perform parametric learning of the functional network.

6.2   Experimental Results

We have tested our results for different values of the polynomial degree ρ, starting with the simplest case ($\rho = 1$). The best solution obtained for the approximating function is given by the expression:

$$\varphi(\xi) = 0.234 \, B_0^1(\xi) + B_1^1(\xi)$$

Table 1. Benchmark used for parametric learning of the associate functional network.

ξ	ζ	$\Phi(\xi,\zeta)$	ξ	ζ	$\Phi(\xi,\zeta)$	ξ	ζ	$\Phi(\xi,\zeta)$	ξ	ζ	$\Phi(\xi,\zeta)$
0.376	0.608	1.240	0.869	0.981	1.660	0.934	0.897	1.650	0.566	0.641	1.350
0.230	0.811	1.300	0.174	0.291	0.984	0.439	0.848	1.400	0.911	0.858	1.620
0.569	0.860	1.460	0.244	0.173	0.960	0.714	0.106	1.200	0.952	0.163	1.365
0.240	0.682	1.230	0.093	0.742	1.210	0.170	0.075	0.876	0.816	0.606	1.460
0.762	0.778	1.510	0.305	0.997	1.450	0.686	0.089	1.170	0.116	0.293	0.959
0.377	0.138	1.010	0.337	0.565	1.201	0.938	0.101	1.340	0.184	0.684	1.210
0.598	0.152	1.150	0.355	0.867	1.380	0.903	0.041	1.301	0.951	0.412	1.460
0.995	0.468	1.501	0.140	0.189	0.916	0.190	0.669	1.202	0.818	0.512	1.421
0.907	0.521	1.470	0.960	0.687	1.570	0.252	0.199	0.976	0.517	0.541	1.269
0.726	0.714	1.461	0.739	0.274	1.280	0.710	0.527	1.360	0.296	0.724	1.280
0.073	0.649	1.151	0.402	0.607	1.252	0.619	0.964	1.541	0.031	0.512	1.050
0.471	0.121	1.061	0.518	0.882	1.462	0.902	0.983	1.681	0.835	0.373	1.368
0.123	0.292	0.962	0.145	0.057	0.854	0.426	0.199	1.071	0.299	0.331	1.060
0.144	0.206	0.926	0.437	0.888	1.432	0.032	0.050	0.794	0.383	0.070	0.988
0.371	0.172	1.031	0.696	0.754	1.471	0.192	0.241	0.967	0.767	0.012	1.200
0.558	0.791	1.420	0.543	0.852	1.450	0.793	0.762	1.521	0.797	0.382	1.350
0.928	0.192	1.372	0.208	0.819	1.301	0.182	0.858	1.319	0.397	0.285	1.089
0.883	0.427	1.421	0.472	0.984	1.499	0.641	0.755	1.440	0.664	0.146	1.180
0.001	0.126	0.818	0.468	0.397	1.181	0.703	0.884	1.539	0.723	0.187	1.230
0.425	0.718	1.330	0.694	0.623	1.401	0.905	0.348	1.398	0.480	0.956	1.482
0.454	0.719	1.341	0.261	0.490	1.131	0.181	0.250	0.966	0.331	0.516	1.172
0.628	0.314	1.230	0.352	0.681	1.269	0.823	0.388	1.370	0.357	0.442	1.151
0.570	0.698	1.381	0.894	0.434	1.429	0.832	0.831	1.570	0.826	0.324	1.350
0.555	0.755	1.402	0.283	0.612	1.209	0.431	0.427	1.180	0.185	0.892	1.340
0.742	0.178	1.240	0.248	0.191	0.971	0.536	0.584	1.299	0.471	0.121	1.061

Proceeding in the same way, we also tested our method for larger values of η. The approximate models associated with $\rho = 2$ and $\rho = 3$ are given by:

$$\varphi(\xi) = 0.423B_0^2(\xi) + 0.521B_1^2(\xi) + B_2^2(\xi)$$

and

$$\varphi(\xi) = 0.395B_0^3(\xi) + 0.5276B_1^3(\xi) + 2.06B_2^3(\xi) + 1.049B_2^3(\xi),$$

respectively. The performance of these models can be better measured in terms of the RMSE (root mean squared error), given by: $\text{RMSE} = \sqrt{\dfrac{\Lambda}{\alpha + 1}}$, which takes into account not only the approximation error but also the sample size used for training.

Table 2 reports our experimental results. The table shows (in columns): the number of approximating functions, ρ, the number of free parameters, $\alpha + 1$, and the averaged RMSE and maximum error values for the 100 training points (columns 3 and 4) and testing points (columns 5 and 6) in Table 1. As the reader can see, the RMSE shown in third column decreases as the number of

Table 2. RMSE and maximum errors for five different approximate models.

		Training phase		Testing phase	
β	par.	RMSE	Max.	RMSE	Max.
2	3	1.93×10^{-1}	4.57×10^{-1}	1.98×10^{-1}	5.16×10^{-1}
3	4	1.49×10^{-2}	4.33×10^{-2}	1.51×10^{-2}	7.63×10^{-2}
4	5	1.28×10^{-3}	3.47×10^{-3}	1.51×10^{-3}	5.72×10^{-3}
6	7	7.56×10^{-6}	2.81×10^{-5}	8.38×10^{-6}	4.56×10^{-5}
11	12	4.98×10^{-8}	1.09×10^{-7}	5.71×10^{-8}	2.63×10^{-7}

approximating functions increases. We also performed cross-validation of our results to check for over-fitting. To this aim, we predicted the values of 500 data points and computed the prediction errors. The results for the RMSE are shown in columns 5 and 6 of Table 2. We can see that the RMSE and the maximum errors for the training and testing data are comparable. As a result, we can conclude that no over-fitting happens and our results can be safely validated.

All the computational work in this paper has been carried out on a 3.4 GHz Intel Core i7 processor, with 16 GB of RAM. The programming code has been implemented by the authors in the programming language of the popular scientific program *Matlab*, particularly on its version 2018b.

7 Conclusions and Future Work

In this paper we addressed the parametric learning problem of functional networks by considering a classical model: the associative functional network, which represents the associativity operator. We showed that learning this functional network for an unknown functions can be transformed into the problem of learning the parameters of an approximating function, leading to a multivariate nonlinear continuous minimization problem. We solved it by applying a modified memetic self-adaptive version of the firefly algorithm. The experimental results on an illustrative example used as a benchmark show that the method performs well and is able to obtain the learn all parameters of the model and hence, replicate the approximating model with good accuracy. Our results also show that the accuracy increases with the number of approximating functions. We also performed cross-validation by using two sets of data for training and testing, respectively. Since the obtained results are comparable, we concluded that no over-fitting occurs.

Our future work includes extending this methodology to other types of functional networks and to more sophisticated approximating functions. For instance, it is still an open problem to determine whether or not the method improves for other choices of the basis functions, such as shifted step functions, logistic functions, or the like. Applying this approach to some practical problems in different domains of science and engineering is also included in the plans for future work in the field.

Acknowledgments. The first two authors would like to thank the financial support provided by the project PDE-GIR of the European Union's Horizon 2020 research and innovation program under the Marie Sklodowska-Curie grant agreement No 778035, and also from the Agencia Estatal de Investigación (AEI) of the Spanish Ministry of Science, Innovation and Universities (Computer Science National Program) under grant #TIN2017-89275-R and European Funds FEDER (AEI/FEDER, UE). The last two authors wish to thank the Basque Government for its funding support through the EMAITEK and ELKARTEK programs. Javier Del Ser also received funding support from the Consolidated Research Group MATH-MODE (IT1294-19), granted by the Department of Education of the Basque Government.

References

1. Campuzano, A., Iglesias, A., Gálvez, A.: Applying firefly algorithm to data fitting for the Van der Waals equation of state with Bézier curves. In: Proceedings of International Conference on Cyberworlds, CW 2019, Los Alamitos, CA, pp. 211–214. IEEE Computer Society Press (2019)
2. Castillo, E.: Functional networks. Neural Process. Lett. **7**, 151–159 (1998)
3. Castillo, E., Iglesias, A., Ruiz-Cobo, R.: Functional Equations in Applied Sciences. Elsevier, Amsterdam (2005)
4. Del Ser, J., et al.: Bio-inspired computation: where we stand and what's next. Swarm Evol. Comput. **48**, 220–250 (2019)
5. Fister, I., Yang, X.S., Brest, J., Fister Jr., I.: A comprehensive review of firefly algorithms. In: Yang, X.S., Cui, Z., Xiao, R., Gandomi, A.H., Karamanoglu, M. (eds.) Swarm Intelligence and Bio-Inspired Computation, pp. 73–102. Elsevier, Theory and Applications (2013)
6. Gálvez, A., Iglesias, A.: Firefly algorithm for polynomial Bézier surface parameterization. J. Appl. Math. **2013**, 9 (2013). Article ID 237984
7. Gálvez, A., Iglesias, A.: Firefly algorithm for explicit B-spline curve fitting to data points. Math. Prob. Eng. 2013, 12 (2013). Article ID 528215
8. Gálvez, A., Iglesias, A.: Modified memetic self-adaptive firefly algorithm for 2D fractal image reconstruction. Proceedings of IEEE 42nd Annual Computer Software and Applications Conference, IEEE COMPSAC 2019, Los Alamitos, CA, pp. 165–170. IEEE Computer Society Press (2018)
9. Gálvez, A., Fister, I., Osaba, E., Del Ser, J., Iglesias, A.: Hybrid modified firefly algorithm for border detection of skin lesions in medical imaging. In: Proceedings of IEEE Congress on Evolutionary Computation, IEEE CEC 2019, Los Alamitos, CA, pp. 111–118. IEEE Computer Society Press (2019)
10. Iglesias, A., Gálvez, A.: Memetic firefly algorithm for data fitting with rational curves. In: Proceedings of IEEE Congress on Evolutionary Computation, CEC 2015, Los Alamitos, CA, pp. 507–514. IEEE Computer Society Press (2015)
11. Iglesias, A., Gálvez, A.: New memetic self-adaptive firefly algorithm for continuous optimisation. Int. J. Bio-Inspir. Comput. **8**(5), 300–317 (2016)
12. Luus, R., Jaakola, T.H.I.: Optimization by direct search and systematic reduction of the size of search region. Am. Inst. Chem. Eng. J. (AIChE) **19**(4), 760–766 (1973)
13. Tilahun, S.L., Ong, H.C.: Modified firefly algorithm. J. Appl. Math. (2012). Article ID 467631

14. Yang, X.-S.: Firefly algorithms for multimodal optimization. In: Watanabe, O., Zeugmann, T. (eds.) SAGA 2009. LNCS, vol. 5792, pp. 169–178. Springer, Heidelberg (2009). https://doi.org/10.1007/978-3-642-04944-6_14
15. Yang, X.S.: Firefly algorithm, stochastic test functions and design optimisation. Int. J. Bio-Inspir. Comput. **2**(2), 78–84 (2010)
16. Yang, X.-S.: Nature-Inspired Metaheuristic Algorithms, 2nd edn. Luniver Press, Frome (2010)

Dual Formulation of the TV-Stokes Denoising Model for Multidimensional Vectorial Images

Alexander Malyshev[✉]

University of Bergen, Bergen, Norway
alexander.malyshev@uib.no

Abstract. The TV-Stokes denoising model for a vectorial image defines a denoised vector field in the form of the gradient of a scalar function. The dual formulation naturally leads to a Chambolle-type algorithm, where the most time consuming part is application of the orthogonal projector onto the range space of the gradient operator. This application can be efficiently executed by the fast cosine transform taking advantage of the fast Fourier transform. Convergence of the Chambolle-type iteration can be improved by Nesterov's acceleration.

Keywords: Total variation · Denoising of a vector field · Chambolle's algorithm

1   Introduction

Let $\tilde{u}(x)$ be a scalar function, or a continuous grayscale image, defined in a domain $\Omega \subset \mathbb{R}^2$, which is corrupted with an additive noise, i.e., $\tilde{u} = u + \eta$, where $u(x)$ is an unknown true function, or image, and $\eta(x)$ is noise. A classical variational model for image denoising is the Rudin-Osher-Fatemi (ROF) variational model introduced in [8],

$$\min_u \int_\Omega |\nabla u| + \frac{1}{2\lambda}\|u - \tilde{u}\|_2^2, \tag{1}$$

where $\nabla u = (u_{x_1}, u_{x_2})$ is the gradient of $u(x)$ and $|\nabla u| = \sqrt{u_{x_1}^2 + u_{x_2}^2}$. The term $\int_\Omega |\nabla u|$ is called the total variation of $u(x)$ in Ω. The term $\|u - \tilde{u}\|_2^2 = \int_\Omega (u - \tilde{u})^2$ is the data fitting term. A suitable regularization parameter $\lambda > 0$ depends on statistical properties of the noise η. Solution of (1) gives an approximation to the true function such that sufficiently large discontinuities available in $u(x)$ are well preserved. A classical numerical method for solving (1) is Chambolle's algorithm from [2]. A recent survey of the most efficient numerical algorithms for solving (1) is found in [3]. These algorithms belong to the class of local methods.

Supported by the Research Executive Agency of the European Commission, grant 778035 - PDE-GIR - H2020-MSCA-RISE-2017.

V. V. Krzhizhanovskaya et al. (Eds.): ICCS 2020, LNCS 12141, pp. 580–587, 2020.
https://doi.org/10.1007/978-3-030-50426-7_43

Modern image denoising techniques are dominated by the non-local patch-based algorithms; see survey in [6]. Nevertheless, the ROF model should not be entirely discarded because the model and its special variants can be useful in some cases, for example, when smoothing an image until a cartoon-looking result or when denoising an image subject to geometric constraints. Our study below is devoted to the latter case.

The model (1) is trivially extended to the case, when $\tilde{u}(x)$ and $u(x)$ are vector functions, by applying the model (1) separately to each component of $\tilde{u}(x)$ and $u(x)$. Since the trivial extension is not always satisfactory, other approaches to the vectorial images have been proposed. For example, the so called TV-Stokes model, which is restricted to two-dimensional vectorial images $v(x) = (v_1(x), v_2(x))$, satisfies the Stokes constraint $\operatorname{div} v = \partial v_1/\partial x_1 + \partial v_2/\partial x_2 = 0$ and reads

$$\min_{v:\,\operatorname{div} v=0} \int_\Omega \lambda|\nabla v| + \frac{1}{2}\|v - \tilde{v}\|_2^2, \tag{2}$$

where $|\nabla v| = \sqrt{(v_1)_{x_1}^2 + (v_1)_{x_2}^2 + (v_2)_{x_1}^2 + (v_2)_{x_2}^2}$; see [4,7,10] for more details and a great deal of numerical illustrations.

The present note introduces a multidimensional TV-Stokes model and derives its dual formulation similar to that of [4]. The dual formulation allows us to propose a Chambolle-type algorithm for numerical solution of the TV-Stokes model. Most of the arithmetical work at each iteration of this algorithm is required for application of the orthogonal projector to the linear subspace defined by the Stokes constraint. We propose an efficient implementation of this operation via the fast Fourier transform. We also note that the original Chambolle algorithm can be improved by means of Nesterov's acceleration as in [1], and similar acceleration may be applied to the Chambolle-type algorithm following the recipes given in [3].

2   TV-Stokes Model for Multidimensional Images and Its Dual Formulation

We consider real-valued functions defined in $\Omega = [0, L_1] \times \cdots \times [0, L_n] \subset \mathbb{R}^n$ for arbitrary $n = 1, 2, \ldots$. The gradient operator ∇ is applied only to functions with homogeneous Neumann boundary conditions. First of all, we use the gradient field of a scalar function $u(x)$, $x \in \Omega$, which is the vector function $\nabla u(x) = [u_{x_1}(x), u_{x_2}(x), \ldots, u_{x_n}(x)]^T$. We also apply the gradient operator to n-dimensional vector fields $v(x) = [v_1(x), v_2(x), \ldots, v_n(x)]$ in Ω and label it with the bar as $\bar{\nabla} v(x)$ in order to distinguish from the scalar case. The object $\bar{\nabla} v(x)$ is the tensor field $\partial v_i(x)/\partial x_j$, $i, j = 1, 2, \ldots, n$.

Given an n-dimensional vector field $\tilde{v}(x) \in \mathbb{R}^n$ corrupted with an additive noise, a constrained variant of the ROF model defines the gradient field $v(x) = [v_1(x), v_2(x), \ldots, v_n(x)] \in \mathbb{R}^n$ satisfying the variational problem

$$\min_{v=\nabla u} \left(|\bar{\nabla} v|_1 + \frac{1}{2\lambda}\|v - \tilde{v}\|_2^2 \right), \tag{3}$$

where $\lambda > 0$ is a suitable scalar parameter and $\|v - \tilde{v}\|_2^2 = \int_\Omega \sum_{i=1}^n (v_i - \tilde{v}_i)^2(x)$.
The seminorm $|\bar{\nabla} v|_1$ is the total variation

$$|\bar{\nabla} v|_1 = \int_\Omega |\bar{\nabla} v| = \int_\Omega \sqrt{\sum_{i,j=1}^n (\partial v_i(x)/\partial x_j)^2}.$$

The inner product of vector functions $v, w \in \mathbb{R}^n$ is

$$\langle v, w \rangle = \int_\Omega \sum_{i=1}^n v_i(x) w_i(x)$$

so that the norm $\|v\|_2$ satisfies $\|v\|_2^2 = \langle v, v \rangle$.

Solution $v(x)$ of the variational model (3) is constrained to the linear subspace $V = \{v \in \mathbb{R}^n : v = \nabla u,$ where $u(x)$ is a scalar function$\}$. When $n = 2$, the constraint $v = \nabla u$ is equivalent to the constraint $\operatorname{div} v = 0$, which participates in the 2D TV-Stokes model from [4,7,10]. Following this observation, we will call (3) the TV-Stokes model too. More specifically, (3) is a primal formulation of the multidimensional TV-Stokes model.

Note that the continuous functional $\mathcal{F}(v) = |\bar{\nabla} v|_1 + \frac{1}{2\lambda}\|v - \tilde{v}\|_2^2$ is strictly convex. Therefore, its minimum in V is unique and attained in the closed ball $\{v : \|v - \tilde{v}\|_2 \le \|\tilde{v}\|_2\}$.

Let us equip tensor fields $p(x)$ having the components $p_{ij}(x)$, $i, j = 1, 2, \ldots, n$, with the two norms

$$\|p\|_\infty = \left\| \sqrt{\sum_{i,j=1}^n p_{ij}^2(x)} \right\|_\infty , \quad \|p\|_2 = \left\| \sqrt{\sum_{i,j=1}^n p_{ij}^2(x)} \right\|_2 .$$

The total variation can be rewritten in the form

$$|\bar{\nabla} v|_1 = \max_{\|p\|_\infty \le 1} \langle \bar{\nabla} v, p \rangle \tag{4}$$

using the tensor $p(x)$ as a dual variable; see arguments in [3]. Thus, the model (3) is equivalently reduced to the primal-dual formulation

$$\min_{v \in V} \max_{\|p\|_\infty \le 1} F(v, p), \quad \text{where} \quad F(v, p) = \langle \bar{\nabla} v, p \rangle + \frac{1}{2\lambda} \langle v - \tilde{v}, v - \tilde{v} \rangle. \tag{5}$$

The order of the operations min and max in (5) may be interchanged due to

Theorem 1 ([9]). *Let X be a convex subset of a linear topological space, Y be a compact convex subset of a linear topological space, and $f : X \times Y \to \mathbb{R}$ be lower semicontinuous on X and upper semicontinuous on Y. Suppose that f is quasiconvex on X and quasiconcave on Y. Then*

$$\inf_{x \in X} \max_{y \in Y} f(x, y) = \max_{y \in Y} \inf_{x \in X} f(x, y).$$

Owing to Theorem 1, $\min_{v \in V} \max_{|p| \leq 1} F(v, p) = \max_{|p| \leq 1} \min_{v \in V} F(v, p)$, and we arrive at the primal-dual max-min formulation

$$\max_{||p||_\infty \leq 1} \min_{v \in V} \left[\langle \bar{\nabla} v, p \rangle + \frac{1}{2\lambda} \langle v - \tilde{v}, v - \tilde{v} \rangle \right]. \tag{6}$$

Further derivations make use of the conjugate to ∇ operator denoted by ∇^*. In particular, $\langle \bar{\nabla} v, p \rangle = \langle v, \bar{\nabla}^* p \rangle$ and

$$F(v, p) = \langle v, \bar{\nabla}^* p \rangle + \frac{1}{2\lambda} \langle v - \tilde{v}, v - \tilde{v} \rangle. \tag{7}$$

Replacing v by ∇u in (7) and further rearrangements yield

$$F(v, p) = \langle \nabla u, \bar{\nabla}^* p \rangle + \frac{1}{2\lambda} \langle \nabla u - \tilde{v}, \nabla u - \tilde{v} \rangle$$
$$= \langle u, \nabla^* \bar{\nabla}^* p \rangle + \frac{1}{2\lambda} \left[\langle \nabla^* \nabla u, u \rangle - 2 \langle u, \nabla^* \tilde{v} \rangle + \langle \tilde{v}, \tilde{v} \rangle \right].$$

The necessary condition for $\min_{v = \nabla u} F(v, p)$ in terms of the first variation of u is the equality

$$\lambda \nabla^* \bar{\nabla}^* p + \nabla^* \nabla u - \nabla^* \tilde{v} = 0. \tag{8}$$

Solution of (8) is not unique because of the homogeneous Neumann boundary conditions. However, all solutions differ only by a constant, i.e. if $u_I(x)$ and $u_{II}(x)$ are two solutions, then $u_I - u_{II} \equiv$ const.

Let us choose the linear least squares solution

$$u = (\nabla^* \nabla)^\dagger \nabla^* \left(\tilde{v} - \lambda \bar{\nabla}^* p \right), \tag{9}$$

where $\dagger$ denotes the Moore-Penrose pseudoinverse, i.e. the solution of (8) with the minimum 2-norm. The vector field $v = \nabla u$ is determined uniquely as

$$v = \nabla (\nabla^* \nabla)^\dagger \nabla^* \left(\tilde{v} - \lambda \nabla^* p \right) = \Pi \left(\tilde{v} - \lambda \nabla^* p \right). \tag{10}$$

Recall that the symmetric operator $\Pi = \nabla (\nabla^* \nabla)^\dagger \nabla^*$ is an orthogonal projector, i.e., $\Pi^2 = \Pi$ and $\Pi^* = \Pi$.

In order to find $\max_{|p| \leq 1} F(v, p)$ subject to (10), we insert the representation $v = \Pi \left(\tilde{v} - \lambda \bar{\nabla}^* p \right)$ into (7) and perform equivalent transformations:

$$F(v, p) = \frac{1}{2\lambda} \left[2 \langle \Pi \left(\tilde{v} - \lambda \bar{\nabla}^* p \right), \lambda \bar{\nabla}^* p \rangle + \langle \Pi \left(\tilde{v} - \lambda \bar{\nabla}^* p \right) - \tilde{v}, \Pi \left(\tilde{v} - \lambda \bar{\nabla}^* p \right) - \tilde{v} \rangle \right]$$
$$= \frac{1}{2\lambda} \left[2 \langle \Pi \left(\tilde{v} - \lambda \bar{\nabla}^* p \right), \lambda \bar{\nabla}^* p \rangle + \langle \Pi \left(\tilde{v} - \lambda \bar{\nabla}^* p \right), \Pi \left(\tilde{v} - \lambda \bar{\nabla}^* p \right) \rangle \right.$$
$$\left. - 2 \langle \Pi \left(\tilde{v} - \lambda \bar{\nabla}^* p \right), \tilde{v} \rangle + \langle \tilde{v}, \tilde{v} \rangle \right]$$
$$= \frac{1}{2\lambda} \left[\langle \tilde{v}, \tilde{v} \rangle - \langle \Pi \left(\tilde{v} - \lambda \bar{\nabla}^* p \right), \Pi \left(\tilde{v} - \lambda \bar{\nabla}^* p \right) \rangle \right]$$
$$= \frac{1}{2\lambda} \| \tilde{v} \|_2^2 - \frac{1}{2\lambda} \| \Pi (\tilde{v} - \lambda \bar{\nabla}^* p) \|_2^2.$$

Therefore, the problem $\max_{|p| \leq 1} F(v, p)$ subject to (10) is equivalently reduced to the constrained minimum distance problem

$$\max_{||p||_\infty \leq 1} \|\Pi(\bar{\nabla}^* p - \tilde{v}/\lambda)\|_2. \tag{11}$$

We formulate the above proven facts in the form of

Theorem 2 (Dual formulation of the TV-Stokes model). *The unique solution to the TV-Stokes variational problem*

$$\min_{v=\nabla u} \left(|\bar{\nabla}v|_1 + \frac{1}{2\lambda}\|v - \tilde{v}\|_2^2 \right)$$

is the vector field

$$v = \Pi\left(\tilde{v} - \lambda\nabla^* p\right),$$

where $\Pi = \nabla(\nabla^*\nabla)^\dagger\nabla^*$ *is an orthogonal projector, and the tensor field* $p(x)$ *solves the dual variational problem*

$$\max_{||p||_\infty \leq 1} \|\Pi(\bar{\nabla}^* p - \tilde{v}/\lambda)\|_2.$$

3   The Chambolle-Type Iteration

Following the derivation of Chambolle's algorithm in [2], we write the Karush-Kuhn-Tucker conditions for (11) as the equation

$$\bar{\nabla}\Pi\left(\bar{\nabla}^* p - \tilde{v}/\lambda\right) + \|\bar{\nabla}\Pi\left(\bar{\nabla}^* p - \tilde{v}/\lambda\right)\|_\infty p = 0. \tag{12}$$

Hence the solution of (11) can be approximated by the projected gradient iteration

$$p^0 = 0, \quad p^{k+1} = \frac{p^k - \tau\bar{\nabla}\Pi\left(\bar{\nabla}^* p^k - \tilde{v}/\lambda\right)}{\max(1, \left\|p^k - \tau\bar{\nabla}\Pi\left(\bar{\nabla}^* p^k - \tilde{v}/\lambda\right)\right\|_\infty)}. \tag{13}$$

where $\tau > 0$ is a step parameter.

Nesterov's acceleration for the iteration (13) and other numerical methods such as the primal-dual methods can be found in [1,3].

4   The Singular Value Decomposition of the Differentiation Matrix

We approximate the partial differentiation operators $\partial/\partial x_d$ by the $N \times N$ differentiation matrices of order $N = N_d$:

$$D = \begin{bmatrix} -1 & 1 & & & \\ & -1 & 1 & & \\ & & \ddots & \ddots & \\ & & & -1 & 1 \\ & & & & 0 \end{bmatrix}. \tag{14}$$

The discrete cosine transform is defined by the orthogonal $N \times N$ matrix C with the entries

$$C_{1j} = \sqrt{\frac{1}{N}}, \quad C_{ij} = \sqrt{\frac{2}{N}} \cos \frac{\pi(i-1)(2j-1)}{2N}, \quad i = 2, \ldots, N, \quad j = 1, \ldots, N.$$

The discrete sine transform is defined by the orthogonal $(N-1) \times (N-1)$ symmetric matrix S with the entries

$$S_{ij} = \sqrt{\frac{2}{N}} \sin \frac{\pi i j}{N}, \quad i, j = 1, \ldots, N-1.$$

The singular value decomposition (SVD) of D is

$$D = - \begin{bmatrix} 0 & S \\ 1 & 0 \end{bmatrix} \Sigma C,$$

where the diagonal matrix Σ has the entries $\Sigma_{ii} = 2 \sin \frac{\pi(i-1)}{2N}$, $i = 1, \ldots, N$.

5   Discrete Gradient Operators

Discretization of a scalar function $u(x)$ on a rectangular grid, which is equidistant along each of n dimensions, is given by the components $u_{\alpha_1 \alpha_2 \ldots \alpha_n}$, $1 \leq \alpha_d \leq N_d$. The set of components is often called the grid function. The r-norm of a grid function u is defined as $\|u\|_r = \left(\sum_{\alpha_1 \alpha_2 \ldots \alpha_n} |u_{\alpha_1 \alpha_2 \ldots \alpha_n}|^r \right)^{1/r}$.

The product of an $N_d \times N_d$ matrix A with a grid function u along dimension d is denoted by $A_{\times d} u$ such that the product $w = A_{\times d} u$ has the components $w_{\alpha_1 \ldots \alpha_{d-1} \beta \alpha_{d+1} \ldots \alpha_n} = \sum_{\gamma=1}^{N_d} A_{\beta \gamma} u_{\alpha_1 \ldots \alpha_{d-1} \gamma \alpha_{d+1} \ldots \alpha_n}$.

The discrete gradient operator ∇ is defined by means of the differentiation matrices D introduced in the previous section. For example, the gradient of a scalar grid function u is the set of n grid functions

$$\nabla u = \{D_{\times 1} u, D_{\times 2} u, \ldots, D_{\times n} u\}.$$

The discrete gradient $\bar{\nabla}$ is defined for a vectorial grid function $v = [v_1, v_2, \ldots v_n]$ as the set of $n \times n$ grid functions $D_{\times j} v_i$, $i, j = 1, 2, \ldots, n$.

It is rather straightforward to introduce discrete analogs of the norms $\|\nabla u\|_2$, $\|\bar{\nabla} v\|_2$, seminorms $|\bar{\nabla} v|_1$ and so on.

The singular value decomposition of D allows us to prove that

$$\|\bar{\nabla}\|_2 = \|\nabla\|_2 < 2\sqrt{n}. \tag{15}$$

Let us consider the iteration (13) for grid functions and with the discrete gradient operators. The following lemma determines the range of steps τ, for which the iteration (13) is stable.

Lemma 1. *The iteration (13) is 1-Lipschitz if*

$$\tau \leq 2/\|\bar{\nabla} \Pi \bar{\nabla}^*\|_2. \tag{16}$$

Proof. Each step of (13) consists of two mappings: $p \mapsto p - \tau \bar{\nabla} \Pi \left(\bar{\nabla}^* p - \tilde{v}/\lambda \right)$ and $q \mapsto q / \max(1, \|q\|_\infty)$. The first mapping is linear and 1-Lipschitz if and only if $\|I - \tau \bar{\nabla} \Pi \bar{\nabla}^*\|_2 \leq 1$, where I is the identity transformation. The second mapping is always 1-Lipschitz.

The estimate (15) implies that stability of (13) holds when $\tau \leq 1/(2n)$. The fastest convergence occurs for $\tau = 1/(2n)$.

6   Computation of $(\nabla^* \nabla)^\dagger$ by the Fast Cosine Transform

Using the differentiation matrices D of order N_d along each dimension d, a discretization of the operator $\nabla^* \nabla$ is applied to a grid function u as follows,

$$\nabla^* \nabla u = (D^T D)_{\times 1} u + (D^T D)_{\times 2} u + \ldots + (D^T D)_{\times n} u.$$

By the aid of the SVD of each differentiation matrix D, the discretized equation $\nabla^* \nabla u = f$ is equivalently reduced to the diagonal system of linear equations

$$\Sigma_{\times 1}^2 \hat{u} + \Sigma_{\times 2}^2 \hat{u} + \ldots + \Sigma_{\times n}^2 \hat{u} = \hat{f},$$

where $\hat{u} = C_{\times n} \ldots C_{\times 2} C_{\times 1} u$ and $\hat{f} = C_{\times n} \ldots C_{\times 2} C_{\times 1} f$. Recall that $C_{\times d}$ are the matrices of the discrete cosine transform of order N_d. The components of $\hat{u}$ and $\hat{f}$ are related by the equalities

$$\hat{u}_{\alpha_1 \cdots \alpha_n} \Psi_{\alpha_1 \alpha_2 \ldots \alpha_n} = \hat{f}_{\alpha_1 \cdots \alpha_n}, \quad \alpha_d = 1, \ldots, N_d,$$

where $\Psi_{\alpha_1 \alpha_2 \ldots \alpha_n} = \Sigma_{\alpha_1 \alpha_1}^2 + \Sigma_{\alpha_2 \alpha_2}^2 + \ldots + \Sigma_{\alpha_n \alpha_n}^2$ Note that $\Psi_{\alpha_1 \alpha_2 \ldots \alpha_n} = 0$ if and only if $\alpha_1 = \ldots = \alpha_n = 1$ and is positive otherwise. Hence the least squares solution of $\nabla^* \nabla u = f$ has the components

$$\hat{u}_{1 \ldots 1} = 0, \quad \hat{u}_{\alpha_1 \cdots \alpha_n} = \hat{f}_{\alpha_1 \cdots \alpha_n} / \Psi_{\alpha_1 \alpha_2 \ldots \alpha_n} \quad \text{if } \alpha_1 + \ldots + \alpha_n > n.$$

Recall that $u = C_{\times 1}^T C_{\times 2}^T \cdots C_{\times n}^T \hat{u}$. Note that multiplication of grid functions by the matrices C and C^T can be efficiently implemented by means of the fast Fourier transform (FFT); see, e.g. [12].

Alternatively, fast application of the operator $(\nabla^* \nabla)^\dagger$ can also be computed by the multigrid method [11].

7   Conclusion

In this note, we propose the variational model (3) for denoising of multidimensional vectorial images satisfying the Stokes constraint. Theorem 2 gives the dual formulation of this model. The dual formulation is used for construction of the Chambolle-type iteration (13), which solves the problem (3). Faster convergence is achieved by applying Nesterov's acceleration to (13) as in [1].

While the potential applicability of the new model (3) is not wide, we hope that the TV-Stokes model will become a useful instrument for denoising of images of hydrodynamical flows and for image inpainting.

References

1. Beck, A., Teboulle, M.: A fast iterative shrinkage-thresholding algorithm for linear inverse problems. SIAM J. Imaging Sci. **2**(1), 183–202 (2009)
2. Chambolle, A.: An algorithm for total variation minimization and applications. J. Math. Imaging Vis. **20**, 89–97 (2004)
3. Chambolle, A., Pock, T.: An introduction to continuous optimization for imaging. Acta Numerica **25**, 161–319 (2016)
4. Elo, C.A., Malyshev, A., Rahman, T.: A dual formulation of the TV-Stokes algorithm for image denoising. In: Tai, X.-C., Mørken, K., Lysaker, M., Lie, K.-A. (eds.) SSVM 2009. LNCS, vol. 5567, pp. 307–318. Springer, Heidelberg (2009). https://doi.org/10.1007/978-3-642-02256-2_26
5. Jia, Z.-G., Wei, M.: A new TV-Stokes model for image deblurring and denoising with fast algorithms. J. Sci. Comput. **72**, 522–541 (2017)
6. Lebrun, M., Colom, M., Buades, A., Morel, J.: Secrets of image denoising cuisine. Acta Numerica **21**, 475–576 (2012)
7. Rahman, T., Tai, X.-C., Osher, S.: A TV-Stokes denoising algorithm. In: Sgallari, F., Murli, A., Paragios, N. (eds.) SSVM 2007. LNCS, vol. 4485, pp. 473–483. Springer, Heidelberg (2007). https://doi.org/10.1007/978-3-540-72823-8_41
8. Rudin, L.I., Osher, S., Fatemi, E.: Nonlinear total variation based noise removal algorithms. Phys. D **60**, 259–268 (1992)
9. Sion, M.: On general minimax theorems. Pac. J. Math. **8**, 171–176 (1958)
10. Tai, X.C., Osher, S., Holm, R.: Image inpainting using a TV-Stokes equation. In: Tai, X.C., Lie, K.A., Chan, T.F., Osher, S. (eds.) Image Processing Based on Partial Differential Equations, Mathematics and Visualization, pp. 3–22. Springer, Heidelberg (2007). https://doi.org/10.1007/978-3-540-33267-1_1
11. Trottenberg, U., Oosterlee, C.W., Schüller, A.: Multigrid. Academic Press (2001)
12. Van Loan, C.: Computational Frameworks for the Fast Fourier Transform. SIAM, Philadalphia (1992)

Minimizing Material Consumption of 3D Printing with Stress-Guided Optimization

Anzong Zheng[1]([⊠]), Shaojun Bian[1], Ehtzaz Chaudhry[1], Jian Chang[1], Habibollah Haron[2], Lihua You[1], and Jianjun Zhang[1]

[1] The National Center for Computer Animation, Bournemouth University, Poole, UK
azheng@bournemouth.ac.uk
[2] Department of Computer Science, Universiti Teknologi Malaysia, Johor Bahru, Malaysia

Abstract. 3D printing has been widely used in daily life, industry, architecture, aerospace, crafts, art, etc. Minimizing 3D printing material consumption can greatly reduce the costs. Therefore, how to design 3D printed objects with less materials while maintain structural soundness is an important problem. The current treatment is to use thin shells. However, thin shells have low strength. In this paper, we use stiffeners to stiffen 3D thin-shell objects for increasing the strength of the objects and propose a stress guided optimization framework to achieve minimum material consumption. First, we carry out finite element calculations to determine stress distribution in 3D objects and use the stress distribution to guide random generation of some points called seeds. Then we map the 3D objects and seeds to a 2D space and create a Voronoi Diagram from the seeds. The stiffeners are taken to be the edges of the Voronoi Diagram whose intersections with the edges of each of the triangles used to represent the polygon models of the 3D objects are used to define stiffeners. The obtained intersections are mapped back to 3D polygon models and the cross-section size of stiffeners is minimized under the constraint of the required strength. Monte-Carlo simulation is finally introduced to repeat the process from random seed generation to cross-section size optimization of stiffeners. Many experiments are presented to demonstrate the proposed framework and its advantages.

Keywords: 3D printing · Thin-shell stiffened objects · Minimum material consumption · Finite element analysis · Stress-guided optimization

1 Introduction

With quick development of 3D printing technologies, the price of desktop 3D printers has become more affordable to general customers. Nowadays, people can make 3D prints easily with these affordable printers. With more and more widely applications of 3D printing, saving material consumption of 3D printing can significantly reduce the costs which can be achieved by using thin shells. Since thin-shell objects have low strength, we use stiffeners to stiffen thin-shell objects and proposed a stress guide optimization framework to obtain stiffened thin-shell objects with minimum material consumption and required strength.

© Springer Nature Switzerland AG 2020
V. V. Krzhizhanovskaya et al. (Eds.): ICCS 2020, LNCS 12141, pp. 588–603, 2020.
https://doi.org/10.1007/978-3-030-50426-7_44

Our proposed stress guide optimization framework achieves minimum material consumption through optimizing stiffener distribution and minimizing the cross-section size of stiffeners. In order to generate optimal distribution of stiffeners, the stress field of the input thin-shell objects under given loads and boundary conditions is calculated with the Finite Element Analysis (FEA). According to the stress distribution, some points called seeds are placed randomly on the surface of 3D thin-shell objects. The 3D objects and seeds are mapped to a 2D space so that a Voronoi diagram can be generated from these mapped seeds. The generated Voronoi diagram is mapped back to the 3D space and the edges of the mapped Voronoi diagram represent the distribution of stiffeners. After that, cross-section size of stiffeners is optimized to minimize the volume of the stiffeners. Since the generation of seeds uses a uniform random process which may not lead to a global optimal solution of stiffener distribution, Monte-Carlo simulation is introduced and iterated a given number of times to avoid any local minimum.

2 Related Work

The work proposed in this paper is related to 3D printing, finite element analysis, and structural optimization. We briefly review the existing work in these areas.

3D Printing: There are a lot of papers on 3D printing. The deformation problem was investigated in [1]. The articulation of 3D printed models was examined in [2]. Mechanical movements of 3D printed objects were studied in [3, 4]. And the appearance of 3D printed models was discussed in [5, 6].

Finite Element Analysis Enormous publications can be found about finite element analysis. For example, the finite element method in solid and structures was introduced in [7]. The finite element analysis of stiffened plates was given in [8]. The finite element calculations of stiffened shell were presented in [9]. The vibration of stiffened plates was investigated with the finite element method in [10]. Stress analysis of stiffened composited plates was carried out in [11]. The plates and shells with geometrically linear and nonlinear problems were studied in [12]. And mesh distortions of plate and shell finite elements were examined in [13].

Structural Optimization is also a well investigated filed. Here we only briefly review some representative literature on optimization of 3D printing objects. Three approaches: hollowing, thickening, and strut insertion were introduced in [14] to obtain structurally sound and lightweight 3D prints. Thickness parameters of shells were optimized in [15]. The number of struts in a skin-frame structure is minimized in [16]. The material consumption of honeycomb-like 3D models is reduced via a hollowing optimization algorithm in [17]. Stiffened objects were first investigated in [18]. A method to produce optimized structures for any input surface with any load configurations was researched in [22].

3 Overview

The algorithm overview is shown in Fig. 1. For an input thin-shell object, the finite element calculation is first carried out to obtain its stress distribution (Fig. 1(a)). The seeds used to determine the positions of stiffeners are dispersed randomly in high stress areas (Fig. 1(b)). By mapping the object and seeds in a 3D space to a 2D space, a Voronoi diagram is generated (Fig. 1(c)). After determining the intersections between the edges of the V oronoi diagram and the edges of each of the triangles used to represent the thin-shell object and mapping them back to a 3D space, the stiffener distribution is determined (Fig. 1(d)). Having determined the stiffener distribution, cross-section size optimization of stiffeners is performed to obtain the minimum volume of the stiffeners. In order to optimize the seed generation, Monte-Carlo simulation is introduced to refine the stiffener distribution further. The final stress field obtained from finite element calculations is shown in Fig. 1(e) which significantly improves the stress distribution.

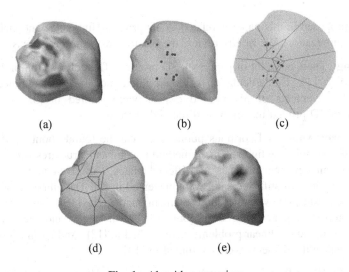

Fig. 1. Algorithm overview.

The finite element formulation of thin-shell objects and stiffened thin-shell objects has been presented in [10, 19]. In what follows, we only investigate the distribution of stiffeners, size optimization of stiffeners, Monte-Carlo simulation, and present the results obtained from our proposed framework.

4   Distribution of Stiffeners

The distribution of stiffeners is obtained through four steps. They are: seed generation, quasi-conformal parameterization, creation of Voronoi diagram, and stiffener extraction.

4.1   Seed Generation

The stress field of an input thin-shell object is first calculated under given boundary conditions and forces. Based on the obtained stress distribution, a given number of seeds are distributed on the object. The seeds are placed through a probability that places more seeds in the areas with a higher stress. By doing so, the areas with higher stresses are stiffened by more stiffeners.

In what follows, n_t stands for the number of triangles of the object mesh, s_i the stress of a randomly selected triangle t_i, σ_s the material strength, n_s the number of expected seeds, and p^* for the probability threshold.

First, a triangle t_i is randomly selected from the n_t triangles, and a probability p is also randomly generated between 0 and 1. If a randomly generated probability p is bigger than the probability threshold p^* but smaller than s_i / σ_s which is the ratio of the stress s_i over the material strength σ_s, the triangle is seeded and marked. If the randomly selected triangle t_i has been seeded and marked, a new triangle is randomly selected. The process is repeated until the number n_s of the expected seeds are reached. This algorithm is shown below.

Algorithm 1: Seed generation

Input: stress field $\{s_i\}$, material strength σ_s, number of expected seeds n_s and probability threshold p^*
Output: markedFaces

count $= 0$
markedFaces$[n_t]$ =false
while $count \leq n_s$ **do**
 | $t_i =$ UniformRandom$(1, \cdots, n_t)$
 | $p =$ UniformRandom$(0, 1)$
 | **if** *not markedfaces$[t_i]$ and $p^* < p < s_i/\sigma_s$* **then**
 | | count++
 | | markedFaces$[t_i] =$ true
 | **end**
end

4.2   Quasi-Conformal Parameterization

Generating a Voronoi diagram from the placed seeds in the 3D thin-shell object and tracing stiffeners from the 3D Voronoi diagram and the 3D mesh is more complicated than in 2D since it requires searching for geodesic lines between arbitrary two points. In order to tackle this problem, we use a quasi-conformal parameterization method called the least square conformal maps (LSCM) [20] to map the 3D mesh to 2D which transforms the problem of tracing stiffeners in 3D space into the one of finding intersections between a segment and mesh edges, which is easier to deal with.

Conformal Maps: As shown in Fig. 2, an application $\mathcal{X}$ mapping a (u, v) domain to a surface is said to be conformal if for each (u, v), the tangent vectors to the iso-u and iso-v curves passing through $\mathcal{X}(u, v)$ are orthogonal and have the same norm, which can be written as:

$$N(u, v) \times \frac{\partial \mathcal{X}(u, v)}{\partial u} = \frac{\partial \mathcal{X}(u, v)}{\partial v} \tag{1}$$

where $N(u, v)$ denotes the unit normal to the surface. In other words, a conformal map is locally isotropic, i.e. maps an elementary circle of the (u, v) domain to an elementary circle of the surface.

Conformality in a Triangulation: Consider a triangulation $\mathcal{G} = \{[1 \cdots n], \mathcal{T}, (p_j) \ 1 \leq j \leq n\}$, where $[1 \cdots n], n \geq 3$ corresponds to the vertices, $\mathcal{T}$ is a set of n' triangles represented by triples of vertices, and $p_j \in \mathbb{R}^3$ denotes the geometric location at the vertex j. Each triangle has a local orthonormal basis, where $(x_1, y_1), (x_2, y_2), (x_3, y_3)$ are the coordinates of its vertices in this basis (i.e., the normal is along the z-axis). The local bases of two triangles sharing an edge are consistently oriented.

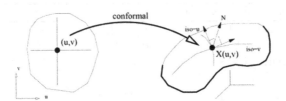

Fig. 2. Conformal map [20]

By considering the restriction of $\mathcal{X}$ to a triangle T and applying the conformality criterion to the inverse map $\mathcal{U}: (x, y) \rightarrow (u, v)$, Eq. (1) becomes:

$$\frac{\partial \mathcal{X}}{\partial u} - i \frac{\partial \mathcal{X}}{\partial v} = 0 \tag{2}$$

where $\mathcal{X}$ has been written in a complex number, i.e. $\mathcal{X} = x + iy$. According to the theorem on the derivatives of inverse functions, this implies that

$$\frac{\partial \mathcal{U}}{\partial x} + i \frac{\partial \mathcal{U}}{\partial y} = 0 \tag{3}$$

where $\mathcal{U} = u + iv$.

Since this equation cannot in general be strictly enforced, the violation of the conformality condition is minimized in the least squares sense, which defines the criterion:

$$C(T) = \int_T \left| \frac{\partial \mathcal{U}}{\partial x} + i \frac{\partial \mathcal{U}}{\partial y} \right|^2 dA = \left| \frac{\partial \mathcal{U}}{\partial x} + i \frac{\partial \mathcal{U}}{\partial y} \right|^2 A_T \qquad (4)$$

where A_T is the area of the triangle and the notation $|z|$ stands for the modulus of the complex number z. Summing over the whole triangulation, the criterion to minimize is then

$$C(T) = \sum_{T \in T} C(T) \qquad (5)$$

After the seeds are obtained on the 3D mesh, they are projected to the 2D space with the above LSCM parameterization for further processing.

4.3   Creation of Voronoi Diagram

A Voronoi diagram is a partition of a plane into regions close to each of a given set of seeds. With the algorithm described in 4.1, the seeds on the 3D mesh shown in Fig. 3 (a) are generated. These seeds are mapped to a 2D space with the algorithm given in 4.2, and the following algorithm is used to generate a Voronoi diagram from the generated seeds as shown in Figs. 3(b) and 3(c).

For the input boundary surface S and a given number n of seeds $\{s_i\}, i \in (1, n)$ defined in the interior domain of S, a Voronoi tessellation of S is defined to be the collection of Voronoi cells Ω_i, $i \in (1, n)$ of these seeds with

$$\Omega_i = \{x \in S \mid \| \, x - s_i \, \| \leq \| \, x - s_j \, \|, \forall j \neq i\} \qquad (6)$$

In the above equation, $\|\cdot\|$ denotes the Euclidean norm. A Voronoi tessellation is called a centroidal Voronoi tessellation (CVT) [21] if each seed coincides with the centroid of its Voronoi cell, where the centroid c_i of its Voronoi cell Ω_i is defined as

$$\mathbf{c}_i = \frac{\int_{x \in \omega_i} \rho(\mathbf{x}) \mathbf{x} d\sigma}{\int_{x_i \in \omega_i} \rho(\mathbf{x}) d\sigma} \qquad (7)$$

where $d\sigma$ is the area differential, and $\rho(\mathbf{x})$ is the density function over the domain S.

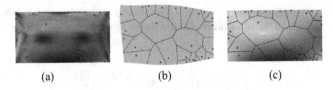

(a)   (b)   (c)

Fig. 3. Generation of Voronoi diagram.

4.4 Stiffener Extraction

Having created the Voronoi diagram in 2D, the next work is to extract stiffeners from the Voronoi diagram. Suppose two ends of an edge of the Voronoi diagram is represented as p_a and p_b respectively. And the edge intersects with the projected input mesh at m_i ($i = 1, \cdots, I$) where I is the number of intersections as shown in Fig. 4.

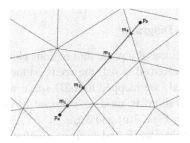

Fig. 4. Stiffener extraction

The stiffener extraction step takes each edge from the Voronoi diagram. All local triangles t_i^l are iterated to detect all intersections p_1, p_2 in all triangles where p_1 stands for m_i, and p_2 stands for m_{i+1} ($i = 1, 2, ..., I - 1$). In order to easily project 2D intersection points back to 3D, the obtained intersections p_1 and p_2 are converted to area coordinates L_1 and L_2 using the local triangle t_i^l. After all edges of the Voronoi diagram have been processed, all intersections represented in local area coordinates are mapped back to 3D coordinates. The algorithm is summarized in Algorithm 2.

5 Size Optimization

With the obtained distribution of stiffeners from previous steps, we further minimize the material consumption by finding optimized cross-section sizes of stiffeners. The objective of the size optimization is to minimize the volume of stiffeners. The constraints of the size optimization consist of 1) user specified lower bound $\underline{w}$ and upper bound $\bar{w}$ for the width of stiffeners, 2) user specified lower bound $\underline{h}$ and upper bound $\bar{h}$ for the height of stiffeners, and 3) the material strength σ_s for both stiffeners and plates.

Algorithm 2: Stiffener Extraction

Input: voronoi diagram, local triangles t_i^l, global triangles t_i^g
Output: stiffeners

for *each edge $\mathfrak{e}_i$ in voronoi diagram* **do**
    for *each t_j^l* **do**
        if $\mathfrak{e}_i$ *intersects with t_j^l* **then**
            $[p_1, p_2] = \text{IntersectSegments}(\mathfrak{e}_i, t_j^l)$
            $[L_1, L_2] = \text{AreaCoords}(p_1, p_2, t_j^l)$
            stiffeners.append(L_1, L_2, t_j^l)
        end
    end
end
Project all 2D stiffeners back to 3D space
for *stiffener in stiffeners* **do**
    $[p_1, p_2] = \text{CartesianCoords}(L_1, L_2, t_i^g)$
end

Considering the above optimization objective and constraints, the problem of the size optimization can be formulated as the following constrained minimum problem:

$$\arg \min_{w,h} \sum vol(\Re_i)$$

s.t.

$$
\begin{aligned}
\underline{w} \leq w \leq \bar{w} \\
\underline{h} \leq h \leq \bar{h} \\
S_{\Re_i} < \sigma_s \\
S_j < \sigma_s
\end{aligned}
\tag{9}
$$

where w is the width of stiffener cross-section, h is the height of stiffener cross-section, $S_{\Re_i}$ stands for the stress of stiffener $\Re_i$, and s_j means the stress of triangle t_j.

6 Monte-Carlo Simulation

As indicated in Algorithm 1, the seeded triangle t_i and probability p are both randomly generated from a uniform distribution. The stiffener distribution relies on the generated seeds from this algorithm which may be a local minimum, not a global optimal solution. In order to tackle this problem, a Monte-Carlo simulation algorithm based on Monte-Carlo stochastic sampling is introduced.

Monte-Carlo sampling is one of the most classic sampling methods used to solve the problems such as evaluation of integrals, physical simulation, optimization and so on. With this sampling algorithm, a number of n_m Monte-Carlo simulation iterations is specified, and then the process of determining the distribution of stiffeners and size optimizations of stiffeners is repeated n_m times with different randomly generated seeds r_s to search for a global optimal solution.

In this research, the number n_m of Monte-Carlo iterations is set to be 100. The experiment indicates 100 Monte-Carlo simulation iterations are large enough to obtain a global optimal solution.

7   Results and Discussions

In this section, we introduce the implementation and parameter setting of the proposed framework, effects of different probability thresholds and Monte-Carlo simulation, and 3D printed objects and the stress comparisons before and after they are stiffened with the method proposed in this paper.

7.1   Implementation and Parameter Setting

The proposed algorithm is implemented in MATLAB with FEM calculations compiled into MEX functions for speed reason. The results are tested on a PC with an Intel Xeon E5 CPU and 32 GB memory, running on Windows OS.

The minimal wall thickness allowed by the used printer is 1 mm. Therefore, both the $\underline{w}$ and $\underline{h}$ are set to be 1 mm. The material strength σ_s of the photosensitive resin used to print all the 3D objects is 42 N/m^2. The upper bounds $\bar{w}$ and $\bar{h}$ are taken to be 4 mm.

7.2   Effect of Different Probability Thresholds

The probability threshold p^* is introduced here to control the spread of the seeds over the geometry. When p^* is set to a low value, the triangles with small probabilities will not be filtered out and marked as seeded ones, causing a wide spread of seeds over all triangles. On the contrary, if p^* is set to a high value, triangles with the stress less than $p^*\sigma_s$ will never be selected which guarantees the concentration of seeds around critical areas.

Figure 5 shows the effect of different probability thresholds p^* on the generated stiffeners. It can be seen a small p^* such as $p^* = 0$ in Fig. 5(a) leads to a more uniform distribution of seeds over the mesh, while a large p^* such as $p^* = 0.5$ in Fig. 5(c) drives seeds towards the areas with higher stress and brings in more stiffeners to enhance them.

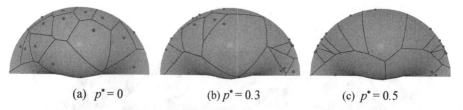

(a) $p^* = 0$ (b) $p^* = 0.3$ (c) $p^* = 0.5$

Fig. 5. Effect of different thresholds p^* on the distribution of seeds.

7.3   Effect of Monte-Carlo Simulations

Figure 6 shows the effect of random number generator seed r_s. With the same stress map and same number of seeds ($n_s = 35$), the distributions of seeds in Figs. 6(a), 6(b) and 6(c) are different, leading to different Voronoi diagrams shown in 6(d), 6(e) and 6(f) and different stiffener distributions shown in 6(g), 6(h), and 6(i), respectively.

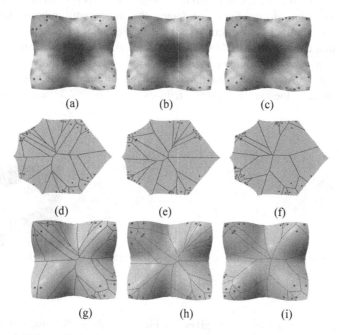

(a) (b) (c)

(d) (e) (f)

(g) (h) (i)

Fig. 6. Effect of Monte-Carlo simulations of a Guscio. The random number generator seeds r_s for each column are 10, 20 and 30 respectively.

7.4   3D Printed Objects and Stress Comparisons

With the optimization algorithm of stress-guided stiffened objects proposed in this paper, the minimum stiffener volumes of some stiffened objects are obtained, their 3D printed models are shown in Fig. 7, and the stress changes with and without the optimized stiffeners are shown in Fig. 8, 9, 10, 11, 12, 13, 14, 15 and 16, respectively.

Fig. 7. All printed 3D objects

Figure 8 shows the stress distributions, stiffeners, and 3D printed model of a stress-guided stiffened plate. In the figure (a) depicts the stress distribution in the flat plate without stiffeners with a maximum stress of 278.198 MPa, (b) shows the optimized stiffeners with a total volume of 418.5148 mm^3, (c) gives the stress distribution in the flat plate stiffened by the optimized stiffeners with a maximum stress 24.6426 MPa, and (d) is a photo of the 3D printed model of the stiffened plate. By applying the optimized stiffeners, the maximum stress reduces from 278.198 MPa to 24.6426 MPa.

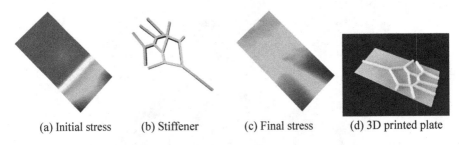

(a) Initial stress   (b) Stiffener   (c) Final stress   (d) 3D printed plate

Fig. 8. Stress-guided stiffened Plate

The example of a Botanic is given in Fig. 9 to show the stress distributions, stiffeners, and 3D printed model. Figure 9(a) shows the initial stress distribution of Botanic without stiffeners with a maximum stress of 90.927 MPa, (b) shows the optimized stiffeners with a total volume of 418.856 mm^3, (c) gives the stress distribution in the Botanic stiffened by the optimized stiffeners with a maximum stress 33.8706 MPa, and (d) is a photo of the 3D printed model of the stiffened Botanic. By applying the optimized stiffeners, the maximum stress reduces from 90.927 MPa to 33.8706 MPa.

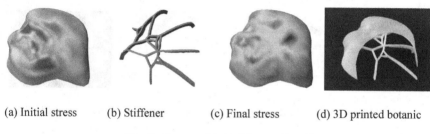

(a) Initial stress   (b) Stiffener   (c) Final stress   (d) 3D printed botanic

Fig. 9. Stress-guided stiffened Botanic

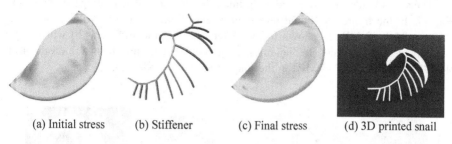

(a) Initial stress   (b) Stiffener   (c) Final stress   (d) 3D printed snail

Fig. 10. Stress-guided stiffened Snail

The stress fields, stiffeners and 3D printed model of a stiffened Snail are shown in Fig. 10. In the figure, the initial maximum stress within the Snail without any stiffeners is 33.273 MPa as shown in (a). After applying the stiffeners (b) with a total volume of 84.0108 mm^3 to the Snail, the maximum stress shown in (c) drops from 33.273 MPa to 28.3634 MPa in the final printed 3D model (d).

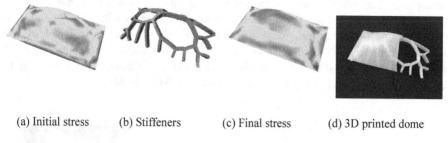

(a) Initial stress   (b) Stiffeners   (c) Final stress   (d) 3D printed dome

Fig. 11. Stress-guided stiffened Dome

Figure 11 shows the stress distributions, stiffeners, and 3D printed object of a Dome. The maximum stress 59.028 MPa in the initial stress distribution (a) without any stiffeners is reduced to the maximum stress 34.3583 MPa in (c) by applying the stiffened stiffeners (b) with a total volume of 754.704 mm^3. (d) is a photo of the 3D printed model of the stiffened Dome.

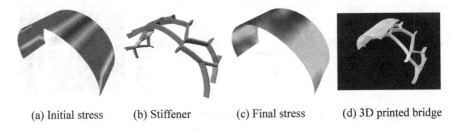

(a) Initial stress   (b) Stiffener   (c) Final stress   (d) 3D printed bridge

Fig. 12. Stress-guided stiffened Bridge

The stress fields, stiffeners and 3D printed model of a stiffened bridge are shown in Fig. 12. In the figure, the initial maximum stress within the bridge without any stiffeners is 94.4982 MPa as shown in (a). After applying the stiffeners (b) with a total volume of 535.109 mm^3 to the bridge, the final maximum stress (c) drops from 94.4982 MPa to 16.8744 MPa in the final printed 3D model (d).

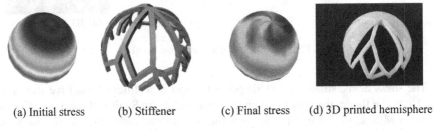

(a) Initial stress   (b) Stiffener   (c) Final stress   (d) 3D printed hemisphere

Fig. 13. Stress-guided stiffened Hemisphere

Figure 13 shows the stress distributions, stiffeners, and 3D printed object of a hemisphere. The initial stress distribution without stiffeners has a maximum stress of 42.0198 MPa shown in (a), (b) shows the optimized stiffeners with a total volume of 1961.93 mm^3, (c) gives the stress distribution in the hemisphere stiffened by the optimized stiffeners with a maximum stress 31.2246 MPa, and (d) is a photo of the 3D printed model of the stiffened hemisphere. The applied optimized stiffeners help to reduce to the maximum stress from 42.0198 MPa to 31.2246 MPa.

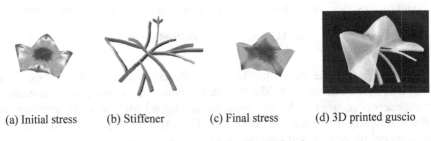

(a) Initial stress   (b) Stiffener   (c) Final stress   (d) 3D printed guscio

Fig. 14. Stress-guided stiffened Guscio

Figure 14 shows the stress distributions, stiffeners, and 3D printed object of a Guscio. The maximum stress 43.8379 MPa in the initial stress distribution (a) without any stiffeners is reduced to the maximum stress 29.5158 MPa in (c) by introducing the stiffened stiffeners (b) with a total volume of 711.483 mm^3. A photo of the 3D printed model of the stiffened Guscio is shown in Fig. 14(d).

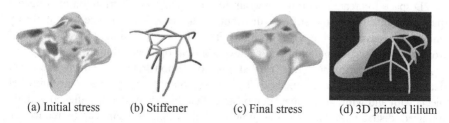

(a) Initial stress   (b) Stiffener   (c) Final stress   (d) 3D printed lilium

Fig. 15. Stress-guided stiffened Lilium

Figure 15 shows the stress distributions, stiffeners, and 3D printed object of a Lilium. The initial stress distribution without stiffeners has a maximum stress of 52.0412 MPa shown in (a), (b) shows the optimized stiffeners with a total volume of 227.294 mm^3, (c) gives the stress distribution in the Lilium stiffened by the optimized stiffeners with a maximum stress 35.3578 MPa, and (d) is a photo of the 3D printed model of the stiffened Lilium. The applied optimized stiffeners help to reduce the maximum stress from 52.0412 MPa to 35.3578 MPa.

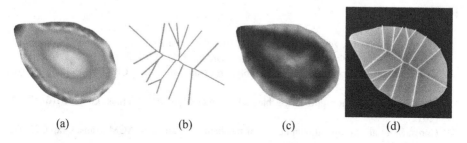

(a)   (b)   (c)   (d)

Fig. 16. Leaf: (a) Initial stress, (b) Stiffener, (c) Final stress (d) 3D printed leaf

The stress fields, stiffeners and 3D printed object of a leaf are shown in Fig. 16. In this example, the initial maximum stress in the leaf without any stiffeners is 54.9437 MPa as shown in (a). After attaching the stiffeners (b) with a total volume of 112.512 mm^3 to the leaf, the final maximum stress drops from 54.9437 MPa to 20.2208 MPa as depicted in (c), and the final printed 3D model is given in (d).

8 Conclusion and Future Work

In this paper, we have developed a stress guided optimization framework to minimize the material consumption of 3D printing. The framework consists of the finite element analysis to obtain the stress distribution in thin-shell objects, random generation of seeds guided by the obtained stress field, mapping the 3D objects and generated seeds to a 2D space to create a Voronoi diagram for optimizing the distribution of stiffeners. Apart from optimizing the stiffener distribution, the cross-section size of stiffeners is minimized to save materials for 3D printing. The Monte-Carlo simulation is introduced to optimize the seed generation and achieve a global optimal solution.

A lot of experiments were carried out to demonstrate the effectiveness and advantages of the proposed method. The stress comparisons between the thin-shell objects with and without stiffeners demonstrate that thin-shell objects stiffened with the optimized distribution and cross-section size of stiffeners significantly reduce the material consumption of 3D printed objects.

This paper assumes the cross-sectional profiles of the stiffeners are the same for each model. To achieve a more efficient structure, it is more reasonable to use various cross-sectional shapes. In the future, one of the aims is to apply various cross-sections and obtain stiffened structures with even less materials.

Acknowledgements. This research is supported by the PDE-GIR project which has received funding from the European Union Horizon 2020 research and innovation programme under the Marie Skodowska-Curie grant agreement No 778035, and Innovate UK (Knowledge Transfer Partnerships Ref: KTP010860). Shaojun Bian is supported by Chinese Scholar Council.

References

1. Skouras, M., Thomaszewski, B., Coros, S., Bickel, B., Gross, M.: Computational design of actuated deformable characters. ACM Trans. Graph. **32**(4), 82 (2013)
2. Calì, J., et al.: 3D-printing of non-assembly, articulated models. ACM Trans. Graph. **31**(6), 1–8 (2012)
3. Zhu, L., et al.: Motion-guided mechanical toy modeling. ACM Trans. Graph **31**(6), 1–10 (2012)
4. Coros, S., et al.: Computational design of mechanical characters. ACM Trans. Graph. **32**(4), 1–12 (2013)
5. Dong, Y., Wang, J., Pellacini, F., Tong, X., Guo, B.: Fabricating spatially-varying subsurface scattering. ACM Trans. Graph. **29**(4), 62 (2013)
6. Chen, D., Levin, D.I., Didyk, P., Sitthi-Amorn, P., Matusik, W.: Spec2Fab: a reducer-tuner model for translating specifications to 3D prints. ACM Trans. Graph. **32**(4), 135 (2013)
7. Zienkiewicz, O.C., Taylor, R.L.: The Finite Element Method For Solid And Structural Mechanics. Elsevier, Amsterdam (2005)
8. Rao, D.V., Sheikh, A.H., Mukhopadhyay, M.: A finite element large displacement analysis of stiffened plates. Comput. Struct. **47**(6), 987–993 (1993)
9. Samanta, A., Mukhopadhyay, M.: Finite element large deflection static analysis of shallow and deep stiffened shells. Finite Elem. Anal. Des. **33**(3), 187–208 (1999)

10. Samanta, A., Mukhopadhyay, M.: Free vibration analysis of stiffened shells by the finite element technique. Eur. J. Mech. -A/Solids **23**(1), 159–179 (2004)
11. Ojeda, R., Prusty, B.G., Lawrence, N., Thomas, G.: A new approach for the large deflection finite element analysis of isotropic and composite plates with arbitrary orientated stiffeners. Finite Elem. Anal. Des. **43**(13), 989–1002 (2007)
12. Cui, X.Y., Liu, G.R., Li, G.Y., Zhao, X., Nguyen-Thoi, T., Sun, G.Y.: A smoothed finite element method (SFEM) for linear and geometrically nonlinear analysis of plates and shells. Comput. Model. Eng. Sci. **28**(2), 109–125 (2008)
13. Nguyen-Van, H., Nguyen-Hoai, N., Chau-Dinh, T., Tran-Cong, T.: Large deflection analysis of plates and cylindrical shells by an efficient four-node flat element with mesh distortions. Acta Mech. **226**(8), 2693–2713 (2015)
14. Stava, O., Vanek, J., Benes, B., Carr, N., Měch, R.: Stress relief: improving structural strength of 3D printable objects. ACM Trans. Graph. **31**(4), 48 (2012)
15. Zhao, H., Xu, W., Zhou, K., Yang, Y., Jin, X., Wu, H.: Stress-constrained thickness optimization for shell object fabrication. Comput. Graph. Forum **36**(6), 368–380 (2017)
16. Wang, W., et al.: Cost-effective printing of 3D objects with skin-frame structures. ACM Trans. Graph. **32**(6), 177 (2013)
17. Lu, L., et al.: Build-to-last: strength to weight 3D printed objects. ACM Trans. Graph. **33**(4), 97 (2014)
18. Li, W., Zheng, A., You, L., Yang, X., Zhang, J., Liu, L.: Rib-reinforced Shell Structure. Comput. Graph. Forum **36**(7), 15–27 (2017)
19. Zheng, A.: Optimally Stiffened Thin Shell Structures in 3D Printing. Ph.D. Thesis, Bournemouth University (2019)
20. Lévy, B., Petitjean, S., Ray, N., Maillot, J.: Least squares conformal maps for automatic texture atlas generation. ACM Trans. Graph. **21**(3), 362–371 (2002)
21. Du, Q., Faber, V., Gunzburger, M.: Centroidal Voronoi tessellations: applications and algorithms. SIAM Rev. **41**(4), 637–676 (1999)
22. Gil-Ureta, F., Pietroni, N., Zorin, D.: Structurally optimized shells. arXiv preprint arXiv: 1904.12240 (2019)

Swarm Intelligence Approach for Rational Global Approximation of Characteristic Curves for the Van der Waals Equation of State

Almudena Campuzano[1,2] , Andrés Iglesias[3,4(✉)] , and Akemi Gálvez[3,4]

[1] Faculty of Social and Behavioural Sciences, University of Amsterdam,
Amsterdam, The Netherlands
campuzanoalmudena@gmail.com
[2] School of Chemical, Biological and Environmental Engineering (CBEE),
College of Engineering, Oregon State University, Corvallis, OR, USA
[3] Department of Applied Mathematics and Computational Sciences,
University of Cantabria, Santander, Spain
{iglesias,galveza}@unican.es
[4] Department of Information Science, Faculty of Sciences, Toho University,
Funabashi-shi, Chiba-ken, Japan

Abstract. The Van der Waals (VdW) equation of state is a popular generalization of the law of ideal gases proposed time ago. In many situations, it is convenient to compute the characteristic curves of the VdW equation of state, called binodal and spinodal curves. Typically, they are constructed through data fitting from a collection of data points represented in the two-dimensional pressure-volume plane. However, the resulting models are still limited and can be further enhanced. In this paper, we propose to extend this polynomial approach by using a rational function as a fitting function. In particular, we consider a rational free-form Bézier curve, which provides a global approximation to the shape of the curve. This rational approach is more flexible than the polynomial one owing to some extra parameters, the weights. Unfortunately, data fitting becomes more difficult as these new parameters have also to be computed. In this paper we address this problem through a powerful nature-inspired swarm intelligence method for continuous optimization called the bat algorithm. Our experimental results show that the method can reconstruct the characteristic curves with very good accuracy.

Keywords: Equation of state · Characteristic curves · Van der Waals equation · Data fitting · Rational curves · Bat algorithm

1 Introduction

Equations of state (EoS) are key in several fields such as physics, thermodynamics, chemical engineering and many others. Roughly speaking, they are algebraic

© Springer Nature Switzerland AG 2020
V. V. Krzhizhanovskaya et al. (Eds.): ICCS 2020, LNCS 12141, pp. 604–616, 2020.
https://doi.org/10.1007/978-3-030-50426-7_45

expressions that describe the relation between physical variables such as temperature, T, pressure, P, and volume, V, for a blend or a component. Therefore, they can be used as predictors of the behavior of thermodynamic systems under different conditions [14].

Different EoS have been defined in the literature. The simplest and most widely known is the ideal gas law, given by:

$$P.V = n.R.T = \frac{m}{M}.R.T \tag{1}$$

where the mass, the molar mass and the number of moles are represented by the variables m, M and n, respectively. As usual, R represents the universal gas constant, whose value was used as R = 0.082 L.atm.mol^{-1}.K^{-1}. This equation is adequate for high temperatures and low pressures (i.e. about 1 atm). Nonetheless, incurs inaccuracies when applied to other conditions. Consecuently, the equation has been modified throughout the years. A modification was proposed in 1873 by Johannes D. Van der Waals. This modification introduces two real, positive parameters, a and b, that account for the forces of interaction between molecules and molecular size, respectively [10]. This equation is referred to as the *Van der Waals (VdW) Equation of State* and is expressed as follows:

$$\left[P + \frac{a}{V_m{}^2}\right](V_m - b) = R.T \tag{2}$$

where V_m is the molar volume. By setting affix the temperature T in (2), and plotting the variables of pressure P vs. volume V, an isotherm can be obtained (go to Sect. 2 for details). A phase diagram can be built by displaying and analysing isotherms for several values of T. This diagram will set the boundaries of the different regions of solid, liquid, and gas phases. These boundaries are defined by curves of non analytic behavior, and indicate the limit in which phase transitions take place. In particular, the gas-liquid transition can be explained by the construction of the characteristic curves: the *binodal curve*, beneath which two different phases can coexist, and the *spinodal* curve, that defines the unstable region in a system. From now on and for convenience reasons these curves will be referred to as *binodal* and *spinodal* respectively.

In general, it is not possible to analytically calculate these curves. They have to be computed by performing data-fitting of a collection of 2-D points previously obtained for distinct isotherms. The sequence of data points that define the binodal consists of the roots placed farthest to the left, the critical point (in which liquid and gas phases are indiscernible), and the roots placed farthest to the right. On the other hand, the spinodal is defined, from right to left, by the local maxima, the critical point, and the local minima of the collection of isotherms. Numerical procedures are used to determine both sequences of points. Then, the characteristic curves are obtained using polynomial data fitting (see [1,2,4,5,12] for details).

The organization of this paper is the following: The problem to be solved is presented in Sect. 2 as a continuous, nonlinear optimization problem. Section 3 offers an insight into the swarm intelligence approach used in this work: the bat algorithm. Section 4 describes in detail the methodology proposed, and the experimental results are reported in Sect. 5. Finally, the main conclusions are presented and some ideas for future work are explored.

2   Problem to Be Solved

2.1   Background

In this paper, we consider the VdW EoS expressed as (2). With some algebraic manipulation and rearranging, it can be obtained an expression for the relation between V and P, for any given T. Hence, for a set of temperatures, being $T_1, T_2, \ldots, T_M$, their respective isotherms can be determined, resulting in a set of curves in the P–V plot. Logically, each temperature will have its associated isotherm. Multiplying the expression by $V_m{}^2/P$, and rearranging, the result is a cubic polynomial:

$$V_m{}^3 - \left(b + \frac{RT}{P}\right) V_m{}^2 + \frac{a}{P}V_m - \frac{ab}{P} = 0 \tag{3}$$

which will have one or three real roots. It will be the case that only one real root exists for values of temperature, T, larger than the critical value of T_c, known as the critical temperature, and being characteristic of each substance. The second case, three real roots, happens for temperatures lower than T_c, when the isotherms oscillate up and down. The isotherm corresponding to $T = T_c$ is associated with a triple root, which defines the *critical point*. The scope of this work will be centered around the case of $T < T_c$; the three real roots linked to each correspondent isotherms will be named as R_1, R_2 and R_3. It is worthwhile to mention that the end roots, R_1 and R_3 are respectively associated to the liquid phase and vapour phase.

Given an scenario in which a temperature $T < T_c$ is raised until meeting the critical value, $T = T_c$, it occurs that the molar volume of the saturated liquid is increased, while in the case of the saturated vapor, the molar volume decreases. This saturated states represent the boundary between the single-phase region (for liquid or vapour respectively), and the coexisting-phases region (liquid/vapour). Mathematically, this can be translated as the two end roots, R_1 and R_3, moving towards each other as the temperature is raised, until they merge in the critical point. This means that at the critical point, liquid and vapour are indistinguishable. The critical values associated with the VdW EoS of a gas are only dependent on the previously-mentioned a and b parameters. This can be proved as follows:

$$V_c = 3.b, \qquad P_c = \frac{a}{27b^2}, \qquad T_c = \frac{8a}{27bR} \tag{4}$$

Working with dimensionless variables by considering the reduced temperature, pressure and volume:

$$(T_r, P_r, V_r) = \left(\frac{T}{T_c}, \frac{P}{P_c}, \frac{V}{V_c}\right) \tag{5}$$

Note that the molar volume, V_m, now is referred to as V for simplicity. Substituting and rearranging terms, Eq. (3) becomes:

$$V_r^3 - \frac{1}{3}\left(1 + \frac{8T_r}{P_r}\right)V_r^2 + \frac{3}{P_r}V_r - \frac{1}{P_r} = 0 \qquad (6)$$

The isotherms for $T < T_c$ exhibit a surprising behavior: if the volume is decreased, then the pressure increases, falls, and then increases again, describing a fluctuation; suggesting that the pressure of certain molar volumes can decrease as a consequence of compressing the fluid, which is associated with a negative isothermal compressibility, and therefore, identified as an unstable phase.

One way to fix this deficiency was proposed by James Clerk-Maxwell in [11], and is now referred to as *Maxwell's construction*, or *Equal area rule*. Basically, it proposes solving the situation by tracing a horizontal line through the fluctuating curve, so that it connects the dew point and the bubble point, in a way that the areas enclosed between the curve and the horizontal line would be equal. This horizontal line is called *tie line*.

2.2   Binodal and Spinodal Curves: Defining the Data Points

Let us examine how are the data-points of the binodal and spinodal obtained. In the case of the binodal, the first step is to define a set of increasing temperatures $T_1 < T_2 < \ldots T_M < T_c$. For every temperature value, there will be a pressure, P_j^*, that will split up the isotherm in two halves of equal dimensions. The value of P_j^* is calculated applying an optimization procedure that, through iteration, identifies the P_j^* that ensures the minimal difference between both areas. To that end, it begins with an initial guess of $\tilde{P}_j$ and iterates until it converges. Once that the right value of pressure has been determined, it is possible to compute the roots, R_k^j, for $(k = 1, 2, 3)$, as the intersect between the isotherm for T_j and the horizontal line $P = P_j^*$. This assemblage of vapour and liquid roots will conform the binodal. Hence, the defining points of the binodal curve, $\mathcal{B}$, can be listed as:

$$\mathcal{B} = \left\{\{(R_1^j, P_j^*)\}_j, (1,1), \{(R_3^{M+1-j}, P_{M+1-j}^*)\}\right\}_{j=1,\ldots,M} \qquad (7)$$

On the other hand, the spinodal is conformed by the collection of points that define the local minima, $\mathbf{l}_j$, the critical point, and local maxima, $\mathbf{L}_l$. Note that vectors appear in bold.

The above-mentioned local optima points can be obtained through various techniques. One of them is solving the derivative $dP/dV = 0$ and examining the second derivative's sign, d^2P/dV^2, at the obtained solutions. If negative, the point corresponds to a maximum; otherwise, it is a minimum. Therefore, the assemblage of the points conforming the spinodal curve, $\mathcal{S}$, is defined by:

$$\mathcal{S} = \{\{\mathbf{l}_j\}_j, (1,1), \{\mathbf{L}_j\}_j\}_{j=1,\ldots,M} \qquad (8)$$

2.3 Characteristic Curves: Data Fitting

With the two sets of data points established, the characteristic curves can be reconstructed using standard numerical routines for data fitting. Taking into account that the obtained data points are influenced by some disturbances such as irregular sampling or noise, approximation is preferable, frequently using least-squares optimization. In such scenario, the function to be minimized is the error functional Ξ, which is the squared sum of residuals. The residual for the i-th data is given by the difference between observed data, μ_i, and the fitted data, $\hat{\mu}_i$:

$$\Xi = \sum_{i=1}^{\chi} (\mu_i - \hat{\mu}_i)^2 \tag{9}$$

Here χ is the total amount of data, and fitted data are procured by a certain fitting model function φ. It is worthwhile to mention that the minimization is carried out on the free variables of φ. Here φ is presumed to be a polynomial of a determined degree. As previously discussed, this choice can be extended by considering rational curves, as described in next section.

2.4 Data Fitting with Rational Bézier Curves

A *free-form rational Bézier curve* $\mathbf{\Phi}(\tau)$ *of degree* η is defined as [12]:

$$\mathbf{\Phi}(\tau) = \frac{\sum\limits_{j=0}^{\eta} \omega_j \mathbf{\Lambda}_j \phi_j^{\eta}(\tau)}{\sum\limits_{j=0}^{\eta} \omega_j \phi_j^{\eta}(\tau)} \tag{10}$$

where $\mathbf{\Lambda}_j$ are vector coefficients called the *poles*, ω_j are their scalar weights, $\phi_j^{\eta}(\tau)$ are the *Bernstein polynomials of index* j *and degree* η, given by:

$$\phi_j^{\eta}(\tau) = \binom{\eta}{j} \tau^j (1-\tau)^{\eta-j}$$

and τ is the *curve parameter*, defined on the finite interval $[0, 1]$. By agreement, $0! = 1$. As mentioned earlier, vectors will be denoted in bold.

Considered a set of data $\{\mathbf{\Delta}_i\}_{i=1,\ldots,\chi}$ in $\mathbb{R}^{\nu}$ (usually $\nu = 2$ or $\nu = 3$), the goal is to achieve the rational Bézier curve $\mathbf{\Phi}(\tau)$ through a discrete approximation of the data $\{\mathbf{\Delta}_i\}_i$. To that end, it is necessary to compute all parameters of the approximating curve $\mathbf{\Phi}(\tau)$, (i.e. weights ω_j, poles $\mathbf{\Lambda}_j$, and parameters τ_i associated with data points $\mathbf{\Delta}_i$, for $i = 1, \ldots, \chi$, $j = 0, \ldots, \eta$), by minimizing the least-squares error, Υ, defined as the sum of squares of the residuals:

$$\Upsilon = \operatorname*{minimize}_{\substack{\{\tau_i\}_i \\ \{\Lambda_j\}_j \\ \{\omega_j\}_j}} \left[\sum_{i=1}^{\chi} \left(\Delta_i - \frac{\displaystyle\sum_{j=0}^{\eta} \omega_j \Lambda_j \phi_j^\eta(\tau_i)}{\displaystyle\sum_{j=0}^{\eta} \omega_j \phi_j^\eta(\tau_i)} \right)^2 \right]. \qquad (11)$$

Now, taking:

$$\varphi_j^\eta(\tau) = \frac{\omega_j \phi_j^\eta(\tau)}{\displaystyle\sum_{k=0}^{\eta} \omega_k \phi_k^\eta(\tau)} \qquad (12)$$

Eq. (11) becomes:

$$\Upsilon = \operatorname*{minimize}_{\substack{\{\tau_i\}_i \\ \{\Lambda_j\}_j \\ \{\omega_j\}_j}} \left[\sum_{i=1}^{\chi} \left(\Delta_i - \sum_{j=0}^{\eta} \Lambda_j \varphi_j^\eta(\tau) \right)^2 \right], \qquad (13)$$

which can be rewritten in matrix form as: $\mathbf{\Omega}.\mathbf{\Lambda} = \mathbf{\Xi}$, where: $\mathbf{\Omega} = [\Omega_{i,j}] = \left[\left(\sum_{k=1}^{\chi} \varphi_i^\eta(\tau_k) \varphi_j^\eta(\tau_k) \right) \right]_{i,j}$, $\mathbf{\Xi} = [\Xi_j] = \left[\left(\sum_{k=1}^{\chi} \Delta_k \varphi_j^\eta(\tau_k) \right) \right]_j$, $\mathbf{\Lambda} = (\Lambda_0, \ldots, \Lambda_\eta)^T$, for $i, j = 0, \ldots, \eta$, and $(.)^T$ means transposition.

Generally, $\chi \gg \eta$, meaning that $\mathbf{\Omega}.\mathbf{\Lambda} = \mathbf{\Xi}$ is an overdetermined system of equations. If τ_i had assigned values, the problem could be solved by standard optimization procedures with coefficients $\{\Lambda_i\}_{i=0,\ldots,\eta}$ as unknowns. However, since τ_i are treated as unknowns, the complexity of the problem escalates. In fact, as the polynomial blending functions $\phi_j^\eta(\tau)$ and the rational blending functions $\varphi_j^\eta(\tau)$, are nonlinear in τ, the least-squares minimization of the residuals turns to be a continuous, nonlinear optimization problem. It can also involve a large number of unknowns, since in reality the problem can present an extremely large amount of data points. Since there may not be only one unique set of parameters leading to the solution, the problem is also multimodal. On the whole, the complicated interplay among all the unknowns (data parameters, poles, and weights) leads to a highly complex overdetermined, continuous, multivariate, multimodal, nonlinear optimization problem. The aim of this work is to solve this problem. Instead of assuming certain values for some free parameters, they are all included in our computations. This problem cannot be solved by applying classical mathematical optimization techniques [4]. In this work we propose the application of the bat algorithm, a high-power evolutionary computational method, already successfully applied to other data-fitting optimization problems in previous works [7–9]. In the next section this algorithm is further discussed.

3   The Bat Algorithm

The *bat algorithm* is a computational intelligence algorithm devised for continuous optimization problems [19,20]. It is inspired by some particular features of the social and motion behavior of small bats (microbats). These microbats use a particular kind of sonar called *echolocation* for different purposes, such as prey detection, obstacle avoidance, or roosting crevices detection, among others. Introduced in 2010, the bat algorithm has found remarkable applications for several problems [9,15–17]. See also [21] for a detailed review of the bat algorithm.

The bat algorithm is a population-based method in which the individuals (bats) are randomly initialized and distributed over the search space and then, they perform extensive exploration searching for the best location, a variable related to the quality of the solution. When a bat i is moving, its dynamics at iteration g is determined by its frequency f_i^g, location $\mathbf{x}_i^g$, and velocity $\mathbf{v}_i^g$. These variables are governed by the following evolution equations:

$$f_i^g = f_{min}^g + \beta(f_{max}^g - f_{min}^g) \tag{14}$$

$$\mathbf{v}_i^g = \mathbf{v}_i^{g-1} + [\mathbf{x}_i^{g-1} - \mathbf{x}^*]f_i^g \tag{15}$$

$$\mathbf{x}_i^g = \mathbf{x}_i^{g-1} + \mathbf{v}_i^g \tag{16}$$

where β is a uniform random variable on $[0,1]$, and $\mathbf{x}^*$ is used to represent the current global best location (solution), obtained by evaluating the fitness function at all bats and then ranking the corresponding fitness values. The method then performs a local search in the neighborhood of the current best solution through a random walk of the form:

$$\mathbf{x}_{new} = \mathbf{x}_{old} + \epsilon \mathcal{A}^g$$

with ϵ being a uniform random number on $[-1,1]$ and where $\mathcal{A}^g =< \mathcal{A}_i^g >$, represents the average loudness of all the bats of the population at generation g. Any new solution that is better than the previous best solution is accepted with a certain probability that depends on the value of the loudness. In case of acceptance, the pulse rate is increased according to the law:

$$r_i^{g+1} = r_i^0[1 - exp(-\gamma g)]$$

where γ is a parameter of the method.

Simultaneously, the loudness is decreased, following an evolution rule:

$$\mathcal{A}_i^{g+1} = \alpha \mathcal{A}_i^g$$

with α being another parameter of the method. This procedure is repeated iteratively for a maximum number of iterations, given by a parameter $\mathcal{G}_{max}$.

It is generally assumed that each bat has different values for the loudness and the pulse emission rate. This is achieved by considering the initial values for the loudness randomly as $\mathcal{A}_i^0 \in (0,2)$. The emission rate takes an initial random value r_i^0 in the interval $[0,1]$. Both parameters are updated only when the new solutions are better than the current ones, which is interpreted as a sign that the bats are advancing towards the optimal global solution.

4   The Method

4.1   Overview of the Method

As explained above, the Van der Waals Equation of State in (2) introduces two parameters, a and b, characteristic of each chemical element. These two parameters, together with a set of temperatures $T_1 < T_2 < \ldots T_M$ below the critical temperature of the substance, T_c, are the starting input of the problem. Our method is comprised of the subsequent steps:

1. Compute V_c, P_c, T_c, the critical values, using (4).
2. Compute the reduced variables V_c, P_c, T_c with (5).
3. Compute isotherms at temperatures T_j from (2).
4. For every isotherm of T_j:
    4a. Contemplate a first guess $\tilde{P}_j$ and obtain the value of P_j^* through optimization, applying Maxwell's construction.
    4b. With P_j^*, compute the roots of (6), as recounted in Sect. 2.2.
    4c. Obtain the local optima of (6), as described in Sect. 2.2.
    The result of (4a.) and (4b.) will be the sets of data points, $\mathcal{B}$ and $\mathcal{S}$, for the binodal and the spinodal curves, found respectively in (7) and (8).
5. Apply rational Bézier curves for data fitting on $\mathcal{B}$ and $\mathcal{S}$ as indicated:
    5a. Obtain data parameterization for $\mathcal{B}$ and $\mathcal{S}$ and weight computation using the bat algorithm (further discussed in Sect. 4.2).
    5b. Compute the poles of the curve applying least-squares optimization. Resolve the equations system applying classical numerical procedures, such as singular value decomposition (SVD), standard LU decomposition, and a modification of the LU decomposition for non-squared sparse problems (see [13] for details).

The most important and crucial part of the method, as well as the key component of this paper is the step (5a), which will be discussed in the next section.

4.2   Bat Algorithm for Data Fitting

This section describes how the bat algorithm, presented in Sect. 3, is used for data parameterization and weight computation with rational Bézier curves. To this purpose, we need to consider:

1. Bat encoding. In our problem, the free variables are going to be represented as follows. Bats, being denoted by $\mathcal{B}_k$, are vectors of real numbers of length $M + \eta + 1$, corresponding to a parameterization of data points and the weights, as follows:

$$\mathcal{B}_k = (\rho_1^k, \rho_2^k, \ldots, \rho_M^k, \omega_0^k, \omega_1^k, \ldots, \omega_\eta^k) \tag{17}$$

All bats $\{\mathcal{B}_k\}_k$ are initialized with uniformly distributed random numbers on the interval $[0, 1]$ for the ρ_j^k and with real positive values on the interval $(0, 20]$ for the ω_j^k. The $\{\rho_i^k\}_i$ are arranged in ascending order to reproduce the orderly form of data parameterizaton.

Table 1. Parameters of the bat algorithm and the values used in this work.

Notation	Explanation	Range	Selected value
$\mathcal{P}$	Size of the population	50–300	100
$\mathcal{G}_{max}$	Max. no. of iterations	200–3,000	1,000
$\mathcal{A}^0$	Initial loudness	$(0, 2)$	0.5
$\mathcal{A}_{min}$	Minimum loudness	$[0, 1]$	0
r^0	Initial pulse rate	$[0, 1]$	0.2
f_{max}	Max. frequency	$[0, 10]$	1.5
α	Multiplicative factor	$(0, 1)$	0.3
γ	Exponential factor	$[0, 1]$	0.2

2. *Fitness function.* It dovetails with the estimation of the least-squares function (11). However, as this function ignores the total number of data points, the RMSE (root-mean squared error) is also computed:

$$RMSE = \sqrt{\frac{\Upsilon}{\chi}} \tag{18}$$

3. *Curve parameters.* There is solely one parameter, which is the degree of the curve, η. This value will influence the amount of weighs and poles. In this work we empirically determined the optimal value, by computing and comparing the RMSE for different values of η, from 2 to 7.

4. *Bat algorithm parameters.* The algorithm has some key parameters that need to be tuned. This is of paramount importance for the proper functioning of the method. The task entails a challenge, because these parameters depend heavily on the problem. In this work the authors chose the best value by comparison from a vast set of empirical results, obtained after performing a large amount of simulations. The adjusted parameters are displayed in rows in Table 1, with their notation, explanation, and range arranged in columns, along with the final selected value. The parameters that are most decisive are the population size, $\mathcal{P}$, and the maximum number of iterations, $\mathcal{G}_{max}$. The size of population is set to $\mathcal{P} = 100$ in all shown cases. More extensive populations were also tested, up to 300, without any significant effect. As for the number of iterations, the bat algorithm results particularly beneficial since a number of $\mathcal{G}_{max} = 1000$ is sufficient to reach convergence, as opposed to other algorithms that typically requires a much larger number.

With the above-mentioned parameters selected, the bat algorithm is run for the fixed number of iterations. Last, the simulation with the top fitness value for (18) is chosen as the problem's best solution.

5   Experiments and Results

5.1   Application to a Real Case: Argon

We have applied our method to the Van der Waals (VdW) Equation of State for the case of argon, Ar. This noble gas is the third most abundant of its kind in the atmosphere, and has countless applications in industrial processes, research, medicine or lighting [18]. Its VdW parameters are $a = 1.355 \ atm.L^2.mol^{-2}$ and $b = 0.03201 \ L.mol^{-1}$. The value of the critical temperature is $T_c = 150.86$ K, with a margin of 0.1 K as reported by [3,6].

First to third steps from our workflow were performed for the following set of temperatures: $\{130, 133, 135, 137, 140, 142, 145, 147, 148, 149, T_c\}$ K and $\{128, 130, 133, 135, 137, 140, 142, 145, 147, 148, 149, 150.2, T_c\}$ K, respectively for the binodal and spinodal. Subsequently, the lists of data points for the characteristic curves, $\mathcal{B}$ and $\mathcal{S}$, were obtained following the fourth step. In the step $4a$, the Vandermonde matrix is used for carrying out the standard polynomial linear fitting of the optimization process. Data parameterization and weight computation are completed following the procedure in Sect. 4.2. The result is a linear system that can be solved using SVD. By doing so, pole computation is accomplished.

5.2   Computational Results

To account for stochastic effects and prevent premature convergence, 30 individual simulations were run for every value of η. The worst 10 executions were dropped to avoid the specious effects of instability. Computational results are reported in Tables 2 and 3 for binodal and spinodal curves respectively, with the degree ranging from $\eta = 2$ to $\eta = 7$ (in rows). We remark that, although our previous experiments for polynomial curves included values up to $\eta = 9$, the values $\eta > 7$ are actually unnecessary because of the extra degrees of freedom given by the weights. They also introduce large numerical errors, so values for η larger than 7 are finally discarded in our discussion. We have also compared our current results for the rational curves with the previous ones with strictly polynomial curves. The comparative results are displayed in Tables 2 and 3. Each table presents, in columns, the curve degree, the best RMSE (for the 30 executions), and the RMSE mean (for the 20 best executions) for the polynomial Bézier curves (columns 2 and 3) and for the rational Bézier curves (columns 4 and 5).

It can be seen from the good values of the fitting errors that the method performs pretty good. The RMSE, best and mean, achieve values of order as low as 10^{-4} for all degrees, except $\eta = 2$, meaning that the fail to be replicated with a quadratic curve, owing to the fact of not being parabolas. The best fitting rational curves are obtained for $\eta = 4$ for the binodal curve (although $\eta = 3$ performs almost equivalently) and for $\eta = 3$ for the spinodal curve (although the errors for $\eta = 4$ are also very similar). The RMSE degrees from $\eta = 4$ to $\eta = 9$, tend to be of the same order. This fact indicates that functions of higher degree are associated with more degrees of freedom (DOFs) and consequently,

Table 2. Computational results for the binodal curve.

Degree	Polynomial approach		Rational approach	
	RMSE (best)	RMSE (mean)	RMSE (best)	RMSE (mean)
$\eta = 2$	3.5349E−2	5.6509E−2	3.3873E−2	5.4522E−2
$\eta = 3$	7.7394E−5	9.8801E−5	6.5902E−5	7.7883E−5
$\eta = 4$	8.5572E−5	1.0833E−4	5.9717E−5	7.4588E−5
$\eta = 5$	1.0365E−4	1.1776E−4	8.4615E−5	1.0513E−4
$\eta = 6$	1.1927E−4	1.3328E−4	9.0631E−5	1.0264E−4
$\eta = 7$	1.1624E−4	1.2684E−4	9.0724E−5	1.0016E−4

Table 3. Computational results for the spinodal curve.

Degree	Polynomial approach		Rational approach	
	RMSE (best)	RMSE (mean)	RMSE (best)	RMSE (mean)
$\eta = 2$	4.9074E−2	7.1226E−2	4.3812E−2	6.2144E−2
$\eta = 3$	9.8649E−5	1.0356E−4	6.3187E−5	8.4551E−5
$\eta = 4$	1.1267E−4	1.2842E−4	7.3063E−5	8.5114E−5
$\eta = 5$	1.2060E−4	1.2953E−4	9.9672E−5	1.0724E−4
$\eta = 6$	1.4971E−4	1.8611E−4	1.1163E−4	1.3974E−4
$\eta = 7$	1.4750E−4	1.6975E−4	1.1708E−4	1.4828E−4

they achieve better fitting. Naturally, this occurs at the price of a higher model complexity, so in the case of numerical errors of a similar order, the values providing the simplest model are more desirable and should be selected with higher priority. In fact, an additional problem is that these extra degrees of freedom may cause over-fitting. Actually, this holds true for the spinodal and binodal respectively, for models of degree $\eta \geq 6$ and $\eta \geq 7$. Hence, the curves that must be considered predictive for other temperature values, are only those of low degree.

Another important observation is the excellent CPU times of around only 2∼4 min. On the other hand, simulations in alternative swarm intelligence methods can take as long as tens of minutes for a single execution. This advantage is owed to the quick convergence of this method. Such competitive computational times are a good indicator of the applicability of our method. We remark, however, that the CPU times for the rational case are still slightly larger (but not dramatically) than for the polynomial case, which is consistent with the fact that some extra free parameters have to be computed, thus requiring extra computation time in our simulations.

Regarding the implementation, the equipment used for all the computations was a 3.4 GHz. Intel Core i7 processor with 8 GB. of RAM. The authors implemented all the source code in *MATLAB*, version 2018b.

6   Conclusions and Future Work

In this manuscript, a new method to construct the characteristic curves of the Van der Waals equation of state through data fitting is presented. The method relies on the use of rational Bézier curves. Considering the parameters a and b of a chemical system as the input for our method, two sets of data points for the binodal and spinodal curves are firstly obtained; then, they are used to perform data parameterization and weight computation by means of the bat algorithm; finally, we apply least-squares optimization with singular value decomposition to compute the poles of the curves. The method is applied to a chemical element, argon. The method performs very well, and reconstructs the curves with high accuracy. Furthermore, it is reasonably fast (although slower than the polynomial case), with CPU times in the range of 2–4 min for each execution.

About the plans for future work, we wish to further improve the accuracy of our method. We also want to reduce the computational time of the method. We are also planning to apply this approach to other chemical components and mixtures, as well as extending this approach to other popular equations of state of interest in the field.

Acknowledgments. The authors would like to thank the financial support provided by the project PDE-GIR of the European Union's Horizon 2020 research and innovation program within the Marie Sklodowska-Curie grant agreement, with reference number 778035, and also from the Agencia Estatal de Investigación (AEI) of the Spanish Ministry of Science, Innovation and Universities (Computer Science National Program), under grant with reference number #TIN2017-89275-R and European Funds FEDER (AEI/FEDER, UE).

References

1. Campuzano, A., Iglesias, A., Gálvez, A.: Applying firefly algorithm to data fitting for the Van der Waals equation of state with Bézier curves. In: Proceedings of the International Conference on Cyberworlds (CW 2019), pp. 211–214. IEEE Computer Society Press, Los Alamitos (2019)
2. Campuzano, A., Iglesias, A., Gálvez, A.: Free-form parametric fitting of van der Waals binodal and spinodal curves with bat algorithm. In: Proceedings of the International Conference on Software, Knowledge, Information Management and Applications (SKIMA 2019), pp. 1–8. IEEE Computer Society Press, Los Alamitos (2019)
3. Angus, S., et al.: International Thermodynamic Tables of the Fluid State - 1 Argon. Butterworths, London (1972)
4. Dierckx, P.: Curve and Surface Fitting with Splines. Oxford University Press, Oxford (1993)
5. Farin, G.: Curves and Surfaces for CAGD, 5th edn. Morgan Kaufmann, San Francisco (2002)
6. Gosman, A.L., McCarty, R.D., Hust, J.G.: Thermodynamic properties of Argon from the triple point to 300 K at pressures to 1000 atmospheres. Data Ser. Nat. Bur. Stand. **27**, 1–146 (1969)

7. Iglesias, A., Gálvez, A., Collantes, M.: Bat algorithm for curve parameterization in data fitting with polynomial Bézier curves. In: Proceedings of the Cyberworlds (CW 2015), pp. 107–114. IEEE Computer Society Press, Los Alamitos (2015)

8. Iglesias, A., Gálvez, A., Collantes, M.: Global-support rational curve method for data approximation with bat algorithm. In: Chbeir, R., Manolopoulos, Y., Maglogiannis, I., Alhajj, R. (eds.) AIAI 2015. IAICT, vol. 458, pp. 191–205. Springer, Cham (2015). https://doi.org/10.1007/978-3-319-23868-5_14

9. Iglesias, A., Gálvez, A., Collantes, M.: Multilayer embedded bat algorithm for B-spline curve reconstruction. Integr. Comput.-Aided Eng. **24**(4), 385–399 (2017)

10. Johnson, D.C.: Advances in Thermodynamics of the Van der Waals Fluid. Morgan & Claypool Publishers, Ames (2014)

11. Maxwell, J.C.: On the dynamical evidence of the molecular constitution of bodies. Nature **11**, 357–359 (1875)

12. Piegl, L., Tiller, W.: The NURBS Book, pp. 146–165. Springer, Heidelberg (1997). https://doi.org/10.1007/978-3-642-59223-2

13. Press, W.H., Teukolsky, S.A., Vetterling, W.T., Flannery, B.P.: Numerical Recipes, 2nd edn. Cambridge University Press, Cambridge (1992)

14. Smith, J.M., Van Ness, H.C., Abbott, M.M.: Introduction to Chemical Engineering Thermodynamics. McGraw-Hill, Boston (2005)

15. Suárez, P., Iglesias, A.: Bat algorithm for coordinated exploration in swarm robotics. In: Del Ser, J. (ed.) ICHSA 2017. AISC, vol. 514, pp. 134–144. Springer, Singapore (2017). https://doi.org/10.1007/978-981-10-3728-3_14

16. Suárez, P., Gálvez, A., Iglesias, A.: Autonomous coordinated navigation of virtual swarm bots in dynamic indoor environments by bat algorithm. In: Tan, Y., Takagi, H., Shi, Y., Niu, B. (eds.) ICSI 2017. LNCS, vol. 10386, pp. 176–184. Springer, Cham (2017). https://doi.org/10.1007/978-3-319-61833-3_19

17. Suárez, P., Iglesias, A., Gálvez, A.: Make robots be bats: specializing robotic swarms to the bat algorithm. Swarm Evol. Comput. **44**, 113–129 (2019)

18. Weast, R.C.: Handbook of Chemistry and Physics, 53rd edn. Chemical Rubber Publishing, Boca Raton (1972)

19. Yang, X.S.: A new metaheuristic bat-inspired algorithm. In: González, J.R., Pelta, D.A., Cruz, C., Terrazas, G., Krasnogor, N. (eds.) NICSO 2010. SCI, vol. 284, pp. 65–74. Springer, Heidelberg (2010). https://doi.org/10.1007/978-3-642-12538-6_6

20. Yang, X.S., Gandomi, A.H.: Bat algorithm: a novel approach for global engineering optimization. Eng. Comput. **29**(5), 464–483 (2012)

21. Yang, X.S.: Bat algorithm: literature review and applications. Int. J. Bio-Inspired Comput. **5**(3), 141–149 (2013)

Author Index

Printed in the United States
by Baker & Taylor Publisher Services